# A Background on Supernovae

" A Source of Cosmic Creation "

Edited by Paul F. Kisak

# Contents

| 1 | **Supernova** | | | **1** |
|---|---|---|---|---|
| | 1.1 | Observation history | | 3 |
| | 1.2 | Discovery | | 4 |
| | 1.3 | Naming convention | | 6 |
| | 1.4 | Classification | | 6 |
| | | 1.4.1 | Type I | 6 |
| | | 1.4.2 | Type II | 7 |
| | | 1.4.3 | Types III, IV, and V | 8 |
| | 1.5 | Current models | | 8 |
| | | 1.5.1 | Thermal runaway | 10 |
| | | 1.5.2 | Core collapse | 11 |
| | | 1.5.3 | Failed | 17 |
| | | 1.5.4 | Light curves | 17 |
| | | 1.5.5 | Asymmetry | 19 |
| | | 1.5.6 | Energy output | 19 |
| | | 1.5.7 | Progenitor | 21 |
| | 1.6 | Interstellar impact | | 22 |
| | | 1.6.1 | Source of heavy elements | 22 |
| | | 1.6.2 | Role in stellar evolution | 22 |
| | | 1.6.3 | Effect on Earth | 24 |
| | 1.7 | Milky Way candidates | | 24 |
| | 1.8 | See also | | 24 |
| | 1.9 | Notes | | 24 |
| | 1.10 | References | | 25 |
| | 1.11 | Further reading | | 33 |
| | 1.12 | External links | | 33 |
| 2 | **Supernova remnant** | | | **34** |
| | 2.1 | Summary of stages | | 34 |
| | 2.2 | Types of supernova remnant | | 36 |

2.3  Origin of cosmic rays . . . . . . . . . . . . . . . . . . . . . . . . . . . . . . . . . . . . . . . . . . .  37

2.4  Gallery . . . . . . . . . . . . . . . . . . . . . . . . . . . . . . . . . . . . . . . . . . . . . . . . . . .  37

2.5  See also . . . . . . . . . . . . . . . . . . . . . . . . . . . . . . . . . . . . . . . . . . . . . . . . . .  38

2.6  References . . . . . . . . . . . . . . . . . . . . . . . . . . . . . . . . . . . . . . . . . . . . . . . . .  38

2.7  External links . . . . . . . . . . . . . . . . . . . . . . . . . . . . . . . . . . . . . . . . . . . . . . .  38

**3  Supernova nucleosynthesis**                                                                               **39**

3.1  Cause . . . . . . . . . . . . . . . . . . . . . . . . . . . . . . . . . . . . . . . . . . . . . . . . . . .  39

   3.1.1  Nuclear fusion sequence and the alpha process . . . . . . . . . . . . . . . . . . . . . . . . .  39

3.2  Products . . . . . . . . . . . . . . . . . . . . . . . . . . . . . . . . . . . . . . . . . . . . . . . . . .  40

3.3  The r-process . . . . . . . . . . . . . . . . . . . . . . . . . . . . . . . . . . . . . . . . . . . . . . .  40

3.4  See also . . . . . . . . . . . . . . . . . . . . . . . . . . . . . . . . . . . . . . . . . . . . . . . . . .  41

3.5  Notes . . . . . . . . . . . . . . . . . . . . . . . . . . . . . . . . . . . . . . . . . . . . . . . . . . .  42

3.6  References . . . . . . . . . . . . . . . . . . . . . . . . . . . . . . . . . . . . . . . . . . . . . . . . .  42

3.7  Other reading . . . . . . . . . . . . . . . . . . . . . . . . . . . . . . . . . . . . . . . . . . . . . . .  42

3.8  External links . . . . . . . . . . . . . . . . . . . . . . . . . . . . . . . . . . . . . . . . . . . . . . .  43

**4  Interstellar medium**                                                                                     **44**

4.1  Interstellar matter . . . . . . . . . . . . . . . . . . . . . . . . . . . . . . . . . . . . . . . . . . . .  46

   4.1.1  The three-phase model . . . . . . . . . . . . . . . . . . . . . . . . . . . . . . . . . . . . . .  46

   4.1.2  Structures . . . . . . . . . . . . . . . . . . . . . . . . . . . . . . . . . . . . . . . . . . . . .  46

   4.1.3  Interaction with interplanetary medium . . . . . . . . . . . . . . . . . . . . . . . . . . . . .  47

   4.1.4  Interstellar extinction . . . . . . . . . . . . . . . . . . . . . . . . . . . . . . . . . . . . . . .  47

4.2  Heating and cooling . . . . . . . . . . . . . . . . . . . . . . . . . . . . . . . . . . . . . . . . . . .  47

   4.2.1  Heating mechanisms . . . . . . . . . . . . . . . . . . . . . . . . . . . . . . . . . . . . . . . .  48

   4.2.2  Cooling mechanisms . . . . . . . . . . . . . . . . . . . . . . . . . . . . . . . . . . . . . . . .  49

4.3  Radiowave propagation . . . . . . . . . . . . . . . . . . . . . . . . . . . . . . . . . . . . . . . . .  49

4.4  The history of knowledge of interstellar space . . . . . . . . . . . . . . . . . . . . . . . . . . . .  50

4.5  See also . . . . . . . . . . . . . . . . . . . . . . . . . . . . . . . . . . . . . . . . . . . . . . . . . .  52

4.6  Notes . . . . . . . . . . . . . . . . . . . . . . . . . . . . . . . . . . . . . . . . . . . . . . . . . . .  52

4.7  References . . . . . . . . . . . . . . . . . . . . . . . . . . . . . . . . . . . . . . . . . . . . . . . . .  53

4.8  External links . . . . . . . . . . . . . . . . . . . . . . . . . . . . . . . . . . . . . . . . . . . . . . .  54

**5  Nuclear fusion**                                                                                          **55**

5.1  Process . . . . . . . . . . . . . . . . . . . . . . . . . . . . . . . . . . . . . . . . . . . . . . . . . .  55

5.2  Nuclear fusion in stars . . . . . . . . . . . . . . . . . . . . . . . . . . . . . . . . . . . . . . . . . .  58

5.3  Requirements . . . . . . . . . . . . . . . . . . . . . . . . . . . . . . . . . . . . . . . . . . . . . . .  58

5.4  Methods for achieving fusion . . . . . . . . . . . . . . . . . . . . . . . . . . . . . . . . . . . . . .  59

   5.4.1  Thermonuclear fusion . . . . . . . . . . . . . . . . . . . . . . . . . . . . . . . . . . . . . . .  59

    5.4.2   Inertial confinement fusion . . . . . . . . . . . . . . . . . . . . . . . . . . . . . 60

    5.4.3   Inertial electrostatic confinement . . . . . . . . . . . . . . . . . . . . . . . . . 60

    5.4.4   Beam-beam or beam-target fusion . . . . . . . . . . . . . . . . . . . . . . . . . 60

    5.4.5   Muon-catalyzed fusion . . . . . . . . . . . . . . . . . . . . . . . . . . . . . . . 60

    5.4.6   Other principles . . . . . . . . . . . . . . . . . . . . . . . . . . . . . . . . . . 60

  5.5   Important reactions . . . . . . . . . . . . . . . . . . . . . . . . . . . . . . . . . . . . 61

    5.5.1   Astrophysical reaction chains . . . . . . . . . . . . . . . . . . . . . . . . . . . 61

    5.5.2   Criteria and candidates for terrestrial reactions . . . . . . . . . . . . . . . . . 61

    5.5.3   Neutronicity, confinement requirement, and power density . . . . . . . . . . . 62

    5.5.4   Bremsstrahlung losses in quasineutral, isotropic plasmas . . . . . . . . . . . . 64

  5.6   See also . . . . . . . . . . . . . . . . . . . . . . . . . . . . . . . . . . . . . . . . . . 65

  5.7   References . . . . . . . . . . . . . . . . . . . . . . . . . . . . . . . . . . . . . . . . . 66

  5.8   Further reading . . . . . . . . . . . . . . . . . . . . . . . . . . . . . . . . . . . . . . 67

  5.9   External links . . . . . . . . . . . . . . . . . . . . . . . . . . . . . . . . . . . . . . . 67

**6  Compact star**   **73**

  6.1   Formation . . . . . . . . . . . . . . . . . . . . . . . . . . . . . . . . . . . . . . . . . 73

  6.2   Lifetime . . . . . . . . . . . . . . . . . . . . . . . . . . . . . . . . . . . . . . . . . . 73

  6.3   White dwarfs . . . . . . . . . . . . . . . . . . . . . . . . . . . . . . . . . . . . . . . 74

  6.4   Neutron stars . . . . . . . . . . . . . . . . . . . . . . . . . . . . . . . . . . . . . . . 75

  6.5   Black holes . . . . . . . . . . . . . . . . . . . . . . . . . . . . . . . . . . . . . . . . 76

    6.5.1   Alternative black hole models . . . . . . . . . . . . . . . . . . . . . . . . . . . 76

  6.6   Exotic stars . . . . . . . . . . . . . . . . . . . . . . . . . . . . . . . . . . . . . . . . 77

    6.6.1   Quark stars and strange stars . . . . . . . . . . . . . . . . . . . . . . . . . . . 78

    6.6.2   Preon stars . . . . . . . . . . . . . . . . . . . . . . . . . . . . . . . . . . . . . 78

    6.6.3   Q stars . . . . . . . . . . . . . . . . . . . . . . . . . . . . . . . . . . . . . . . 78

    6.6.4   Electroweak stars . . . . . . . . . . . . . . . . . . . . . . . . . . . . . . . . . 78

    6.6.5   Other ideas . . . . . . . . . . . . . . . . . . . . . . . . . . . . . . . . . . . . . 78

  6.7   Compact Relativistic Objects and Generalized Uncertainty Principle (GUP) . . . . . . . 78

  6.8   References . . . . . . . . . . . . . . . . . . . . . . . . . . . . . . . . . . . . . . . . . 79

  6.9   Sources . . . . . . . . . . . . . . . . . . . . . . . . . . . . . . . . . . . . . . . . . . 79

**7  Gravitational collapse**   **80**

  7.1   Star formation . . . . . . . . . . . . . . . . . . . . . . . . . . . . . . . . . . . . . . 81

  7.2   Stellar remnants . . . . . . . . . . . . . . . . . . . . . . . . . . . . . . . . . . . . . 81

    7.2.1   White dwarf . . . . . . . . . . . . . . . . . . . . . . . . . . . . . . . . . . . . 81

    7.2.2   Neutron star . . . . . . . . . . . . . . . . . . . . . . . . . . . . . . . . . . . . 81

    7.2.3   Black holes . . . . . . . . . . . . . . . . . . . . . . . . . . . . . . . . . . . . . 82

  7.3   See also . . . . . . . . . . . . . . . . . . . . . . . . . . . . . . . . . . . . . . . . . . 84

7.4   References . . . . . . . . . . . . . . . . . . . . . . . . . . . . . . . . . . . . . . . . . . . . . . 84

7.5   External links . . . . . . . . . . . . . . . . . . . . . . . . . . . . . . . . . . . . . . . . . . 85

**8   White dwarf**                                                                                      **86**

8.1   Discovery . . . . . . . . . . . . . . . . . . . . . . . . . . . . . . . . . . . . . . . . . . . . . 87

8.2   Composition and structure . . . . . . . . . . . . . . . . . . . . . . . . . . . . . . . . . . . . 89

   8.2.1   Mass–radius relationship and mass limit . . . . . . . . . . . . . . . . . . . . . . . . 91

   8.2.2   Radiation and cooling . . . . . . . . . . . . . . . . . . . . . . . . . . . . . . . . . . 94

   8.2.3   Atmosphere and spectra . . . . . . . . . . . . . . . . . . . . . . . . . . . . . . . . . 95

   8.2.4   Magnetic field . . . . . . . . . . . . . . . . . . . . . . . . . . . . . . . . . . . . . . 96

8.3   Variability . . . . . . . . . . . . . . . . . . . . . . . . . . . . . . . . . . . . . . . . . . . . . 96

8.4   Formation . . . . . . . . . . . . . . . . . . . . . . . . . . . . . . . . . . . . . . . . . . . . . 97

   8.4.1   Stars with very low mass . . . . . . . . . . . . . . . . . . . . . . . . . . . . . . . . 97

   8.4.2   Stars with low to medium mass . . . . . . . . . . . . . . . . . . . . . . . . . . . . . 97

   8.4.3   Stars with medium to high mass . . . . . . . . . . . . . . . . . . . . . . . . . . . . . 97

   8.4.4   Type Ia supernovae . . . . . . . . . . . . . . . . . . . . . . . . . . . . . . . . . . . 98

8.5   Fate . . . . . . . . . . . . . . . . . . . . . . . . . . . . . . . . . . . . . . . . . . . . . . . . 98

8.6   Debris disks and planets . . . . . . . . . . . . . . . . . . . . . . . . . . . . . . . . . . . . . 99

8.7   Habitability . . . . . . . . . . . . . . . . . . . . . . . . . . . . . . . . . . . . . . . . . . . . 99

8.8   Binary stars and novae . . . . . . . . . . . . . . . . . . . . . . . . . . . . . . . . . . . . . . 99

   8.8.1   Type Ia supernovae . . . . . . . . . . . . . . . . . . . . . . . . . . . . . . . . . . . 100

   8.8.2   Cataclysmic variables . . . . . . . . . . . . . . . . . . . . . . . . . . . . . . . . . . 100

8.9   Nearest . . . . . . . . . . . . . . . . . . . . . . . . . . . . . . . . . . . . . . . . . . . . . . 100

8.10  See also . . . . . . . . . . . . . . . . . . . . . . . . . . . . . . . . . . . . . . . . . . . . . . 100

8.11  References . . . . . . . . . . . . . . . . . . . . . . . . . . . . . . . . . . . . . . . . . . . . 101

8.12  External links and further reading . . . . . . . . . . . . . . . . . . . . . . . . . . . . . . . . 107

   8.12.1   General . . . . . . . . . . . . . . . . . . . . . . . . . . . . . . . . . . . . . . . . . 107

   8.12.2   Physics . . . . . . . . . . . . . . . . . . . . . . . . . . . . . . . . . . . . . . . . . 107

   8.12.3   Variability . . . . . . . . . . . . . . . . . . . . . . . . . . . . . . . . . . . . . . . 108

   8.12.4   Magnetic field . . . . . . . . . . . . . . . . . . . . . . . . . . . . . . . . . . . . . . 108

   8.12.5   Frequency . . . . . . . . . . . . . . . . . . . . . . . . . . . . . . . . . . . . . . . . 108

   8.12.6   Observational . . . . . . . . . . . . . . . . . . . . . . . . . . . . . . . . . . . . . . 108

   8.12.7   Images . . . . . . . . . . . . . . . . . . . . . . . . . . . . . . . . . . . . . . . . . . 108

**9   Stellar evolution**                                                                                **110**

9.1   Birth of a star . . . . . . . . . . . . . . . . . . . . . . . . . . . . . . . . . . . . . . . . . . 111

   9.1.1   Protostar . . . . . . . . . . . . . . . . . . . . . . . . . . . . . . . . . . . . . . . . . 111

   9.1.2   Brown dwarfs and sub-stellar objects . . . . . . . . . . . . . . . . . . . . . . . . . . 112

   9.1.3   Hydrogen fusion . . . . . . . . . . . . . . . . . . . . . . . . . . . . . . . . . . . . . 112

9.2 Mature stars . . . . . . . . . . . . . . . . . . . . . . . . . . . . . . . . . . . . . . . . 114

    9.2.1 Low-mass stars . . . . . . . . . . . . . . . . . . . . . . . . . . . . . . . . . . . . 114

    9.2.2 Mid-sized stars . . . . . . . . . . . . . . . . . . . . . . . . . . . . . . . . . . . . 115

    9.2.3 Massive stars . . . . . . . . . . . . . . . . . . . . . . . . . . . . . . . . . . . . . 118

9.3 Stellar remnants . . . . . . . . . . . . . . . . . . . . . . . . . . . . . . . . . . . . . . . 121

    9.3.1 White and black dwarfs . . . . . . . . . . . . . . . . . . . . . . . . . . . . . . . . 121

    9.3.2 Neutron stars . . . . . . . . . . . . . . . . . . . . . . . . . . . . . . . . . . . . . 123

    9.3.3 Black holes . . . . . . . . . . . . . . . . . . . . . . . . . . . . . . . . . . . . . . 123

9.4 Models . . . . . . . . . . . . . . . . . . . . . . . . . . . . . . . . . . . . . . . . . . . 123

9.5 See also . . . . . . . . . . . . . . . . . . . . . . . . . . . . . . . . . . . . . . . . . . . 125

9.6 Further reading . . . . . . . . . . . . . . . . . . . . . . . . . . . . . . . . . . . . . . . 125

9.7 External links . . . . . . . . . . . . . . . . . . . . . . . . . . . . . . . . . . . . . . . . 125

9.8 References . . . . . . . . . . . . . . . . . . . . . . . . . . . . . . . . . . . . . . . . . . 125

**10 Gravitational energy**      **128**

10.1 Newtonian mechanics . . . . . . . . . . . . . . . . . . . . . . . . . . . . . . . . . . . . 128

10.2 General relativity . . . . . . . . . . . . . . . . . . . . . . . . . . . . . . . . . . . . . . 128

10.3 See also . . . . . . . . . . . . . . . . . . . . . . . . . . . . . . . . . . . . . . . . . . . 129

10.4 References . . . . . . . . . . . . . . . . . . . . . . . . . . . . . . . . . . . . . . . . . . 129

**11 History of supernova observation**      **130**

11.1 Early history . . . . . . . . . . . . . . . . . . . . . . . . . . . . . . . . . . . . . . . . . 130

11.2 Telescope observation . . . . . . . . . . . . . . . . . . . . . . . . . . . . . . . . . . . . 132

11.3 1970–1999 . . . . . . . . . . . . . . . . . . . . . . . . . . . . . . . . . . . . . . . . . . 134

11.4 2000 to present . . . . . . . . . . . . . . . . . . . . . . . . . . . . . . . . . . . . . . . . 136

11.5 Future . . . . . . . . . . . . . . . . . . . . . . . . . . . . . . . . . . . . . . . . . . . . 138

11.6 See also . . . . . . . . . . . . . . . . . . . . . . . . . . . . . . . . . . . . . . . . . . . 138

11.7 References . . . . . . . . . . . . . . . . . . . . . . . . . . . . . . . . . . . . . . . . . . 139

11.8 External links . . . . . . . . . . . . . . . . . . . . . . . . . . . . . . . . . . . . . . . . 142

**12 Type Ia supernova**      **143**

12.1 Consensus model . . . . . . . . . . . . . . . . . . . . . . . . . . . . . . . . . . . . . . 144

12.2 Formation . . . . . . . . . . . . . . . . . . . . . . . . . . . . . . . . . . . . . . . . . . 146

    12.2.1 Single degenerate progenitors . . . . . . . . . . . . . . . . . . . . . . . . . . . . . 148

    12.2.2 Double degenerate progenitors . . . . . . . . . . . . . . . . . . . . . . . . . . . . . 148

    12.2.3 Type Iax . . . . . . . . . . . . . . . . . . . . . . . . . . . . . . . . . . . . . . . . 148

12.3 Observation . . . . . . . . . . . . . . . . . . . . . . . . . . . . . . . . . . . . . . . . . 149

    12.3.1 Light curve . . . . . . . . . . . . . . . . . . . . . . . . . . . . . . . . . . . . . . 149

12.4 Two distinct types . . . . . . . . . . . . . . . . . . . . . . . . . . . . . . . . . . . . . . 150

12.5   See also . . . . . . . . . . . . . . . . . . . . . . . . . . . . . . . . . . . . . . . . . . . 150

12.6   References . . . . . . . . . . . . . . . . . . . . . . . . . . . . . . . . . . . . . . . . . . 150

12.7   External links . . . . . . . . . . . . . . . . . . . . . . . . . . . . . . . . . . . . . . . . 153

**13  Type Ib and Ic supernovae**                                                      **154**

13.1   Spectra . . . . . . . . . . . . . . . . . . . . . . . . . . . . . . . . . . . . . . . . . . . . 154

13.2   Formation . . . . . . . . . . . . . . . . . . . . . . . . . . . . . . . . . . . . . . . . . . . 155

13.3   Light curves . . . . . . . . . . . . . . . . . . . . . . . . . . . . . . . . . . . . . . . . . 156

13.4   See also . . . . . . . . . . . . . . . . . . . . . . . . . . . . . . . . . . . . . . . . . . . 156

13.5   References . . . . . . . . . . . . . . . . . . . . . . . . . . . . . . . . . . . . . . . . . . 156

**14  Type II supernova**                                                              **158**

14.1   Formation . . . . . . . . . . . . . . . . . . . . . . . . . . . . . . . . . . . . . . . . . . . 159

14.2   Core collapse . . . . . . . . . . . . . . . . . . . . . . . . . . . . . . . . . . . . . . . . . 159

14.3   Theoretical models . . . . . . . . . . . . . . . . . . . . . . . . . . . . . . . . . . . . . . 162

14.4   Light curves for Type II-L and Type II-P supernovae . . . . . . . . . . . . . . . . . . . . . 162

14.5   Type IIn supernovae . . . . . . . . . . . . . . . . . . . . . . . . . . . . . . . . . . . . . . 163

14.6   Type IIb supernovae . . . . . . . . . . . . . . . . . . . . . . . . . . . . . . . . . . . . . . 164

14.7   Hypernovae (collapsars) . . . . . . . . . . . . . . . . . . . . . . . . . . . . . . . . . . . . 164

14.8   See also . . . . . . . . . . . . . . . . . . . . . . . . . . . . . . . . . . . . . . . . . . . 164

14.9   References . . . . . . . . . . . . . . . . . . . . . . . . . . . . . . . . . . . . . . . . . . 165

14.10 External links . . . . . . . . . . . . . . . . . . . . . . . . . . . . . . . . . . . . . . . . . 167

**15  Light curve**                                                                    **168**

15.1   Astronomy . . . . . . . . . . . . . . . . . . . . . . . . . . . . . . . . . . . . . . . . . . 169

15.2   Planetology . . . . . . . . . . . . . . . . . . . . . . . . . . . . . . . . . . . . . . . . . . 169

15.3   Botany . . . . . . . . . . . . . . . . . . . . . . . . . . . . . . . . . . . . . . . . . . . . 169

15.4   References . . . . . . . . . . . . . . . . . . . . . . . . . . . . . . . . . . . . . . . . . . 169

15.5   External links . . . . . . . . . . . . . . . . . . . . . . . . . . . . . . . . . . . . . . . . 169

**16  Apparent magnitude**                                                             **170**

16.1   History . . . . . . . . . . . . . . . . . . . . . . . . . . . . . . . . . . . . . . . . . . . . 171

16.2   Calculations . . . . . . . . . . . . . . . . . . . . . . . . . . . . . . . . . . . . . . . . . 171

16.2.1   Example: Sun and Moon . . . . . . . . . . . . . . . . . . . . . . . . . . . . . . . 173

16.2.2   Magnitude addition . . . . . . . . . . . . . . . . . . . . . . . . . . . . . . . . . 173

16.3   Standard reference values . . . . . . . . . . . . . . . . . . . . . . . . . . . . . . . . . . 173

16.4   Table of notable celestial objects . . . . . . . . . . . . . . . . . . . . . . . . . . . . . . 174

16.5   See also . . . . . . . . . . . . . . . . . . . . . . . . . . . . . . . . . . . . . . . . . . . 174

16.6   References . . . . . . . . . . . . . . . . . . . . . . . . . . . . . . . . . . . . . . . . . . 175

16.7   External links . . . . . . . . . . . . . . . . . . . . . . . . . . . . . . . . . . . . . . . . 176

**17 Chandrasekhar limit**     **177**

17.1 Physics . . . . . . . . . . . . . . . . . . . . . . . . . . . . . . . . . . . . . . . . . 177

17.2 History . . . . . . . . . . . . . . . . . . . . . . . . . . . . . . . . . . . . . . . . 179

17.3 Applications . . . . . . . . . . . . . . . . . . . . . . . . . . . . . . . . . . . . . 179

17.4 Super-Chandrasekhar mass supernovae . . . . . . . . . . . . . . . . . . . . . . . 180

17.5 Tolman–Oppenheimer–Volkoff limit . . . . . . . . . . . . . . . . . . . . . . . . 180

17.6 References . . . . . . . . . . . . . . . . . . . . . . . . . . . . . . . . . . . . . . 181

17.7 Further reading . . . . . . . . . . . . . . . . . . . . . . . . . . . . . . . . . . . 182

**18 Electron degeneracy pressure**     **183**

18.1 References . . . . . . . . . . . . . . . . . . . . . . . . . . . . . . . . . . . . . . 184

**19 Carbon-burning process**     **185**

19.1 Fusion reactions . . . . . . . . . . . . . . . . . . . . . . . . . . . . . . . . . . . 185

19.2 Reaction products . . . . . . . . . . . . . . . . . . . . . . . . . . . . . . . . . . 185

19.3 Neutrino losses . . . . . . . . . . . . . . . . . . . . . . . . . . . . . . . . . . . 186

19.4 Stellar evolution . . . . . . . . . . . . . . . . . . . . . . . . . . . . . . . . . . . 186

19.5 See also . . . . . . . . . . . . . . . . . . . . . . . . . . . . . . . . . . . . . . . 187

19.6 References . . . . . . . . . . . . . . . . . . . . . . . . . . . . . . . . . . . . . . 187

**20 Main sequence**     **189**

20.1 History . . . . . . . . . . . . . . . . . . . . . . . . . . . . . . . . . . . . . . . . 189

20.2 Formation . . . . . . . . . . . . . . . . . . . . . . . . . . . . . . . . . . . . . . 192

20.3 Properties . . . . . . . . . . . . . . . . . . . . . . . . . . . . . . . . . . . . . . 194

20.4 Dwarf terminology . . . . . . . . . . . . . . . . . . . . . . . . . . . . . . . . . . 194

20.5 Parameters . . . . . . . . . . . . . . . . . . . . . . . . . . . . . . . . . . . . . . 194

     20.5.1 Sample parameters . . . . . . . . . . . . . . . . . . . . . . . . . . . . . . 195

20.6 Energy generation . . . . . . . . . . . . . . . . . . . . . . . . . . . . . . . . . . 195

20.7 Structure . . . . . . . . . . . . . . . . . . . . . . . . . . . . . . . . . . . . . . . 195

20.8 Luminosity-color variation . . . . . . . . . . . . . . . . . . . . . . . . . . . . . 196

20.9 Lifetime . . . . . . . . . . . . . . . . . . . . . . . . . . . . . . . . . . . . . . . 198

20.10 Evolutionary tracks . . . . . . . . . . . . . . . . . . . . . . . . . . . . . . . . . 199

20.11 See also . . . . . . . . . . . . . . . . . . . . . . . . . . . . . . . . . . . . . . . 200

20.12 Notes . . . . . . . . . . . . . . . . . . . . . . . . . . . . . . . . . . . . . . . . . 200

20.13 References . . . . . . . . . . . . . . . . . . . . . . . . . . . . . . . . . . . . . . 201

20.14 Further reading . . . . . . . . . . . . . . . . . . . . . . . . . . . . . . . . . . . 203

     20.14.1 General . . . . . . . . . . . . . . . . . . . . . . . . . . . . . . . . . . . . 203

     20.14.2 Technical . . . . . . . . . . . . . . . . . . . . . . . . . . . . . . . . . . . 203

**21 Black hole**     **205**

21.1 History . . . . . . . . . . . . . . . . . . . . . . . . . . . . . . . . . . . . . . . 207

    21.1.1 General relativity . . . . . . . . . . . . . . . . . . . . . . . . . . . . . 208

    21.1.2 Golden age . . . . . . . . . . . . . . . . . . . . . . . . . . . . . . . . 208

21.2 Properties and structure . . . . . . . . . . . . . . . . . . . . . . . . . . . . . . 209

    21.2.1 Physical properties . . . . . . . . . . . . . . . . . . . . . . . . . . . . . 209

    21.2.2 Event horizon . . . . . . . . . . . . . . . . . . . . . . . . . . . . . . . 211

    21.2.3 Singularity . . . . . . . . . . . . . . . . . . . . . . . . . . . . . . . . 211

    21.2.4 Photon sphere . . . . . . . . . . . . . . . . . . . . . . . . . . . . . . . 212

    21.2.5 Ergosphere . . . . . . . . . . . . . . . . . . . . . . . . . . . . . . . . 212

21.3 Formation and evolution . . . . . . . . . . . . . . . . . . . . . . . . . . . . . . 212

    21.3.1 Gravitational collapse . . . . . . . . . . . . . . . . . . . . . . . . . . . 213

    21.3.2 High-energy collisions . . . . . . . . . . . . . . . . . . . . . . . . . . . 214

    21.3.3 Growth . . . . . . . . . . . . . . . . . . . . . . . . . . . . . . . . . . 215

    21.3.4 Evaporation . . . . . . . . . . . . . . . . . . . . . . . . . . . . . . . . 215

21.4 Observational evidence . . . . . . . . . . . . . . . . . . . . . . . . . . . . . . . 216

    21.4.1 Accretion of matter . . . . . . . . . . . . . . . . . . . . . . . . . . . . 216

    21.4.2 X-ray binaries . . . . . . . . . . . . . . . . . . . . . . . . . . . . . . . 217

    21.4.3 Galactic nuclei . . . . . . . . . . . . . . . . . . . . . . . . . . . . . . 220

    21.4.4 Effects of strong gravity . . . . . . . . . . . . . . . . . . . . . . . . . . 222

    21.4.5 Alternatives . . . . . . . . . . . . . . . . . . . . . . . . . . . . . . . . 222

21.5 Open questions . . . . . . . . . . . . . . . . . . . . . . . . . . . . . . . . . . 224

    21.5.1 Entropy and thermodynamics . . . . . . . . . . . . . . . . . . . . . . . 225

    21.5.2 Information loss paradox . . . . . . . . . . . . . . . . . . . . . . . . . . 226

21.6 See also . . . . . . . . . . . . . . . . . . . . . . . . . . . . . . . . . . . . . . 226

21.7 Notes . . . . . . . . . . . . . . . . . . . . . . . . . . . . . . . . . . . . . . . 227

21.8 References . . . . . . . . . . . . . . . . . . . . . . . . . . . . . . . . . . . . . 227

21.9 Further reading . . . . . . . . . . . . . . . . . . . . . . . . . . . . . . . . . . 233

21.10 External links . . . . . . . . . . . . . . . . . . . . . . . . . . . . . . . . . . . 234

**22 Neutron star**                                                                           **235**

22.1 Formation . . . . . . . . . . . . . . . . . . . . . . . . . . . . . . . . . . . . . 235

22.2 Properties . . . . . . . . . . . . . . . . . . . . . . . . . . . . . . . . . . . . . 237

22.3 Structure . . . . . . . . . . . . . . . . . . . . . . . . . . . . . . . . . . . . . . 238

22.4 History of discoveries . . . . . . . . . . . . . . . . . . . . . . . . . . . . . . . 239

22.5 Rotation . . . . . . . . . . . . . . . . . . . . . . . . . . . . . . . . . . . . . . 241

22.6 Population and distances . . . . . . . . . . . . . . . . . . . . . . . . . . . . . . 242

22.7 Binary neutron stars . . . . . . . . . . . . . . . . . . . . . . . . . . . . . . . . 243

22.8 Subtypes . . . . . . . . . . . . . . . . . . . . . . . . . . . . . . . . . . . . . . 244

22.9 Giant nucleus . . . . . . . . . . . . . . . . . . . . . . . . . . . . . . . . . . . 244

22.10 Examples of neutron stars . . . . . . . . . . . . . . . . . . . . . . . . . . . . . . . . . . . . 244

22.11 Gallery . . . . . . . . . . . . . . . . . . . . . . . . . . . . . . . . . . . . . . . . . . . . . . . 245

22.12 See also . . . . . . . . . . . . . . . . . . . . . . . . . . . . . . . . . . . . . . . . . . . . . . 245

22.13 Notes . . . . . . . . . . . . . . . . . . . . . . . . . . . . . . . . . . . . . . . . . . . . . . . . 245

22.14 References . . . . . . . . . . . . . . . . . . . . . . . . . . . . . . . . . . . . . . . . . . . . . . 246

22.15 External links . . . . . . . . . . . . . . . . . . . . . . . . . . . . . . . . . . . . . . . . . . . 247

**23 Electron capture**     **248**

23.1 History . . . . . . . . . . . . . . . . . . . . . . . . . . . . . . . . . . . . . . . . . . . . . . . 250

23.2 Reaction details . . . . . . . . . . . . . . . . . . . . . . . . . . . . . . . . . . . . . . . . . . 250

23.3 Common examples . . . . . . . . . . . . . . . . . . . . . . . . . . . . . . . . . . . . . . . . . 250

23.4 References . . . . . . . . . . . . . . . . . . . . . . . . . . . . . . . . . . . . . . . . . . . . . . 250

23.5 External links . . . . . . . . . . . . . . . . . . . . . . . . . . . . . . . . . . . . . . . . . . . 251

**24 Pair-instability supernova**     **252**

24.1 Physics . . . . . . . . . . . . . . . . . . . . . . . . . . . . . . . . . . . . . . . . . . . . . . . 253

    24.1.1 Photon pressure . . . . . . . . . . . . . . . . . . . . . . . . . . . . . . . . . . . . . . 253

    24.1.2 Pair creation and annihilation . . . . . . . . . . . . . . . . . . . . . . . . . . . . . . 253

    24.1.3 Pair-instability . . . . . . . . . . . . . . . . . . . . . . . . . . . . . . . . . . . . . . 253

24.2 Stellar susceptibility . . . . . . . . . . . . . . . . . . . . . . . . . . . . . . . . . . . . . . . . 253

24.3 Stellar behavior . . . . . . . . . . . . . . . . . . . . . . . . . . . . . . . . . . . . . . . . . . 254

    24.3.1 Below 100 solar masses . . . . . . . . . . . . . . . . . . . . . . . . . . . . . . . . . . 254

    24.3.2 100 to 130 solar masses . . . . . . . . . . . . . . . . . . . . . . . . . . . . . . . . . . 254

    24.3.3 130 to 250 solar masses . . . . . . . . . . . . . . . . . . . . . . . . . . . . . . . . . . 254

    24.3.4 250 solar masses or more . . . . . . . . . . . . . . . . . . . . . . . . . . . . . . . . . 255

24.4 Appearance . . . . . . . . . . . . . . . . . . . . . . . . . . . . . . . . . . . . . . . . . . . . . 255

    24.4.1 Luminosity . . . . . . . . . . . . . . . . . . . . . . . . . . . . . . . . . . . . . . . . . 255

    24.4.2 Spectrum . . . . . . . . . . . . . . . . . . . . . . . . . . . . . . . . . . . . . . . . . . 256

    24.4.3 Light curves . . . . . . . . . . . . . . . . . . . . . . . . . . . . . . . . . . . . . . . . 256

    24.4.4 Remnant . . . . . . . . . . . . . . . . . . . . . . . . . . . . . . . . . . . . . . . . . . 256

24.5 See also . . . . . . . . . . . . . . . . . . . . . . . . . . . . . . . . . . . . . . . . . . . . . . 256

24.6 References . . . . . . . . . . . . . . . . . . . . . . . . . . . . . . . . . . . . . . . . . . . . . . 257

**25 Photodisintegration**     **258**

25.1 Photodisintegration of deuterium . . . . . . . . . . . . . . . . . . . . . . . . . . . . . . . . . 258

25.2 Photodisintegration of beryllium . . . . . . . . . . . . . . . . . . . . . . . . . . . . . . . . . 258

25.3 Hypernovae . . . . . . . . . . . . . . . . . . . . . . . . . . . . . . . . . . . . . . . . . . . . . 258

25.4 Photofission . . . . . . . . . . . . . . . . . . . . . . . . . . . . . . . . . . . . . . . . . . . . 258

25.5 References . . . . . . . . . . . . . . . . . . . . . . . . . . . . . . . . . . . . . . . . . . . . . . 259

**26 Metallicity**   **260**

  26.1 Definition . . . . . . . . . . . . . . . . . . . . . . . . . . . . . . . . . . . . . . . . 260

    26.1.1 Calculation . . . . . . . . . . . . . . . . . . . . . . . . . . . . . . . . . . 262

    26.1.2 Relation between Z and [Fe/H] . . . . . . . . . . . . . . . . . . . . . . 262

  26.2 See also . . . . . . . . . . . . . . . . . . . . . . . . . . . . . . . . . . . . . . . . . 263

  26.3 References . . . . . . . . . . . . . . . . . . . . . . . . . . . . . . . . . . . . . . . 263

  26.4 Sources . . . . . . . . . . . . . . . . . . . . . . . . . . . . . . . . . . . . . . . . . 264

**27 Supergiant**   **265**

  27.1 Properties . . . . . . . . . . . . . . . . . . . . . . . . . . . . . . . . . . . . . . . . 266

    27.1.1 Categorisation of stars . . . . . . . . . . . . . . . . . . . . . . . . . . . . 266

    27.1.2 Variability . . . . . . . . . . . . . . . . . . . . . . . . . . . . . . . . . . 267

  27.2 Evolution . . . . . . . . . . . . . . . . . . . . . . . . . . . . . . . . . . . . . . . . 268

  27.3 Supernova progenitors . . . . . . . . . . . . . . . . . . . . . . . . . . . . . . . . . 268

  27.4 Well known examples . . . . . . . . . . . . . . . . . . . . . . . . . . . . . . . . . 269

  27.5 See also . . . . . . . . . . . . . . . . . . . . . . . . . . . . . . . . . . . . . . . . . 269

  27.6 References . . . . . . . . . . . . . . . . . . . . . . . . . . . . . . . . . . . . . . . 269

**28 Hypergiant**   **271**

  28.1 Formation . . . . . . . . . . . . . . . . . . . . . . . . . . . . . . . . . . . . . . . . 273

  28.2 Stability . . . . . . . . . . . . . . . . . . . . . . . . . . . . . . . . . . . . . . . . . 273

  28.3 Relationships with Ofpe, WNL, LBV, and other supergiant stars . . . . . . . . . . 275

  28.4 Known hypergiants . . . . . . . . . . . . . . . . . . . . . . . . . . . . . . . . . . 275

    28.4.1 Luminous blue variables . . . . . . . . . . . . . . . . . . . . . . . . . . . 275

    28.4.2 Blue hypergiants . . . . . . . . . . . . . . . . . . . . . . . . . . . . . . . 276

    28.4.3 Yellow hypergiants . . . . . . . . . . . . . . . . . . . . . . . . . . . . . . 278

    28.4.4 Red hypergiants . . . . . . . . . . . . . . . . . . . . . . . . . . . . . . . 279

  28.5 See also . . . . . . . . . . . . . . . . . . . . . . . . . . . . . . . . . . . . . . . . . 280

  28.6 References . . . . . . . . . . . . . . . . . . . . . . . . . . . . . . . . . . . . . . . 280

**29 Supernova impostor**   **282**

  29.1 Appearance, origin and mass loss . . . . . . . . . . . . . . . . . . . . . . . . . . . 283

  29.2 Examples . . . . . . . . . . . . . . . . . . . . . . . . . . . . . . . . . . . . . . . . 283

  29.3 References . . . . . . . . . . . . . . . . . . . . . . . . . . . . . . . . . . . . . . . 283

**30 Wolf–Rayet star**   **284**

  30.1 Observation history . . . . . . . . . . . . . . . . . . . . . . . . . . . . . . . . . . 284

  30.2 Classification . . . . . . . . . . . . . . . . . . . . . . . . . . . . . . . . . . . . . 286

  30.3 Nomenclature . . . . . . . . . . . . . . . . . . . . . . . . . . . . . . . . . . . . . 289

  30.4 Nebulae . . . . . . . . . . . . . . . . . . . . . . . . . . . . . . . . . . . . . . . . 290

30.5 Properties . . . . . . . . . . . . . . . . . . . . . . . . . . . . . . . . . . . . . . . . . . 290

    30.5.1 Metallicity . . . . . . . . . . . . . . . . . . . . . . . . . . . . . . . . . . . . . 292

    30.5.2 Rotation . . . . . . . . . . . . . . . . . . . . . . . . . . . . . . . . . . . . . . 292

    30.5.3 Binaries . . . . . . . . . . . . . . . . . . . . . . . . . . . . . . . . . . . . . . 292

30.6 Evolution . . . . . . . . . . . . . . . . . . . . . . . . . . . . . . . . . . . . . . . . . . 293

    30.6.1 Early ideas . . . . . . . . . . . . . . . . . . . . . . . . . . . . . . . . . . . . . 293

    30.6.2 Current models . . . . . . . . . . . . . . . . . . . . . . . . . . . . . . . . . . 294

    30.6.3 Supernovae . . . . . . . . . . . . . . . . . . . . . . . . . . . . . . . . . . . . 296

30.7 Examples . . . . . . . . . . . . . . . . . . . . . . . . . . . . . . . . . . . . . . . . . . 297

30.8 References . . . . . . . . . . . . . . . . . . . . . . . . . . . . . . . . . . . . . . . . . 297

30.9 Further reading . . . . . . . . . . . . . . . . . . . . . . . . . . . . . . . . . . . . . . 301

30.10 External links . . . . . . . . . . . . . . . . . . . . . . . . . . . . . . . . . . . . . . . 301

**31 Gamma-ray burst**     **302**

31.1 History . . . . . . . . . . . . . . . . . . . . . . . . . . . . . . . . . . . . . . . . . . . 303

    31.1.1 Counterpart objects as candidate sources . . . . . . . . . . . . . . . . . . . . 303

    31.1.2 Afterglow . . . . . . . . . . . . . . . . . . . . . . . . . . . . . . . . . . . . . 304

31.2 Classification . . . . . . . . . . . . . . . . . . . . . . . . . . . . . . . . . . . . . . . 306

    31.2.1 Short gamma-ray bursts . . . . . . . . . . . . . . . . . . . . . . . . . . . . . . 308

    31.2.2 Long gamma-ray bursts . . . . . . . . . . . . . . . . . . . . . . . . . . . . . . 309

    31.2.3 Ultra-long gamma-ray bursts . . . . . . . . . . . . . . . . . . . . . . . . . . . 309

31.3 Energetics and beaming . . . . . . . . . . . . . . . . . . . . . . . . . . . . . . . . . . 309

31.4 Progenitors . . . . . . . . . . . . . . . . . . . . . . . . . . . . . . . . . . . . . . . . 310

    31.4.1 Tidal disruption events . . . . . . . . . . . . . . . . . . . . . . . . . . . . . . 312

31.5 Emission mechanisms . . . . . . . . . . . . . . . . . . . . . . . . . . . . . . . . . . . 312

31.6 Rate of occurrence and potential effects on life on Earth . . . . . . . . . . . . . . . . 312

    31.6.1 Hypothetical effects of gamma-ray bursts in the past . . . . . . . . . . . . . . 313

    31.6.2 Hypothetical effects of gamma-ray bursts in the future . . . . . . . . . . . . . 313

    31.6.3 Effects after exposure to the gamma-ray burst on Earth's atmosphere . . . . . . 313

31.7 See also . . . . . . . . . . . . . . . . . . . . . . . . . . . . . . . . . . . . . . . . . . 314

31.8 Footnotes . . . . . . . . . . . . . . . . . . . . . . . . . . . . . . . . . . . . . . . . . 314

31.9 Notes . . . . . . . . . . . . . . . . . . . . . . . . . . . . . . . . . . . . . . . . . . . 314

31.10 Books . . . . . . . . . . . . . . . . . . . . . . . . . . . . . . . . . . . . . . . . . . . 318

31.11 References . . . . . . . . . . . . . . . . . . . . . . . . . . . . . . . . . . . . . . . . . 318

31.12 External links . . . . . . . . . . . . . . . . . . . . . . . . . . . . . . . . . . . . . . . 323

**32 Hypernova**     **325**

32.1 History of the term . . . . . . . . . . . . . . . . . . . . . . . . . . . . . . . . . . . . 325

32.2 Gamma-ray bursts . . . . . . . . . . . . . . . . . . . . . . . . . . . . . . . . . . . . . 325

32.3  Causes of hypernovae . . . . . . . . . . . . . . . . . . . . . . . . . . . . . . . . . . . . 326

    32.3.1  Collapsar model . . . . . . . . . . . . . . . . . . . . . . . . . . . . . . . . . . . 327

    32.3.2  CSM model (circumstellar material) . . . . . . . . . . . . . . . . . . . . . . . . 328

    32.3.3  Pair-instability supernova . . . . . . . . . . . . . . . . . . . . . . . . . . . . . 328

    32.3.4  Magnetar energy release . . . . . . . . . . . . . . . . . . . . . . . . . . . . . . 328

    32.3.5  Other models . . . . . . . . . . . . . . . . . . . . . . . . . . . . . . . . . . . . 329

32.4  See also . . . . . . . . . . . . . . . . . . . . . . . . . . . . . . . . . . . . . . . . . . . 329

32.5  References . . . . . . . . . . . . . . . . . . . . . . . . . . . . . . . . . . . . . . . . . . 329

32.6  Further reading . . . . . . . . . . . . . . . . . . . . . . . . . . . . . . . . . . . . . . . 330

**33  Astrophysical jet**                                                                          **332**

33.1  Relativistic jet . . . . . . . . . . . . . . . . . . . . . . . . . . . . . . . . . . . . . . . 332

    33.1.1  Rotating black hole as energy source . . . . . . . . . . . . . . . . . . . . . . . . 333

33.2  Other images . . . . . . . . . . . . . . . . . . . . . . . . . . . . . . . . . . . . . . . . 334

33.3  See also . . . . . . . . . . . . . . . . . . . . . . . . . . . . . . . . . . . . . . . . . . . 335

33.4  References . . . . . . . . . . . . . . . . . . . . . . . . . . . . . . . . . . . . . . . . . . 335

33.5  External links . . . . . . . . . . . . . . . . . . . . . . . . . . . . . . . . . . . . . . . . 336

33.6  Videos . . . . . . . . . . . . . . . . . . . . . . . . . . . . . . . . . . . . . . . . . . . . 336

**34  Binary star**                                                                                **337**

34.1  Discovery . . . . . . . . . . . . . . . . . . . . . . . . . . . . . . . . . . . . . . . . . . 339

34.2  Classifications . . . . . . . . . . . . . . . . . . . . . . . . . . . . . . . . . . . . . . . . 340

    34.2.1  Methods of observation . . . . . . . . . . . . . . . . . . . . . . . . . . . . . . . 340

    34.2.2  Configuration of the system . . . . . . . . . . . . . . . . . . . . . . . . . . . . . 344

    34.2.3  Cataclysmic variables and X-ray binaries . . . . . . . . . . . . . . . . . . . . . . 345

34.3  Orbital period . . . . . . . . . . . . . . . . . . . . . . . . . . . . . . . . . . . . . . . . 345

    34.3.1  Variations in period . . . . . . . . . . . . . . . . . . . . . . . . . . . . . . . . . 345

34.4  Designations . . . . . . . . . . . . . . . . . . . . . . . . . . . . . . . . . . . . . . . . . 345

    34.4.1  A and B . . . . . . . . . . . . . . . . . . . . . . . . . . . . . . . . . . . . . . . 345

    34.4.2  Discoverer designations . . . . . . . . . . . . . . . . . . . . . . . . . . . . . . . 346

    34.4.3  Hot and cold . . . . . . . . . . . . . . . . . . . . . . . . . . . . . . . . . . . . . 346

34.5  Evolution . . . . . . . . . . . . . . . . . . . . . . . . . . . . . . . . . . . . . . . . . . . 346

    34.5.1  Formation . . . . . . . . . . . . . . . . . . . . . . . . . . . . . . . . . . . . . . 346

    34.5.2  Mass transfer and accretion . . . . . . . . . . . . . . . . . . . . . . . . . . . . . 346

    34.5.3  Runaways and novae . . . . . . . . . . . . . . . . . . . . . . . . . . . . . . . . . 347

34.6  Astrophysics . . . . . . . . . . . . . . . . . . . . . . . . . . . . . . . . . . . . . . . . . 347

    34.6.1  Calculating the center of mass in binary stars . . . . . . . . . . . . . . . . . . . 347

    34.6.2  Center of mass animations . . . . . . . . . . . . . . . . . . . . . . . . . . . . . . 348

    34.6.3  Research findings . . . . . . . . . . . . . . . . . . . . . . . . . . . . . . . . . . . 348

34.7  Examples . . . . . . . . . . . . . . . . . . . . . . . . . . . . . . . . . . . . . . . . . . . . . 350

34.8  Multiple star examples . . . . . . . . . . . . . . . . . . . . . . . . . . . . . . . . . . . . . 350

34.9  See also . . . . . . . . . . . . . . . . . . . . . . . . . . . . . . . . . . . . . . . . . . . . . 351

34.10 Notes and references . . . . . . . . . . . . . . . . . . . . . . . . . . . . . . . . . . . . . . . 352

34.11 External links . . . . . . . . . . . . . . . . . . . . . . . . . . . . . . . . . . . . . . . . . . . 355

**35  Quark-nova**                                                                                       **356**

35.1  See also . . . . . . . . . . . . . . . . . . . . . . . . . . . . . . . . . . . . . . . . . . . . . 356

35.2  References . . . . . . . . . . . . . . . . . . . . . . . . . . . . . . . . . . . . . . . . . . . . 356

35.3  External links . . . . . . . . . . . . . . . . . . . . . . . . . . . . . . . . . . . . . . . . . . 357

**36  Galaxy morphological classification**                                                              **358**

36.1  Hubble sequence . . . . . . . . . . . . . . . . . . . . . . . . . . . . . . . . . . . . . . . . 359

36.2  De Vaucouleurs system . . . . . . . . . . . . . . . . . . . . . . . . . . . . . . . . . . . . . 360

36.2.1  Numerical Hubble stage . . . . . . . . . . . . . . . . . . . . . . . . . . . . . . . . 362

36.3  Yerkes (or Morgan) scheme . . . . . . . . . . . . . . . . . . . . . . . . . . . . . . . . . . 363

36.4  See also . . . . . . . . . . . . . . . . . . . . . . . . . . . . . . . . . . . . . . . . . . . . . 363

36.5  References . . . . . . . . . . . . . . . . . . . . . . . . . . . . . . . . . . . . . . . . . . . . 364

36.6  External links . . . . . . . . . . . . . . . . . . . . . . . . . . . . . . . . . . . . . . . . . . 365

**37  Near-Earth supernova**                                                                             **366**

37.1  Effects on Earth . . . . . . . . . . . . . . . . . . . . . . . . . . . . . . . . . . . . . . . . . 366

37.2  Risk by supernova type . . . . . . . . . . . . . . . . . . . . . . . . . . . . . . . . . . . . . 366

37.3  Past events . . . . . . . . . . . . . . . . . . . . . . . . . . . . . . . . . . . . . . . . . . . . 367

37.4  See also . . . . . . . . . . . . . . . . . . . . . . . . . . . . . . . . . . . . . . . . . . . . . 368

37.5  Footnotes . . . . . . . . . . . . . . . . . . . . . . . . . . . . . . . . . . . . . . . . . . . . 368

37.6  References . . . . . . . . . . . . . . . . . . . . . . . . . . . . . . . . . . . . . . . . . . . . 368

**38  List of supernova candidates**                                                                     **370**

38.1  Notes . . . . . . . . . . . . . . . . . . . . . . . . . . . . . . . . . . . . . . . . . . . . . . . 370

38.2  References . . . . . . . . . . . . . . . . . . . . . . . . . . . . . . . . . . . . . . . . . . . . 370

**39  Timeline of white dwarfs, neutron stars, and supernovae**                                          **373**

39.1  Text and image sources, contributors, and licenses . . . . . . . . . . . . . . . . . . . . . . . 375

39.1.1  Text . . . . . . . . . . . . . . . . . . . . . . . . . . . . . . . . . . . . . . . . . . . 375

39.1.2  Images . . . . . . . . . . . . . . . . . . . . . . . . . . . . . . . . . . . . . . . . . . 387

39.1.3  Content license . . . . . . . . . . . . . . . . . . . . . . . . . . . . . . . . . . . . . 397

**40  Afterword- Editor'sComments**

40.1  SN 1997ff and its Impact on Fundamental CosmologyTheory . . . . . . . . . . . . . . . . . . . . . . . 398

# Chapter 1

# Supernova

This article is about the astronomical event. For other uses, see Supernova (disambiguation).

A **supernova** is a stellar explosion that briefly outshines an entire galaxy, radiating as much energy as the Sun or any ordinary star is expected to emit over its entire life span, before fading from view over several weeks or months.[1] The extremely luminous burst of radiation expels much or all of a star's material[2] at a velocity of up to 30,000 km/s (10% of the speed of light), driving a shock wave[3] into the surrounding interstellar medium. This shock wave sweeps up an expanding shell of gas and dust called a supernova remnant. Supernovae are potentially strong galactic sources of gravitational waves.[4] A great proportion of primary cosmic rays comes from supernovae.[5]

Supernovae are more energetic than novae. *Nova* means "new" in Latin, referring to what appears to be a very bright new star shining in the celestial sphere; the prefix "super-" distinguishes supernovae from ordinary novae, which are far less luminous. The word *supernova* was coined by Walter Baade and Fritz Zwicky in 1931.[6] It is pronounced /ˌsuːpərˈnoʊvə/ with the plural **supernovae** /ˌsuːpərˈnoʊviː/ or **supernovas** (abbreviated *SN*, plural *SNe* after "supernovae").

Supernovae can be triggered in one of two ways: by the sudden re-ignition of nuclear fusion in a degenerate star; or by the gravitational collapse of the core of a massive star. In the first case, a degenerate white dwarf may accumulate sufficient material from a companion, either through accretion or via a merger, to raise its core temperature, ignite carbon fusion, and trigger runaway nuclear fusion, completely disrupting the star. In the second case, the core of a massive star may undergo sudden gravitational collapse, releasing gravitational potential energy that can create a supernova explosion.

The most recent directly observed supernova in the Milky Way was Kepler's Star of 1604 (SN 1604); remnants of two more recent supernovae have been found retrospectively.[7] Observations in other galaxies indicate that supernovae should occur on average about three times every century in the Milky Way, and that any galactic supernova would almost certainly be observable in modern astronomical equipment.[8] Supernovae play a significant role in enriching the interstellar medium with higher mass elements.[9] Furthermore, the expanding shock waves from supernova explosions can trigger the formation of new stars.[10][11]

*SN 1994D (bright spot on the lower left), a type Ia supernova in the NGC 4526 galaxy*

*In this much speeded-up artist's impression showing a collection of distant galaxies, the occasional supernova can be seen. Each of these exploding stars briefly rivals the brightness of its host galaxy.*

## 1.1 Observation history

Main article: History of supernova observation

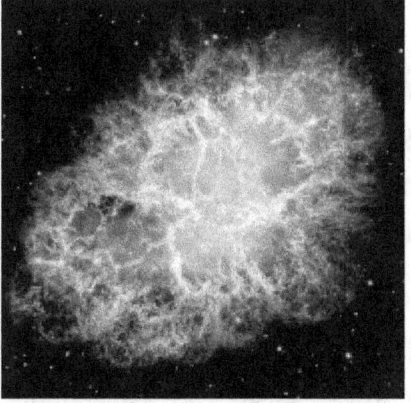

The Crab Nebula is a pulsar wind nebula associated with the 1054 supernova

The highlighted passages refer to the Chinese observation of SN 1054

Hipparchus' interest in the fixed stars may have been inspired by the observation of a supernova (according to Pliny).[12] The earliest recorded supernova, SN 185, was viewed by Chinese astronomers in 185 AD. The brightest recorded supernova was the SN 1006, which was described in detail by Chinese and Islamic astronomers.[13] The widely observed supernova SN 1054 produced the Crab Nebula. Supernovae SN 1572 and SN 1604, the latest to be observed with the naked eye in the Milky Way galaxy, had notable effects on the development of astronomy in Europe because they were used to argue against the Aristotelian idea that the universe beyond the Moon and planets was immutable.[14] Johannes Kepler began observing SN 1604 at its peak on October 17, 1604, and continued to make estimates of its brightness until it faded from naked eye view a year later.[15] It was the second supernova to be observed in a generation (after SN 1572 seen by Tycho Brahe in Cassiopeia).[12]

Before the development of the telescope, there have only been five supernovae seen in the last millennium. In the perspective of how long a star's lifetime is, its death is very brief. In fact, a star's death may only last a few months. Due to this, a typical human will only experience this rarity, on average, once in their lifetime. This is a microscopic fraction in comparison to the 100 billion stars that compose a galaxy.[16]

Since the development of the telescope, the field of supernova discovery has extended to other galaxies, starting with the 1885 observation of supernova S Andromedae in the Andromeda galaxy. American astronomers Rudolph Minkowski and Fritz Zwicky developed the modern supernova classification scheme beginning in 1941.[17] In the 1960s, astronomers found that the maximum intensities of supernova explosions could be used as standard candles, hence indicators of astronomical distances.[18] Some of the most distant supernovae recently observed appeared dimmer than expected. This supports the view that the expansion of the universe is accelerating.[19] Techniques were developed for reconstructing supernova explosions that have no written records of being observed. The date of the Cassiopeia A supernova event was determined from light echoes off nebulae,[20] while the age of supernova remnant RX J0852.0-4622 was estimated from temperature measurements[21] and the gamma ray emissions from the decay of titanium-44.[22] In 2009, nitrates were discovered in Antarctic ice deposits that matched the times of past supernova events.[23]

## 1.2   Discovery

Main article: History of supernova observation § Telescope observation

Early work on what was originally believed to be simply a new category of novae was performed during the 1930s by Walter Baade and Fritz Zwicky at Mount Wilson Observatory.[24] The name *super-novae* was first used during 1931 lectures held at Caltech by Baade and Zwicky, then used publicly in 1933 at a meeting of the American Physical Society.[6] By 1938, the hyphen had been lost and the modern name was in use.[25] Because supernovae are relatively rare events within a galaxy, occurring about three times a century in the Milky Way,[7] obtaining a good sample of supernovae to study requires regular monitoring of many galaxies.

Supernovae in other galaxies cannot be predicted with any meaningful accuracy. Normally, when they are discovered, they are already in progress.[26] Most scientific interest in supernovae—as standard candles for measuring distance, for example—require an observation of their peak luminosity. It is therefore important to discover them well before they reach their maximum. Amateur astronomers, who greatly outnumber professional astronomers, have played an important role in finding supernovae, typically by looking at some of the closer galaxies through an optical telescope and comparing them to earlier photographs.[27]

Toward the end of the 20th century astronomers increasingly turned to computer-controlled telescopes and CCDs for hunting supernovae. While such systems are popular with amateurs, there are also professional installations such as the Katzman Automatic Imaging Telescope.[28] Recently the Supernova Early Warning System (SNEWS) project has begun using a network of neutrino detectors to give early warning of a supernova in the Milky Way galaxy.[29][30] Neutrinos are particles that are produced in great quantities by a supernova explosion,[31] and they are not significantly absorbed by the interstellar gas and dust of the galactic disk.

Supernova searches fall into two classes: those focused on relatively nearby events and those looking for explosions farther away. Because of the expansion of the universe, the distance to a remote object with a known emission spectrum can be estimated by measuring its Doppler shift (or redshift); on average, more distant objects recede with greater velocity than those nearby, and so have a higher redshift. Thus the search is split between high redshift and low redshift, with

*The remnant of a star gone supernova.*

the boundary falling around a redshift range of $z = 0.1$–$0.3$[33]—where $z$ is a dimensionless measure of the spectrum's frequency shift.

High redshift searches for supernovae usually involve the observation of supernova light curves. These are useful for standard or calibrated candles to generate Hubble diagrams and make cosmological predictions. Supernova spectroscopy, used to study the physics and environments of supernovae, is more practical at low than at high redshift.[34][35] Low redshift observations also anchor the low-distance end of the Hubble curve, which is a plot of distance versus redshift for visible galaxies.[36][37] (See also Hubble's law).

*A star set to explode.*[32]

## 1.3   Naming convention

Supernova discoveries are reported to the International Astronomical Union's Central Bureau for Astronomical Telegrams, which sends out a circular with the name it assigns to that supernova. The name is the marker *SN* followed by the year of discovery, suffixed with a one or two-letter designation. The first 26 supernovae of the year are designated with a capital letter from *A* to *Z*. Afterward pairs of lower-case letters are used: *aa*, *ab*, and so on. Hence, for example, *SN 2003C* designates the third supernova reported in the year 2003.[38] The last supernova of 2005 was SN 2005nc, indicating that it was the 367th[nb 1] supernova found in 2005. Since 2000, professional and amateur astronomers have been finding several hundreds of supernovae each year (572 in 2007, 261 in 2008, 390 in 2009; 231 in 2013).[39][40]

Historical supernovae are known simply by the year they occurred: SN 185, SN 1006, SN 1054, SN 1572 (called *Tycho's Nova*) and SN 1604 (*Kepler's Star*). Since 1885 the additional letter notation has been used, even if there was only one supernova discovered that year (e.g. SN 1885A, SN 1907A, etc.) — this last happened with SN 1947A. *SN*, for SuperNova, is a standard prefix. Until 1987, two-letter designations were rarely needed; since 1988, however, they have been needed every year.

## 1.4   Classification

As part of the attempt to understand supernovae, astronomers have classified them according to their light curves and the absorption lines of different chemical elements that appear in their spectra. The first element for division is the presence or absence of a line caused by hydrogen. If a supernova's spectrum contains lines of hydrogen (known as the Balmer series in the visual portion of the spectrum) it is classified *Type II*; otherwise it is *Type I*. In each of these two types there are subdivisions according to the presence of lines from other elements or the shape of the light curve (a graph of the supernova's apparent magnitude as a function of time).[42][43]

### 1.4.1   Type I

The type I supernovae are subdivided on the basis of their spectra, with type Ia showing a strong ionised silicon absorption line. Type I supernovae without this strong line are classified as types Ib and Ic, with type Ib showing strong neutral helium

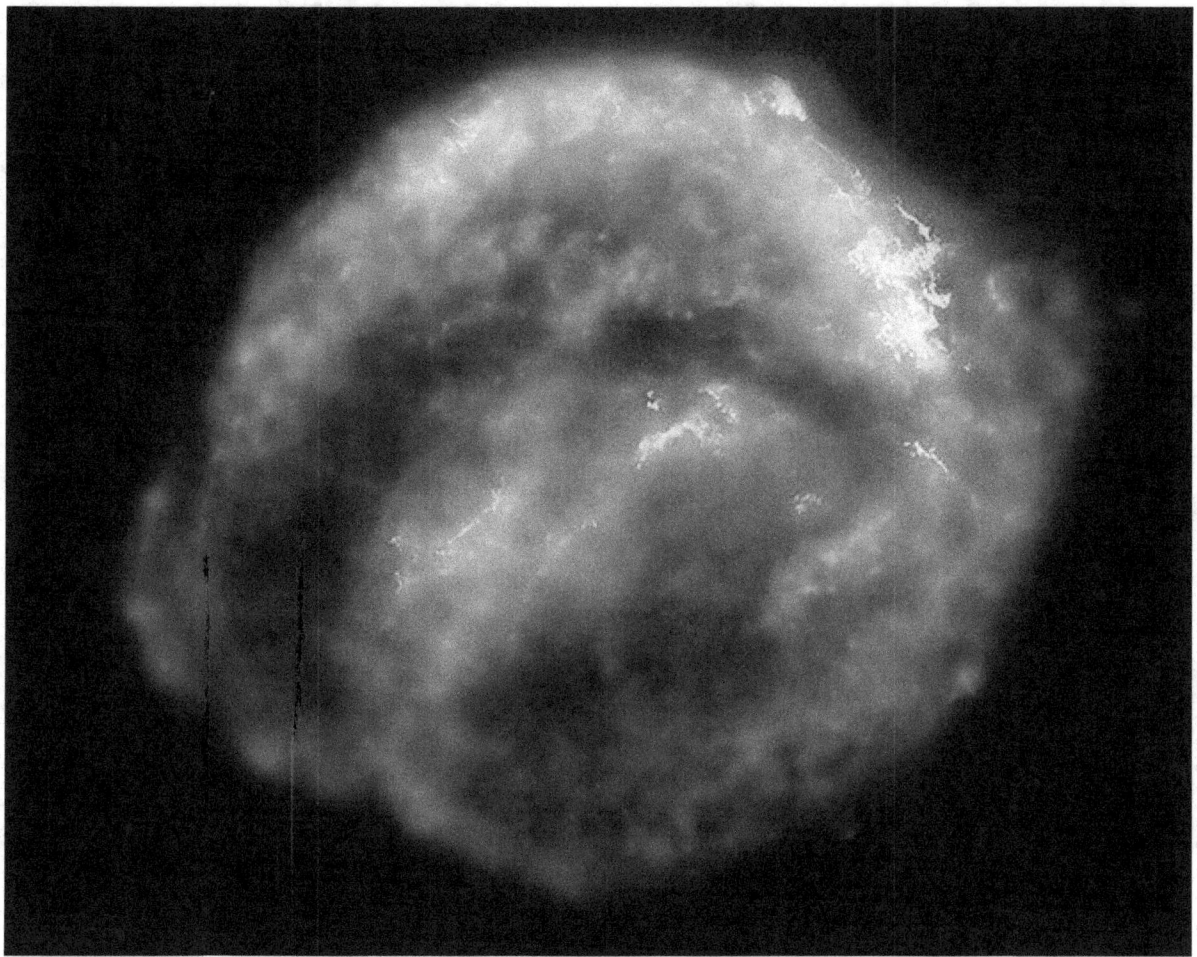

*Multiwavelength X-ray, infrared, and optical compilation image of Kepler's supernova remnant, SN 1604.*

lines and type Ic lacking them. The light curves are all similar, although type Ia are generally brighter at peak luminosity, but the light curve is not important for classification of type I supernovae.

A small number of type Ia supernovae exhibit unusual features such as non-standard luminosity or broadened light curves, and these are typically classified by referring to the earliest example showing similar features. For example, the sub-luminous SN 2008ha is often referred to as SN 2002cx-like or class Ia-2002cx.

## 1.4.2   Type II

The supernovae of Type II can also be sub-divided based on their spectra. While most Type II supernovae show very broad emission lines which indicate expansion velocities of many thousands of kilometres per second, some, such as SN 2005gl, have relatively narrow features in their spectra. These are called Type IIn, where the 'n' stands for 'narrow'.

A few supernovae, such as SN 1987K and SN 1993J, appear to change types: they show lines of hydrogen at early times, but, over a period of weeks to months, become dominated by lines of helium. The term "Type IIb" is used to describe the combination of features normally associated with Types II and Ib.[43]

Type II supernovae with normal spectra dominated by broad hydrogen lines that remain for the life of the decline are classified on the basis of their light curves. The most common type shows a distinctive "plateau" in the light curve shortly after peak brightness where the visual luminosity stays relatively constant for several months before the decline resumes. These are called type II-P referring to the plateau. Less common are type II-L supernovae that lack a distinct plateau. The "L" signifies "linear" although the light curve is not actually a straight line.

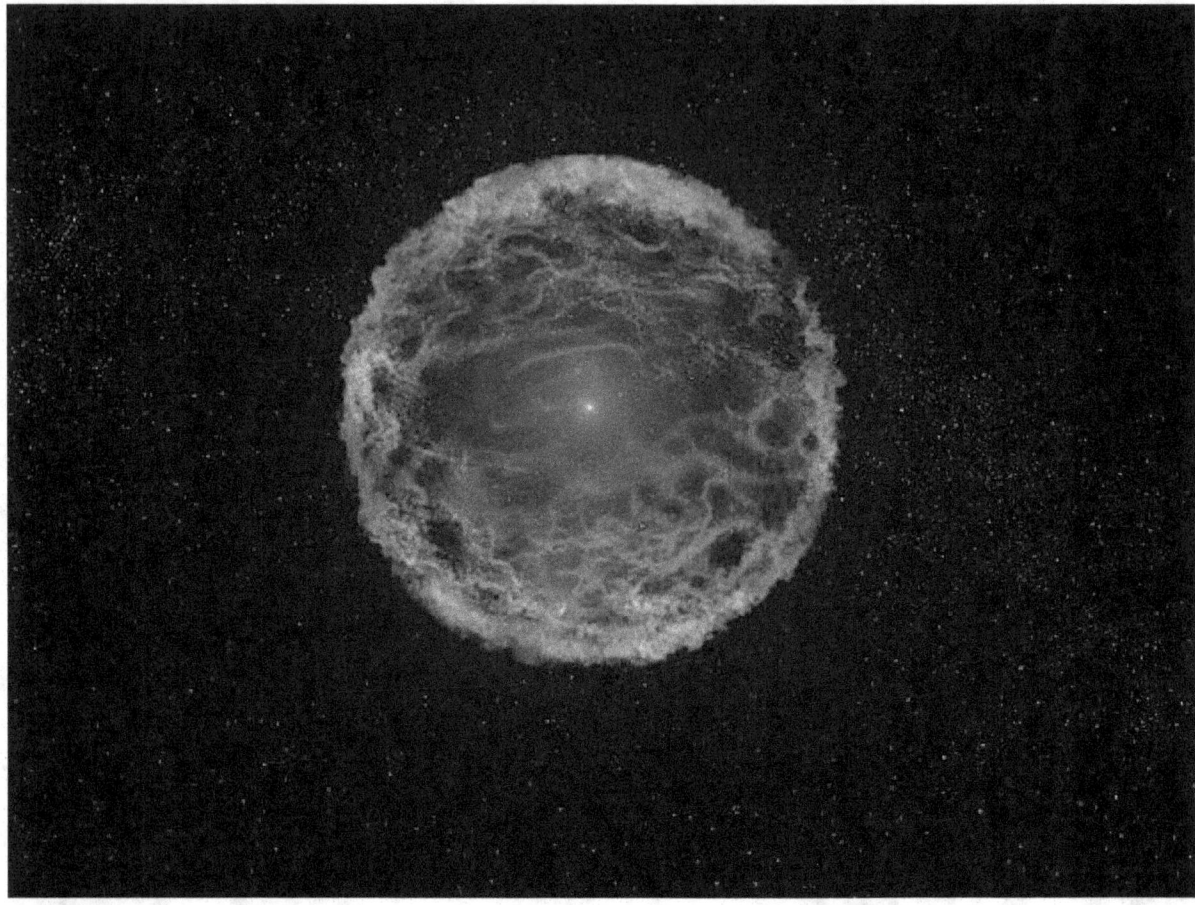

*Artist's impression of supernova 1993J.*[41]

Supernovae that do not fit into the normal classifications are designated peculiar, or 'pec'.[43]

### 1.4.3   Types III, IV, and V

Fritz Zwicky defined additional supernovae types, although based on a very few examples that didn't cleanly fit the parameters for a type I or type II supernova. SN 1961i in NGC 4303 was the prototype and only member of the type III
supernova class, noted for its broad light curve maximum and broad hydrogen Balmer lines that were slow to develop in
the spectrum. SN 1961f in NGC 3003 was the prototype and only member of the type IV class, with a light curve similar
to a type II-P supernova, with hydrogen absorption lines but weak hydrogen emission lines. The type V class was coined
for SN 1961V in NGC 1058, an unusual faint supernova or supernova imposter with a slow rise to brightness, a maximum
lasting many months, and an unusual emission spectrum. The similarity of SN 1961V to the Eta Carinae Great Outburst
was noted.[45] Supernovae in M101 (1909) and M83 (1923 and 1957) were also suggested as possible type IV or type V
supernovae.[46]

These types would now all be treated as peculiar type II supernovae, of which many more examples have been discovered,
although it is still debated whether SN 1961V was a true supernova following an LBV outburst or an imposter.[44]

## 1.5   Current models

The type codes described above that astronomers give to supernovae are *taxonomic* in nature: the type number describes
the light observed from the supernova, not necessarily its cause. For example, type Ia supernovae are produced by runaway

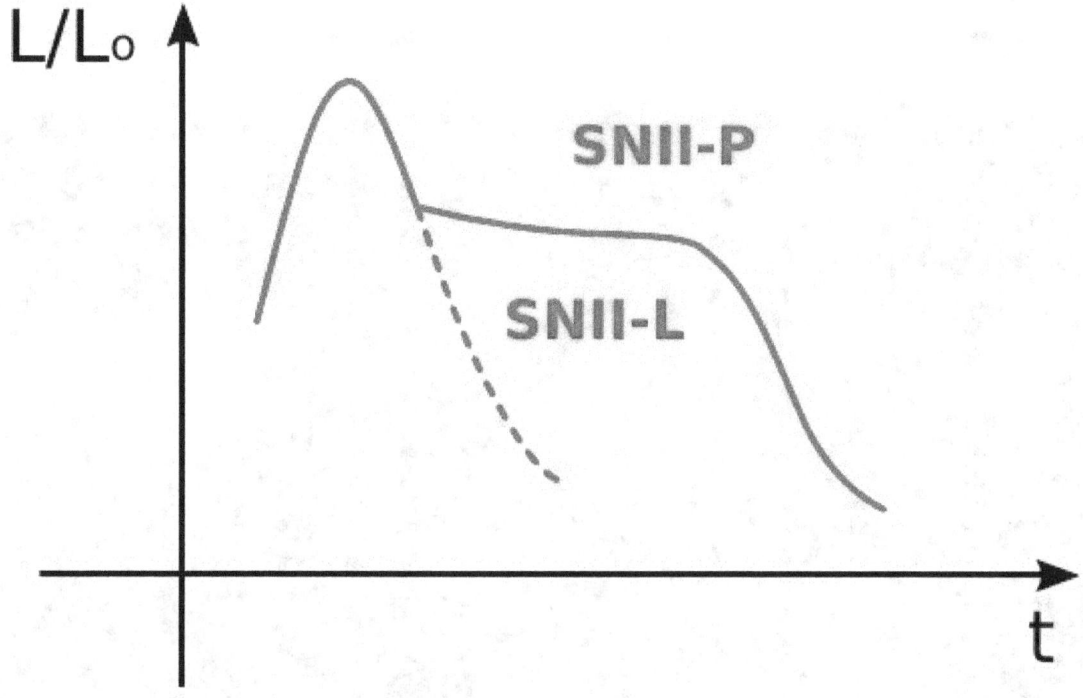

*Light curves are used to classify type II-P and type II-L supernovae*

*Sequence shows the rapid brightening and slower fading of a supernova explosion in the galaxy NGC 1365[47]*

fusion ignited on degenerate white dwarf progenitors while the spectrally similar type Ib/c are produced from massive Wolf-Rayet progenitors by core collapse. The following summarizes what astronomers currently believe are the most plausible explanations for supernovae.

## 1.5.1  Thermal runaway

Main article: Type Ia supernova

A white dwarf star may accumulate sufficient material from a stellar companion to raise its core temperature enough to

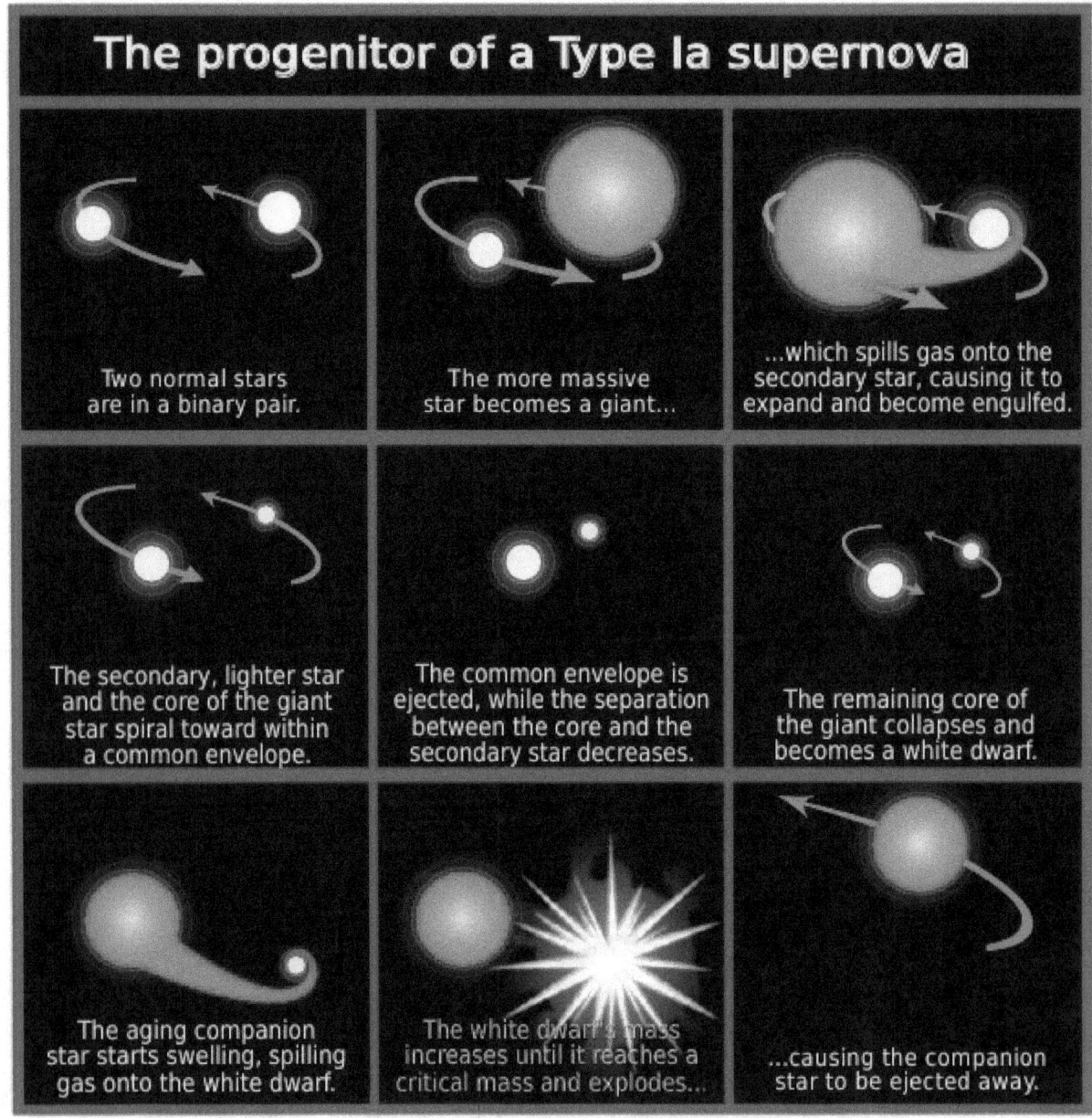

*Formation of a type Ia supernova*

ignite carbon fusion, at which point it undergoes runaway nuclear fusion, completely disrupting it. There are three avenues by which this detonation is theorized to happen: stable accretion of material from a companion, the collision of two white dwarfs, or accretion that causes ignition in a shell that then ignites. The dominant mechanism by which Type Ia supernovae are produced remains unclear.[48] Despite this uncertainty in how Type Ia supernovae are produced, Type Ia supernovae have very uniform properties, and are useful standard candles over intergalactic distances. Some calibrations are required to compensate for the gradual change in properties or different frequencies of abnormal luminosity supernovae at high red shift, and for small variations in brightness identified by light curve shape or spectrum.[49][50]

## Normal Type Ia

There are several means by which a supernova of this type can form, but they share a common underlying mechanism. If a carbon-oxygen[nb 2] white dwarf accreted enough matter to reach the Chandrasekhar limit of about 1.44 solar masses $(M\odot)$[51] (for a non-rotating star), it would no longer be able to support the bulk of its plasma through electron degeneracy pressure[52][53] and would begin to collapse. However, the current view is that this limit is not normally attained; increasing temperature and density inside the core ignite carbon fusion as the star approaches the limit (to within about 1%[54]), before collapse is initiated.[51]

Within a few seconds, a substantial fraction of the matter in the white dwarf undergoes nuclear fusion, releasing enough energy ($1-2\times10^{44}$ J)[55] to unbind the star in a supernova explosion.[56] An outwardly expanding shock wave is generated, with matter reaching velocities on the order of 5,000–20,000 km/s, or roughly 3% of the speed of light. There is also a significant increase in luminosity, reaching an absolute magnitude of −19.3 (or 5 billion times brighter than the Sun), with little variation.[57]

The model for the formation of this category of supernova is a closed binary star system. The larger of the two stars is the first to evolve off the main sequence, and it expands to form a red giant. The two stars now share a common envelope, causing their mutual orbit to shrink. The giant star then sheds most of its envelope, losing mass until it can no longer continue nuclear fusion. At this point it becomes a white dwarf star, composed primarily of carbon and oxygen.[58] Eventually the secondary star also evolves off the main sequence to form a red giant. Matter from the giant is accreted by the white dwarf, causing the latter to increase in mass. Despite widespread acceptance of the basic model, the exact details of initiation and of the heavy elements produced in the explosion are still unclear.

Type Ia supernovae follow a characteristic light curve—the graph of luminosity as a function of time—after the explosion. This luminosity is generated by the radioactive decay of nickel−56 through cobalt−56 to iron−56.[57] The peak luminosity of the light curve is extremely consistent across normal Type Ia supernovae, having a maximum absolute magnitude of about −19.3. This allows them to be used as a secondary[59] standard candle to measure the distance to their host galaxies.[60]

## Non-standard Type Ia

Another model for the formation of a Type Ia explosion involves the merger of two white dwarf stars, with the combined mass momentarily exceeding the Chandrasekhar limit.[61] There is much variation in this type of explosion,[62] and in many cases there may be no supernova at all, but it is expected that they will have a broader and less luminous light curve than the more normal Type Ia explosions.

Abnormally bright Type Ia supernovae are expected when the white dwarf already has a mass higher than the Chandrasekhar limit,[63] possibly enhanced further by asymmetry,[64] but the ejected material will have less than normal kinetic energy.

There is no formal sub-classification for the non-standard Type Ia supernovae. It has been proposed that a group of sub-luminous supernovae that occur when helium accretes onto a white dwarf should be classified as **type Iax**.[65][66] This type of supernova may not always completely destroy the white dwarf progenitor and could leave behind a zombie star.[67]

One specific type of non-standard Type Ia supernova develops hydrogen, and other, emission lines and gives the appearance of mixture between a normal Type Ia and a Type IIn supernova. Examples are SN 2002ic and SN 2005gj. These supernova have been dubbed **Type Ia/IIn**, **Type Ian**, **Type IIa** and **Type IIan**.[68]

## 1.5.2 Core collapse

Very massive stars can undergo core collapse when nuclear fusion suddenly becomes unable to sustain the core against its own gravity; this is the cause of all types of supernova except type Ia. The collapse may cause violent expulsion of the outer layers of the star resulting in a supernova, or the release of gravitational potential energy may be insufficient and the star may collapse into a black hole or neutron star with little radiated energy.

Core collapse can be caused by several different mechanisms: electron capture; exceeding the Chandrasekhar limit; pair-instability; or photodisintegration.[2][69] When a massive star develops an iron core larger than the Chandrasekhar mass it

# Supernovae / mass-metallicity

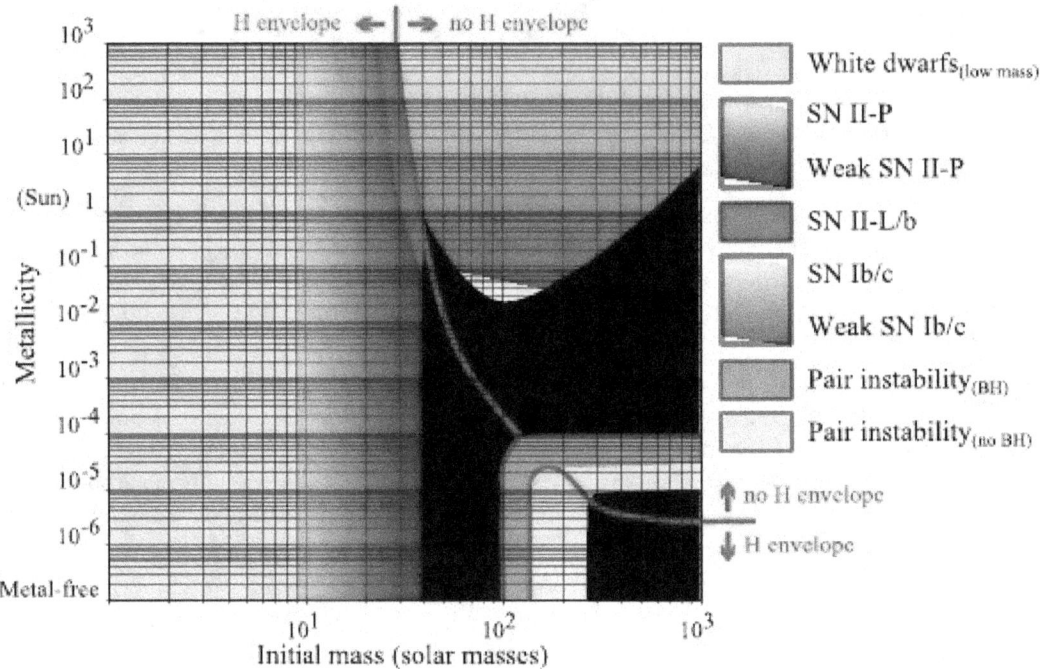

*Supernovae as initial mass-metallicity*

will no longer be able to support itself by electron degeneracy pressure and will collapse further to a neutron star or black hole. Electron capture by magnesium in a degenerate O/Ne/Mg core causes gravitational collapse followed by explosive oxygen fusion, with very similar results. Electron-positron pair production in a large post-helium burning core removes thermodynamic support and causes initial collapse followed by runaway fusion, resulting in a pair-instability supernova. A sufficiently large and hot stellar core may generate gamma-rays energetic enough to initiate photodisintegration directly, which will cause a complete collapse of the core.

The table below lists the known reasons for core collapse in massive stars, the types of star that they occur in, their associated supernova type, and the remnant produced. The metallicity is the proportion of elements other than hydrogen or helium, as compared to the Sun. The initial mass is the mass of the star prior to the supernova event, given in multiples of the Sun's mass, although the mass at the time of the supernova may be much lower.

Type IIn supernovae are not listed in the table. They can potentially be produced by various types of core collapse in different progenitor stars, possibly even by type Ia white dwarf ignitions, although it seems that most will be from iron core collapse in luminous supergiants or hypergiants (including LBVs). The narrow spectral lines for which they are named occur because the supernova is expanding into a small dense cloud of circumstellar material.[70] It appears that a significant proportion of supposed type IIn supernovae are actually supernova imposters, massive eruptions of LBV-like stars similar to the Great Eruption Eta Carinae. In these events, material previously ejected from the star creates the narrow absorption lines and causes a shock wave through interaction with the newly ejected material.[71]

When a stellar core is no longer supported against gravity it collapses in on itself with velocities reaching 70,000 km/s (0.23c),[72] resulting in a rapid increase in temperature and density. What follows next depends on the mass and structure of the collapsing core, with low mass degenerate cores forming neutron stars, higher mass degenerate cores mostly collapsing completely to black holes, and non-degenerate cores undergoing runaway fusion.

The initial collapse of degenerate cores is accelerated by beta decay, photodisintegration and electron capture, which causes a burst of electron neutrinos. As the density increases, neutrino emission is cut off as they become trapped in the core. The inner core eventually reaches typically 30 km diameter[73] and a density comparable to that of an atomic nucleus, and neutron degeneracy pressure tries to halt the collapse. If the core mass is more than about 15 $M\odot$ then

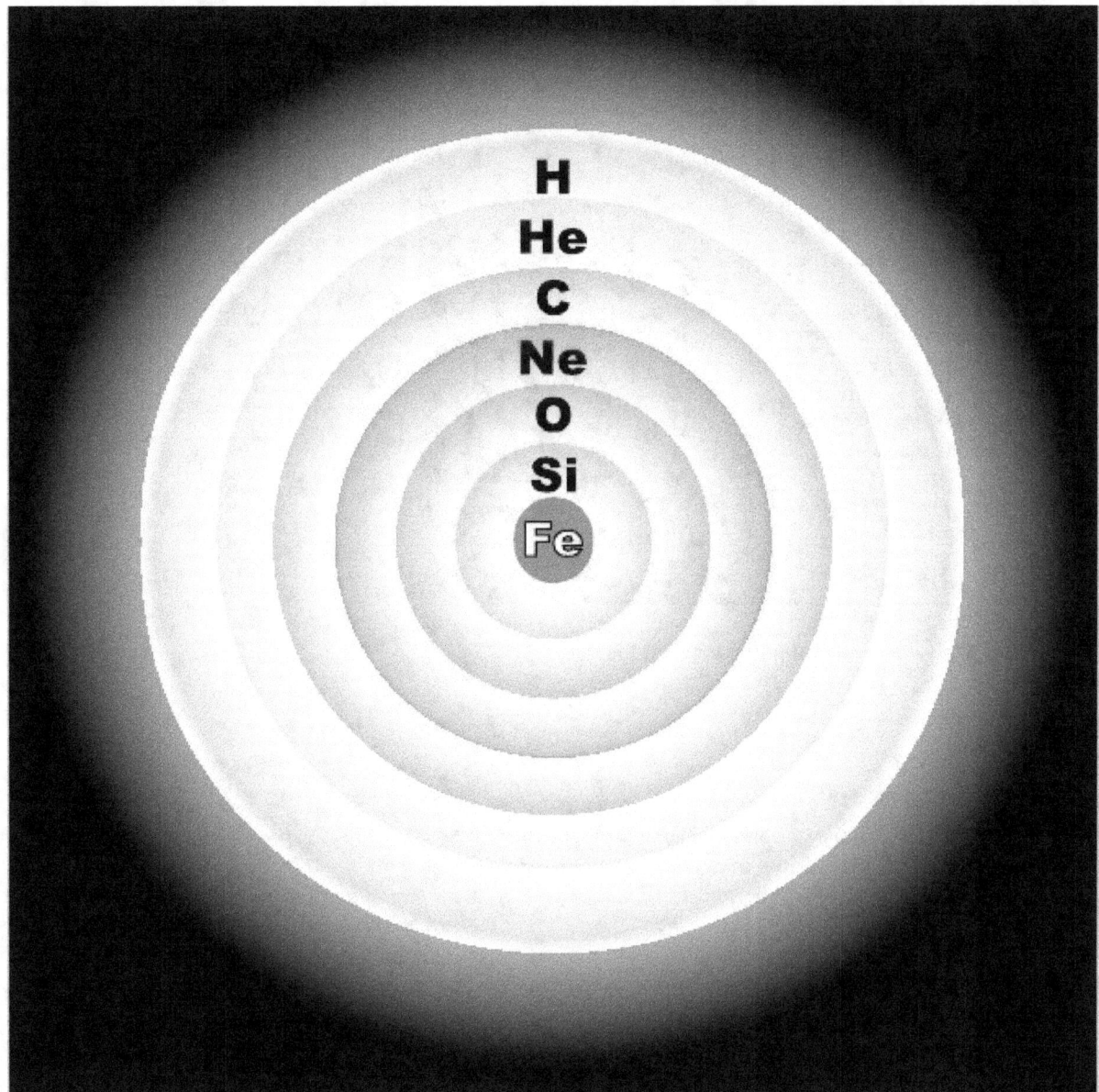

*The onion-like layers of a massive, evolved star just prior to core collapse (Not to scale)*

neutron degeneracy is insufficient to stop the collapse and a black hole forms directly with no supernova explosion.

In lower mass cores the collapse is stopped and the newly formed neutron core has an initial temperature of about 100 billion kelvin, 6000 times the temperature of the sun's core.[74] 'Thermal' neutrinos form as neutrino-antineutrino pairs of all flavors, and total several times the number of electron-capture neutrinos.[75] About $10^{46}$ joules, approximately 10% of the star's rest mass, is converted into a ten-second burst of neutrinos which is the main output of the event.[73][76] The suddenly halted core collapse rebounds and produces a shock wave that stalls within milliseconds[77] in the outer core as energy is lost through the dissociation of heavy elements. A process that is not clearly understood is necessary to allow the outer layers of the core to reabsorb around $10^{44}$ joules[76] (1 foe) from the neutrino pulse, producing the visible explosion, although there are also other theories on how to power the explosion.[73]

Some material from the outer envelope falls back onto the neutron star, and for cores beyond about 8 $M\odot$ there is sufficient fallback to form a black hole. This fallback will reduce the kinetic energy of the explosion and the mass of expelled radioactive material, but in some situations it may also generate relativistic jets that result in a gamma-ray burst

# Remnants of massive single stars

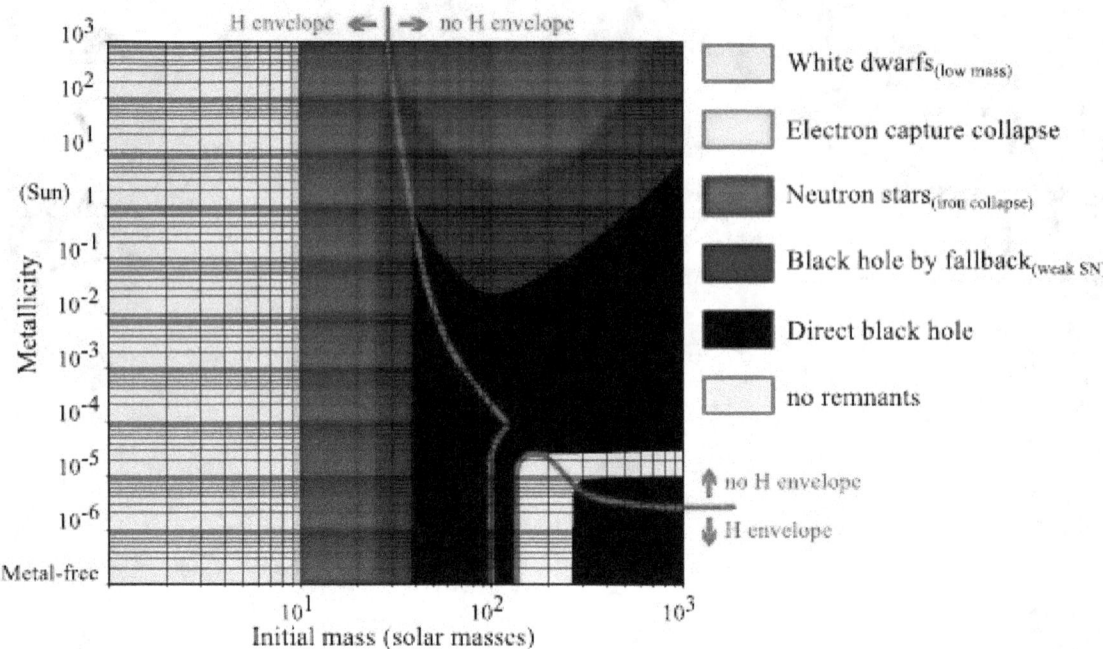

*Remnants of single massive stars*

or an exceptionally luminous supernova.

Collapse of massive non-degenerate cores will ignite further fusion. When the core collapse is initiated by pair instability, oxygen fusion begins and the collapse may be halted. For core masses of 40–60 $M\odot$, the collapse halts and the star remains intact, but core collapse will occur again when a larger core has formed. For cores of around 60–130 $M\odot$, the fusion of oxygen and heavier elements is so energetic that the entire star is disrupted, causing a supernova. At the upper end of the mass range, the supernova is unusually luminous and extremely long-lived due to many solar masses of ejected $Ni_{56}$. For even larger core masses, the core temperature becomes high enough to allow photodisintegration and the core collapses completely into a black hole.[78]

## Type II

Main article: Type II supernova
 Stars with initial masses less than about eight times the sun never develop a core large enough to collapse and they eventually lose their atmospheres to become white dwarfs. Stars with at least 9 $M\odot$ (possibly as much as 12 $M\odot$[79]) evolve in a complex fashion, progressively burning heavier elements at hotter temperatures in their cores.[73][80] The star becomes layered like an onion, with the burning of more easily fused elements occurring in larger shells.[2][81] Although popularly described as an onion with an iron core, the least massive supernova progenitors only have oxygen-neon(-magnesium) cores. These super AGB stars may form the majority of core collapse supernovae, although less luminous and so less commonly observed than those from more massive progenitors.[79]

If core collapse occurs during a supergiant phase when the star still has a hydrogen envelope, the result is a type II supernova. The rate of mass loss for luminous stars depends on the metallicity and luminosity. Extremely luminous stars at near solar metallicity will lose all their hydrogen before they reach core collapse and so will not form a type II supernova. At low metallicity, all stars will reach core collapse with a hydrogen envelope but sufficiently massive stars collapse directly to a black hole without producing a visible supernova.

Stars with an initial mass up to about 90 times the sun, or a little less at high metallicity, are expected to result in a type II-P supernova which is the most commonly observed type. At moderate to high metallicity, stars near the upper end of that

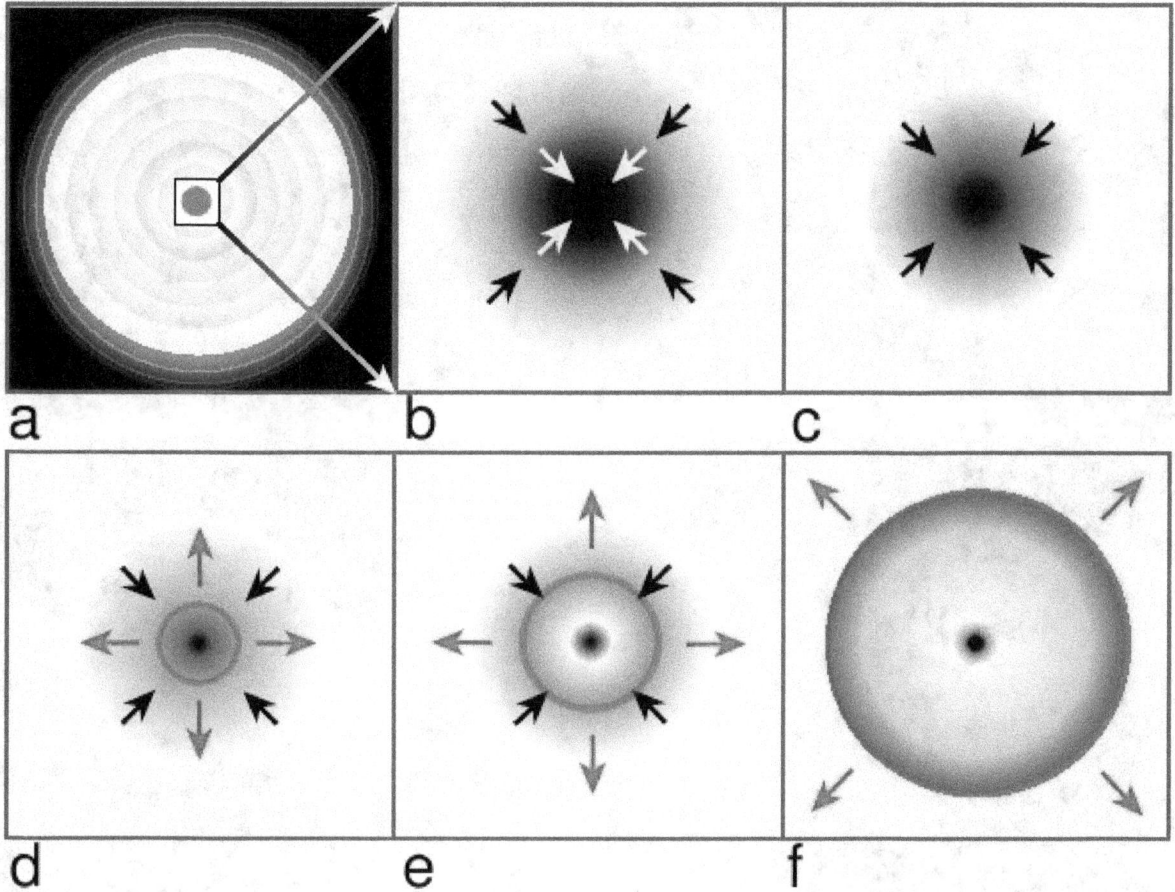

*Within a massive, evolved star (a) the onion-layered shells of elements undergo fusion, forming an iron core (b) that reaches Chandrasekhar-mass and starts to collapse. The inner part of the core is compressed into neutrons (c), causing infalling material to bounce (d) and form an outward-propagating shock front (red). The shock starts to stall (e), but it is re-invigorated by a process that may include neutrino interaction. The surrounding material is blasted away (f), leaving only a degenerate remnant.*

mass range will have lost most of their hydrogen when core collapse occurs and the result will be a type II-L supernova. At very low metallicity, stars of around 140–250 $M\odot$ will reach core collapse by pair instability while they still have a hydrogen atmosphere and an oxygen core and the result will be a supernova with type II characteristics but a very large mass of ejected $^{56}$Ni and high luminosity.

## Type Ib and Ic

Main article: Type Ib and Ic supernovae

These supernovae, like those of Type II, are massive stars that undergo core collapse. However the stars which become Types Ib and Ic supernovae have lost most of their outer (hydrogen) envelopes due to strong stellar winds or else from interaction with a companion.[84] These stars are known as Wolf-Rayet stars, and they occur at moderate to high metallicity where continuum driven winds cause sufficiently high mass loss rates. Observations of type Ib/c supernova do not match the observed or expected occurrence of Wolf Rayet stars and alternate explanations for this type of core collapse supernova involve stars stripped of their hydrogen by binary interactions. Binary models provide a better match for the observed supernovae, with the proviso that no suitable binary helium stars have ever been observed.[85] Since a supernova explosion can occur whenever the mass of the star at the time of core collapse is low enough not to cause complete fallback to a black hole, any massive star may result in a supernova if it loses enough mass before core collapse occurs.

Type Ib supernovae are the more common and result from Wolf-Rayet stars of type WC which still have helium in their

*The atypical subluminous type II SN 1997D*

atmospheres. For a narrow range of masses, stars evolve further before reaching core collapse to become WO stars with very little helium remaining and these are the progenitors of type Ic supernovae.

A few percent of the Type Ic supernovae are associated with gamma-ray bursts (GRB), though it is also believed that any hydrogen-stripped Type Ib or Ic supernova could produce a GRB, depending on the geometry of the explosion.[86] The mechanism for producing this type of GRB is the jets produced by the magnetic field of the rapidly spinning magnetar formed at the collapsing core of the star. The jets would also transfer energy into the expanding outer shell of the explosion to produce a super-luminous supernova.[87]

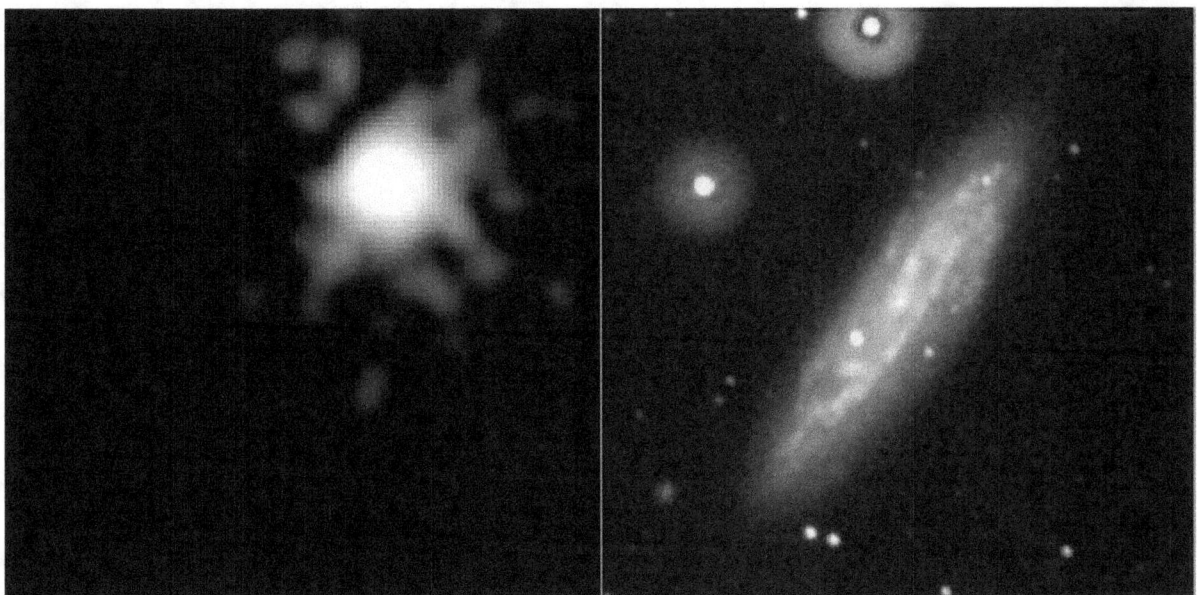

*SN 2008D, a Type Ib[82] supernova, shown in X-ray (left) and visible light (right) at the far upper end of the galaxy[83]*

Ultra-stripped supernovae occur when the exploding star has been stripped (almost) all the way to the metal core, via mass transfer in a close binary.[88] As a result, very little material is ejected from the exploding star (~0.1 MS$_{un}$). In the most extreme cases, ultra-stripped supernovae can occur in naked metal cores, barely above the Chandrasekhar mass limit. SN 2005ek [89] might be an observational example of an ultra-stripped supernova, giving rise to a relatively dim and fast decaying light curve. The nature of ultra-stripped supernovae can be both iron core-collapse and electron capture supernovae, depending on the mass of the collapsing core.

### 1.5.3 Failed

The core collapse of some massive stars may not result in a visible supernova. The main model for this is a sufficiently massive core that the explosion is insufficient to reverse the infall of the outer layers onto a black hole. These events are difficult to detect, but large surveys have detected possible candidates.[90][91]

### 1.5.4 Light curves

The visual light curves of the different supernova types vary in shape and amplitude, based on the underlying mechanisms of the explosion, the way that visible radiation is produced, and the transparency of the ejected material. The light curves can be significantly different at other wavelengths. For example, at UV and shorter wavelengths there is an extremely luminous peak lasting just a few hours, corresponding to the shock breakout of the initial explosion, which is hardly detectable at longer wavelengths.

The light curves for type Ia are mostly very uniform, with a consistent maximum absolute magnitude and a relatively steep decline in luminosity. The energy output is driven by radioactive decay of nickel-56 (half life 6 days), which then decays to radioactive cobalt-56 (half life 77 days). These radioisotopes from material ejected in the explosion excite surrounding material to incandescence. The initial phases of the light curve decline steeply as the effective size of the photosphere decreases and trapped electromagnetic radiation is depleted. The light curve continues to decline in the B band while it may show a small shoulder in the visual at about 40 days, but this is only a hint of a secondary maximum that occurs in the infra-red as certain ionised heavy elements recombine to produce infra-red radiation and the ejecta become transparent to it. The visual light curve continues to decline at a rate slightly greater than the decay rate of the radioactive cobalt (which has the longer half life and controls the later curve), because the ejected material becomes more diffuse and less able to convert the high energy radiation into visual radiation. After several months, the light curve changes its decline rate again

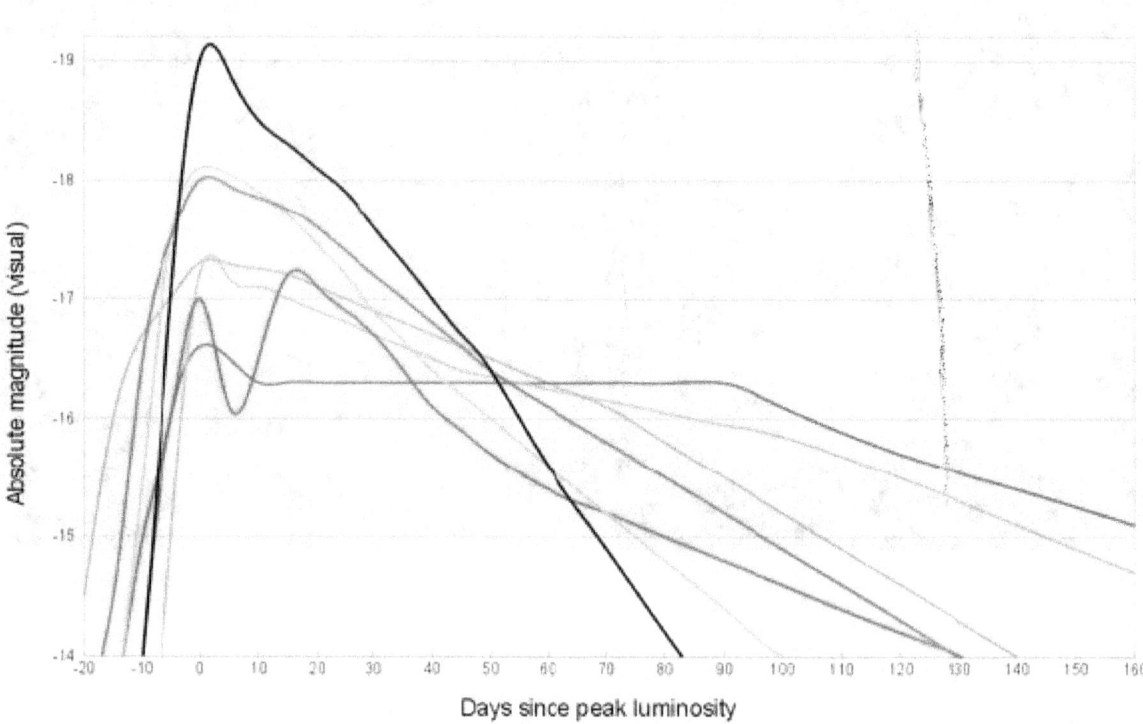

Comparative supernova type light curves

as positron emission becomes dominant from the remaining cobalt-56, although this portion of the light curve has been little-studied.

Type Ib and Ic light curves are basically similar to type Ia although with a lower average peak luminosity. The visual light output is again due to radioactive decay being converted into visual radiation, but there is a much lower mass of nickel-56 produced in these types of explosion. The peak luminosity varies considerably and there are even occasional type Ib/c supernovae orders of magnitude more and less luminous than the norm. The most luminous type Ic supernovae are referred to as hypernovae and tend to have broadened light curves in addition to the increases peak luminosity. The source of the extra energy is thought to be relativistic jets driven by the formation of a rotating black hole, which also produce gamma-ray bursts.

The light curves for type II supernovae are characterised by a much slower decline than type I, on the order of 0.05 magnitudes per day,[92] excluding the plateau phase. The visual light output is dominated by kinetic energy rather than radioactive decay for several months, due primarily to the existence of hydrogen in the ejecta from the atmosphere of the supergiant progenitor star. In the initial explosion this hydrogen becomes heated and ionised. The majority of type II supernovae show a prolonged plateau in their light curves as this hydrogen recombines, emitting visible light and becoming more transparent. This is then followed by a declining light curve driven by radioactive decay although slower than in type I supernovae, due to the efficiency of conversion into light by all the hydrogen.[44]

In type II-L the plateau is absent because the progenitor had relatively little hydrogen left in its atmosphere, sufficient to appear in the spectrum but insufficient to produce a noticeable plateau in the light output. In type IIb supernovae the hydrogen atmosphere of the progenitor is so depleted (thought to be due to tidal stripping by a companion star) that the light curve is closer to a type I supernova and the hydrogen even disappears from the spectrum after several weeks.[44]

Type IIn supernovae are characterised by additional narrow spectral lines produced in a dense shell of circumstellar material. Their light curves are generally very broad and extended, occasionally also extremely luminous and referred to as a hypernova. These light curves are produced by the highly efficient conversion of kinetic energy of the ejecta into electromagnetic radiation by interaction with the dense shell of material. This only occurs when the material is sufficiently dense and compact, indicating that it has been produced by the progenitor star itself only shortly before the supernova occurs.

Large numbers of supernovae have been catalogued and classified to provide distance candles and test models. Average characteristics vary somewhat with distance and type of host galaxy, but can broadly be specified for each supernova type.

Notes:

- a. ^ Faint types may be a distinct sub-class. Bright types may be a continuum from slightly over-luminous to hypernovae.

- b. ^ These magnitudes are measured in the R band. Measurements in V or B bands are common and will be around half a magnitude brighter for supernovae.

- c. ^ Order of magnitude kinetic energy. Total electromagnetic radiated energy is usually lower, (theoretical) neutrino energy much higher.

- d. ^ Probably a heterogeneous group, any of the other types embedded in nebulosity.

## 1.5.5   Asymmetry

A long-standing puzzle surrounding Type II supernovae is why the compact object remaining after the explosion is given a large velocity away from the epicentre;[96] pulsars, and thus neutron stars, are observed to have high velocities, and black holes presumably do as well, although they are far harder to observe in isolation. The initial impetus can be substantial, propelling an object of more than a solar mass at a velocity of 500 km/s or greater. This indicates an asymmetry in the explosion, but the mechanism by which momentum is transferred to the compact object remains a puzzle. Proposed explanations for this kick include convection in the collapsing star and jet production during neutron star formation.

One possible explanation for the asymmetry in the explosion is large-scale convection above the core. The convection can create variations in the local abundances of elements, resulting in uneven nuclear burning during the collapse, bounce and resulting explosion.[97]

Another possible explanation is that accretion of gas onto the central neutron star can create a disk that drives highly directional jets, propelling matter at a high velocity out of the star, and driving transverse shocks that completely disrupt the star. These jets might play a crucial role in the resulting supernova explosion.[98][99] (A similar model is now favored for explaining long gamma-ray bursts.)

Initial asymmetries have also been confirmed in Type Ia supernova explosions through observation. This result may mean that the initial luminosity of this type of supernova depends on the viewing angle. However, the explosion becomes more symmetrical with the passage of time. Early asymmetries are detectable by measuring the polarization of the emitted light.[100]

## 1.5.6   Energy output

Although we are used to thinking of supernovae primarily as luminous visible events, the electromagnetic radiation they produce is almost a minor side-effect of the explosion. Particularly in the case of core collapse supernovae, the emitted electromagnetic radiation is a tiny fraction of the total event energy.

There is a fundamental difference between the balance of energy production in the different types of supernova. In type Ia white dwarf detonations, most of the explosion energy is directed into heavy element synthesis and kinetic energy of the ejecta. In core collapse supernovae, the vast majority of the energy is directed into neutrino emission, and while some of this apparently powers the main explosion 99%+ of the neutrinos escape in the first few minutes following the start of the collapse.

Type Ia supernovae derive their energy from runaway nuclear fusion of a carbon-oxygen white dwarf. Details of the energetics are still not fully modelled, but the end result is the ejection of the entire mass of the original star with high kinetic energy. Around half a solar mass of this is $Ni_{56}$ generated from silicon burning. $Ni_{56}$ is radioactive and generates $Co_{56}$ by beta plus decay with a half life of six days, plus gamma rays. $Co_{56}$ itself decays by the beta plus path with a

*The pulsar in the Crab nebula is travelling at 375 km/s relative to the nebula.*[95]

half life of 77 days to stable $Fe_{56}$. These two processes are responsible for the electromagnetic radiation from type Ia supernovae. In combination with the changing transparency of the ejected material, they produce the rapidly declining light curve.[101]

Core collapse supernovae are on average visually fainter than type Ia supernovae, but the total energy released is far higher. This is driven by gravitational potential energy from the core collapse, initially producing electron neutrinos from disintegrating nucleons, followed by all flavours of thermal neutrinos from the super-heated neutron star core. Around 1% of these neutrinos are thought to deposit sufficient energy into the outer layers of the star to drive the resulting explosion, but again the details cannot be reproduced exactly in current models. Kinetic energies and nickel yields are somewhat lower than type Ia supernovae, hence the reduced visual luminosity, but energy from the ionisation of the many solar masses of remaining hydrogen can contribute to a much slower decline in luminosity and produce the plateau phase seen in the majority of core collapse supernovae.

In some core collapse supernovae, fallback onto a black hole drives relativistic jets which may produce a brief energetic

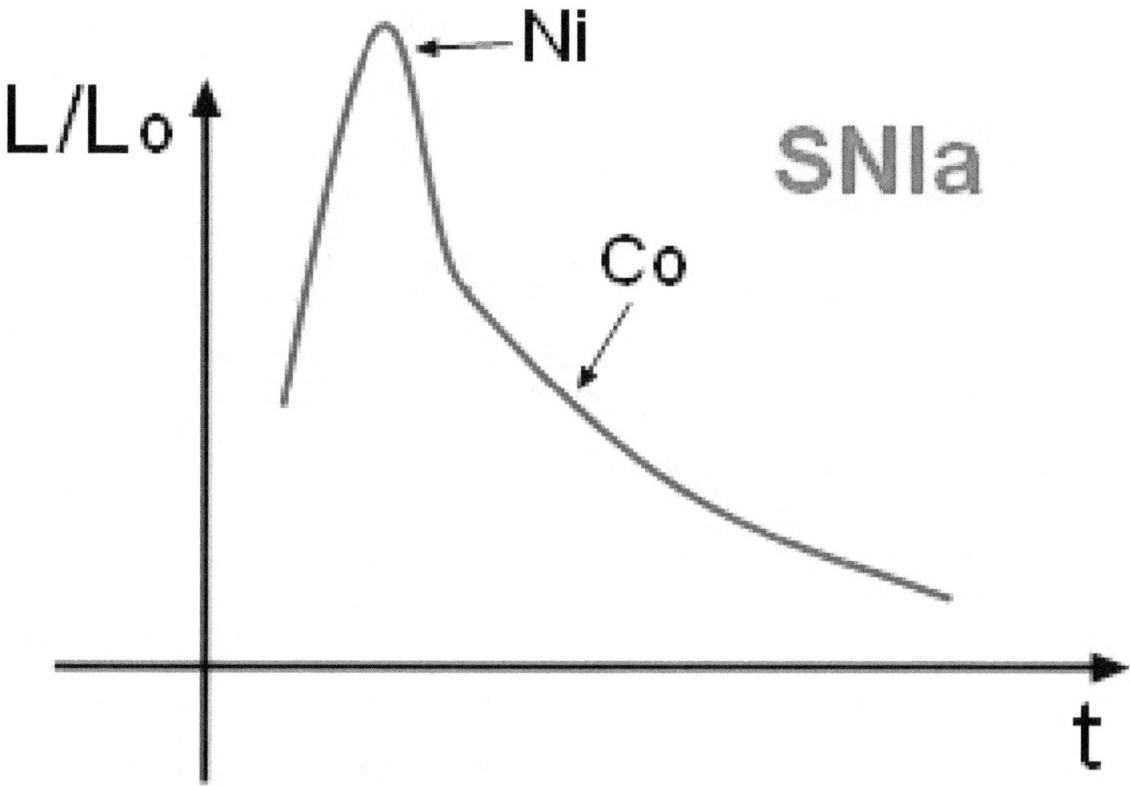

*The radioactive decays of nickel-56 and cobalt-56 that produce a supernova visible light curve*

and directional burst of gamma rays and also transfers substantial further energy into the ejected material. This is one scenario for producing high luminosity supernovae and is thought to be the cause of type Ic hypernovae and long duration gamma-ray bursts. If the relativistic jets are too brief and fail to penetrate the stellar envelope then a low luminosity gamma-ray burst may be produced and the supernova may be sub-luminous.

When a supernova occurs inside a small dense cloud of circumstellar material then it will produce a shock wave that can efficiently convert a high fraction of the kinetic energy into electromagnetic radiation. Even though the initial explosion energy was entirely normal the resulting supernova will have high luminosity and extended duration since it does not rely on exponential radioactive decay. This type of event may cause type IIn hypernovae.

Although pair-instability supernovae are core collapse supernovae with spectra and light curves similar to type II-P, the nature of the explosion following core collapse is more like a giant type Ia with runaway fusion of carbon, oxygen, and silicon. The total energy released by the highest mass events is comparable to other core collapse supernovae but neutrino production is thought to be very low, hence the kinetic and electromagnetic energy is very high. The cores of these stars are much larger than any white dwarf and the amount of radioactive nickel and other heavy elements ejected can be orders of magnitude higher, with consequently high visual luminosity.

## 1.5.7 Progenitor

The supernova classification type is closely tied to the type of star at the time of the explosion. The occurrence of each type of supernova depends dramatically on the metallicity and hence the age of the host galaxy.

Type Ia supernovae are produced from white dwarf stars in binary systems and occur in all galaxy types. Core collapse supernovae are only found in galaxies undergoing current or very recent star formation, since they result from short-lived massive stars. They are most commonly found in type Sc spirals, but also in the arms of other spiral galaxies and in irregular galaxies, especially starburst galaxies.

Type Ib/c and II-L, and possibly most type IIn, supernovae are only thought to be produced from stars having near-solar metallicity levels that result in high mass loss from massive stars, hence they are less common in older more distant galaxies. The table shows the expected progenitor for the main types of core collapse supernova, and the approximate proportions of each in the local neighbourhood.

There are a number of difficulties reconciling modelled and observed stellar evolution leading up to core collapse supernovae. Red supergiants are the expected progenitors for the vast majority of core collapse supernovae, and these have been observed but only at relatively low masses and luminosities, below about 18 $M\odot$ and 100,000 $L\odot$ respectively. Most progenitors of type II supernovae are not detected and must be considerably fainter, and presumably less massive. It is now proposed that higher mass red supergiants do not explode as supernovae, but instead evolve back towards hotter temperatures. Several progenitors of type IIb supernovae have been confirmed, and these were K and G supergiants, plus one A supergiant.[106] Yellow hypergiants or LBVs are proposed progenitors for type IIb supernovae, and almost all type IIb supernovae near enough to observe have shown such progenitors.[107][108]

Until just a few decades ago, hot supergiants were not considered likely to explode, but observations have shown otherwise. Blue supergiants form an unexpectedly high proportion of confirmed supernova progenitors, partly due to their high luminosity and easy detection, while not a single Wolf-Rayet progenitor has yet been clearly identified.[106][109] Models have had difficulty showing how blue supergiants lose enough mass to reach supernova without progressing to a different evolutionary stage. One study has shown a possible route for low-luminosity post-red supergiant luminous blue variables to collapse, most likely as a type IIn supernova.[110]

The expected progenitors of type Ib supernovae, luminous WC stars, are not observed at all. Instead WC stars are found at lower luminosities, apparently post-red supergiant stars. WO stars are extremely rare and visually relatively faint, so it is difficult to say whether such progenitors are missing or just yet to be observed. Very luminous progenitors, despite numerous supernovae being observed near enough that such progenitors would have been clearly imaged.[111] Several examples of hot luminous progenitors of type IIn supernovae have been detected: SN 2005gy and SN 2010jl were both apparently massive luminous stars, but are very distant; and SN 2009ip had a highly luminous progenitor likely to have been an LBV, but is a peculiar supernova whose exact nature is disputed.[106]

## 1.6    Interstellar impact

### 1.6.1    Source of heavy elements

Main article: Supernova nucleosynthesis

Supernovae are a key source of elements heavier than oxygen.[112] These elements are produced by nuclear fusion (for iron$-56$ and lighter elements), and by nucleosynthesis during the supernova explosion for elements heavier than iron.[113] Supernovae are the most likely, although not undisputed, candidate sites for the r-process, which is a rapid form of nucleosynthesis that occurs under conditions of high temperature and high density of neutrons. The reactions produce highly unstable nuclei that are rich in neutrons. These forms are unstable and rapidly beta decay into more stable forms.

The r-process reaction, which is likely to occur in type II supernovae, produces about half of all the element abundance beyond iron, including plutonium and uranium.[114] The only other major competing process for producing elements heavier than iron is the s-process in large, old red giant stars, which produces these elements much more slowly, and which cannot produce elements heavier than lead.[115]

### 1.6.2    Role in stellar evolution

Main article: Supernova remnant

The remnant of a supernova explosion consists of a compact object and a rapidly expanding shock wave of material. This cloud of material sweeps up the surrounding interstellar medium during a free expansion phase, which can last for up to two centuries. The wave then gradually undergoes a period of adiabatic expansion, and will slowly cool and mix with the surrounding interstellar medium over a period of about 10,000 years.[116]

*Supernova remnant N 63A lies within a clumpy region of gas and dust in the Large Magellanic Cloud.*

The Big Bang produced hydrogen, helium, and traces of lithium, while all heavier elements are synthesized in stars and supernovae. Supernovae tend to enrich the surrounding interstellar medium with *metals*—elements other than hydrogen and helium.

These injected elements ultimately enrich the molecular clouds that are the sites of star formation.[117] Thus, each stellar generation has a slightly different composition, going from an almost pure mixture of hydrogen and helium to a more metal-rich composition. Supernovae are the dominant mechanism for distributing these heavier elements, which are formed in a star during its period of nuclear fusion. The different abundances of elements in the material that forms a star have important influences on the star's life, and may decisively influence the possibility of having planets orbiting it.

The kinetic energy of an expanding supernova remnant can trigger star formation due to compression of nearby, dense molecular clouds in space.[118] The increase in turbulent pressure can also prevent star formation if the cloud is unable to lose the excess energy.[10]

Evidence from daughter products of short-lived radioactive isotopes shows that a nearby supernova helped determine the composition of the Solar System 4.5 billion years ago, and may even have triggered the formation of this system.[119] Supernova production of heavy elements over astronomic periods of time ultimately made the chemistry of life on Earth possible.

### 1.6.3 Effect on Earth

Main article: Near-Earth supernova

A **near-Earth supernova** is a supernova close enough to the Earth to have noticeable effects on its biosphere. Depending upon the type and energy of the supernova, it could be as far as 3000 light-years away. Gamma rays from a supernova would induce a chemical reaction in the upper atmosphere converting molecular nitrogen into nitrogen oxides, depleting the ozone layer enough to expose the surface to harmful solar radiation. This has been proposed as the cause of the Ordovician–Silurian extinction, which resulted in the death of nearly 60% of the oceanic life on Earth.[120] In 1996 it was theorized that traces of past supernovae might be detectable on Earth in the form of metal isotope signatures in rock strata. Iron-60 enrichment was later reported in deep-sea rock of the Pacific Ocean.[121][122][123] In 2009, elevated levels of nitrate ions were found in Antarctic ice, which coincided with the 1006 and 1054 supernovae. Gamma rays from these supernovae could have boosted levels of nitrogen oxides, which became trapped in the ice.[124]

Type Ia supernovae are thought to be potentially the most dangerous if they occur close enough to the Earth. Because these supernovae arise from dim, common white dwarf stars, it is likely that a supernova that can affect the Earth will occur unpredictably and in a star system that is not well studied. The closest known candidate is IK Pegasi (see below).[125] Recent estimates predict that a Type II supernova would have to be closer than eight parsecs (26 light-years) to destroy half of the Earth's ozone layer.[126]

## 1.7 Milky Way candidates

Main article: List of supernova candidates

Several large stars within the Milky Way have been suggested as possible supernovae within the next million years. These include Rho Cassiopeiae,[128] Eta Carinae,[129] RS Ophiuchi,[130] U Scorpii,[131] VY Canis Majoris,[132] Betelgeuse, and Antares.[133] Many Wolf–Rayet stars, such as Gamma Velorum,[134] WR 104,[135] and those in the Quintuplet Cluster,[136] are also considered possible precursor stars to a supernova explosion in the 'near' future.

The nearest supernova candidate is IK Pegasi (HR 8210), located at a distance of 150 light-years. This closely orbiting binary star system consists of a main sequence star and a white dwarf 31 million kilometres apart. The dwarf has an estimated mass 1.15 times that of the Sun.[137] It is thought that several million years will pass before the white dwarf can accrete the critical mass required to become a Type Ia supernova.[138]

## 1.8 See also

- List of supernovae
- List of supernova remnants
- Quark-nova
- Supernova impostor
- Supernovae in fiction
- Timeline of white dwarfs, neutron stars, and supernovae

## 1.9 Notes

[1] The value is obtained by converting the suffix "nc" from bijective base-26, with $a = 1$, $b = 2$, $c = 3$, ... $z = 26$. Thus $nc = n \times 26 + c = 14 \times 26 + 3 = 367$.

[2] For a core primarily composed of oxygen, neon and magnesium, the collapsing white dwarf will typically form a neutron star. In this case, only a fraction of the star's mass will be ejected during the collapse.

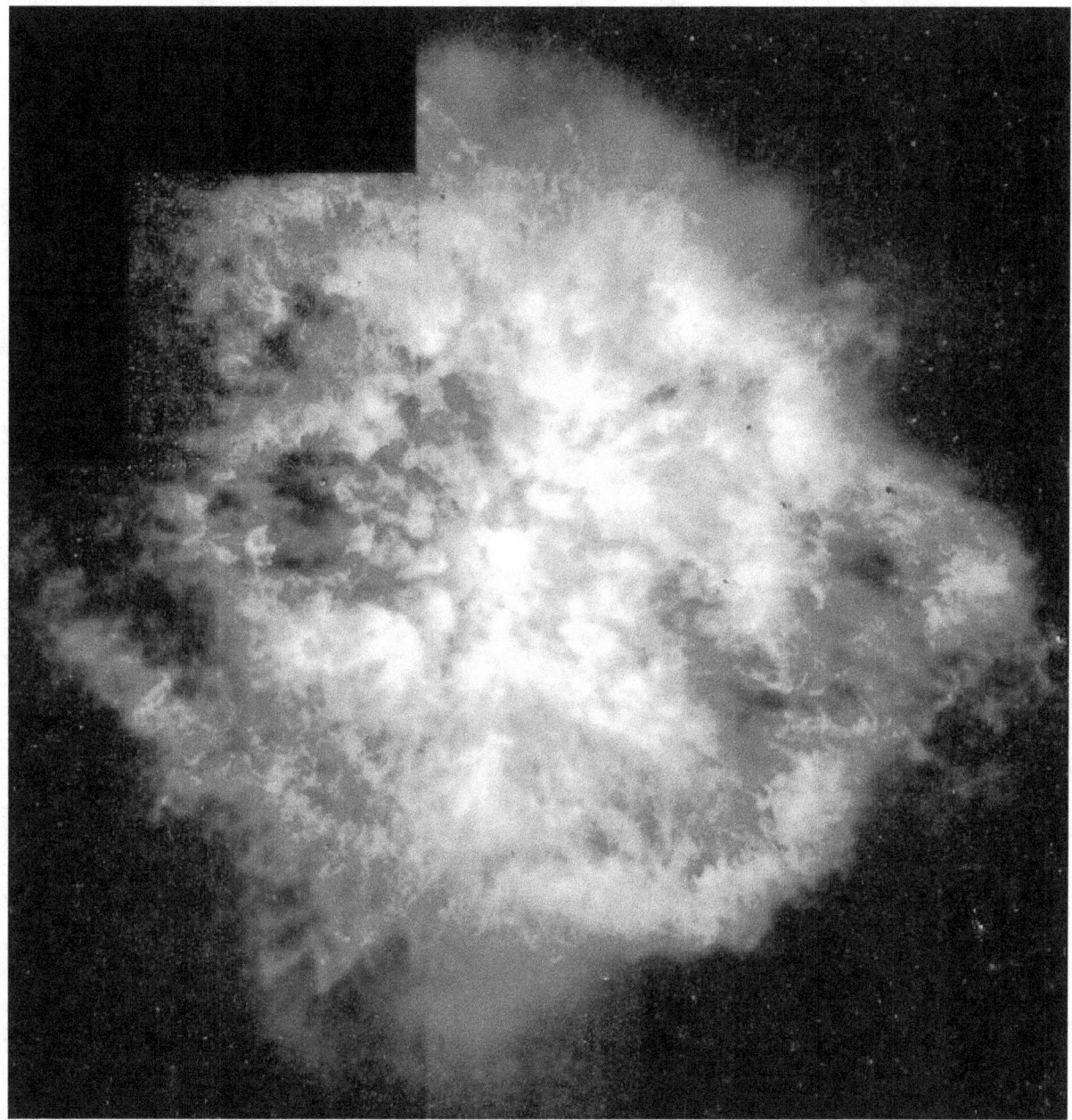

*The nebula around Wolf–Rayet star WR124, which is located at a distance of about 21,000 light years.[127]*

## 1.10 References

[1] Giacobbe, Frederick W. (2005). "How a Type II Supernova Explodes". *Electronic Journal of Theoretical Physics* **2** (6): 30–38. Bibcode:2005EJTP....2f..30G.

[2] Heger, Alexander; Fryer, Chris L.; Woosley, Stanford E.; Langer, Norbert; Hartmann, Dieter H. (2003). "How Massive Single Stars End Their Life". *Astrophysical Journal* **591**: 288. arXiv:astro-ph/0212469. Bibcode:2003ApJ...591..288H. doi:10.1086/375341.

[3] Schawinski, Kevin; et al. (2008). "Supernova Shock Breakout from a Red Supergiant". *Science* **321** (5886): 223–226. arXiv:0803.3596. Bibcode:2008Sci...321..223S. doi:10.1126/science.1160456. PMID 18556514.

[4] Ott, Christian D.; et al. (2012). "Core-Collapse Supernovae, Neutrinos, and Gravitational Waves". *Nuclear Physics B: Proceed-

*ings Supplement* **235**: 381. arXiv:1212.4250. Bibcode:2013NuPhS.235..381O. doi:10.1016/j.nuclphysbps.2013.04.036.

[5] Ackermann, M.; et al. (2013). "Detection of the Characteristic Pion-Decay Signature in Supernova Remnants". *Science* **339** (6121): 807–11. arXiv:1302.3307. Bibcode:2013Sci...339..807A. doi:10.1126/science.1231160. PMID 23413352.

[6] Osterbrock, Donald E. (2001). "Who Really Coined the Word Supernova? Who First Predicted Neutron Stars?". *Bulletin of the American Astronomical Society* **33**: 1330. Bibcode:2001AAS...199.1501O.

[7] Reynolds, Stephen P.; et al. (2008). "The Youngest Galactic Supernova Remnant: G1.9+0.3". *The Astrophysical Journal Letters* **680** (1): L41–L44. arXiv:0803.1487. Bibcode:2008ApJ...680L..41R. doi:10.1086/589570.

[8] Adams, Scott M.; Kochanek, Christopher S.; Beacom, John F.; Vagins, Mark R.; Stanek, Krzysztof Z. (2013). "Observing the Next Galactic Supernova". *The Astrophysical Journal* **778** (2): 164. arXiv:1306.0559. Bibcode:2013ApJ...778..164A. doi:10.1088/0004-637X/778/2/164.

[9] Whittet, Doug C. B. (2003). *Dust in the Galactic Environment*. CRC Press. pp. 45–46. ISBN 0-7503-0624-6.

[10] Krebs, J.; Hillebrandt, Wolfgang (1983). "The interaction of supernova shockfronts and nearby interstellar clouds". *Astronomy and Astrophysics* **128**: 411. Bibcode:1983A&A...128..411K.

[11] Boss, A. P.; Ipatov, S. I.; Keiser, S. A.; Myhill, E. A.; Vanhala, H. A. T. (2008). "Simultaneous Triggered Collapse of the Presolar Dense Cloud Core and Injection of Short-Lived Radioisotopes by a Supernova Shock Wave". *The Astrophysical Journal Letters* **686** (2): L119–L122. arXiv:0809.3045. Bibcode:2008ApJ...686L.119B. doi:10.1086/593057.

[12] Motz, Lloyd; Weaver, Jefferson Hane (2001). *The Story of Astronomy*. Basic Books. p. 76. ISBN 0-7382-0586-9.

[13] Winkler, P. Frank; Gupta, Gaurav; Long, Knox S. (2003). "The SN 1006 Remnant: Optical Proper Motions, Deep Imaging, Distance, and Brightness at Maximum". *Astrophysical Journal* **585** (1): 324. arXiv:astro-ph/0208415. Bibcode:2003ApJ...585..324W.doi:10.1086/345985.

[14] Clark, David H.; Stephenson, Francis Richard (1982). "The Historical Supernovae". *Supernovae: A survey of current research; Proceedings of the Advanced Study Institute, Cambridge, England, June 29 – July 10, 1981*. Dordrecht: D. Reidel. pp. 355–370. Bibcode:1982ssrc.conf..355C.

[15] Baade, Walter (1943). "No. 675. Nova Ophiuchi of 1604 as a supernova". *Contributions from the Mount Wilson Observatory / Carnegie Institution of Washington* **675**: 1–9. Bibcode:1943CMWCI.675....1B.

[16] Murdin, Paul & Lesley (1978). *Supernovae*. New York, NY: Press Syndicate of the University of Cambridge. pp. 1–3. ISBN 052130038X.

[17] da Silva, Luiz Augusto L. (1993). "The Classification of Supernovae". *Astrophysics and Space Science* **202** (2): 215–236. Bibcode:1993Ap&SS.202..215D. doi:10.1007/BF00626878.

[18] Kowal, Charles T. (1968). "Absolute magnitudes of supernovae". *Astronomical Journal* **73**: 1021–1024. Bibcode:1968AJ.....73. doi:10.1086/110763.

[19] Leibundgut, B.; et al. (2003). "A cosmological surprise: The universe accelerates". *Europhysics News* **32**(4): 121. Bibcode: doi:10.1086/378560.

[20] Fabian, Andrew C. (2008). "A Blast from the Past". *Science* **320** (5880): 1167–1168. doi:10.1126/science.1158538. PMID 18511676.

[21] Aschenbach, Bernd (1998). "Discovery of a young nearby supernova remnant". *Nature* **396**(6707): 141–142. Bibcode:1998 doi:10.1038/24103.

[22] Iyudin, A. F.; et al. (1998). "Emission from $^{44}$Ti associated with a previously unknown Galactic supernova". *Nature* **396** (6707): 142–144. Bibcode:1998Natur.396..142I. doi:10.1038/24106.

[23] Motizuki, Y.; et al. (2009). "An Antarctic ice core recording both supernovae and solar cycles". arXiv:0902.3446 [astro-ph.HE].

[24] Baade, Walter; Zwicky, Fritz (1934). "On Super-novae". *Proceedings of the National Academy of Sciences* **20** (5): 254–259. Bibcode:1934PNAS...20..254B. doi:10.1073/pnas.20.5.254. PMC 1076395. PMID 16587881.

[25] Murdin, Paul; Murdin, Lesley (1985). *Supernovae* (2nd ed.). Cambridge University Press. p. 42. ISBN 0-521-30038-X.

[26] Colgate, Stirling A.; McKee, Chester (1969). "Early Supernova Luminosity". *The Astrophysical Journal* **157**: 623. Bibcode: doi:10.1086/150102.

[27] Zuckerman, Ben; Malkan, Matthew A. (1996). *The Origin and Evolution of the Universe*. Jones & Bartlett Learning. p. 68. ISBN 0-7637-0030-4.

[28] Filippenko, Alexei V.; Li, Wei-Dong; Treffers, Richard R.; Modjaz, Maryam (2001). "The Lick Observatory Supernova Search with the Katzman Automatic Imaging Telescope". In Paczynski, B.; Chen, W.-P.; Lemme, C. *Small Telescope Astronomy on Global Scale*. ASP Conference Series **246**. San Francisco: Astronomical Society of the Pacific. p. 121. Bibcode:2001ASPC ISBN978-1-58381-084-2.

[29] Antonioli, P.; et al. (2004). "SNEWS: The SuperNova Early Warning System". *New Journal of Physics* **6**: 114. arXiv:astro-ph/0406214. Bibcode:2004NJPh....6..114A. doi:10.1088/1367-2630/6/1/114.

[30] Scholberg, Kate (2000). "SNEWS: The supernova early warning system". *AIP Conference Proceedings* **523**: 355. arXiv:astro-ph/9911359. Bibcode:2000AIPC..523..355. doi:10.1063/1.1291879.

[31] Beacom, John F. (1999). "Supernova neutrinos and the neutrino masses". *Revista Mexicana de Fisica* **45** (2): 36. arXiv:hep-ph/9901300. Bibcode:1999RMxF...45...36B.

[32] "A star set to explode". *SpaceTelescope.org*. Retrieved 2014-01-07.

[33] Frieman, J. A.; et al. (2008). "The Sloan Digital Sky Survey-Ii Supernova Survey: Technical Summary". *The Astronomical Journal* **135**: 338. arXiv:0708.2749. Bibcode:2008AJ....135..338F. doi:10.1088/0004-6256/135/1/338.

[34] Perlmutter, Saul A.; et al. (1997). "Scheduled discovery of 7+ high-redshift SNe: First cosmology results and bounds on $q_0$". In Ruiz-Lapuente, P.; Canal, R.; Isern, J. *Thermonuclear Supernovae, Proceedings of the NATO Advanced Study Institute*. NATO Advanced Science Institutes Series C **486**. Dordrecht: Kluwer Academic Publishers. p. 749. arXiv:astro-ph/9602122. Bibcode:1997ASIC..486..749P.

[35] Linder, Eric V.; Huterer, Dragan (2003). "Importance of supernovae at $z > 1.5$ to probe dark energy". *Physical Review D* **67** (8): 081303. arXiv:astro-ph/0208138. Bibcode:2002astro.ph..8138L. doi:10.1103/PhysRevD.67.081303.

[36] Perlmutter, Saul A.; et al. (1997). "Measurements of the Cosmological Parameters $\Omega$ and $\Lambda$ from the First Seven Supernovae at $z \geq 0.35$". *The Astrophysical Journal* **483** (2): 565. arXiv:astro-ph/9608192. Bibcode:1997ApJ...483..565P. doi:10.1086/304265.

[37] Copin, Y.; et al. (2006). "The Nearby Supernova Factory". *New Astronomy Review* **50**(4–5): 436. Bibcode:2006NewAR..50. doi:10.1016/j.newar.2006.02.035.

[38] Kirshner, Robert P. (1980). "Type I supernovae: An observer's view". *AIP Conference Proceedings* **63**: 33. Bibcode:1980AIPC doi:10.1063/1.32212.

[39] "List of Supernovae". IAU Central Bureau for Astronomical Telegrams. Retrieved 2010-10-25.

[40] "The Padova-Asiago supernova catalogue". Osservatorio Astronomico di Padova. Retrieved 2014-01-10.

[41] "Artist's impression of supernova 1993J". *SpaceTelescope.org*. Retrieved 2014-09-12.

[42] Cappellaro, Enrico; Turatto, Massimo (2001). "Supernova Types and Rates". *Influence of Binaries on Stellar Population Studies* **264**. Dordrecht: Kluwer Academic Publishers. p. 199. arXiv:astro-ph/0012455. Bibcode:2001ASSL..264..199C. doi:10.1007/978-94-015-9723-4_16. ISBN 978-0-7923-7104-5.

[43] Turatto, Massimo (2003). "Classification of Supernovae". *Supernovae and Gamma-Ray Bursters*. Lecture Notes in Physics **598**. p. 21. doi:10.1007/3-540-45863-8_3. ISBN 978-3-540-44053-6.

[44] Doggett, Jesse B.; Branch, David (1985). "A comparative study of supernova light curves". *The Astronomical Journal* **90**: 2303. Bibcode:1985AJ.....90.2303D. doi:10.1086/113934.

[45] Zwicky, Fritz (1964). "NGC 1058 and its Supernova 1961". *The Astrophysical Journal* **139**: 514. Bibcode:1964ApJ...139..514Z. doi:10.1086/147779.

[46] Zwicky, Fritz (1962). "New Observations of Importance to Cosmology". In McVittie, G. C. *Problems of Extra-Galactic Research, Proceedings from IAU Symposium* **15**. New York: Macmillan Press. p. 347. Bibcode:1962IAUS...15..347Z.

[47]  "The Rise and Fall of a Supernova". *ESO Picture of the Week*. Retrieved 2013-06-14.

[48]  Piro, Anthony L.; Thompson, Todd A.; Kochanek, Christopher S. (2014). "Reconciling 56Ni production in Type Ia super-novae with double degenerate scenarios". *Monthly Notices of the Royal Astronomical Society* **438** (4): 3456. arXiv:1308.0334. Bibcode:2014MNRAS.438.3456P. doi:10.1093/mnras/stt2451.

[49]  Chen, Wen-Cong; Li, Xiang-Dong (2009). "On the Progenitors of Super-Chandrasekhar Mass Type Ia Supernovae". *The Astrophysical Journal* **702**: 686. arXiv:0907.0057. Bibcode:2009ApJ...702..686C. doi:10.1088/0004-637X/702/1/686.

[50]  Howell, D. Andrew; Sullivan, Mark; Conley, Alexander J.; Carlberg, Raymond G. (2007). "Predicted and Observed Evolution in the Mean Properties of Type Ia Supernovae with Redshift". *Astrophysical Journal Letters* **667** (1): L37–L40. arXiv:astro-ph/0701912. Bibcode:2007ApJ...667L..37H. doi:10.1086/522030.

[51]  Mazzali, Paolo A.; Röpke, Friedrich K.; Benetti, Stefano; Hillebrandt, Wolfgang (2007). "A Common Explosion Mech-anism for Type Ia Supernovae". *Science* **315** (5813): 825–828. arXiv:astro-ph/0702351. Bibcode:2007Sci...315..825M. doi:10.1126/science.1136259. PMID 17289993.

[52]  Lieb, Elliott H.; Yau, Horng-Tzer (1987). "A rigorous examination of the Chandrasekhar theory of stellar collapse". *The Astrophysical Journal* **323** (1): 140–144. Bibcode:1987ApJ...323..140L. doi:10.1086/165813.

[53]  Canal, Ramon; Gutiérrez, Jordi L. (1997). "The possible white dwarf-neutron star connection". In Isern, J.; Hernanz, M.; Gracia-Berro, E. *Proceedings of the 10th European Workshop on White Dwarfs* **214**. Dordrecht: Kluwer Academic Publishers. p. 49. Bibcode:1997astro.ph..1225C. ISBN 978-0-7923-4585-5.

[54]  Wheeler, J. Craig (2000). *Cosmic Catastrophes: Supernovae, Gamma-Ray Bursts, and Adventures in Hyperspace*. Cambridge University Press. p. 96. ISBN 978-0-521-65195-0.

[55]  Khokhlov, Alexei M.; Mueller, Ewald; Höflich, Peter A. (1993). "Light curves of Type IA supernova models with different explosion mechanisms". *Astronomy and Astrophysics* **270** (1–2): 223–248. Bibcode:1993A&A...270..223K.

[56]  Röpke, Friedrich K.; Hillebrandt, Wolfgang (2004). "The case against the progenitor's carbon-to-oxygen ratio as a source of peak luminosity variations in Type Ia supernovae". *Astronomy and Astrophysics Letters* **420** (1): L1–L4. arXiv:astro-ph/0403509. Bibcode:2004A&A...420L...1R. doi:10.1051/0004-6361:20040135.

[57]  Hillebrandt, Wolfgang; Niemeyer, Jens C. (2000). "Type IA Supernova Explosion Models". *Annual Review of Astronomy and Astrophysics* **38** (1): 191–230. arXiv:astro-ph/0006305. Bibcode:2000ARA&A..38..191H. doi:10.1146/annurev.astro.38.1.191.

[58]  Paczyński, Bohdan (1976). "Common Envelope Binaries". In Eggleton, P.; Mitton, S.; Whelan, J. *Structure and Evolution of Close Binary Systems*. IAU Symposium No. 73. Dordrecht: D. Reidel. pp. 75–80. Bibcode:1976IAUS...73...75P.

[59]  Macri, Lucas M.; Stanek, Krzysztof Z.; Bersier, David; Greenhill, Lincoln J.; Reid, Mark J. (2006). "A New Cepheid Distance to the Maser-Host Galaxy NGC 4258 and Its Implications for the Hubble Constant". *The Astrophysical Journal* **652** (2): 1133–1149. arXiv:astro-ph/0608211. Bibcode:2006ApJ...652.1133M. doi:10.1086/508530.

[60]  Colgate, Stirling A. (1979). "Supernovae as a standard candle for cosmology". *The Astrophysical Journal* **232** (1): 404–408. Bibcode:1979ApJ...232..404C. doi:10.1086/157300.

[61]  Ruiz-Lapuente, P.; et al. (2000). "Type IA supernova progenitors". *Memorie della Societa Astronomica Italiana* **71**: 435. Bibcode:2000MmSAI..71..435R.

[62]  Dan, Marius; Rosswog, Stephan; Guillochon, James; Ramirez-Ruiz, Enrico (2012). "How the merger of two white dwarfs depends on their mass ratio: Orbital stability and detonations at contact". *Monthly Notices of the Royal Astronomical Society* **422** (3): 2417. arXiv:1201.2406. Bibcode:2012MNRAS.422.2417D. doi:10.1111/j.1365-2966.2012.20794.x.

[63]  Howell, D. Andrew; et al. (2006). "The type Ia supernova SNLS-03D3bb from a super-Chandrasekhar-mass white dwarf star". *Nature* **443** (7109): 308–311. arXiv:astro-ph/0609616. Bibcode:2006Natur.443..308H. doi:10.1038/nature05103. PMID 16988705.

[64]  Tanaka, M.; et al. (2010). "Spectropolarimetry of Extremely Luminous Type Ia Supernova 2009dc: Nearly Spherical Explosion of Super-Chandrasekhar Mass White Dwarf". *The Astrophysical Journal* **714** (2): 1209. arXiv:0908.2057. Bibcode:2010ApJ... doi:10.1088/0004-637X/714/2/1209.

[65]  Wang, B.; Liu, D.; Jia, S.; Han, Z. (2014). "Helium double-detonation explosions for the progenitors of type Ia supernovae". *Proceedings of the International Astronomical Union* **9** (S298): 442. arXiv:1301.1047. Bibcode:2014IAUS..298..442W. doi:10 .1017/S1743921313007072.

[66] Foley, R. J.; et al. (2013). "Type Iax Supernovae: A New Class of Stellar Explosion". *The Astrophysical Journal* **767**: 57. arXiv:1212.2209. Bibcode:2013ApJ...767...57F. doi:10.1088/0004-637X/767/1/57.

[67] McCully, Curtis; et al. (2014). "A luminous, blue progenitor system for the type Iax supernova 2012Z". *Nature* **512** (7512): 54–56. arXiv:1408.1089. Bibcode:2014Natur.512...54M. doi:10.1038/nature13615. PMID 25100479.

[68] Silverman, J. M.; et al. (2013). "Type Ia Supernovae strongle interaction with their circumstellar medium". *The Astrophysical Journal Supplement Series* **207** (1): 3. arXiv:1304.0763. Bibcode:2013ApJS..207....3S. doi:10.1088/0067-0049/207/1/3.

[69] Nomoto, Ken'ichi; Tanaka, Masaomi; Tominaga, Nozomu; Maeda, Keiichi (2010). "Hypernovae, gamma-ray bursts, and first stars". *New Astronomy Reviews* **54** (3–6): 191. Bibcode:2010NewAR..54..191N. doi:10.1016/j.newar.2010.09.022.

[70] Moriya, Takashi J. (2012). "Progenitors of Recombining Supernova Remnants". *The Astrophysical Journal* **750** (1): L13. arXiv:1203.5799. Bibcode:2012ApJ...750L..13M. doi:10.1088/2041-8205/750/1/L13.

[71] Smith, N.; et al. (2009). "Sn 2008S: A Cool Super-Eddington Wind in a Supernova Impostor". *The Astrophysical Journal* **697**: L49. arXiv:0811.3929. Bibcode:2009ApJ...697L..49S. doi:10.1088/0004-637X/697/1/L49.

[72] Fryer, Chris L.; New, Kimberly C. B. (2003). "Gravitational Waves from Gravitational Collapse". *Living Reviews in Relativity* **6**. doi:10.12942/lrr-2003-2.

[73] Woosley, Stanford E.; Janka, Hans-Thomas (2005). "The Physics of Core-Collapse Supernovae". *Nature Physics* **1** (3): 147–154. arXiv:astro-ph/0601261. Bibcode:2005NatPh...1..147W. doi:10.1038/nphys172.

[74] Janka, Hans-Thomas; Langanke, Karlheinz; Marek, Andreas; Martínez-Pinedo, Gabriel; Müller, Bernhard (2007). "Theory of core-collapse supernovae". *Physics Reports* **442**: 38. arXiv:astro-ph/0612072. Bibcode:2007PhR...442...38J. doi:10.1016/j

[75] Gribbin, J. R.; Gribbin, M. (2000). *Stardust: Supernovae and Life – The Cosmic Connection*. Yale University Press. p. 173. ISBN 978-0-300-09097-0.

[76] Barwick, S. W.; et al. (2004). "APS Neutrino Study: Report of the Neutrino Astrophysics and Cosmology Working Group". arXiv:astro-ph/0412544 [astro-ph].

[77] Myra, Eric S.; Burrows, Adam (1990). "Neutrinos from type II supernovae- The first 100 milliseconds". *Astrophysical Journal* **364**: 222–231. Bibcode:1990ApJ...364..222M. doi:10.1086/169405.

[78] Kasen, D.; Woosley, Stanford E.; Heger, Alexander (2011). "Pair Instability Supernovae: Light Curves, Spectra, and Shock Breakout" (PDF). *The Astrophysical Journal* **734** (2): 102. arXiv:1101.3336. Bibcode:2011ApJ...734..102K. doi:10.1088/0004-637X/734/2/102.

[79] Poelarends, Arend J. T.; Herwig, Falk; Langer, Norbert; Heger, Alexander (2008). "The Supernova Channel of Super-AGB Stars". *The Astrophysical Journal* **675**: 614. arXiv:0705.4643. Bibcode:2008ApJ...675..614P. doi:10.1086/520872.

[80] Gilmore, Gerry (2004). "ASTRONOMY: The Short Spectacular Life of a Superstar". *Science* **304** (5679): 1915–1916. doi:10.1126/science.1100370. PMID 15218132.

[81] Faure, Gunter; Mensing, Teresa M. (2007). "Life and Death of Stars". *Introduction to Planetary Science*. pp. 35–48. doi:10.1007/978-1-4020-5544-7_4. ISBN 978-1-4020-5233-0.

[82] Malesani, D.; et al. (2009). "Early Spectroscopic Identification of SN 2008D". *The Astrophysical Journal Letters* **692** (2): L84. arXiv:0805.1188. Bibcode:2009ApJ...692L..84M. doi:10.1088/0004-637X/692/2/L84.

[83] Svirski, Gilad; Nakar, Ehud (2014). "Sn 2008D: A Wolf-Rayet Explosion Through a Thick Wind". *The Astrophysical Journal* **788**: L14. arXiv:1403.3400. Bibcode:2014ApJ...788L..14S. doi:10.1088/2041-8205/788/1/L14.

[84] Pols, Onno (1997). "Close Binary Progenitors of Type Ib/Ic and IIb/II-L Supernovae". In Leung, K.-C. *Proceedings of The Third Pacific Rim Conference on Recent Development on Binary Star Research*. ASP Conference Series **130**. pp. 153–158. Bibcode:1997rdbs.conf..153P.

[85] Eldridge, John J.; Fraser, Morgan; Smartt, Stephen J.; Maund, Justyn R.; Crockett, R. Mark (2013). "The death of massive stars – II. Observational constraints on the progenitors of Type Ibc supernovae". *Monthly Notices of the Royal Astronomical Society* **436**: 774. arXiv:1301.1975. Bibcode:2013MNRAS.436..774E. doi:10.1093/mnras/stt1612.

[86] Ryder, Stuart D.; et al. (2004). "Modulations in the radio light curve of the Type IIb supernova 2001ig: evidence for a Wolf-Rayet binary progenitor?". *Monthly Notices of the Royal Astronomical Society* **349** (3): 1093–1100. arXiv:astro-ph/0401135. Bibcode:2004MNRAS.349.1093R. doi:10.1111/j.1365-2966.2004.07589.x.

[87] Nicholl, M.; et al. (2013). "Slowly fading super-luminous supernovae that are not pair-instability explosions". *Nature* **502** (7471): 346–349. arXiv:1310.4446. Bibcode:2013Natur.502..346N. doi:10.1038/nature12569. PMID 24132291.

[88] Tauris, Thomas M.; Langer, Norbert; Moriya, Takashi J.; Podsiadlowski, Philipp; Yoon, Sung-Chul; Blinnikov, Sergey I. (2013). "Ultra-stripped Type Ic supernovae from close binary evolution". *Astrophysical Journal Letters* **778**. arXiv:1310.6356. Bibcode:2013ApJ...778L..23T. doi:10.1088/2041-8205/778/2/L23.

[89] Drout, M. R.; Soderberg, A. M.; Mazzali, P. A.; Parrent, J. T.; Margutti, R.; Milisavljevic, D.; Sanders, N. E.; Chornock, R.; Foley, R. J.; Kirshner, Robert P.; Filippenko, Alexei V.; Li, W.; Brown, P. J.; Cenko, S. B.; Chakraborti, S.; Challis, P.; Friedman, A.; Ganeshalingam, M.; Hicken, M.; Jensen, C.; Modjaz, Maryam; Perets, H. B.; Silverman, J. M.; Wong, D. S. (2013). "The Fast and Furious Decay of the Peculiar Type Ic Supernova 2005ek". *Astrophysical Journal* **774** (58): 44. arXiv:1306.2337. Bibcode:2013ApJ...774...58D. doi:10.1088/0004-637X/774/1/58.

[90] Reynolds, Thomas M.; Fraser, Morgan; Gilmore, Gerard (2015). "Gone without a bang: an archival HST survey for disappearing massive stars". *Monthly Notices of the Royal Astronomical Society* **453** (3): 2886–2901. doi:10.1093/mnras/stv1809. ISSN 0035-8711.

[91] Gerke, J. R.; Kochanek, Christopher S.; Stanek, Krzysztof Z. (2015). "The search for failed supernovae with the Large Binocular Telescope: first candidates". *Monthly Notices of the Royal Astronomical Society* **450** (3): 3289–3305. doi:10.1093/mnras/stv776. ISSN 0035-8711.

[92] Barbon, Roberto; Ciatti, Franco; Rosino, Leonida (1979). "Photometric properties of type II supernovae". *Astronomy and Astrophysics* **72**: 287. Bibcode:1979A&A....72..287B.

[93] Li, W.; Leaman, J.; Chornock, R.; Filippenko, A. V.; Poznanski, D.; Ganeshalingam, M.; Wang, X.; Modjaz, M.; Jha, S.; Foley, R. J.; Smith, N. (2011). "Nearby supernova rates from the Lick Observatory Supernova Search – II. The observed luminosity functions and fractions of supernovae in a complete sample". *Monthly Notices of the Royal Astronomical Society* **412** (3): 1441. arXiv:1006.4612. Bibcode:2011MNRAS.412.1441L. doi:10.1111/j.1365-2966.2011.18160.x.

[94] Richardson, D.; Branch, D.; Casebeer, D.; Millard, J.; Thomas, R. C.; Baron, E. (2002). "A Comparative Study of the Absolute Magnitude Distributions of Supernovae". *The Astronomical Journal* **123** (2): 745. arXiv:astro-ph/0112051. Bibcode:2002AJ....123..745R. doi:10.1086/338318.

[95] Frail, Dale A.; Giacani, Elsa B.; Goss, W. Miller; Dubner, Gloria M. (1996). "The Pulsar Wind Nebula Around PSR B1853+01 in the Supernova Remnant W44". *Astrophysical Journal Letters* **464** (2): L165–L168. arXiv:astro-ph/9604121. Bibcode:1996ApJ...464L.165F. doi:10.1086/310103.

[96] Höflich, Peter A.; Kumar, Pawan; Wheeler, J. Craig (2004). "Neutron star kicks and supernova asymmetry". *Cosmic explosions in three dimensions: Asymmetries in supernovae and gamma-ray bursts*. Cambridge University Press. p. 276. Bibcode:2004cetd

[97] Fryer, Chris L. (2004). "Neutron Star Kicks from Asymmetric Collapse". *Astrophysical Journal* **601** (2): L175–L178. arXiv:astro-ph/0312265. Bibcode:2004ApJ...601L.175F. doi:10.1086/382044.

[98] Gilkis, Avishai; Soker, Noam (2014). "Implications of turbulence for jets in core-collapse supernova explosions" **1412**. p. 4984. arXiv:1412.4984. Bibcode:2014arXiv1412.4984G.

[99] Khokhlov, Alexei M.; et al. (1999). "Jet-induced Explosions of Core Collapse Supernovae". *The Astrophysical Journal* **524** (2): L107. arXiv:astro-ph/9904419. Bibcode:1999ApJ...524L.107K. doi:10.1086/312305.

[100] Wang, L.; et al. (2003). "Spectropolarimetry of SN 2001el in NGC 1448: Asphericity of a Normal Type Ia Supernova". *The Astrophysical Journal* **591** (2): 1110. arXiv:astro-ph/0303397. Bibcode:2003ApJ...591.1110W. doi:10.1086/375444.

[101] Mazzali, P. A.; Nomoto, K. I.; Cappellaro, E.; Nakamura, T.; Umeda, H.; Iwamoto, K. (2001). "Can Differences in the Nickel Abundance in Chandrasekhar-Mass Models Explain the Relation between the Brightness and Decline Rate of Normal Type Ia Supernovae?". *The Astrophysical Journal* **547** (2): 988. arXiv:astro-ph/0009490. Bibcode:2001ApJ...547..988M. doi:10.1086/318428.

[102] Iwamoto, K. (2006). "Neutrino Emission from Type Ia Supernovae". *AIP Conference Proceedings* **847**. p. 406. doi:10.1063/1

[103] Hayden, B. T.; Garnavich, P. M.; Kessler, R.; Frieman, J. A.; Jha, S. W.; Bassett, B.; Cinabro, D.; Dilday, B.; Kasen, D.; Marriner, J.; Nichol, R. C.; Riess, A. G.; Sako, M.; Schneider, D. P.; Smith, M.; Sollerman, J. (2010). "The Rise and Fall of Type Ia Supernova Light Curves in the SDSS-II Supernova Survey". *The Astrophysical Journal* **712**: 350. arXiv:1001.3428. Bibcode:2010ApJ...712..350H. doi:10.1088/0004-637X/712/1/350.

[104] Janka, Hans-Thomas (2012). "Explosion Mechanisms of Core-Collapse Supernovae". *Annual Review of Nuclear and Particle Science* **62**: 407. arXiv:1206.2503. Bibcode:2012ARNPS..62..407J. doi:10.1146/annurev-nucl-102711-094901.

[105] Smartt (2009). "Progenitors of core-collapse supernovae". *Annual Review of Astronomy and Astrophysics* **47**: 63–106. arXiv: Bibcode:2009ARA&A..47...63S. doi:10.1146/annurev-astro-082708-101737.

[106] Smartt, Stephen J. (2009). "Progenitors of Core-Collapse Supernovae". *Annual Review of Astronomy & Astrophysics* **47**: 63. arXiv:0908.0700. Bibcode:2009ARA&A..47...63S. doi:10.1146/annurev-astro-082708-101737.

[107] Walmswell, J. J.; Eldridge, J. J. (2012). "Circumstellar dust as a solution to the red supergiant supernova progenitor problem". *Monthly Notices of the Royal Astronomical Society* **419** (3): 2054. arXiv:1109.4637. Bibcode:2012MNRAS.419.2054W. doi:10.1111/j.1365-2966.2011.19860.x.

[108] Georgy, C. (2012). "Yellow supergiants as supernova progenitors: An indication of strong mass loss for red supergiants?". *Astronomy & Astrophysics* **538**: L8–L2. arXiv:1111.7003. Bibcode:2012A&A...538L...8G. doi:10.1051/0004-6361/201118372.

[109] Yoon, S. -C.; Gräfener, G.; Vink, J. S.; Kozyreva, A.; Izzard, R. G. (2012). "On the nature and detectability of Type Ib/c supernova progenitors". *Astronomy & Astrophysics* **544**: L11. arXiv:1207.3683. Bibcode:2012A&A...544L..11Y. doi:10.1051/0004-6361/201219790.

[110] Groh, J. H.; Meynet, G.; Ekström, S. (2013). "Massive star evolution: Luminous blue variables as unexpected supernova progenitors". *Astronomy & Astrophysics* **550**: L7. arXiv:1301.1519. Bibcode:2013A&A...550L...7G. doi:10.1051/0004-6361/201220741.

[111] Yoon, S.-C.; Gräfener, G.; Vink, J. S.; Kozyreva, A.; Izzard, R. G. (2012). "On the nature and detectability of Type Ib/c supernova progenitors". *Astronomy & Astrophysics* **544**: L11. arXiv:1207.3683. Bibcode:2012A&A...544L..11Y. doi:10.1051/0004-6361/201219790.

[112] François, P.; et al. (2004). "The evolution of the Milky Way from its earliest phases: Constraints on stellar nucleosynthesis". *Astronomy and Astrophysics* **421** (2): 613–621. arXiv:astro-ph/0401499. Bibcode:2004A&A...421..613F. doi:10.1051/0004-6361:20034140.

[113] Woosley, Stanford E.; Arnett, W. D.; Clayton, D. D. (1973). "The Explosive Burning of Oxygen and Silicon". *Astrophysical Journal Supplement* **26**: 231–312. Bibcode:1973ApJS...26..231W. doi:10.1086/190282.

[114] Qian, Y.-Z.; Vogel, P.; Wasserburg, G. J. (1998). "Diverse Supernova Sources for the r-Process". *Astrophysical Journal* **494** (1): 285–296. arXiv:astro-ph/9706120. Bibcode:1998ApJ...494..285Q. doi:10.1086/305198.

[115] Gonzalez, Guillermo; Brownlee, Donald; Ward, Peter (2001). "The Galactic Habitable Zone: Galactic Chemical Evolution". *Icarus* **152**: 185. arXiv:astro-ph/0103165. Bibcode:2001Icar..152..185G. doi:10.1006/icar.2001.6617.

[116] Cox, Donald P. (1972). "Cooling and Evolution of a Supernova Remnant". *Astrophysical Journal* **178**: 159. Bibcode:1972ApJ...178..159C. doi:10.1086/151775.

[117] Sandstrom, Karin M.; Bolatto, Alberto D.; Stanimirović, Snežana; Van Loon, Jacco Th.; Smith, J. D. T. (2009). "Measuring Dust Production in the Small Magellanic Cloud Core-Collapse Supernova Remnant 1E 0102.2−7219". *The Astrophysical Journal* **696** (2): 2138. arXiv:0810.2803. Bibcode:2009ApJ...696.2138S. doi:10.1088/0004-637X/696/2/2138.

[118] Preibisch, T.; Zinnecker, H. (2001). "Triggered Star Formation in the Scorpius-Centaurus OB Association (Sco OB2)". *ASP Conference Proceedings, From Darkness to Light: Origin and Evolution of Young Stellar Clusters* **243**. San Francisco: Astronomical Society of the Pacific. p. 791. Bibcode:2001ASPC..243..791P.

[119] Cameron, A.G.W.; Truran, J.W. (1977). "The supernova trigger for formation of the solar system". *Icarus* **30** (3): 447. Bibcode:1977Icar...30..447C. doi:10.1016/0019-1035(77)90101-4.

[120] Melott, A.; et al. (2004). "Did a gamma-ray burst initiate the late Ordovician mass extinction?". *International Journal of Astrobiology* **3** (2): 55–61. arXiv:astro-ph/0309415. Bibcode:2004IJAsB...3...55M. doi:10.1017/S1473550404001910.

[121] Fields, Brian D.; Hochmuth, Kathrin A.; Ellis, John (2005). "Deep-Ocean Crusts as Telescopes: Using Live Radioisotopes to Probe Supernova Nucleosynthesis". *The Astrophysical Journal* **621** (2): 902. arXiv:astro-ph/0410525. Bibcode:2005ApJ...621 doi:10.1086/427797.

[122] Knie, K.; et al. (2004). "$^{60}$Fe Anomaly in a Deep-Sea Manganese Crust and Implications for a Nearby Supernova Source". *Physical Review Letters* **93** (17): 171103–171106. Bibcode:2004PhRvL..93q1103K. doi:10.1103/PhysRevLett.93.171103.

[123] Fields, B. D.; Ellis, J. (1999). "On Deep-Ocean Fe-60 as a Fossil of a Near-Earth Supernova". *New Astronomy* **4** (6): 419–430. arXiv:astro-ph/9811457. Bibcode:1999NewA....4..419F. doi:10.1016/S1384-1076(99)00034-2.

[124] "In Brief". *Scientific American* **300** (5): 28. 2009. doi:10.1038/scientificamerican0509-28a.

[125] Gorelick, M. (2007). "The Supernova Menace". *Sky & Telescope* **113**: 26. Bibcode:2007S&T...113c..26G.

[126] Gehrels, Neil; et al. (2003). "Ozone Depletion from Nearby Supernovae". *Astrophysical Journal* **585** (2): 1169–1176. arXiv:astro-ph/0211361. Bibcode:2003ApJ...585.1169G. doi:10.1086/346127.

[127] Van Der Sluys, M. V.; Lamers, H. J. G. L. M. (2003). "The dynamics of the nebula M1-67 around the run-away Wolf-Rayet star WR 124". *Astronomy and Astrophysics* **398**: 181. arXiv:astro-ph/0211326. Bibcode:2003A&A...398..181V. doi:10.1051/0004-6361:20021634.

[128] Lobel, A.; et al. (2004). "Spectroscopy of the Millennium Outburst and Recent Variability of the Yellow Hypergiant Rho Cassiopeiae". *Stars as suns : activity* **219**: 903. arXiv:astro-ph/0312074. Bibcode:2004IAUS..219..903L.

[129] Van Boekel, R.; et al. (2003). "Direct measurement of the size and shape of the present-day stellar wind of eta Carinae". *Astronomy and Astrophysics* **410** (3): L37. arXiv:astro-ph/0310399. Bibcode:2003A&A...410L..37V. doi:10.1051/0004-6361:20031500.

[130] Bode, M. F.; et al. (2006). "Swift *Observations* of the 2006 Outburst of the Recurrent Nova RS Ophiuchi. 1. Early X-Ray Emission from the Shocked Ejecta and Red Giant Wind". *The Astrophysical Journal* **652**: 629. arXiv:astro-ph/0604618. Bibcode:2006ApJ...652..629B. doi:10.1086/507980.

[131] Thoroughgood, T. D.; et al. (2002). "The recurrent nova U Scorpii — A type Ia supernova progenitor". *The Physics of Cataclysmic Variables and Related Objects* **261**. San Francisco, CA:Astronomical Society of the Pacific. Bibcode:2002ASPC..261.

[132] Humphreys, Roberta M.; Helton, L. Andrew; Jones, Terry J. (2007). "The Three-Dimensional Morphology of VY Canis Majoris. 1. The Kinematics of the Ejecta". *The Astronomical Journal* **133** (6): 2716. arXiv:astro-ph/0702717. Bibcode:2007AJ ....133.2716H.doi:10.1086/517609.

[133] Inglis, Michael (2015). "Star Death: Supernovae, Neutron Stars & Black Holes". *Astrophysics is Easy!*. The Patrick Moore Practical Astronomy Series. p. 203. doi:10.1007/978-3-319-11644-0_12. ISBN 978-3-319-11643-3.

[134] Thielemann, F.-K.; Hirschi, R.; Liebendörfer, M.; Diehl, R. (2011). "Massive Stars and Their Supernovae". *Astronomy with Radioactivities*. Lecture Notes in Physics **812**. p. 153. doi:10.1007/978-3-642-12698-7_4. ISBN 978-3-642-12697-0.

[135] Tuthill, Peter G.; et al. (2008). "The Prototype Colliding-Wind Pinwheel WR 104". *The Astrophysical Journal* **675**: 698. arXiv:0712.2111. Bibcode:2008ApJ...675..698T. doi:10.1086/527286.

[136] Tuthill, Peter G.; et al. (2006). "Pinwheels in the Quintuplet Cluster". *Science* **313** (5789): 935. arXiv:astro-ph/0608427. Bibcode:2006Sci...313..935T. doi:10.1126/science.1128731. PMID 16917053.

[137] Landsman, W.; Simon, T.; Bergeron, P. (1999). "The hot white-dwarf companions of HR 1608, HR 8210, and HD 15638". *Astronomical Society of the Pacific* **105** (690): 841–847. Bibcode:1993PASP..105..841L. doi:10.1086/133242.

[138] Vennes, S.; Kawka, A. (2008). "On the empirical evidence for the existence of ultramassive white dwarfs". *Monthly Notices of the Royal Astronomical Society* **389** (3): 1367. arXiv:0806.4742. Bibcode:2008MNRAS.389.1367V. doi:10.1111/j.1365-2966.2008.13652.x.

## 1.11 Further reading

- "Introduction to Supernova Remnants". NASA/GSFC. 2007-10-04. Retrieved 2011-03-15.

- Bethe, Hans A. (1990). "Supernovae".*Physics Today (ISSN 0031-9228)***43**(9): 736–739. Bibcode:1990PhT....43 doi:10.1063/1.881256. PMID 10035857.

- Croswell, Ken (1996). *The Alchemy of the Heavens: Searching for Meaning in the Milky Way*. Anchor Books. ISBN 0-385-47214-5. A popular-science account.

- Filippenko, Alexei V. (1997). "Optical Spectra of Supernovae". *Annual Review of Astronomy and Astrophysics* **35**: 309. Bibcode:1997ARA&A..35..309F. doi:10.1146/annurev.astro.35.1.309. An article describing spectral classes of supernovae.

- Takahashi, K.; Sato, K.; Burrows, Adam; Thompson, Todd A. (2003). "Supernova Neutrinos, Neutrino Oscillations, and the Mass of the Progenitor Star". *Physical Review D* **68** (11): 77–81. arXiv:hep-ph/0306056. Bibcode:2003PhRvD..68k3009T. doi:10.1103/PhysRevD.68.113009. A good review of supernova events.

- Hillebrandt, Wolfgang; Janka, Hans-Thomas; Müller, Ewald (2006). "How to Blow Up a Star". *Scientific American* **295** (4): 42–49. doi:10.1038/scientificamerican1006-42.

- Woosley, Stanford E.; Janka, Hans-Thomas (2005). "The Physics of Core-Collapse Supernovae". *Nature Physics* **1** (3): 147–154. arXiv:astro-ph/0601261. Bibcode:2005NatPh...1..147W. doi:10.1038/nphys172.

## 1.12 External links

- "RSS news feed" (RSS). The Astronomer's Telegram. Retrieved 2006-11-28.

- Tsvetkov, D. Yu.; Pavlyuk, N. N.; Bartunov, O. S.; Pskovskii, Yu. P. "Sternberg Astronomical Institute Supernova Catalogue". Sternberg Astronomical Institute, Moscow University. Retrieved 2006-11-28. A searchable catalog.

- "List of Supernovae with IAU Designations". IAU: Central Bureau for Astronomical Telegrams. Retrieved 2010-10-25.

- Overbye, Dennis (2008-05-21). "Scientists See Supernova in Action". *The New York Times*. Retrieved 2008-05-21.(subscription required)

# Chapter 2

# Supernova remnant

A **supernova remnant** (**SNR**) is the structure resulting from the explosion of a star in a supernova. The supernova remnant is bounded by an expanding shock wave, and consists of ejected material expanding from the explosion, and the interstellar material it sweeps up and shocks along the way.

There are two common routes to a supernova: either a massive star may run out of fuel, ceasing to generate fusion energy in its core, and collapsing inward under the force of its own gravity to form a neutron star or a black hole; or a white dwarf star may accumulate (accrete) material from a companion star until it reaches a critical mass and undergoes a thermonuclear explosion.

In either case, the resulting supernova explosion expels much or all of the stellar material with velocities as much as 10% the speed of light, that is, about 30,000 km/s. These ejecta are highly supersonic: assuming a typical temperature of the interstellar medium of 10,000 K, the Mach number can initially be > 1000. Therefore, a strong shock wave forms ahead of the ejecta, that heats the upstream plasma up to temperatures well above millions of K. The shock continuously slows down over time as it sweeps up the ambient medium, but it can expand over hundreds or thousands of years and over tens of parsecs before its speed falls below the local sound speed.

One of the best observed young supernova remnants was formed by SN 1987A, a supernova in the Large Magellanic Cloud that was observed in February 1987. Other well-known supernova remnants include the Crab Nebula, Tycho, the remnant of SN 1572, named after Tycho Brahe who recorded the brightness of its original explosion, and Kepler, the remnant of SN 1604, named after Johannes Kepler. The youngest known remnant in our galaxy is G1.9+0.3, discovered in the galactic center.[1]

## 2.1  Summary of stages

An SNR passes through the following stages as it expands:[2]

1. Free expansion of the ejecta, until they sweep up their own weight in circumstellar or interstellar medium. This can last tens to a few hundred years depending on the density of the surrounding gas.

2. Sweeping up of a shell of shocked circumstellar and interstellar gas. This begins the Sedov-Taylor phase, which can be well modeled by a self-similar analytic solution (see Blast wave#Astronomy). Strong X-ray emission traces the strong shock waves and hot shocked gas.

3. Cooling of the shell, to form a thin (< 1 pc), dense (1-100 million atoms per cubic metre) shell surrounding the hot (few million kelvin) interior. This is the pressure-driven snowplow phase. The shell can be clearly seen in optical emission from recombining ionized hydrogen and ionized oxygen atoms.

4. Cooling of the interior. The dense shell continues to expand from its own momentum. This stage is best seen in the radio emission from neutral hydrogen atoms.

*SN 1054 remnant* (Crab Nebula).

5. Merging with the surrounding interstellar medium. When the supernova remnant slows to the speed of the random velocities in the surrounding medium, after roughly 30,000 years, it will merge into the general turbulent flow, contributing its remaining kinetic energy to the turbulence.

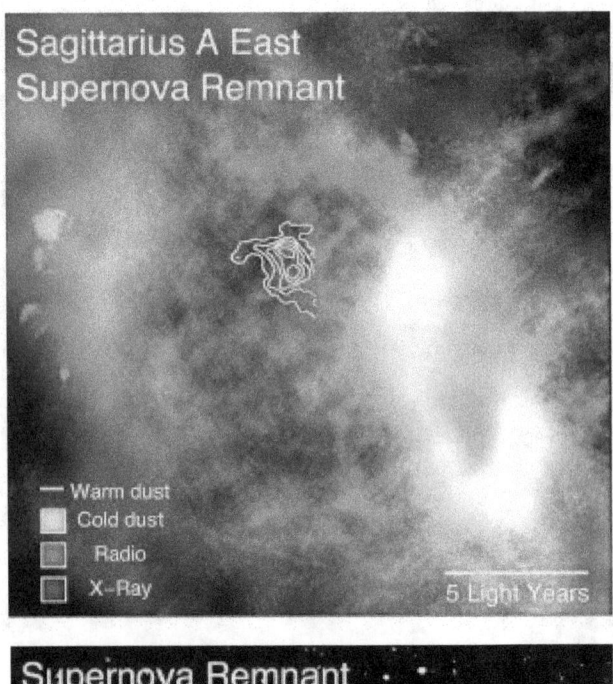

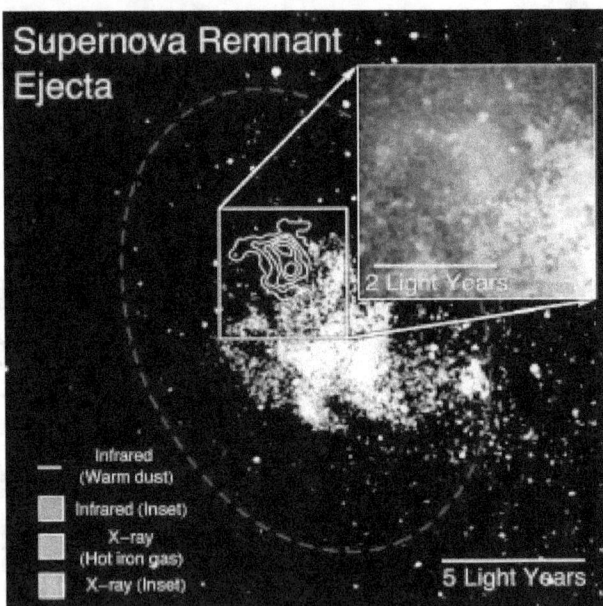

Supernova remnant ejecta producing planet-forming material.

## 2.2   Types of supernova remnant

There are three types of supernova remnant:

- Shell-like, such as Cassiopeia A

- Composite, in which a shell contains a central pulsar wind nebula, such as G11.2-0.3 or G21.5-0.9.

- Mixed-morphology (also called "thermal composite") remnants, in which central thermal X-ray emission is seen, enclosed by a radio shell. The thermal X-rays are primarily from swept-up interstellar material, rather than supernova ejecta. Examples of this class include the SNRs W28 and W44. (Confusingly, W44 additionally contains a pulsar and pulsar wind nebula; so it is simultaneously both a "classic" composite and a thermal composite.)

## 2.3 Origin of cosmic rays

Supernova remnants are considered the major source of galactic cosmic rays.[3][4][5] The connection between cosmic rays and supernovas was first suggested by Walter Baade and Fritz Zwicky in 1934. Vitaly Ginzburg and Sergei Syrovatskii in 1964 remarked that if the efficiency of cosmic ray acceleration in supernova remnants is about 10 percent, the cosmic ray losses of the Milky Way are compensated. This hypothesis is supported by a specific mechanism called "shock wave acceleration" based on Enrico Fermi's ideas, which is still under development.

Indeed, Enrico Fermi proposed in 1949 a model for the acceleration of cosmic rays through particle collisions with magnetic clouds in the interstellar medium.[6] This process, known as the "Second Order Fermi Mechanism", increases particle energy during head-on collisions, resulting in a steady gain in energy. A later model to produce Fermi Acceleration was generated by a powerful shock front moving through space. Particles that repeatedly cross the front of the shock can gain significant increases in energy. This became known as the "First Order Fermi Mechanism".[7]

Supernova remnants can provide the energetic shock fronts required to generate ultra-high energy cosmic rays. Observation of the SN 1006 remnant in the X-ray has shown synchrotron emission consistent with it being a source of cosmic rays.[3] However, for energies higher than about $10^{18}$ eV a different mechanism is required as supernova remnants cannot provide sufficient energy.[7]

It is still unclear whether supernova remnants accelerate cosmic rays up to PeV energies. The future telescope CTA will help to answer this question.

## 2.4 Gallery

- Cygnus Loop

- GK Persei remnant

- G299 remnant

- N49 remnant
  (Large Magellanic Cloud)

- Puppis A

- SN 185 remnant

- SN 1006 remnant

- SN 1054 remnant (*Crab Nebula*)

- SN 1572 remnant (*Tycho Supernova*

- SN 1604 remnant (*Kepler's Supernova*)

- SN 1680 remnant (*Cassiopeia A*)

- SN 1987A remnant

- SNR 0519-69.0 remnant

- T Pyxidis remnant

## 2.5   See also

- List of supernova remnants

- Local Bubble

- Nova remnant

- Planetary nebula

- Superbubble

## 2.6   References

[1] Discovery of most recent supernova in our galaxy May 14, 2008

[2] Reynolds, Stephen P. (2008). "Supernova Remnants at High Energy". *Annual Review of Astronomy and Astrophysics* **46** (46): 89–126. Bibcode:2008ARA&A..46...89R. doi:10.1146/annurev.astro.46.060407.145237.

[3] K. Koyama; R. Petre; E.V. Gotthelf; U. Hwang; et al. (1995). "Evidence for shock acceleration of high-energy electrons in the supernova remnant SN1006". *Nature* **378** (6554): 255–258. Bibcode:1995Natur.378..255K. doi:10.1038/378255a0.

[4] "Supernova produces cosmic rays". BBC News. November 4, 2004. Retrieved 2006-11-28.

[5] "SNR and Cosmic Ray Acceleration". NASA Goddard Space Flight Center. Retrieved 2007-02-08.

[6] E. Fermi (1949). "On the Origin of the Cosmic Radiation". *Physical Review* **75** (8): 1169–1174. Bibcode:1949PhRv...75.1169F. doi:10.1103/PhysRev.75.1169.

[7] "Ultra-High Energy Cosmic Rays". University of Utah. Retrieved 2006-08-10.

## 2.7   External links

- Galactic SNR Catalogue (D. A. Green, University of Cambridge)

- Chandra observations of supernova remnants: catalog, photo album, selected picks

- 2MASS images of Supernova Remnants

- NASA: Introduction to Supernova Remnants

- NASA's Imagine: Supernova Remnants

- Afterlife of a Supernova on UniverseToday.com

- Supernova remnant on arxiv.org

- Supernova Remnants, SEDS

# Chapter 3

# Supernova nucleosynthesis

**Supernova nucleosynthesis** is a theory of the production of many different chemical elements in supernova explosions, first advanced by Fred Hoyle in 1954.[1] The nucleosynthesis, or fusion of lighter elements into heavier ones, occurs during explosive oxygen burning and silicon burning.[2] Those fusion reactions create the elements silicon, sulfur, chlorine, argon, sodium, potassium, calcium, scandium, titanium and iron peak elements: vanadium, chromium, manganese, iron, cobalt, and nickel. These are called "primary elements", in that they can be fused from pure hydrogen and helium in massive stars. As a result of their ejection from supernovae, their abundances increase within the interstellar medium. Elements heavier than nickel are created primarily by a rapid capture of neutrons in a process called the r-process. However, these are much less abundant than the primary chemical elements. Other processes thought to be responsible for some of the nucleosynthesis of underabundant heavy elements, notably a proton capture process known as the rp-process and a photodisintegration process known as the gamma (or p) process. The latter synthesizes the lightest, most neutron-poor, isotopes of the heavy elements.

## 3.1 Cause

Main article: Supernova

A supernova is a massive explosion of a star that occurs under two principal scenarios. The first is that a white dwarf star undergoes a nuclear-based explosion after it reaches its Chandrasekhar limit after absorbing mass from a neighboring star (usually a red giant). The second, and more common, cause is when a massive star, usually a supergiant, reaches nickel-56 in its nuclear fusion (or burning) processes. This isotope undergoes radioactive decay into iron-56, which has one of the highest binding energies of all of the isotopes, and is the last element that produces a net release of energy by nuclear fusion, exothermically.

All nuclear fusion reactions that produce heavier elements cause the star to lose energy and are said to be endothermic reactions. The pressure that supports the star's outer layers drops sharply. As the outer envelope is no longer sufficiently supported by the radiation pressure, the star's gravity pulls its outer layers rapidly inward. As the star collapses, these outer layers collide with the incompressible stellar core, producing a shockwave that expands outward through the unfused material of the outer shell. The pressures and densities in the shockwave are sufficient to induce fusion in that material, and the energy released leads to the star's explosion, dispersing material from the star into interstellar space.

### 3.1.1 Nuclear fusion sequence and the alpha process

After a star completes the oxygen burning process, its core is composed primarily of silicon and sulfur.[3] If it has sufficiently high mass, it further contracts until its core reaches temperatures in the range of 2.7–3.5 GK (230–300 keV). At these temperatures, silicon and other elements can photodisintegrate, emitting a proton or alpha particle.[3] Silicon burning entails the *alpha process*, which creates new elements by adding one of these alpha particles[3] (the equivalent of a helium nucleus, two protons plus two neutrons) per step in the following sequence:

The entire silicon-burning sequence lasts about one day and stops when nickel-56 has been produced. The star can no longer release energy via nuclear fusion because a nucleus with 56 nucleons has the lowest mass per nucleon (any proton or neutron) of all the elements in the alpha process sequence. Although iron-58 and nickel-62 have slightly higher binding energies per nucleon than iron-56,[4] the next step up in the alpha process would be zinc−60, which has slightly *more* mass per nucleon and thus, is less thermodynamically favorable. Nickel-56 (which has 28 protons) has a half-life of 6.02 days and decays via $\beta^+$ decay to cobalt−56 (27 protons), which in turn has a half-life of 77.3 days as it decays to iron-56 (26 protons). However, only minutes are available for the nickel-56 to decay within the core of a massive star. The star has run out of nuclear fuel and within minutes begins to contract.

During this phase of the contraction, the potential energy of gravitational contraction heats the interior to 5 GK (430 keV) and this opposes and delays the contraction. However, since no additional heat energy can be generated via new fusion reactions, the final unopposed contraction rapidly accelerates into a collapse lasting only a few seconds. The central portion of the star is now crushed into either a neutron star or, if the star is massive enough, a black hole. The outer layers of the star are blown off in an explosion known as a Type II supernova that lasts days to months. The supernova explosion releases a large burst of neutrons, which synthesizes, in about one second while-inside the star, roughly half of the supply of elements in the universe that are heavier than iron, via a neutron-capture mechanism known as the *r-process* (where the "r" stands for rapid neutron capture).

## 3.2   Products

The maximum weight for an element produced by fusion in a normal star is that of iron, reaching an isotope with an atomic mass of 56 (see Stellar nucleosynthesis). Prior to a supernova, fusion of elements between silicon and iron occurs only in the largest of stars, in the silicon burning process. (A slow neutron capture process, known as the s-process which also occurs during normal stellar nucleosynthesis can create elements up to bismuth with an atomic mass of approximately 209. However, the s-process occurs primarily in low-mass stars that evolve more slowly.) Once the core fails to produce enough energy to support the outer envelope of gases the star explodes as a supernova producing the bulk of elements beyond iron. Production of elements from iron to uranium occurs within seconds in a supernova explosion. Due to the large amounts of energy released, much higher temperatures and densities are reached than at normal stellar temperatures. These conditions allow for an environment where transuranium elements might be formed.

## 3.3   The r-process

Main article: r-process

 During supernova nucleosynthesis, the r-process (r for rapid) creates very neutron-rich heavy isotopes, which decay after the event to the first stable isotope, thereby creating the neutron-rich stable isotopes of all heavy elements. This neutron capture process occurs in high neutron density with high temperature conditions. In the r-process, any heavy nuclei are bombarded with a large neutron flux to form highly unstable neutron rich nuclei which very rapidly undergo beta decay to form more stable nuclei with higher atomic number and the same atomic mass. The neutron flux is astonishingly high, about $10^{22}$ neutrons per square centimeter per second. First calculation of a dynamic r-process, showing the evolution of calculated results with time,[5] also suggested that the r-process abundances are a superposition of differing neutron fluences. Small fluence produces the first r-process abundance peak near atomic weight A=130 but no actinides, whereas large fluence produces the actinides uranium and thorium but no longer contains the A=130 abundance peak. These processes occur in a fraction of a second to a few seconds, depending on details. Hundreds of subsequent papers published have utilized this time-dependent approach. Interestingly, the only modern nearby supernova, 1987A, has not revealed r-process enrichments. Modern thinking is that the r-process yield may be ejected from some supernovae but swallowed up in others as part of the residual neutron star or black hole.

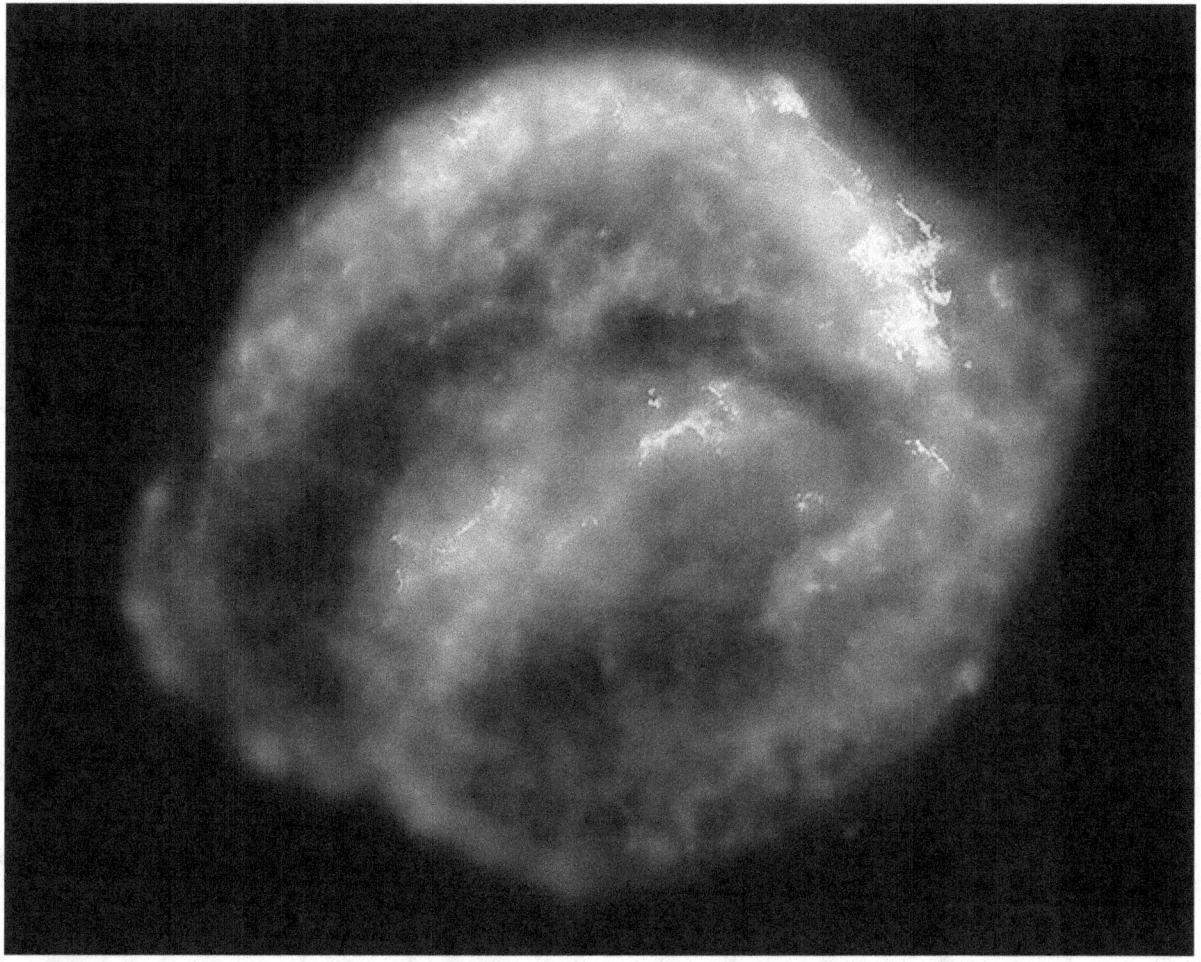

*Composite image of Kepler's supernova from pictures by the Spitzer Space Telescope, Hubble Space Telescope, and Chandra X-ray Observatory.*

## 3.4   See also

- Big bang nucleosynthesis

- Critical mass

- Nuclear decay

- Nuclear fission

- Nuclear fusion

- Nucleosynthesis

- Primordial nuclide

- Stellar nucleosynthesis

- Supernova

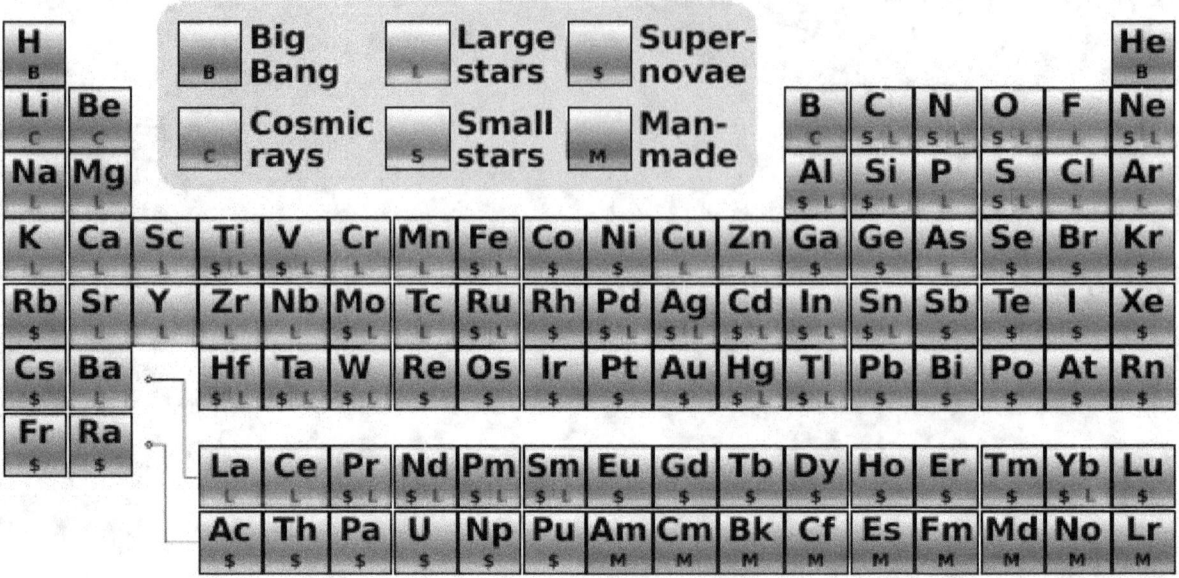

*A version of the periodic table indicating the origins – including supernova nucleosynthesis – of the elements. All elements above 103 (lawrencium) are also manmade and are not included.*

## 3.5   Notes

[1] Energy is produced in the isolated fusion reaction of nickel-56 with helium-4, but production of the latter (by photodisintegration of heavier nuclei) is costly, and consumes energy, causing alpha buildup of nickel to be shut off due to the essential fact that nickel-56 has nucleon binding energy less zinc-60.

## 3.6   References

[1] "Synthesis of the laments from carbon to nickel" Astrophys. J. Suppl. 1, 121 (1954)

[2] Woosley, S.E.; W. D. Arnett & D. D. Clayton (1973). "Explosive burning of oxygen and silicon". *The Astrophysical Journal Supplement* **26**: 231–312. Bibcode:1973ApJS...26..231W. doi:10.1086/190282.

[3] Clayton, Donald D. (1983). *Principles of Stellar Evolution and Nucleosynthesis*. University of Chicago Press. pp. 519–524. ISBN 9780226109534.

[4] Citation: *The atomic nuclide with the highest mean binding energy*, Fewell, M. P., American Journal of Physics, Volume 63, Issue 7, pp. 653–658 (1995). Click here for a high-resolution graph, *The Most Tightly Bound Nuclei*, which is part of the Hyperphysics project at Georgia State University.

[5] P. A. Seeger; W.A. Fowler; D. D. Clayton (1965). "Nucleosynthesis of heavy elements by neutron capture". *The Astrophysical Journal Supplement* **11**: 121–166. Bibcode:1965ApJS...11..121S. doi:10.1086/190111.

## 3.7   Other reading

- E. M. Burbidge, G. R. Burbidge, W. A. Fowler, F. Hoyle, *Synthesis of the Elements in Stars*, Rev. Mod. Phys. 29 (1957) 547 (article at the Physical Review Online Archive).

- D. D. Clayton. "Handbook of Isotopes in the Cosmos", Cambridge University Press, 2003, ISBN 0-521-82381-1.

## 3.8 External links

- Atom Smashers Shed Light on Supernovae, Big Bang *Sky & Telescope Online*, April 22, 2005

- G. Gonzalez; D. Brownlee; P. Ward (2001). "The Galactic Habitable Zone: Galactic Chemical Evolution" (PDF). *Icarus* **152**: 185–200. arXiv:astro-ph/0103165. Bibcode:2001Icar..152..185G. doi:10.1006/icar.2001.6617.

# Chapter 4

# Interstellar medium

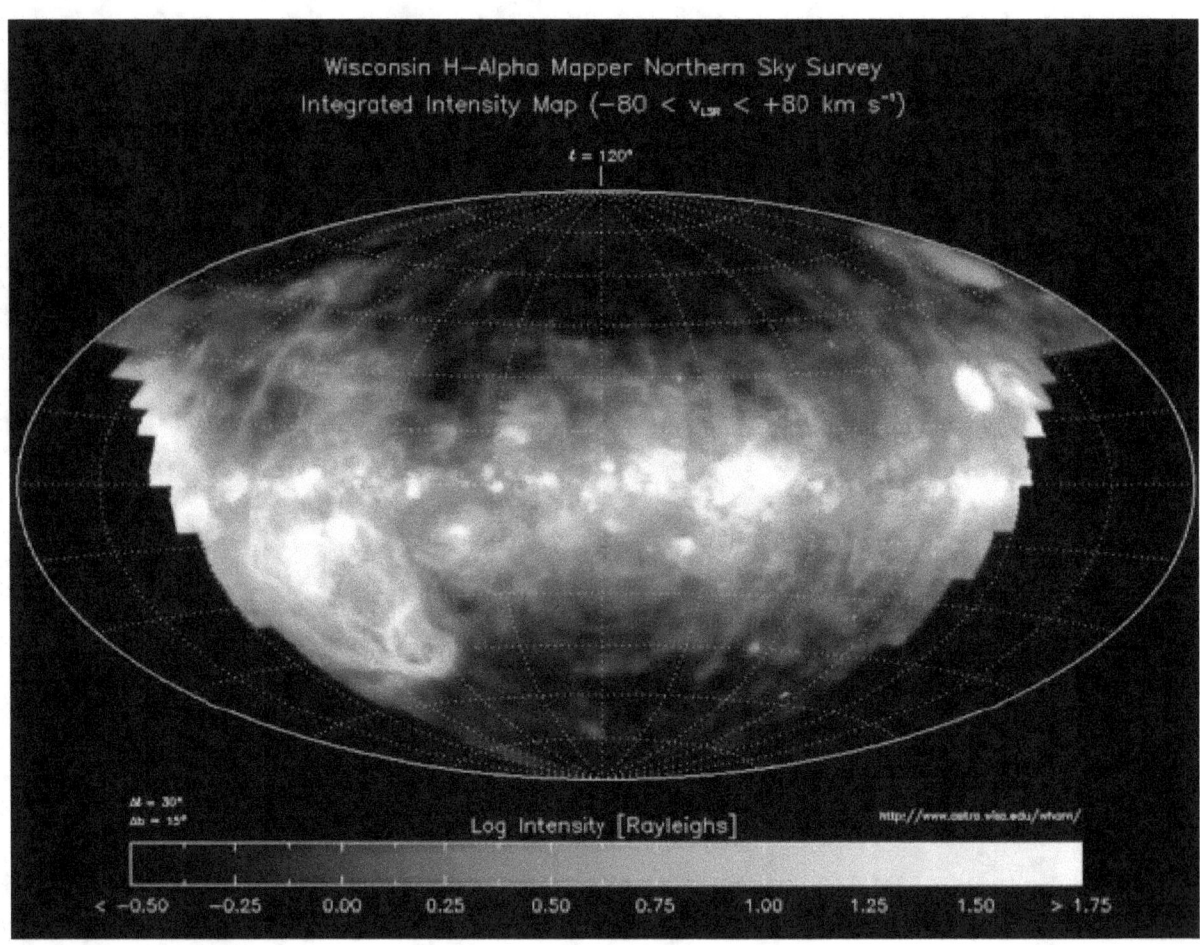

*The distribution of ionized hydrogen (known by astronomers as H II from old spectroscopic terminology) in the parts of the Galactic interstellar medium visible from the Earth's northern hemisphere as observed with the Wisconsin Hα Mapper (Haffner et al. 2003).*

In astronomy, the **interstellar medium** (**ISM**) is the matter that exists in the space between the star systems in a galaxy. This matter includes gas in ionic, atomic, and molecular form, as well as dust and cosmic rays. It fills interstellar space and blends smoothly into the surrounding intergalactic space. The energy that occupies the same volume, in the form of electromagnetic radiation, is the **interstellar radiation field**.

The interstellar medium is composed of multiple phases, distinguished by whether matter is ionic, atomic, or molecular, and the temperature and density of the matter. The interstellar medium is composed primarily by hydrogen followed by helium with trace amounts of carbon, oxygen, and nitrogen comparatively to hydrogen.[1] The thermal pressures of these phases are in rough equilibrium with one another. Magnetic fields and turbulent motions also provide pressure in the ISM, and are typically more important dynamically than the thermal pressure is.

In all phases, the interstellar medium is extremely tenuous by terrestrial standards. In cool, dense regions of the ISM, matter is primarily in molecular form, and reaches number densities of $10^6$ molecules per $cm^3$. In hot, diffuse regions of the ISM, matter is primarily ionized, and the density may be as low as $10^{-4}$ ions per $cm^3$. Compare this with a number density of roughly $10^{19}$ molecules per $cm^3$ for air, and $10^{10}$ molecules per $cm^3$ for a laboratory high-vacuum chamber. By mass, 99% of the ISM is gas in any form, and 1% is dust.[2] Of the gas in the ISM, by number 91% of atoms are hydrogen and 9% are helium, with 0.1% being atoms of elements heavier than hydrogen or helium,[3] known as "metals" in astronomical parlance. By mass this amounts to 70% hydrogen, 28% helium, and 1.5% heavier elements. The hydrogen and helium are primarily a result of primordial nucleosynthesis, while the heavier elements in the ISM are mostly a result of enrichment in the process of stellar evolution.

The ISM plays a crucial role in astrophysics precisely because of its intermediate role between stellar and galactic scales. Stars form within the densest regions of the ISM, molecular clouds, and replenish the ISM with matter and energy through planetary nebulae, stellar winds, and supernovae. This interplay between stars and the ISM helps determine the rate at which a galaxy depletes its gaseous content, and therefore its lifespan of active star formation.

On September 12, 2013, NASA officially announced that Voyager 1 had reached the ISM on August 25, 2012, making it the first artificial object to do so. Interstellar plasma and dust will be studied until the mission's end in 2025.

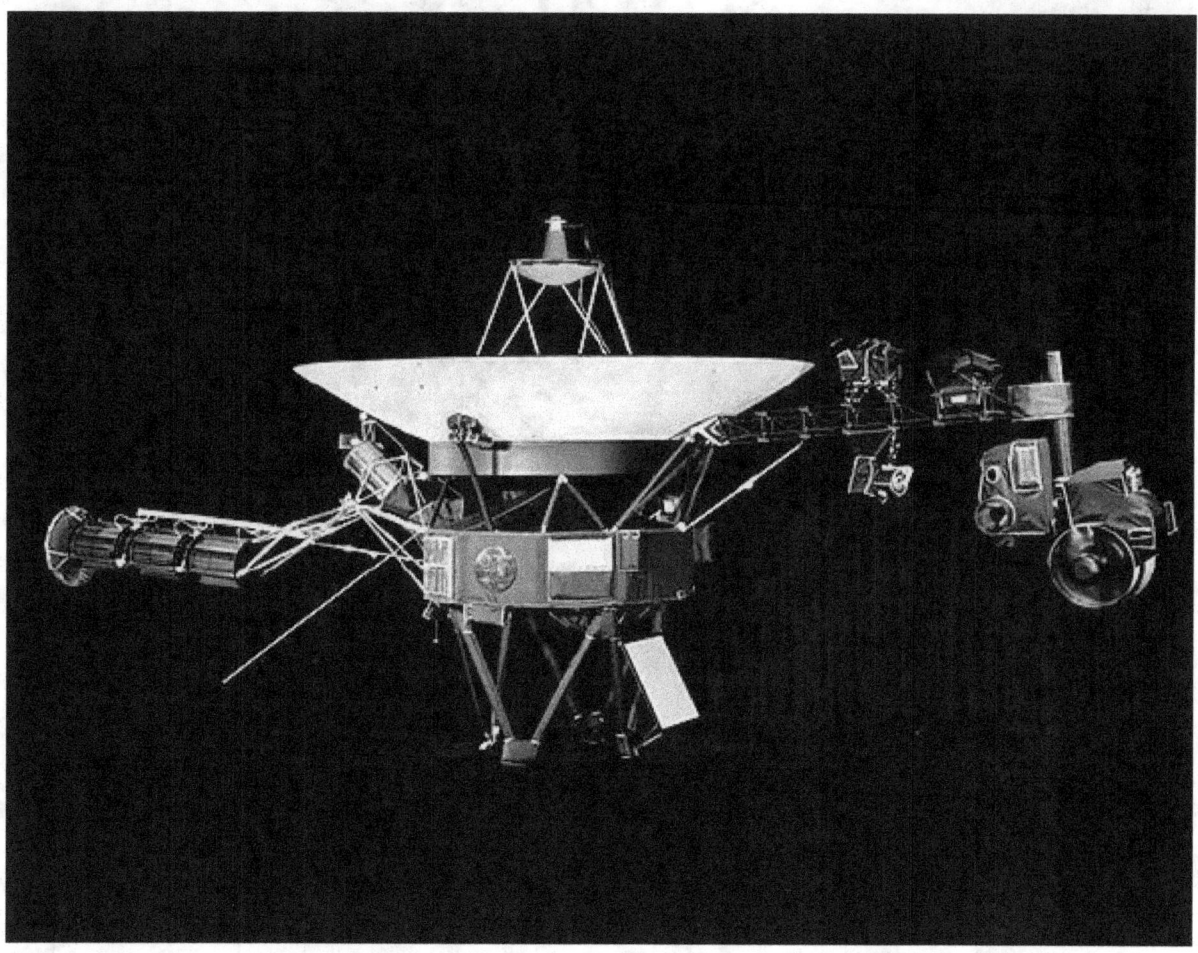

*Voyager 1 is the first artificial object to reach the ISM*

# 4.1   Interstellar matter

Table 1 shows a breakdown of the properties of the components of the ISM of the Milky Way.

## 4.1.1   The three-phase model

Field, Goldsmith & Habing (1969) put forward the static two *phase* equilibrium model to explain the observed properties of the ISM. Their modeled ISM consisted of a cold dense phase (T < 300 K), consisting of clouds of neutral and molecular hydrogen, and a warm intercloud phase (T ~ $10^4$ K), consisting of rarefied neutral and ionized gas. McKee & Ostriker (1977) added a dynamic third phase that represented the very hot (T ~ $10^6$ K) gas which had been shock heated by supernovae and constituted most of the volume of the ISM. These phases are the temperatures where heating and cooling can reach a stable equilibrium. Their paper formed the basis for further study over the past three decades. However, the relative proportions of the phases and their subdivisions are still not well known.[3]

## 4.1.2   Structures

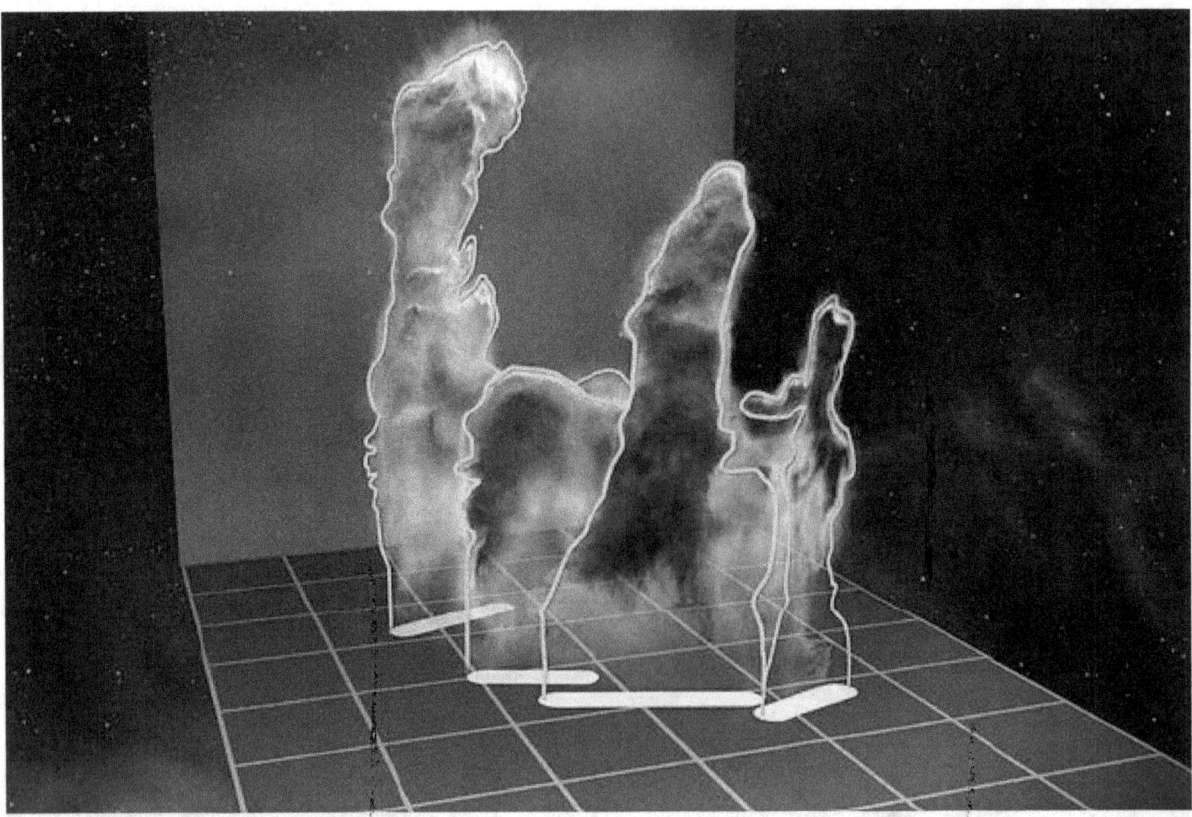

*Three-dimensional structure in Pillars of Creation.*[4]

The ISM is turbulent and therefore full of structure on all spatial scales. Stars are born deep inside large complexes of molecular clouds, typically a few parsecs in size. During their lives and deaths, stars interact physically with the ISM.

Stellar winds from young clusters of stars (often with giant or supergiant HII regions surrounding them) and shock waves created by supernovae inject enormous amounts of energy into their surroundings, which leads to hypersonic turbulence. The resultant structures – of varying sizes – can be observed, such as stellar wind bubbles and superbubbles of hot gas, seen by X-ray satellite telescopes or turbulent flows observed in radio telescope maps.

The Sun is currently traveling through the Local Interstellar Cloud, a denser region in the low-density Local Bubble.

### 4.1.3 Interaction with interplanetary medium

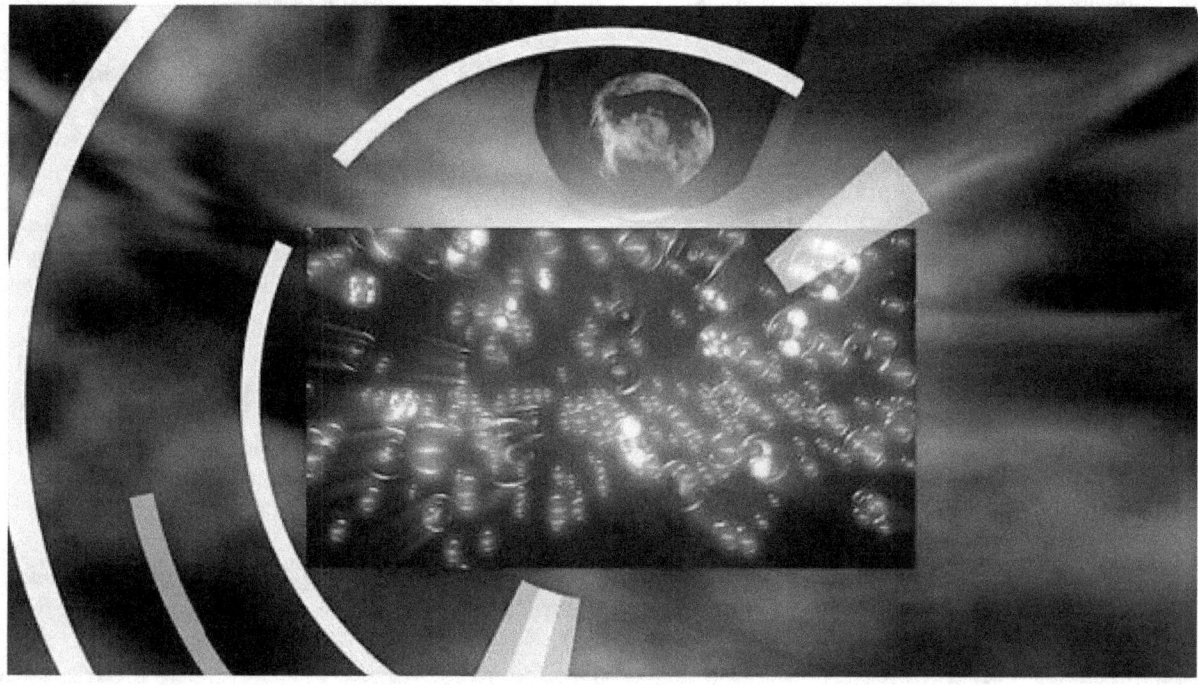

*Short, narrated video about IBEX's interstellar matter observations.*

The interstellar medium begins where the interplanetary medium of the Solar System ends. The solar wind slows to subsonic velocities at the termination shock, 90—100 astronomical units from the Sun. In the region beyond the termination shock, called the heliosheath, interstellar matter interacts with the solar wind. Voyager 1, the farthest human-made object from the Earth (after 1998[5]), crossed the termination shock December 16, 2004 and later entered interstellar space when it crossed the heliopause on August 25, 2012, providing the first direct probe of conditions in the ISM (Stone et al. 2005).

### 4.1.4 Interstellar extinction

The ISM is also responsible for extinction and reddening, the decreasing light intensity and shift in the dominant observable wavelengths of light from a star. These effects are caused by scattering and absorption of photons and allow the ISM to be observed with the naked eye in a dark sky. The apparent rifts that can be seen in the band of the Milky Way— a uniform disk of stars— are caused by absorption of background starlight by molecular clouds within a few thousand light years from Earth.

Far ultraviolet light is absorbed effectively by the neutral components of the ISM. For example, a typical absorption wavelength of atomic hydrogen lies at about 121.5 nanometers, the Lyman-alpha transition. Therefore, it is nearly impossible to see light emitted at that wavelength from a star farther than a few hundred light years from Earth, because most of it is absorbed during the trip to Earth by intervening neutral hydrogen.

## 4.2 Heating and cooling

The ISM is usually far from thermodynamic equilibrium. Collisions establish a Maxwell–Boltzmann distribution of velocities, and the 'temperature' normally used to describe interstellar gas is the 'kinetic temperature', which describes the temperature at which the particles would have the observed Maxwell–Boltzmann velocity distribution in thermodynamic equilibrium. However, the interstellar radiation field is typically much weaker than a medium in thermodynamic equilibrium; it is most often roughly that of an A star (surface temperature of ~10,000 K) highly diluted. Therefore, bound levels within an atom or molecule in the ISM are rarely populated according to the Boltzmann formula (Spitzer 1978, § 2.4).

Depending on the temperature, density, and ionization state of a portion of the ISM, different heating and cooling mechanisms determine the temperature of the gas.

### 4.2.1   Heating mechanisms

**Heating by low-energy cosmic rays**   The first mechanism proposed for heating the ISM was heating by low-energy cosmic rays.  Cosmic rays are an efficient heating source able to penetrate in the depths of molecular clouds. Cosmic rays transfer energy to gas through both ionization and excitation and to free electrons through Coulomb interactions.  Low-energy cosmic rays (a few MeV) are more important because they are far more numerous than high-energy cosmic rays.

**Photoelectric heating in grains**   The ultraviolet radiation emitted by hot stars can remove electrons from dust grains. The photon hits the dust grain, and some of its energy is used in overcoming the potential energy barrier (due to the possible positive charge of the grain) to remove the electron from the grain. The remainder of the photon's energy heats the grain and gives the ejected electron kinetic energy.  Since the size distribution of dust grains is $n(r) \propto r^{-3.5}$, where r is the size of the dust particle, the grain area distribution is $r^2 n \propto r^{-1.5}$. This indicates that the smallest dust grains dominate this method of heating.

**Photoionization**   When an electron is freed from an atom (typically from absorption of a UV photon) it carries kinetic energy away of the order: $E_{photon} - E_{ionization}$. This heating mechanism dominates in HII regions, but is negligible in the diffuse ISM due to the relative lack of neutral carbon atoms.

**X-ray heating**   X-rays remove electrons from atoms and ions, and those photoelectrons can provoke secondary ionizations. As the intensity is often low, this heating is only efficient in warm, less dense atomic medium (as the column density is small). For example in molecular clouds only hard x-rays can penetrate and x-ray heating can be ignored. This is assuming the region is not near an x-ray source such as a supernova remnant.

**Chemical heating**   Molecular hydrogen ($H_2$) can be formed on the surface of dust grains when two H atoms (which can travel over the grain) meet. This process yields 4.48 eV of energy distributed over the rotational and vibrational modes, kinetic energy of the $H_2$ molecule, as well as heating the dust grain. This kinetic energy, as well as the energy transferred from de-excitation of the hydrogen molecule through collisions, heats the gas.

**Grain-gas heating**   Collisions at high densities between gas atoms and molecules with dust grains can transfer thermal energy. This is not important in HII regions because UV radiation is more important. It is also not important in diffuse ionized medium due to the low density. In the neutral diffuse medium grains are always colder, but do not effectively cool the gas due to the low densities.

Grain heating by thermal exchange is very important in supernova remnants where densities and temperatures are very high.

Gas heating via grain-gas collisions is dominant deep in giant molecular clouds (especially at high densities). Far infrared radiation penetrates deeply due to the low optical depth. Dust grains are heated via this radiation and can transfer thermal energy during collisions with the gas. A measure of efficiency in the heating is given by the accommodation coefficient:

$$\alpha = \frac{T_2 - T}{T_d - T}$$

where $T$ is the gas temperature, $T_d$ the dust temperature, and $T_2$ the post-collision temperature of the gas atom/molecule. This coefficient was measured by (Burke & Hollenbach 1983) as $\alpha = 0.35$.

**Other heating mechanisms** A variety of macroscopic heating mechanisms are present including:

- Gravitational collapse of a cloud
- Supernova explosions
- Stellar winds
- Expansion of H II regions
- Magnetohydrodynamic waves created by supernova remnants

### 4.2.2 Cooling mechanisms

**Fine structure cooling** The process of fine structure cooling is dominant in most regions of the Interstellar Medium, except regions of hot gas and regions deep in molecular clouds. It occurs most efficiently with abundant atoms having fine structure levels close to the fundamental level such as: CII and OI in the neutral medium and OII, OIII, NII, NIII, NeII and NeIII in HII regions. Collisions will excite these atoms to higher levels, and they will eventually de-excite through photon emission, which will carry the energy out of the region.

**Cooling by permitted lines** At lower temperatures, more levels than fine structure levels can be populated via collisions. For example, collisional excitation of the n=2 level of hydrogen will release a Ly $\alpha$ photon upon de-excitation. In molecular clouds, excitation of rotational lines of CO is important. Once a molecule is excited, it eventually returns to a lower energy state, emitting a photon which can leave the region, cooling the cloud.

## 4.3 Radiowave propagation

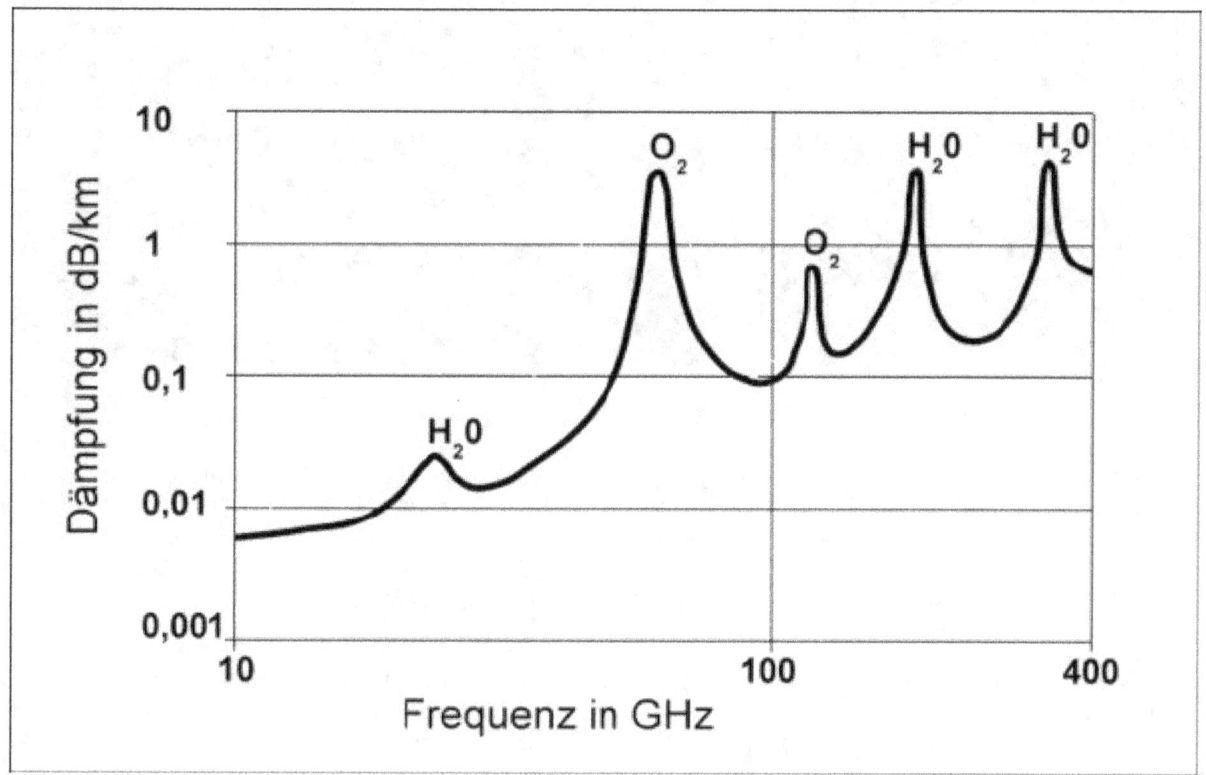

*Atmospheric attenuation in dB/km as a function of frequency over the EHF band. Peaks in absorption at specific frequencies are a problem, due to atmosphere constituents such as water vapor ($H_2O$) and carbon dioxide ($CO_2$).*

Radio waves from ≈10 kHz (very low frequency) to ≈300 GHz (extremely high frequency) propagate differently in interstellar space than on the Earth's surface. There are many sources of interference and signal distortion that do not exist on Earth. A great deal of radio astronomy depends on compensating for the different propagation effects to uncover the desired signal.[6][7]

## 4.4   The history of knowledge of interstellar space

*Herbig–Haro 110 object ejects gas through interstellar space.*[8]

The nature of the interstellar medium has received the attention of astronomers and scientists over the centuries, and understanding of the ISM has developed. However, they first had to acknowledge the basic concept of "interstellar" space. The term appears to have been first used in print by Bacon (1626, § 354–5): "The Interstellar Skie.. hath .. so much Affinity with the Starre, that there is a Rotation of that, as well as of the Starre." Later, natural philosopher Robert Boyle (1674) discussed "The inter-stellar part of heaven, which several of the modern Epicureans would have to be empty."

Before modern electromagnetic theory, early physicists postulated that an invisible luminiferous aether existed as a medium to carry lightwaves. It was assumed that this aether extended into interstellar space, as Patterson (1862) wrote, "this efflux occasions a thrill, or vibratory motion, in the ether which fills the interstellar spaces."

The advent of deep photographic imaging allowed Edward Barnard to produce the first images of dark nebulae silhouetted against the background star field of the galaxy, while the first actual detection of cold diffuse matter in interstellar space was made by Johannes Hartmann in 1904[9] through the use of absorption line spectroscopy. In his historic study of the spectrum and orbit of Delta Orionis, Hartmann observed the light coming from this star and realized that some of this light was being absorbed before it reached the Earth. Hartmann reported that absorption from the "K" line of calcium appeared "extraordinarily weak, but almost perfectly sharp" and also reported the "quite surprising result that the calcium line at 393.4 nanometres does not share in the periodic displacements of the lines caused by the orbital motion of the spectroscopic binary star". The stationary nature of the line led Hartmann to conclude that the gas responsible for the absorption was not present in the atmosphere of Delta Orionis, but was instead located within an isolated cloud of matter residing somewhere along the line-of-sight to this star. This discovery launched the study of the Interstellar Medium.

In the series of investigations, Viktor Ambartsumian introduced the now commonly accepted notion that interstellar matter occurs in the form of clouds.[10]

Following Hartmann's identification of interstellar calcium absorption, interstellar sodium was detected by Heger (1919) through the observation of stationary absorption from the atom's "D" lines at 589.0 and 589.6 nanometres towards Delta Orionis and Beta Scorpii.

Subsequent observations of the "H" and "K" lines of calcium by Beals (1936) revealed double and asymmetric profiles in the spectra of Epsilon and Zeta Orionis. These were the first steps in the study of the very complex interstellar sightline towards Orion. Asymmetric absorption line profiles are the result of the superposition of multiple absorption lines, each corresponding to the same atomic transition (for example the "K" line of calcium), but occurring in interstellar clouds with different radial velocities. Because each cloud has a different velocity (either towards or away from the observer/Earth) the absorption lines occurring within each cloud are either Blue-shifted or Red-shifted (respectively) from the lines' rest wavelength, through the Doppler Effect. These observations confirming that matter is not distributed homogeneously were the first evidence of multiple discrete clouds within the ISM.

*This light-year-long knot of interstellar gas and dust resembles a caterpillar.*[11]

The growing evidence for interstellar material led Pickering (1912) to comment that "While the interstellar absorbing medium may be simply the ether, yet the character of its selective absorption, as indicated by Kapteyn, is characteristic of a gas, and free gaseous molecules are certainly there, since they are probably constantly being expelled by the Sun and stars."

The same year Victor Hess's discovery of cosmic rays, highly energetic charged particles that rain onto the Earth from space, led others to speculate whether they also pervaded interstellar space. The following year the Norwegian explorer and physicist Kristian Birkeland wrote: "It seems to be a natural consequence of our points of view to assume that the whole of space is filled with electrons and flying electric ions of all kinds. We have assumed that each stellar system in evolutions throws off electric corpuscles into space. It does not seem unreasonable therefore to think that the greater part of the material masses in the universe is found, not in the solar systems or nebulae, but in 'empty' space" (Birkeland 1913).

Thorndike (1930) noted that "it could scarcely have been believed that the enormous gaps between the stars are completely void. Terrestrial aurorae are not improbably excited by charged particles from the Sun emitted by the Sun. If the millions of other stars are also ejecting ions, as is undoubtedly true, no absolute vacuum can exist within the galaxy."

In September 2012, NASA scientists reported that polycyclic aromatic hydrocarbons (PAHs), subjected to *interstellar medium (ISM)* conditions, are transformed, through hydrogenation, oxygenation and hydroxylation, to more complex organics - "a step along the path toward amino acids and nucleotides, the raw materials of proteins and DNA, respectively".[12][13] Further, as a result of these transformations, the PAHs lose their spectroscopic signature which could be one of the reasons "for the lack of PAH detection in interstellar ice grains, particularly the outer regions of cold, dense clouds or the upper molecular layers of protoplanetary disks."[12][13]

In February 2014, NASA announced a greatly upgraded database for tracking polycyclic aromatic hydrocarbons (PAHs) in the universe. According to scientists, more than 20% of the carbon in the universe may be associated with PAHs, possible starting materials for the formation of life. PAHs seem to have been formed shortly after the Big Bang, are widespread throughout the universe, and are associated with new stars and exoplanets.[14]

## 4.5   See also

- Astrophysical maser

- Diffuse interstellar band

- Fossil stellar magnetic field

- Heliosphere

- List of interstellar and circumstellar molecules

- List of plasma physics articles

- Molecular Clouds

- Photodissociation region

## 4.6   Notes

[1] Herdst, Eric (1995). "Chemistry in The Interstellar Medium". *Annual Review Physical Chemistry*. Bibcode:1995ARPC...46...27H. doi:10.1146/annurev.pc.46.100195.000331. Retrieved 2014-10-24.

[2] Boulanger, F.; Cox, P.; and Jones, A. P. (2000). "Course 7: Dust in the Interstellar Medium". In F. Casoli, J. Lequeux, & F. David. *Infrared Space Astronomy, Today and Tomorrow*. p. 251. Bibcode:2000isat.conf..251B.

[3] Ferriere (2001)

[4] "The Pillars of Creation Revealed in 3D". Retrieved 14 June 2015.

[5] Voyager: Fast Facts

[6] Samantha Blair. "Interstellar Medium Interference (video)". *SETI Talks*.

[7] "Voyager 1 Experiences Three Tsunami Waves in Interstellar Space (video)". *JPL*.

[8] "A geyser of hot gas flowing from a star". *ESA/Hubble Press Release*. Retrieved 3 July 2012.

[9] Asimov, Isaac, *Asimov's Biographical Encyclopedia of Science and Technology* (2nd ed.)

[10] S. Chandrasekhar (1989), "To Victor Ambartsumian on his 80th birthday", *Journal of Astrophysics and Astronomy* **18**: 3, Bibcode:1988Ap.....29..408C, doi:10.1007/BF01005852

[11] "Hubble sees a cosmic caterpillar". *Image Archive*. ESA/Hubble. Retrieved 9 September 2013.

[12] Staff (September 20, 2012), *NASA Cooks Up Icy Organics to Mimic Life's Origins*, Space.com, retrieved September 22, 2012

[13] Gudipati, Murthy S.; Yang, Rui (September 1, 2012), "In-Situ Probing Of Radiation-Induced Processing Of Organics In Astrophysical Ice Analogs—Novel Laser Desorption Laser Ionization Time-Of-Flight Mass Spectroscopic Studies", *The Astrophysical Journal Letters* **756** (1): L24, Bibcode:2012ApJ...756L..24G, doi:10.1088/2041-8205/756/1/L24, retrieved September 22, 2012

[14] Hoover, Rachel (February 21, 2014). "Need to Track Organic Nano-Particles Across the Universe? NASA's Got an App for That". *NASA*. Retrieved February 22, 2014.

## 4.7 References

- Bacon, Francis (1626), *Sylva* (354–5 ed.)

- Beals, C. S. (1936), "On the interpretation of interstellar lines", *Monthly Notices of the Royal Astronomical Society* **96**: 661, Bibcode:1936MNRAS..96..661B, doi:10.1093/mnras/96.7.661

- Birkeland, Kristian (1913), "Polar Magnetic Phenomena and Terrella Experiments", *The Norwegian Aurora Polaris Expedition, 1902-03 (section 2)*, New York: Christiania (now Oslo), H. Aschelhoug & Co., p. 720 out-of-print, full text online

- Boyle, Robert (1674), *The Excellency of Theology Compar'd with Natural Philosophy*, ii. iv., p. 178

- Burke, J. R.; Hollenbach, D.J. (1983), "The gas-grain interaction in the interstellar medium - Thermal accommodation and trapping", *Astrophysical Journal* **265**: 223, Bibcode:1983ApJ...265..223B, doi:10.1086/160667

- Dyson, J. (1997), *Physics of the Interstellar Medium*, London: Taylor & Francis

- Field, G. B.; Goldsmith, D. W.; Habing, H. J. (1969), "Cosmic-Ray Heating of the Interstellar Gas", *Astrophysical Journal* **155**: L149, Bibcode:1969ApJ...155L.149F, doi:10.1086/180324

- Ferriere, K. (2001), "The Interstellar Environment of our Galaxy", *Reviews of Modern Physics* **73** (4): 1031–1066, arXiv:astro-ph/0106359, Bibcode:2001RvMP...73.1031F, doi:10.1103/RevModPhys.73.1031

- Haffner, L. M.; Reynolds, R. J.; Tufte, S. L.; Madsen, G. J.; Jaehnig, K. P.; Percival, J. W. (2003), "The Wisconsin Hα Mapper Northern Sky Survey", *Astrophysical Journal Supplement* **145** (2): 405, arXiv:astro-ph/0309117, Bibcode:2003ApJS..149..405H, doi:10.1086/378850. The Wisconsin Hα Mapper is funded by the National Science Foundation.

- Heger, Mary Lea (1919), "Stationary Sodium Lines in Spectroscopic Binaries", *Publications of the Astronomical Society of the Pacific* **31** (184): 304, Bibcode:1919PASP...31..304H, doi:10.1086/122890

- Lequeux, J. *The Interstellar Medium*. Springer 2005.

- McKee, C. F.; Ostriker, J. P. (1977), "A theory of the interstellar medium - Three components regulated by supernova explosions in an inhomogeneous substrate", *Astrophysical Journal* **218**: 148, Bibcode:1977ApJ...218..148M, doi:10.1086/155667

- Patterson, Robert Hogarth (1862), "Colour in nature and art", *Essays in History and Art* **10** Reprinted from *Blackwood's Magazine*.

- Pickering, W. H. (1912), "The Motion of the Solar System relatively to the Interstellar Absorbing Medium", *Monthly Notices of the Royal Astronomical Society* **72**: 740, Bibcode:1912MNRAS..72..740P, doi:10.1093/mnras/72.9.740

- Spitzer, L. (1978), *Physical Processes in the Interstellar Medium*, Wiley, ISBN 0-471-29335-0

- Stone, E. C.; Cummings, A. C.; McDonald, F. B.; Heikkila, B. C.; Lal, N.; Webber, W. R. (2005), "Voyager 1 Explores the Termination Shock Region and the Heliosheath Beyond", *Science* **309** (5743): 2017–20, Bibcode:2005Sci...309.2017S, doi:10.1126/science.1117684, PMID 16179468

- Thorndike, S. L. (1930), "Interstellar Matter", *Publications of the Astronomical Society of the Pacific* **42** (246): 99, Bibcode:1930PASP...42...99T, doi:10.1086/124007

## 4.8   External links

- Freeview Video 'Chemistry of Interstellar Space' William Klemperer, Harvard University. A Royal Institution Discourse by the Vega Science Trust.

- The interstellar medium: an online tutorial

# Chapter 5

# Nuclear fusion

In nuclear physics, **nuclear fusion** is a nuclear reaction in which two or more atomic nuclei come very close and then collide at a very high speed and join to form a new nucleus. During this process, matter is not conserved because some of the matter of the fusing nuclei is converted to photons (energy). Fusion is the process that powers active or "main sequence" stars.

The fusion of two nuclei with lower masses than iron-56 (which, along with nickel-62, has the largest binding energy per nucleon) generally releases energy, while the fusion of nuclei heavier than iron *absorbs* energy. The opposite is true for the reverse process, nuclear fission. This means that fusion generally occurs for lighter elements only, and likewise, that fission normally occurs only for heavier elements. There are extreme astrophysical events that can lead to short periods of fusion with heavier nuclei. This is the process that gives rise to nucleosynthesis, the creation of the heavy elements during events such as supernova.

Following the discovery of quantum tunneling by Friedrich Hund, in 1929 Robert Atkinson and Fritz Houtermans used the measured masses of light elements to predict that large amounts of energy could be released by fusing small nuclei. Building upon the nuclear transmutation experiments by Ernest Rutherford, carried out several years earlier, the laboratory fusion of hydrogen isotopes was first accomplished by Mark Oliphant in 1932. During the remainder of that decade the steps of the main cycle of nuclear fusion in stars were worked out by Hans Bethe. Research into fusion for military purposes began in the early 1940s as part of the Manhattan Project. Fusion was accomplished in 1951 with the Greenhouse Item nuclear test. Nuclear fusion on a large scale in an explosion was first carried out on November 1, 1952, in the Ivy Mike hydrogen bomb test.

Research into developing controlled thermonuclear fusion for civil purposes also began in earnest in the 1950s, and it continues to this day. The present article is about the theory of fusion. For details of the quest for controlled fusion and its history, see the article Fusion power.

## 5.1 Process

The origin of the energy released in fusion of light elements is due to interplay of two opposing forces, the nuclear force which combines together protons and neutrons, and the Coulomb force which causes protons to repel each other. The protons are positively charged and repel each other but they nonetheless stick together, demonstrating the existence of another force referred to as nuclear attraction. This force, called the strong nuclear force, overcomes electric repulsion in a very close range. The effect of this force is not observed outside the nucleus, hence the force has a strong dependence on distance, making it a short-range force. The same force also pulls the nucleons together, or neutrons and protons together.[2] Because the nuclear force is stronger than the Coulomb force for atomic nuclei smaller than iron and nickel, building up these nuclei from lighter nuclei by **fusion** releases the extra energy from the net attraction of these particles. For larger nuclei, however, no energy is released, since the nuclear force is short-range and cannot continue to act across still larger atomic nuclei. Thus, energy is no longer released when such nuclei are made by fusion; instead, energy is absorbed in such processes.

*The Sun is a main-sequence star, and thus generates its energy by nuclear fusion of hydrogen nuclei into helium. In its core, the Sun fuses 620 million metric tons of hydrogen each second.*

Fusion reactions of light elements power the stars and produce virtually all elements in a process called nucleosynthesis. The fusion of lighter elements in stars releases energy (and the mass that always accompanies it). For example, in the fusion of two hydrogen nuclei to form helium, 0.7% of the mass is carried away from the system in the form of kinetic energy or other forms of energy (such as electromagnetic radiation).[3]

Research into controlled fusion, with the aim of producing fusion power for the production of electricity, has been conducted for over 60 years. It has been accompanied by extreme scientific and technological difficulties, but has resulted in progress. At present, controlled fusion reactions have been unable to produce break-even (self-sustaining) controlled fusion reactions.[4] Workable designs for a reactor that theoretically will deliver ten times more fusion energy than the amount needed to heat up plasma to required temperatures are in development (see ITER). The ITER facility is expected to finish its construction phase in 2019. It will start commissioning the reactor that same year and initiate plasma experiments in 2020, but is not expected to begin full deuterium-tritium fusion until 2027.[5]

It takes considerable energy to force nuclei to fuse, even those of the lightest element, hydrogen. This is because all nuclei have a positive charge due to their protons, and as like charges repel, nuclei strongly resist being put close together.

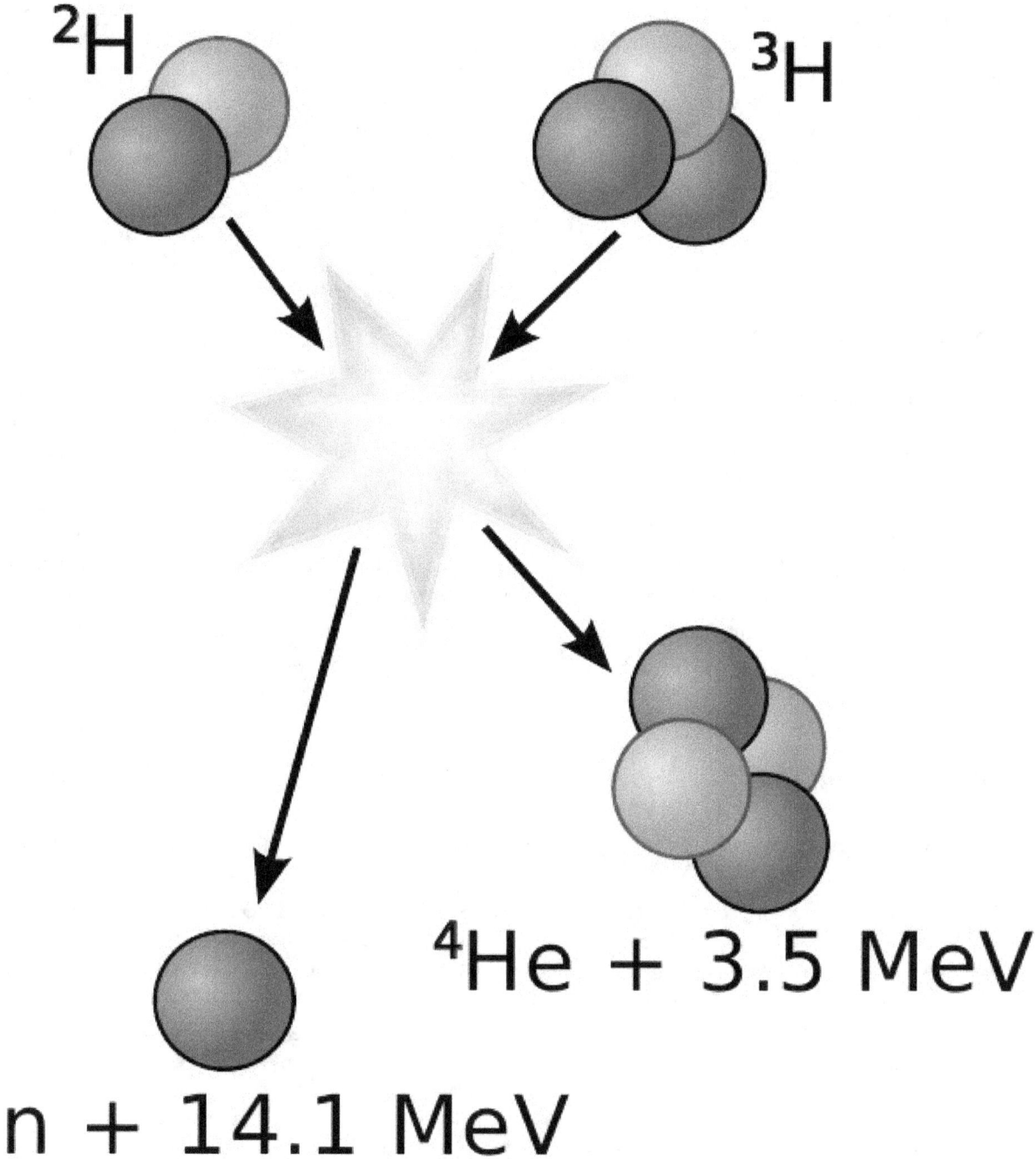

$^2$H $^3$H

$^4$He + 3.5 MeV

n + 14.1 MeV

*Fusion of deuterium with tritium creating helium-4, freeing a neutron, and releasing 17.59 MeV of energy, as an appropriate amount of mass changing forms to appear as the kinetic energy of the products, in agreement with kinetic E = $\Delta$mc$^2$, where $\Delta$m is the change in rest mass of particles.*[1]

Accelerated to high speeds, they can overcome this electrostatic repulsion and be forced close enough for the attractive nuclear force to be sufficiently strong to achieve fusion. The fusion of lighter nuclei, which creates a heavier nucleus and often a free neutron or proton, generally releases more energy than it takes to force the nuclei together; this is an exothermic process that can produce self-sustaining reactions. The US National Ignition Facility, which uses laser-driven inertial confinement fusion, is thought to be capable of break-even fusion.

The first large-scale laser target experiments were performed in June 2009 and ignition experiments began in early 2011.[6][7]

Energy released in most nuclear reactions is much larger than in chemical reactions, because the binding energy that holds a nucleus together is far greater than the energy that holds electrons to a nucleus. For example, the ionization energy gained by adding an electron to a hydrogen nucleus is 13.6 eV—less than one-millionth of the 17.6 MeV released in the deuterium–tritium (D–T) reaction shown in the diagram to the right (one gram of matter would release 339 GJ of energy). Fusion reactions have an energy density many times greater than nuclear fission; the reactions produce far greater energy per unit of mass even though *individual* fission reactions are generally much more energetic than *individual* fusion ones, which are themselves millions of times more energetic than chemical reactions. Only direct conversion of mass into energy, such as that caused by the annihilatory collision of matter and antimatter, is more energetic per unit of mass than nuclear fusion.

## 5.2  Nuclear fusion in stars

The most important fusion process in nature is the one that powers stars. In the 20th century, it was realized that the energy released from nuclear fusion reactions accounted for the longevity of the Sun and other stars as a source of heat and light. The fusion of nuclei in a star, starting from its initial hydrogen and helium abundance, provides that energy and synthesizes new nuclei as a byproduct of that fusion process. The prime energy producer in the Sun is the fusion of hydrogen to form helium, which occurs at a solar-core temperature of 14 million kelvin. The net result is the fusion of four protons into one alpha particle, with the release of two positrons, two neutrinos (which changes two of the protons into neutrons), and energy. Different reaction chains are involved, depending on the mass of the star. For stars the size of the sun or smaller, the proton-proton chain dominates. In heavier stars, the CNO cycle is more important.

As a star uses up a substantial fraction of its hydrogen, it begins to synthesize heavier elements, as part of stellar nucleosynthesis. However the heaviest elements are synthesized by fusion that occurs as a more massive star undergoes a violent supernova at the end of its life, a process known as supernova nucleosynthesis.

## 5.3  Requirements

Details and supporting references on the material in this section can be found in textbooks on nuclear physics or nuclear fusion.[8]

A substantial energy barrier of electrostatic forces must be overcome before fusion can occur. At large distances, two naked nuclei repel one another because of the repulsive electrostatic force between their positively charged protons. If two nuclei can be brought close enough together, however, the electrostatic repulsion can be overcome by the attractive nuclear force, which is stronger at close distances.

When a nucleon such as a proton or neutron is added to a nucleus, the nuclear force attracts it to other nucleons, but primarily to its immediate neighbours due to the short range of the force. The nucleons in the interior of a nucleus have more neighboring nucleons than those on the surface. Since smaller nuclei have a larger surface area-to-volume ratio, the binding energy per nucleon due to the nuclear force generally increases with the size of the nucleus but approaches a limiting value corresponding to that of a nucleus with a diameter of about four nucleons. It is important to keep in mind that the above picture is a toy model because nucleons are quantum objects, and so, for example, since two neutrons in a nucleus are identical to each other, distinguishing one from the other, such as which one is in the interior and which is on the surface, is in fact meaningless, and the inclusion of quantum mechanics is necessary for proper calculations.

The electrostatic force, on the other hand, is an inverse-square force, so a proton added to a nucleus will feel an electrostatic repulsion from *all* the other protons in the nucleus. The electrostatic energy per nucleon due to the electrostatic force thus increases without limit as nuclei get larger.

The net result of these opposing forces is that the binding energy per nucleon generally increases with increasing size, up to the elements iron and nickel, and then decreases for heavier nuclei. Eventually, the binding energy becomes negative and very heavy nuclei (all with more than 208 nucleons, corresponding to a diameter of about 6 nucleons) are not stable. The four most tightly bound nuclei, in decreasing order of binding energy per nucleon, are 62Ni, 58Fe, 56Fe, and 60Ni.[9] Even though the nickel isotope, 62Ni, is more stable, the iron isotope 56Fe is an order of magnitude more common. This is due to the fact that there is no easy way for stars to create 62Ni through the alpha process.

An exception to this general trend is the helium-4 nucleus, whose binding energy is higher than that of lithium, the next heaviest element. This is because protons and neutrons are fermions, which according to the Pauli exclusion principle cannot exist in the same nucleus in exactly the same state. Each proton or neutron energy state in a nucleus can accommodate both a spin up particle and a spin down particle. Helium-4 has an anomalously large binding energy because its nucleus consists of two protons and two neutrons, so all four of its nucleons can be in the ground state. Any additional nucleons would have to go into higher energy states. Indeed, the helium-4 nucleus is so tightly bound that it is commonly treated as a single particle in nuclear physics, namely, the alpha particle.

The situation is similar if two nuclei are brought together. As they approach each other, all the protons in one nucleus repel all the protons in the other. Not until the two nuclei actually come in contact can the strong nuclear force take over. Consequently, even when the final energy state is lower, there is a large energy barrier that must first be overcome. It is called the Coulomb barrier.

The Coulomb barrier is smallest for isotopes of hydrogen, as their nuclei contain only a single positive charge. A diproton is not stable, so neutrons must also be involved, ideally in such a way that a helium nucleus, with its extremely tight binding, is one of the products.

Using deuterium-tritium fuel, the resulting energy barrier is about 0.1 MeV. In comparison, the energy needed to remove an electron from hydrogen is 13.6 eV, about 7500 times less energy. The (intermediate) result of the fusion is an unstable $^5$He nucleus, which immediately ejects a neutron with 14.1 MeV. The recoil energy of the remaining $^4$He nucleus is 3.5 MeV, so the total energy liberated is 17.6 MeV. This is many times more than what was needed to overcome the energy barrier.

The reaction **cross section** σ is a measure of the probability of a fusion reaction as a function of the relative velocity of the two reactant nuclei. If the reactants have a distribution of velocities, e.g. a thermal distribution, then it is useful to perform an average over the distributions of the product of cross section and velocity. This average is called the 'reactivity', denoted <σv>. The reaction rate (fusions per volume per time) is <σv> times the product of the reactant number densities:

$$f = n_1 n_2 \langle \sigma v \rangle .$$

If a species of nuclei is reacting with itself, such as the DD reaction, then the product $n_1 n_2$ must be replaced by $(1/2)n^2$
.

$\langle \sigma v \rangle$ increases from virtually zero at room temperatures up to meaningful magnitudes at temperatures of 10–100 keV. At these temperatures, well above typical ionization energies (13.6 eV in the hydrogen case), the fusion reactants exist in a plasma state.

The significance of $\langle \sigma v \rangle$ as a function of temperature in a device with a particular energy confinement time is found by considering the Lawson criterion. This is an extremely challenging barrier to overcome on Earth, which explains why fusion research has taken many years to reach the current high state of technical prowess.[10]

# 5.4 Methods for achieving fusion

Main article: Fusion power

## 5.4.1 Thermonuclear fusion

Main article: Thermonuclear fusion

If the matter is sufficiently heated (hence being plasma), the fusion reaction may occur due to collisions with extreme thermal kinetic energies of the particles. In the form of thermonuclear weapons, thermonuclear fusion is the only fusion technique so far to yield undeniably large amounts of useful fusion energy. Usable amounts of thermonuclear fusion energy released in a controlled manner have yet to be achieved. In nature, this is what produces energy in stars through stellar nucleosynthesis.

## 5.4.2   Inertial confinement fusion

Main article: Inertial confinement fusion

Inertial confinement fusion (**ICF**) is a type of fusion energy research that attempts to initiate nuclear fusion reactions by heating and compressing a fuel target, typically in the form of a pellet that most often contains a mixture of deuterium and tritium.

## 5.4.3   Inertial electrostatic confinement

Main article: Inertial electrostatic confinement

Inertial electrostatic confinement is a set of devices that use an electric field to heat ions to fusion conditions. The most well known is the fusor. Starting in 1999, a number of amateurs have been able to do amateur fusion using these homemade devices.[11][12][13][14][15] Other IEC devices include: the Polywell, MIX POPS[16] and Marble concepts.[17]

## 5.4.4   Beam-beam or beam-target fusion

If the energy to initiate the reaction comes from accelerating one of the nuclei, the process is called *beam-target* fusion; if both nuclei are accelerated, it is *beam-beam* fusion.

Accelerator-based light-ion fusion is a technique using particle accelerators to achieve particle kinetic energies sufficient to induce light-ion fusion reactions. Accelerating light ions is relatively easy, and can be done in an efficient manner—all it takes is a vacuum tube, a pair of electrodes, and a high-voltage transformer; fusion can be observed with as little as 10 kV between electrodes. The key problem with accelerator-based fusion (and with cold targets in general) is that fusion cross sections are many orders of magnitude lower than Coulomb interaction cross sections. Therefore, the vast majority of ions end up expending their energy on bremsstrahlung and ionization of atoms of the target. Devices referred to as sealed-tube neutron generators are particularly relevant to this discussion. These small devices are miniature particle accelerators filled with deuterium and tritium gas in an arrangement that allows ions of these nuclei to be accelerated against hydride targets, also containing deuterium and tritium, where fusion takes place. Hundreds of neutron generators are produced annually for use in the petroleum industry where they are used in measurement equipment for locating and mapping oil reserves.

## 5.4.5   Muon-catalyzed fusion

Muon-catalyzed fusion is a well-established and reproducible fusion process that occurs at ordinary temperatures. It was studied in detail by Steven Jones in the early 1980s. Net energy production from this reaction cannot occur because of the high energy required to create muons, their short 2.2 µs half-life, and the high chance that a muon will bind to the new alpha particle and thus stop catalyzing fusion.[18]

## 5.4.6   Other principles

Some other confinement principles have been investigated.

Antimatter-initialized fusion uses small amounts of antimatter to trigger a tiny fusion explosion. This has been studied primarily in the context of making nuclear pulse propulsion, and pure fusion bombs feasible. This is not near becoming a practical power source, due to the cost of manufacturing antimatter alone.

Pyroelectric fusion was reported in April 2005 by a team at UCLA. The scientists used a pyroelectric crystal heated from −34 to 7 °C (−29 to 45 °F), combined with a tungsten needle to produce an electric field of about 25 gigavolts per meter to ionize and accelerate deuterium nuclei into an erbium deuteride target. At the estimated energy levels,[19] the D-D fusion reaction may occur, producing helium-3 and a 2.45 MeV neutron. Although it makes a useful neutron generator, the apparatus is not intended for power generation since it requires far more energy than it produces.[20][21][22][23]

Hybrid nuclear fusion-fission (hybrid nuclear power) is a proposed means of generating power by use of a combination of nuclear fusion and fission processes. The concept dates to the 1950s, and was briefly advocated by Hans Bethe during the 1970s, but largely remained unexplored until a revival of interest in 2009, due to the delays in the realization of pure fusion.[24] Project PACER, carried out at Los Alamos National Laboratory (LANL) in the mid-1970s, explored the possibility of a fusion power system that would involve exploding small hydrogen bombs (fusion bombs) inside an underground cavity. As an energy source, the system is the only fusion power system that could be demonstrated to work using existing technology. However it would also require a large, continuous supply of nuclear bombs, making the economics of such a system rather questionable.

## 5.5 Important reactions

### 5.5.1 Astrophysical reaction chains

At the temperatures and densities in stellar cores the rates of fusion reactions are notoriously slow. For example, at solar core temperature ($T \approx 15$ MK) and density (160 g/cm$^3$), the energy release rate is only 276 $\mu$W/cm$^3$—about a quarter of the volumetric rate at which a resting human body generates heat.[25] Thus, reproduction of stellar core conditions in a lab for nuclear fusion power production is completely impractical. Because nuclear reaction rates strongly depend on temperature ($\exp(-E/kT)$), achieving reasonable power levels in terrestrial fusion reactors requires 10–100 times higher temperatures (compared to stellar interiors): $T \approx 0.1$–$1.0$ GK.

### 5.5.2 Criteria and candidates for terrestrial reactions

Main article: Fusion power § Fuels

In artificial fusion, the primary fuel is not constrained to be protons and higher temperatures can be used, so reactions with larger cross-sections are chosen. Another concern is the production of neutrons, which activate the reactor structure radiologically, but also have the advantages of allowing volumetric extraction of the fusion energy and tritium breeding. Reactions that release no neutrons are referred to as *aneutronic*.

To be a useful energy source, a fusion reaction must satisfy several criteria. It must:

- **Be exothermic**: This limits the reactants to the low $Z$ (number of protons) side of the curve of binding energy. It also makes helium 4He the most common product because of its extraordinarily tight binding, although 3He and 3H also show up.

- **Involve low Z nuclei**: This is because the electrostatic repulsion must be overcome before the nuclei are close enough to fuse.

- **Have two reactants**: At anything less than stellar densities, three body collisions are too improbable. In inertial confinement, both stellar densities and temperatures are exceeded to compensate for the shortcomings of the third parameter of the Lawson criterion, ICF's very short confinement time.

- **Have two or more products**: This allows simultaneous conservation of energy and momentum without relying on the electromagnetic force.

- **Conserve both protons and neutrons**: The cross sections for the weak interaction are too small.

Few reactions meet these criteria. The following are those with the largest cross sections:

For reactions with two products, the energy is divided between them in inverse proportion to their masses, as shown. In most reactions with three products, the distribution of energy varies. For reactions that can result in more than one set of products, the branching ratios are given.

Some reaction candidates can be eliminated at once.[26] The D-$^6$Li reaction has no advantage compared to p$^+$-11

5B because it is roughly as difficult to burn but produces substantially more neutrons through 2

1D-2

1D side reactions. There is also a p$^+$-7

3Li reaction, but the cross section is far too low, except possibly when $T_i > 1$ MeV, but at such high temperatures an endothermic, direct neutron-producing reaction also becomes very significant. Finally there is also a p$^+$-9

4Be reaction, which is not only difficult to burn, but 9

4Be can be easily induced to split into two alpha particles and a neutron.

In addition to the fusion reactions, the following reactions with neutrons are important in order to "breed" tritium in "dry" fusion bombs and some proposed fusion reactors:

The latter of the two equations was unknown when the U.S. conducted the Castle Bravo fusion bomb test in 1954. Being just the second fusion bomb ever tested (and the first to use lithium), the designers of the Castle Bravo "Shrimp" had understood the usefulness of Lithium-6 in tritium production, but had failed to recognize that Lithium-7 fission would greatly increase the yield of the bomb. While Li-7 has a small neutron cross-section for low neutron energies, it has a higher cross section above 5 MeV.[27] Li-7 also undergoes a chain reaction due to its release of a neutron after fissioning. The 15 Mt yield was 150% greater than the predicted 6 Mt and caused casualties from the fallout generated.

To evaluate the usefulness of these reactions, in addition to the reactants, the products, and the energy released, one needs to know something about the cross section. Any given fusion device has a maximum plasma pressure it can sustain, and an economical device would always operate near this maximum. Given this pressure, the largest fusion output is obtained when the temperature is chosen so that $<\sigma v>/T^2$ is a maximum. This is also the temperature at which the value of the triple product $nT\tau$ required for ignition is a minimum, since that required value is inversely proportional to $<\sigma v>/T^2$ (see Lawson criterion). (A plasma is "ignited" if the fusion reactions produce enough power to maintain the temperature without external heating.) This optimum temperature and the value of $<\sigma v>/T^2$ at that temperature is given for a few of these reactions in the following table.

Note that many of the reactions form chains. For instance, a reactor fueled with 3

1T and 3

2He creates some 2

1D, which is then possible to use in the 2

1D-3

2He reaction if the energies are "right". An elegant idea is to combine the reactions (8) and (9). The 3

2He from reaction (8) can react with 6

3Li in reaction (9) before completely thermalizing. This produces an energetic proton, which in turn undergoes reaction (8) before thermalizing. Detailed analysis shows that this idea would not work well, but it is a good example of a case where the usual assumption of a Maxwellian plasma is not appropriate.

### 5.5.3   Neutronicity, confinement requirement, and power density

Any of the reactions above can in principle be the basis of fusion power production. In addition to the temperature and cross section discussed above, we must consider the total energy of the fusion products $E_{fus}$, the energy of the charged fusion products $E_{ch}$, and the atomic number $Z$ of the non-hydrogenic reactant.

Specification of the 2

1D-2

1D reaction entails some difficulties, though. To begin with, one must average over the two branches (2i) and (2ii). More difficult is to decide how to treat the 3

1T and 3

$^3_2$He products. $^3_1$T burns so well in a deuterium plasma that it is almost impossible to extract from the plasma. The $^2_1$D-$^3_2$He reaction is optimized at a much higher temperature, so the burnup at the optimum $^2_1$D-$^2_1$D temperature may be low, so it seems reasonable to assume the $^3_1$T but not the $^3_2$He gets burned up and adds its energy to the net reaction. Thus the total reaction would be the sum of (2i), (2ii), and (1):

$$5 \, ^2_1\text{D} \rightarrow 4 \, ^4_2\text{He} + 2 \, n^0 + 3 \, ^3_2\text{He} + p^+, \quad E_{\text{fus}} = 4.03+17.6+3.27 = 24.9 \text{ MeV}, \quad E_{\text{ch}} = 4.03+3.5+0.82 = 8.35 \text{ MeV}.$$

We count the $^2_1$D-$^2_1$D fusion energy *per D-D reaction* (not per pair of deuterium atoms) as $E_{\text{fus}}$ = (4.03 MeV + 17.6 MeV)×50% + (3.27 MeV)×50% = 12.5 MeV and the energy in charged particles as $E_{\text{ch}}$ = (4.03 MeV + 3.5 MeV)×50% + (0.82 MeV)×50% = 4.2 MeV. (Note: if the tritium ion reacts with a deuteron while it still has a large kinetic energy, then the kinetic energy of the helium-4 produced may be quite different from 3.5 MeV, so this calculation of energy in charged particles is only approximate.)

Another unique aspect of the $^2_1$D-$^2_1$D reaction is that there is only one reactant, which must be taken into account when calculating the reaction rate.

With this choice, we tabulate parameters for four of the most important reactions

The last column is the **neutronicity** of the reaction, the fraction of the fusion energy released as neutrons. This is an important indicator of the magnitude of the problems associated with neutrons like radiation damage, biological shielding, remote handling, and safety. For the first two reactions it is calculated as $(E_{\text{fus}}-E_{\text{ch}})/E_{\text{fus}}$. For the last two reactions, where this calculation would give zero, the values quoted are rough estimates based on side reactions that produce neutrons in a plasma in thermal equilibrium.

Of course, the reactants should also be mixed in the optimal proportions. This is the case when each reactant ion plus its associated electrons accounts for half the pressure. Assuming that the total pressure is fixed, this means that density of the non-hydrogenic ion is smaller than that of the hydrogenic ion by a factor $2/(Z+1)$. Therefore, the rate for these reactions is reduced by the same factor, on top of any differences in the values of $<\sigma v>/T^2$. On the other hand, because the $^2_1$D-$^2_1$D reaction has only one reactant, its rate is twice as high as when the fuel is divided between two different hydrogenic species, thus creating a more efficient reaction.

Thus there is a "penalty" of $(2/(Z+1))$ for non-hydrogenic fuels arising from the fact that they require more electrons, which take up pressure without participating in the fusion reaction. (It is usually a good assumption that the electron temperature will be nearly equal to the ion temperature. Some authors, however discuss the possibility that the electrons could be maintained substantially colder than the ions. In such a case, known as a "hot ion mode", the "penalty" would not apply.) There is at the same time a "bonus" of a factor 2 for $^2_1$D-$^2_1$D because each ion can react with any of the other ions, not just a fraction of them.

We can now compare these reactions in the following table.

The maximum value of $<\sigma v>/T^2$ is taken from a previous table. The "penalty/bonus" factor is that related to a non-hydrogenic reactant or a single-species reaction. The values in the column "reactivity" are found by dividing $1.24\times10^{-24}$ by the product of the second and third columns. It indicates the factor by which the other reactions occur more slowly than the $^2_1$D-$^3_1$T.

1T reaction under comparable conditions. The column "Lawson criterion" weights these results with $E_{ch}$ and gives an indication of how much more difficult it is to achieve ignition with these reactions, relative to the difficulty for the 2

1D-3

1T reaction. The last column is labeled "power density" and weights the practical reactivity with $E_{fus}$. It indicates how much lower the fusion power density of the other reactions is compared to the 2

1D-3

1T reaction and can be considered a measure of the economic potential.

## 5.5.4   Bremsstrahlung losses in quasineutral, isotropic plasmas

The ions undergoing fusion in many systems will essentially never occur alone but will be mixed with electrons that in aggregate neutralize the ions' bulk electrical charge and form a plasma. The electrons will generally have a temperature comparable to or greater than that of the ions, so they will collide with the ions and emit x-ray radiation of 10–30 keV energy, a process known as Bremsstrahlung.

The huge size of the Sun and stars means that the x-rays produced in this process will not escape and will deposit their energy back into the plasma. They are said to are opaque to x-rays. But any terrestrial fusion reactor will be optically thin for x-rays of this energy range. X-rays are difficult to reflect but they are effectively absorbed (and converted into heat) in less than mm thickness of stainless steel (which is part of a reactor's shield). This means the bremsstrahlung process is carrying energy out of the plasma, cooling it.

The ratio of fusion power produced to x-ray radiation lost to walls is an important figure of merit. This ratio is generally maximized at a much higher temperature than that which maximizes the power density (see the previous subsection). The following table shows estimates of the optimum temperature and the power ratio at that temperature for several reactions.[26]

The actual ratios of fusion to Bremsstrahlung power will likely be significantly lower for several reasons. For one, the calculation assumes that the energy of the fusion products is transmitted completely to the fuel ions, which then lose energy to the electrons by collisions, which in turn lose energy by Bremsstrahlung. However, because the fusion products move much faster than the fuel ions, they will give up a significant fraction of their energy directly to the electrons. Secondly, the ions in the plasma are assumed to be purely fuel ions. In practice, there will be a significant proportion of impurity ions, which will then lower the ratio. In particular, the fusion products themselves *must* remain in the plasma until they have given up their energy, and *will* remain some time after that in any proposed confinement scheme. Finally, all channels of energy loss other than Bremsstrahlung have been neglected. The last two factors are related. On theoretical and experimental grounds, particle and energy confinement seem to be closely related. In a confinement scheme that does a good job of retaining energy, fusion products will build up. If the fusion products are efficiently ejected, then energy confinement will be poor, too.

The temperatures maximizing the fusion power compared to the Bremsstrahlung are in every case higher than the temperature that maximizes the power density and minimizes the required value of the fusion triple product. This will not change the optimum operating point for 2

1D-3

1T very much because the Bremsstrahlung fraction is low, but it will push the other fuels into regimes where the power density relative to 2

1D-3

1T is even lower and the required confinement even more difficult to achieve. For 2

1D-2

1D and 2

1D-3

2He, Bremsstrahlung losses will be a serious, possibly prohibitive problem. For 3

2He-3

2He, p⁺-6

3Li and p⁺-11

5B the Bremsstrahlung losses appear to make a fusion reactor using these fuels with a quasineutral, isotropic plasma impossible. Some ways out of this dilemma are considered—and rejected—in *Fundamental limitations on plasma fusion systems not in thermodynamic equilibrium* by Todd Rider.[28] This limitation does not apply to non-neutral and anisotropic plasmas; however, these have their own challenges to contend with.

## 5.6    See also

- Aneutronic fusion

- CNO cycle

- Direct energy conversion

- Inertial electrostatic confinement

- Focus fusion

- Fusenet

- Fusion power

- Fusion rocket

- Helium-3

- Impulse generator

- ITER

- Joint European Torus

- List of fusion experiments

- List of plasma (physics) articles

- National Ignition Facility

- Nuclear fission

- Nuclear physics

- Nuclear reactor

- Nucleosynthesis

- Neutron generator

- Neutron source

- Periodic table

- Polywell

- Proton-proton chain

- Pulsed power

- Teller–Ulam design

- Thermonuclear fusion

- Timeline of nuclear fusion

- Triple-alpha process

# 5.7 References

[1] Shultis, J.K. and Faw, R.E. (2002). *Fundamentals of nuclear science and engineering*. CRC Press. p. 151. ISBN 0-8247-0834-2.

[2] Physics Flexbook. Ck12.org. Retrieved on 2012-12-19.

[3] Bethe, Hans A. "The Hydrogen Bomb", *Bulletin of the Atomic Scientists*, April 1950, p. 99.

[4] "Progress in Fusion". ITER. Retrieved 2010-02-15.

[5] "ITER – the way to new energy". *ITER*. 2014.

[6] "The National Ignition Facility: Ushering in a new age for high energy density science". National Ignition Facility. Retrieved 2014-03-27.

[7] "DOE looks again at inertial fusion as potential clean-energy source", David Kramer, *Physics Today*, March 2011, p 26

[8] S. Atzeni, J. Meyer-ter-Vehn (2004). Chapter 1: "Nuclear fusion reactions". The Physics of Inertial Fusion. University of Oxford Press. ISBN 978-0-19-856264-1

[9] The Most Tightly Bound Nuclei. Hyperphysics.phy-astr.gsu.edu. Retrieved on 2011-08-17.

[10] What Is The Lawson Criteria, Or How to Make Fusion Power Viable

[11] "Fusor Forums • Index page". Fusor.net. Retrieved 2014-08-24.

[12] "Build a Nuclear Fusion Reactor? No Problem". Clhsonline.net. 2012-03-23. Retrieved 2014-08-24.

[13] "Extreme DIY: Building a homemade nuclear reactor in NYC". *BBC News*. Retrieved 30 October 2014.

[14] Schechner, Sam (2008-08-18). "Nuclear Ambitions: Amateur Scientists Get a Reaction From Fusion – WSJ". Online.wsj.com. Retrieved 2014-08-24.

[15] "Will's Amateur Science and Engineering: Fusion Reactor's First Light!". Tidbit77.blogspot.com. 2010-02-09. Retrieved 2014-08-24.

[16] Park J, Nebel RA, Stange S, Murali SK (2005). "Experimental Observation of a Periodically Oscillating Plasma Sphere in a Gridded Inertial Electrostatic Confinement Device". *Phys Rev Lett* **95** (1): 015003. Bibcode:2005PhRvL..95a5003P. doi:10.1103/PhysRevLett.95.015003. PMID 16090625.

[17] "The Multiple Ambipolar Recirculating Beam Line Experiment" Poster presentation, 2011 US-Japan IEC conference, Dr. Alex Klein

[18] Jones, S.E. (1986). "Muon-Catalysed Fusion Revisited". *Nature* **321** (6066): 127–133. Bibcode:1986Natur.321..127J. doi:10.1038/321127a0

[19] Supplementary methods for "Observation of nuclear fusion driven by a pyroelectric crystal". Main article Naranjo, B.; Gimzewski, J.K.; Putterman, S. (2005). "Observation of nuclear fusion driven by a pyroelectric crystal". *Nature* **434** (7037): 1115–1117. Bibcode:2005Natur.434.1115N. doi:10.1038/nature03575. PMID 15858570.

[20] UCLA Crystal Fusion. Rodan.physics.ucla.edu. Retrieved on 2011-08-17. Archived 8 June 2015 at the Wayback Machine

[21] Schewe, Phil and Stein, Ben (2005). "Pyrofusion: A Room-Temperature, Palm-Sized Nuclear Fusion Device". *Physics News Update* **729** (1). Archived from the original on 12 November 2013.

[22] Coming in out of the cold: nuclear fusion, for real. Christiansciencemonitor.com (2005-06-06). Retrieved on 2011-08-17.

[23] fusion on the desktop ... really!. MSNBC (2005-04-27). Retrieved on 2011-08-17.

[24] Gerstner, E. (2009). "Nuclear energy: The hybrid returns". *Nature* **460** (7251): 25–8. doi:10.1038/460025a. PMID 19571861.

[25] FusEdWeb | Fusion Education. Fusedweb.pppl.gov (1998-11-09). Retrieved on 2011-08-17.

[26] Archived January 3, 2006 at the Wayback Machine. Retrieved on 2012-12-19.

[27] Subsection 4.7.4c. Kayelaby.npl.co.uk. Retrieved on 2012-12-19.

[28] Portable Document Format (PDF) Archived 26 March 2006 at the Wayback Machine

## 5.8 Further reading

- "What is Nuclear Fusion?". NuclearFiles.org.

- S. Atzeni, J. Meyer-ter-Vehn (2004). "Nuclear fusion reactions". *The Physics of Inertial Fusion* (PDF). University of Oxford Press. ISBN 978-0-19-856264-1.

- G. Brumfiel (22 May 2006). "Chaos could keep fusion under control". *Nature*. doi:10.1038/news060522-2.

- R.W. Bussard (9 November 2006). "Should Google Go Nuclear? Clean, Cheap, Nuclear Power". *Google TechTalks*.

- A. Wenisch, R. Kromp, D. Reinberger (November 2007). "Science of Fiction: Is there a Future for Nuclear?" (PDF). Austrian Institute of Ecology.

- W.J. Nuttall (September 2008). "Fusion as an Energy Source: Challenges and Opportunities" (PDF). *Institute of Physics Report*. Institute of Physics.

## 5.9 External links

- NuclearFiles.org—A repository of documents related to nuclear power.

- Annotated bibliography for nuclear fusion from the Alsos Digital Library for Nuclear Issues

- -NRL Fusion Formulary

**Organizations**

- Fusion for Energy website

- ITER (International Thermonuclear Experimental Reactor) website

- CCFE (Culham Centre for Fusion Energy) website

- JET (Joint European Torus) website

- Naka Fusion Institute at JAEA (Japan Atomic Energy Agency) website

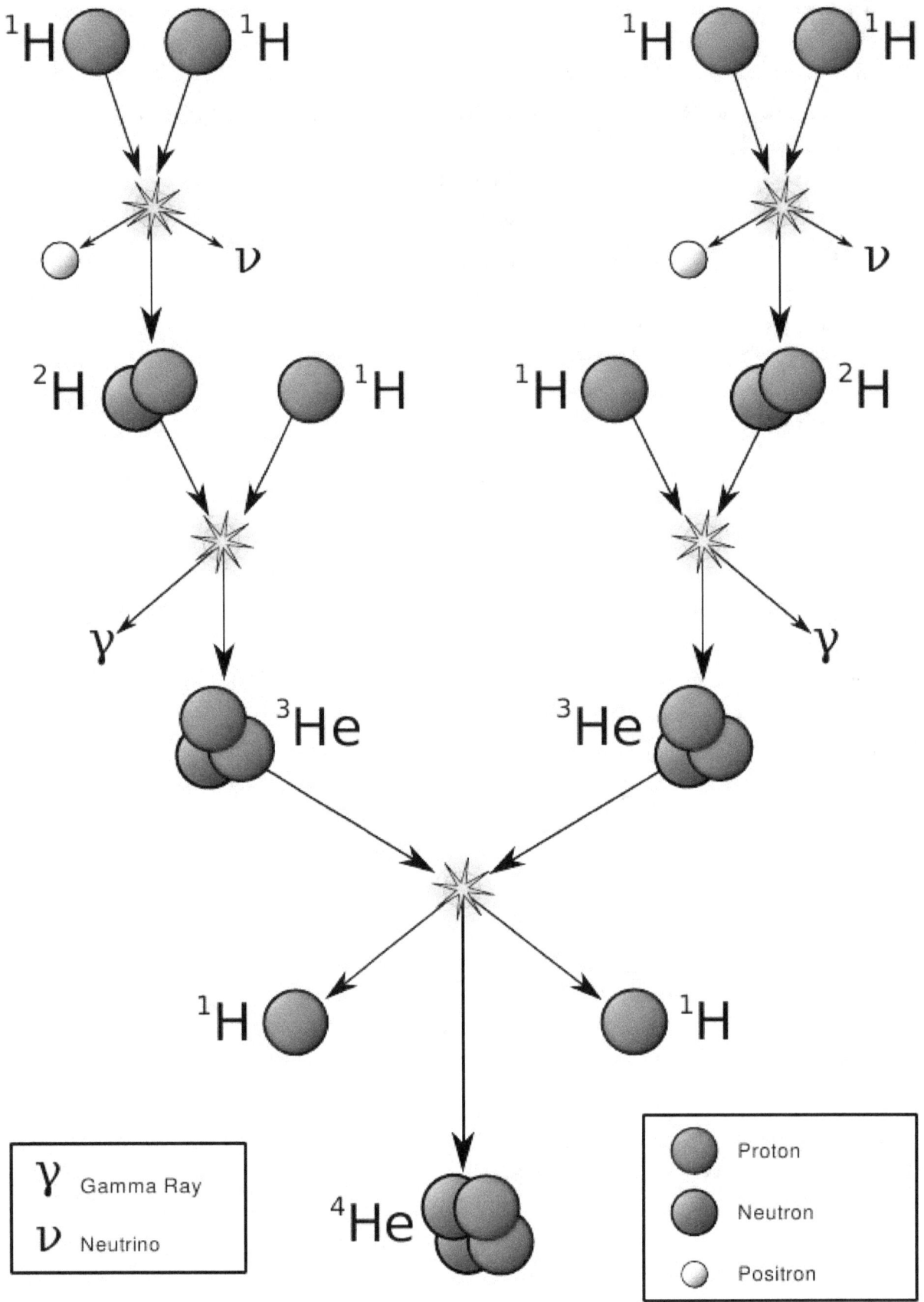

The proton-proton chain dominates in stars the size of the Sun or smaller.

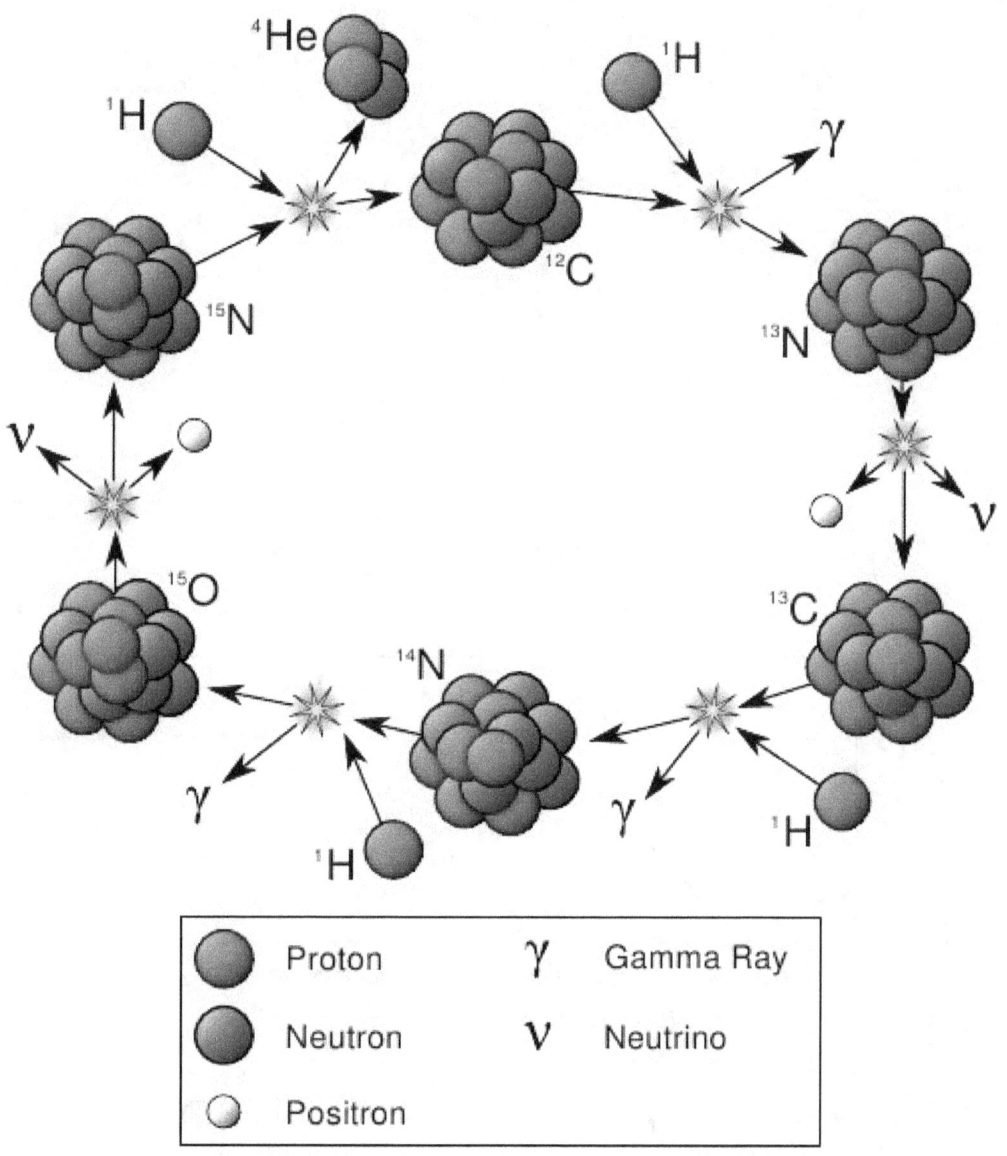

The CNO cycle dominates in stars heavier than the Sun.

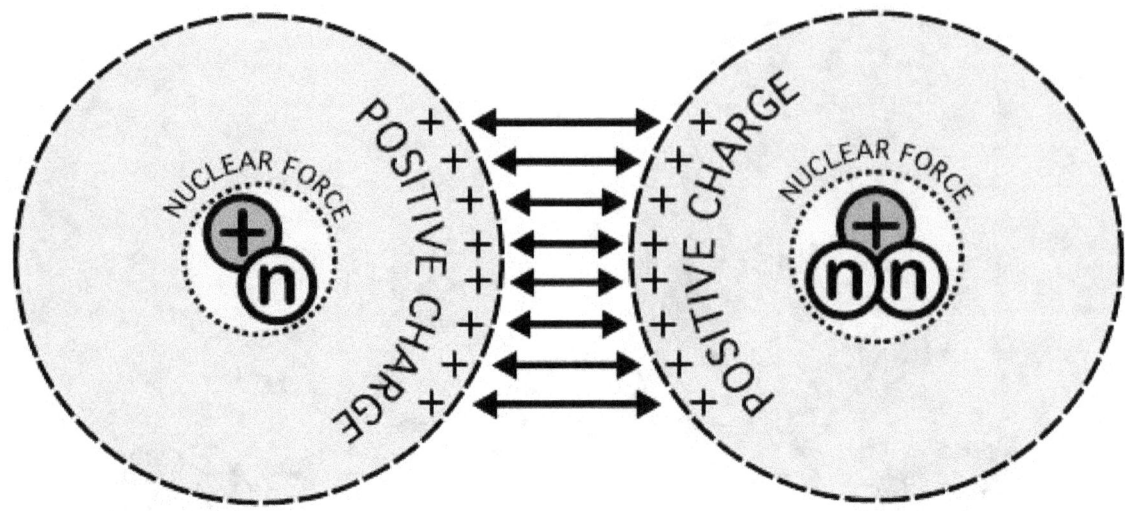

*The electrostatic force between the positively charged nuclei is repulsive, but when the separation is small enough, the attractive nuclear force is stronger. Therefore, the prerequisite for fusion is that the nuclei have enough kinetic energy that they can approach each other despite the electrostatic repulsion.*

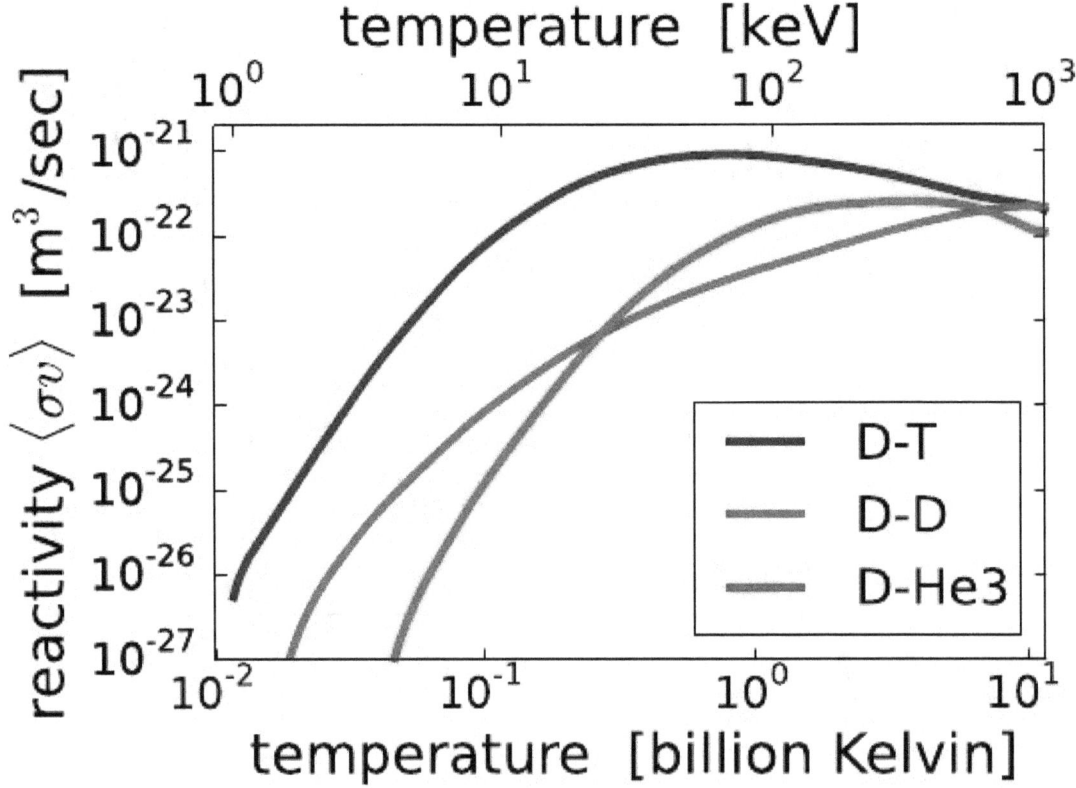

*The fusion reaction rate increases rapidly with temperature until it maximizes and then gradually drops off. The DT rate peaks at a lower temperature (about 70 keV, or 800 million kelvin) and at a higher value than other reactions commonly considered for fusion energy.*

*The* Tokamak à configuration variable, *research fusion reactor, at the École Polytechnique Fédérale de Lausanne (Switzerland).*

*The only man-made fusion device to achieve ignition to date is the hydrogen bomb. The detonation of the first device, codenamed Ivy Mike, occurred in 1952 and is shown here.*

# Chapter 6

# Compact star

In astronomy, the term **compact star** (sometimes **compact object**) is used to refer collectively to white dwarfs, neutron stars, other exotic dense stars, and black holes.

Most compact stars are the endpoints of stellar evolution and are thus often referred to as **stellar remnants**, the form of the remnant depending primarily on the mass of the star when it formed. These objects are all small in volume for their mass, giving them a very high density. The term *compact star* is often used when the exact nature of the star is not known, but evidence suggests that it is very massive and has a small radius, thus implying one of the above-mentioned categories. A compact star that is not a black hole may be called a **degenerate star**.

## 6.1  Formation

The usual endpoint of stellar evolution is the formation of a compact star.

Most stars will eventually come to a point in their evolution, when the outward radiation pressure from the nuclear fusions in its interior can no longer resist the ever present gravitational forces. When this happens, the star collapses under its own weight and undergo the process of stellar death. For most stars, this will result in the formation of a very dense and compact stellar remnant, also known as a compact star.

Compact stars have no internal energy production, but will - with the exception of black holes - usually radiate for millions of years with excess heat left from the collapse itself.[1]

According to the most recent understanding, compact stars could also form during the phase separations of the early Universe following the Big Bang.[2] Primordial origins of known compact objects have not been determined with certainty.

## 6.2  Lifetime

Although compact stars may radiate, and thus cool off and lose energy, they do not depend on high temperatures to maintain their structure, as ordinary stars do. Barring external disturbances and baryonic decay, they can persist virtually forever. Black holes are however generally believed to finally evaporate from Hawking radiation after trillions of years. According to our current standard models of physical cosmology, all stars will eventually evolve into cool and dark compact stars, by the time the Universe enters the so-called degenerate era in a very distant future.

The somewhat wider definition of *compact objects* often includes smaller solid objects such as planets, asteroids, and comets. There is a remarkable variety of stars and other clumps of hot matter, but all matter in the Universe must eventually end as some form of compact object, according to the theory of thermodynamics.

## 6.3    White dwarfs

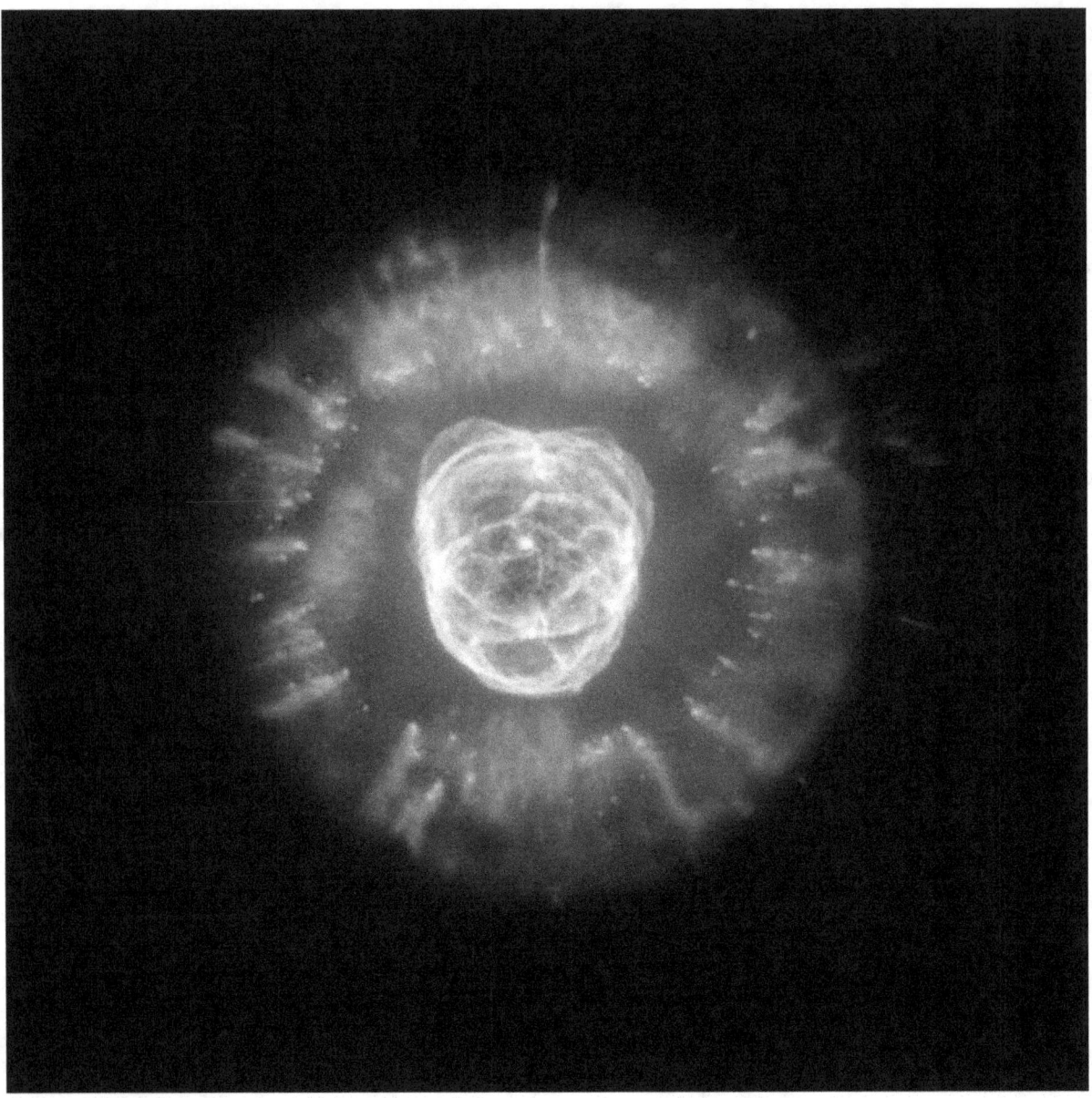

*The Eskimo Nebula is illuminated by the white dwarf at its center.*

Main article: White dwarf

The stars called *degenerate dwarfs* or, more usually, *white dwarfs* are made up mainly of degenerate matter—typically, carbon and oxygen nuclei in a sea of degenerate electrons. White dwarfs arise from the cores of main-sequence stars and are therefore very hot when they are formed. As they cool they will redden and dim until they eventually become dark *black dwarfs*. White dwarfs were observed in the 19th century, but the extremely high densities and pressures they contain were not explained until the 1920s.

The equation of state for degenerate matter is "soft", meaning that adding more mass will result in a smaller object. Continuing to add mass to what is now a white dwarf, the object shrinks and the central density becomes even larger, with higher degenerate-electron energies. The star's radius has now shrunk to only a few thousand kilometers, and the mass is approaching the theoretical upper limit of the mass of a white dwarf, the Chandrasekhar limit, about 1.4 times the mass of the Sun ($M\odot$).

If we were to take matter from the center of our white dwarf and slowly start to compress it, we would first see electrons forced to combine with nuclei, changing their protons to neutrons by inverse beta decay. The equilibrium would shift towards heavier, neutron-richer nuclei that are not stable at everyday densities. As the density increases, these nuclei become still larger and less well-bound. At a critical density of about $4 \cdot 10^{14}$ kg/m$^3$, called the neutron drip line, the atomic nucleus would tend to fall apart into protons and neutrons. Eventually we would reach a point where the matter is on the order of the density (~$2 \cdot 10^{17}$ kg/m$^3$) of an atomic nucleus. At this point the matter is chiefly free neutrons, with a small amount of protons and electrons.

## 6.4 Neutron stars

*The Crab Nebula is a supernova remnant containing the Crab Pulsar, a neutron star.*

Main article: Neutron star

In certain binary stars containing a white dwarf, mass is transferred from the companion star onto the white dwarf, eventually pushing it over the Chandrasekhar limit. Electrons react with protons to form neutrons and thus no longer supply the necessary pressure to resist gravity, causing the star to collapse. If the center of the star is composed mostly of carbon and oxygen then such a gravitational collapse will ignite runaway fusion of the carbon and oxygen, resulting in a Type Ia supernova that entirely blows apart the star before the collapse can become irreversible. If the center is composed mostly of magnesium or heavier elements, the collapse continues.[3][4][5] As the density further increases, the remaining electrons react with the protons to form more neutrons. The collapse continues until (at higher density) the neutrons become degenerate. A new equilibrium is possible after the star shrinks by three orders of magnitude, to a radius between 10 and 20 km. This is a *neutron star*.

Although the first neutron star was not observed until 1967 when the first radio pulsar was discovered, neutron stars were proposed by Baade and Zwicky in 1933, only one year after the neutron was discovered in 1932. They realized that because neutron stars are so dense, the collapse of an ordinary star to a neutron star would liberate a large amount of gravitational potential energy, providing a possible explanation for supernovae.[6][7][8] This is the explanation for supernovae of types Ib, Ic, and II. Such supernovae occur when the iron core of a massive star exceeds the Chandrasekhar limit and collapses to a neutron star.

Like electrons, neutrons are fermions. They therefore provide neutron degeneracy pressure to support a neutron star against collapse. In addition, repulsive neutron-neutron interactions provide additional pressure. Like the Chandrasekhar limit for white dwarfs, there is a limiting mass for neutron stars: the Tolman-Oppenheimer-Volkoff limit, where these forces are no longer sufficient to hold up the star. As the forces in dense hadronic matter are not well understood, this limit is not known exactly but is thought to be between 2 and 3 $M\odot$. If more mass accretes onto a neutron star, eventually this mass limit will be reached. What happens next is not completely clear.

## 6.5   Black holes

Main article: Stellar black hole

As more mass is accumulated, equilibrium against gravitational collapse reaches its breaking point. The star's pressure is insufficient to counterbalance gravity and a catastrophic gravitational collapse occurs in milliseconds. The escape velocity at the surface, already at least 1/3 light speed, quickly reaches the velocity of light. No energy or matter can escape: a black hole has formed. All light will be trapped within an event horizon, and so a black hole appears truly black, except for the possibility of Hawking radiation. It is presumed that the collapse will continue.

In the classical theory of general relativity, a gravitational singularity occupying no more than a point will form. There may be a new halt of the catastrophic gravitational collapse at a size comparable to the Planck length, but at these lengths there is no known theory of gravity to predict what will happen. Adding any extra mass to the black hole will cause the radius of the event horizon to increase linearly with the mass of the central singularity. This will induce certain changes in the properties of the black hole, such as reducing the tidal stress near the event horizon, and reducing the gravitational field strength at the horizon. However, there will not be any further qualitative changes in the structure associated with any mass increase.

### 6.5.1   Alternative black hole models

- Fuzzball

- Gravastar

- Dark energy star

- Black star

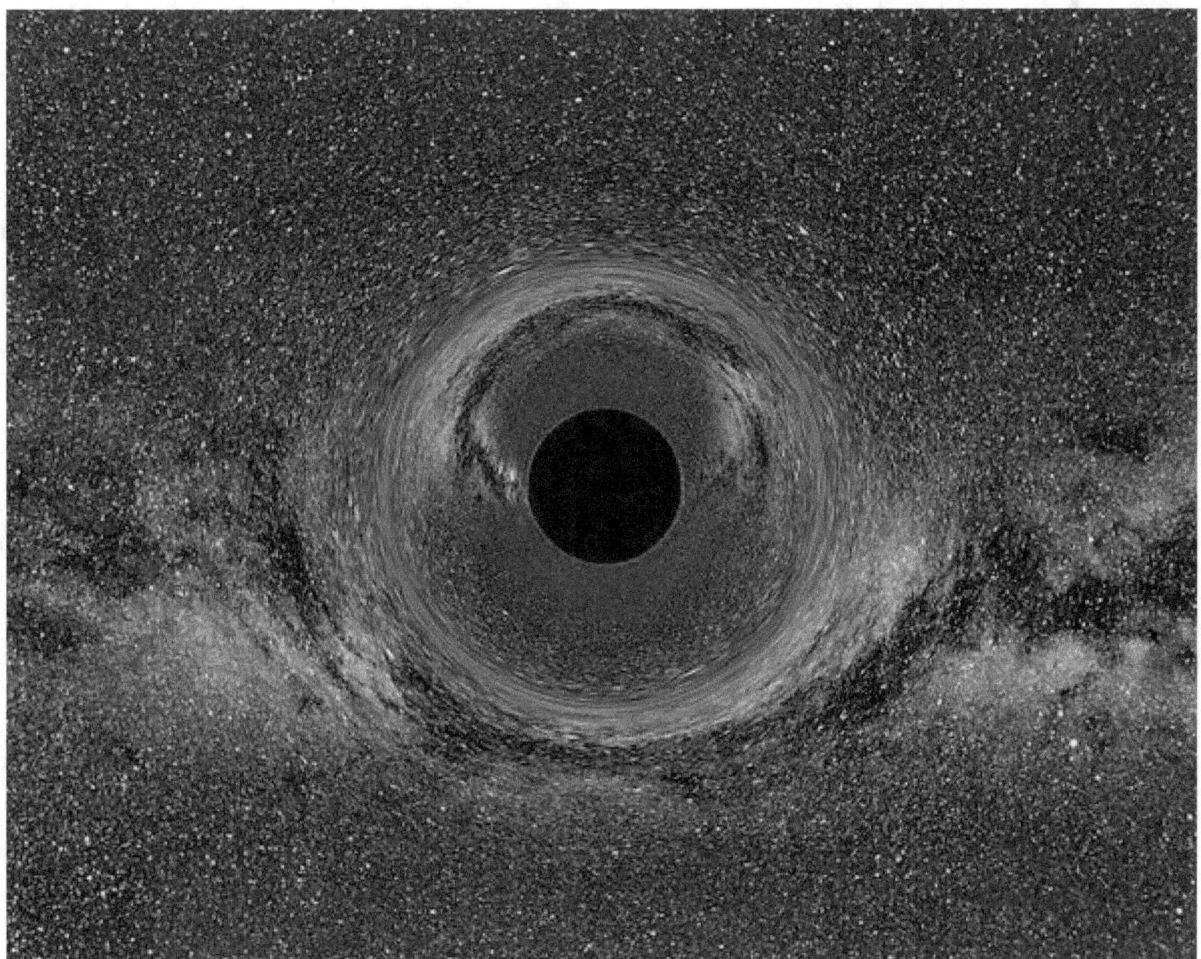

*A simulated black hole of ten solar masses, at a distance of 600km.*

- Magnetospheric eternally collapsing object

- Dark star

- Primordial black holes

## 6.6 Exotic stars

Main article: Exotic star

An *exotic star* is a hypothetical compact star composed of something other than electrons, protons, and neutrons balanced against gravitational collapse by degeneracy pressure or other quantum properties. These include strange stars (composed of strange matter) and the more speculative preon stars (composed of preons).

Exotic stars are hypothetical, but observations released by the Chandra X-Ray Observatory on April 10, 2002 detected two candidate strange stars, designated RX J1856.5-3754 and 3C58, which had previously been thought to be neutron stars. Based on the known laws of physics, the former appeared much smaller and the latter much colder than they should, suggesting that they are composed of material denser than neutronium. However, these observations are met with skepticism by researchers who say the results were not conclusive.

### 6.6.1   Quark stars and strange stars

Main article: Quark star

If neutrons are squeezed enough at a high temperature, they will decompose into their component quarks, forming what is known as a quark matter. In this case, the star will shrink further and become denser, but instead of a total collapse into a black hole, it is possible, that the star may stabilize itself and survive in this state indefinitely, as long as no extra mass is added. It has, to some extent, become a very large nucleon. A-type star in this hypothetical state is called a *quark star* or more specifically a *strange star*. The pulsars RX J1856.5-3754 and 3C58 have been suggested as possible quark stars. Most neutron stars are thought to hold a core of quark matter, but it has proven hard to determine observationally.

### 6.6.2   Preon stars

A *preon star* is a proposed type of compact star made of preons, a group of hypothetical subatomic particles. Preon stars would be expected to have huge densities, exceeding $10^{23}$ kilogram per cubic meter – intermediate between quark stars and black holes. Preon stars could originate from supernova explosions or the big bang; however, current observations from particle accelerators speak against the existence of preons.

### 6.6.3   Q stars

Main article: Q star

*Q stars* are hypothetical compact, heavier neutron stars with an exotic state of matter where particle numbers are preserved. Q stars are also called "gray holes".

### 6.6.4   Electroweak stars

Main article: Electroweak star

An *electroweak star* is a theoretical type of exotic star, whereby the gravitational collapse of the star is prevented by radiation pressure resulting from electroweak burning, that is, the energy released by conversion of quarks to leptons through the electroweak force. This process occurs in a volume at the star's core approximately the size of an apple, containing about two Earth masses.[9]

### 6.6.5   Other ideas

[10][11]

- Boson star

## 6.7   Compact Relativistic Objects and Generalized Uncertainty Principle

Based on the generalized uncertainty principle (GUP), proposed by some approaches to quantum gravity such as String Theory and Doubly Special Relativity Theories, the effect of GUP on the thermodynamic properties of compact stars with two different components has been studied, recently.[12] Tawfik et al. noted that the existence of quantum gravity correction tends to resist the collapse of stars if the GUP parameter is taking values between Planck scale and electroweak scale. Comparing with other approaches, it was found that the radii of compact stars should be smaller and increasing energy decreases the radii of the compact stars.

## 6.8 References

[1] Tauris, T. M.; J. van den Heuvel, E. P. (20 Mar 2003). "Formation and Evolution of Compact Stellar X-ray Sources". *arXiv*. Bibcode:2006csxs.book..623T.

[2] Witten, Edward (1984). "Cosmic separation of phases". *Physical Review D* **30** (2): 272–285. Bibcode:1984PhRvD..30..272W. doi:10.1103/PhysRevD.30.272.

[3] Hashimoto, M.; Iwamoto, K.; Nomoto, K. (1993). "Type II supernovae from 8–10 solar mass asymptotic giant branch stars". *The Astrophysical Journal* **414**: L105. Bibcode:1993ApJ...414L.105H. doi:10.1086/187007.

[4] Ritossa, C.; Garcia-Berro, E.; Iben, I., Jr. (1996). "On the Evolution of Stars That Form Electron-degenerate Cores Processed by Carbon Burning. II. Isotope Abundances and Thermal Pulses in a 10 $M_{sun}$ Model with an ONe Core and Applications to Long-Period Variables, Classical Novae, and Accretion-induced Collapse". *The Astrophysical Journal* **460**: 489. Bibcode:1996ApJ...460..489R. doi:10.1086/176987.

[5] Wanajo, S.; et al. (2003). "Ther-Process in Supernova Explosions from the Collapse of O-Ne-Mg Cores". *The Astrophysical Journal* **593** (2): 968. arXiv:astro-ph/0302262. Bibcode:2003ApJ...593..968W. doi:10.1086/376617.

[6] Osterbrock, D. E. (2001). "Who Really Coined the Word Supernova? Who First Predicted Neutron Stars?". *Bulletin of the American Astronomical Society* **33**: 1330. Bibcode:2001AAS...199.1501O.

[7] Baade, W.; Zwicky, F. (1934). "On Super-Novae".*Proceedings of the National Academy of Sciences***20**(5): 254–9. Bibcode:193 doi:10.1073/pnas.20.5.254. PMC 1076395. PMID 16587881.

[8] Baade, W.; Zwicky, F. (1934). "Cosmic Rays from Super-Novae". *Proceedings of the National Academy of Sciences* **20** (5): 259. Bibcode:1934PNAS...20..259B. doi:10.1073/pnas.20.5.259.

[9] Shiga, D. (4 January 2010). "Exotic stars may mimic big bang". *New Scientist*. Retrieved 2010-02-18.

[10] Visser, M.; Barcelo, C.; Liberati, S.; Sonego, S. (2009). "Small, dark, and heavy: But is it a black hole?". arXiv:0902.0346 [hep-th].

[11] Barcelo, C.; Liberati, S.; Sonego, S.; Visser, M. (30 September 2009). "How Quantum Effects Could Create Black Stars, Not Holes". *Scientific American*.

[12] Ahmed Farag Ali and A. Tawfik. Int. J. Mod. Phys. D22 (2013) 1350020

## 6.9 Sources

- Blaschke, D.; Fredriksson, S.; Grigorian, H.; Öztaş, A.; Sandin, F. (2005). "Phase diagram of three-flavor quark matter under compact star constraints". *Physical Review D* **72** (6). arXiv:hep-ph/0503194. Bibcode:2005PhRvD..72 f5020B.doi:10.1103/PhysRevD.72.065020.

- Sandin, F. (2005). "Compact stars in the standard model – and beyond". *European Physical Journal C* **40**: 15. arXiv:astro-ph/0410407. Bibcode:2005EPJC...40...15S. doi:10.1140/epjcd/s2005-03-003-y.

- Sandin, F. (2005). *Exotic Phases of Matter in Compact Stars* (PDF) (Thesis). Luleå University of Technology.

# Chapter 7

# Gravitational collapse

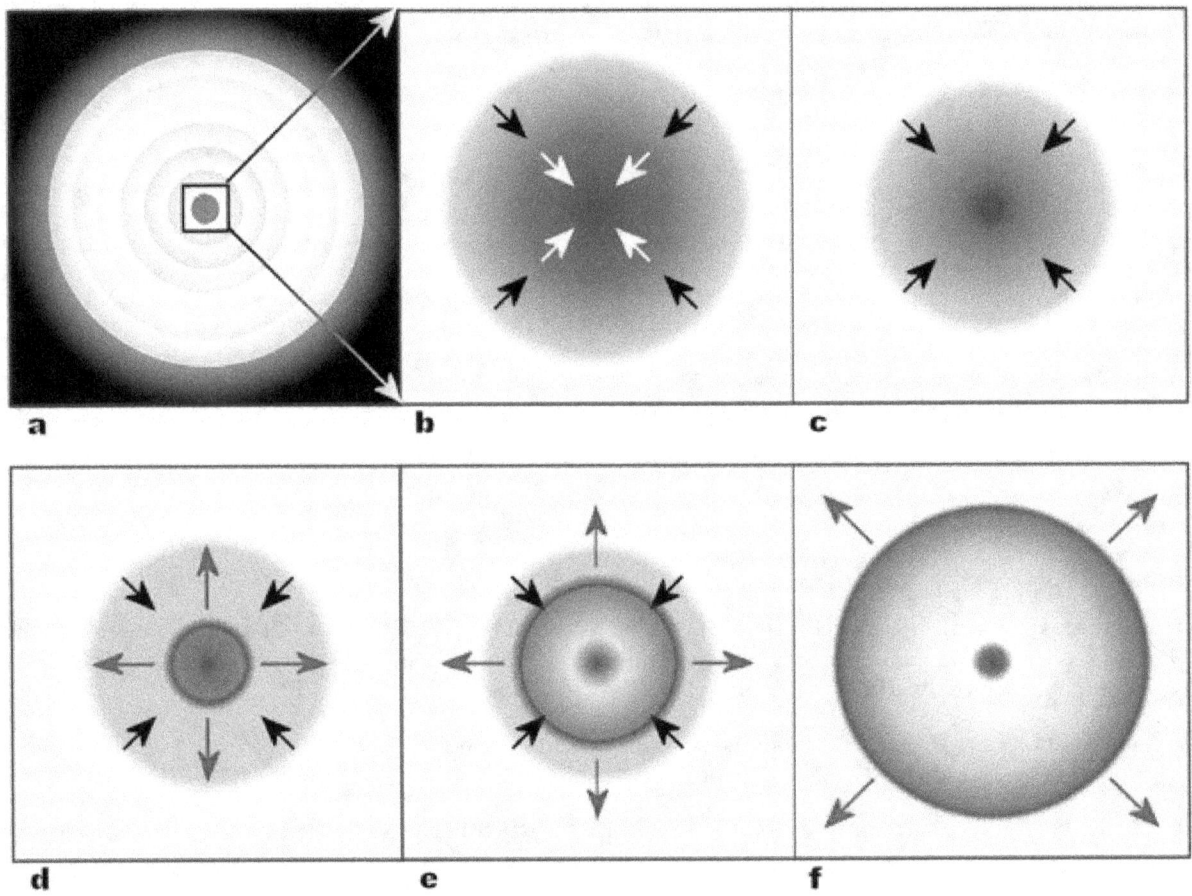

*Gravitational collapse of a star*

**Gravitational collapse** is the inward fall of an astronomical object due to the influence of its own gravity which tends to draw the object toward its center of mass.

Gravitational collapse is at the heart of structure formation in the universe. An initial smooth distribution of matter will eventually collapse and cause a hierarchy of structures, such as clusters of galaxies, stellar groups, stars and planets.

A star is born through the gradual gravitational collapse of a cloud of interstellar matter. The compression caused by the collapse raises the temperature until nuclear fuel ignites in the center of the star and the collapse comes to a halt due to the outward thermal pressure balances the gravitational forces and the star is in dynamic equilibrium. And, when all its energy sources are exhausted, a star will again collapse until it reaches a new equilibrium state.

# 7.1 Star formation

Main article: Star formation

An interstellar cloud of gas will remain in hydrostatic equilibrium as long as the kinetic energy of the gas pressure is in balance with the potential energy of the internal gravitational force. Mathematically this is expressed using the virial theorem, which states that, to maintain equilibrium, the gravitational potential energy must equal twice the internal thermal energy.[1] If an interstellar cloud of gas is massive enough that the gas pressure is insufficient to support it, the cloud will undergo gravitational collapse. The mass above which a cloud will undergo such collapse is called the Jeans mass. The Jeans mass depends on the temperature and density of the cloud, but is typically thousands to tens of thousands of solar masses.[2]

# 7.2 Stellar remnants

At what is called the death of the star when a star has burned out its fuel supply, it will undergo a contraction that can be halted only if it reaches a new state of equilibrium. Depending on the mass during its lifetime, these stellar remnants can take one of three forms:

- White dwarfs, in which gravity is opposed by electron degeneracy pressure[3]

- Neutron stars, in which gravity is opposed by neutron degeneracy pressure and short-range repulsive neutron–neutron interactions mediated by the strong force

- Black hole

## 7.2.1 White dwarf

Main article: White dwarf

The collapse to a white dwarf takes place over tens of thousands of years, while the star blows off its outer envelope to form a planetary nebula. If it has a companion star, a white dwarf-sized object can accrete matter from the companion star until it reaches the Chandrasekhar limit (about one and a half times the mass of our Sun) at which point gravitational collapse takes over again. While it might seem that the white dwarf might collapse to the next stage (neutron star), they instead undergo runaway carbon fusion, blowing completely apart in a Type Ia supernova.

## 7.2.2 Neutron star

Main article: Neutron star

Neutron stars are formed by gravitational collapse of larger stars, the remnant of other types of supernova. They are so compact that a Newtonian description is inadequate for an accurate treatment, which requires the use of Einstein's general relativity.

*NGC 6745 produces material densities sufficiently extreme to trigger star formation through gravitational collapse*

### 7.2.3   Black holes

Main article: Black hole

According to Einstein's theory, for even larger stars, above the Landau-Oppenheimer-Volkoff limit, also known as the Tolman–Oppenheimer–Volkoff limit (roughly double the mass of our Sun) no known form of cold matter can provide the force needed to oppose gravity in a new dynamical equilibrium. Hence, the collapse continues with nothing to stop it.

Once a body collapses to within its Schwarzschild radius it forms what is called a black hole, meaning a space-time region from which not even light can escape. It follows from a theorem of Roger Penrose[5] that the subsequent formation of some kind of singularity is inevitable. Nevertheless, according to Penrose's cosmic censorship hypothesis, the singularity will be confined within the event horizon bounding the black hole, so the space-time region outside will still have a well behaved geometry, with strong but finite curvature, that is expected[6] to evolve towards a rather simple form describable by the historic Schwarzschild metric in the spherical limit and by the more recently discovered Kerr metric if angular momentum is present.

On the other hand, the nature of the kind of singularity to be expected inside a black hole remains rather controversial.

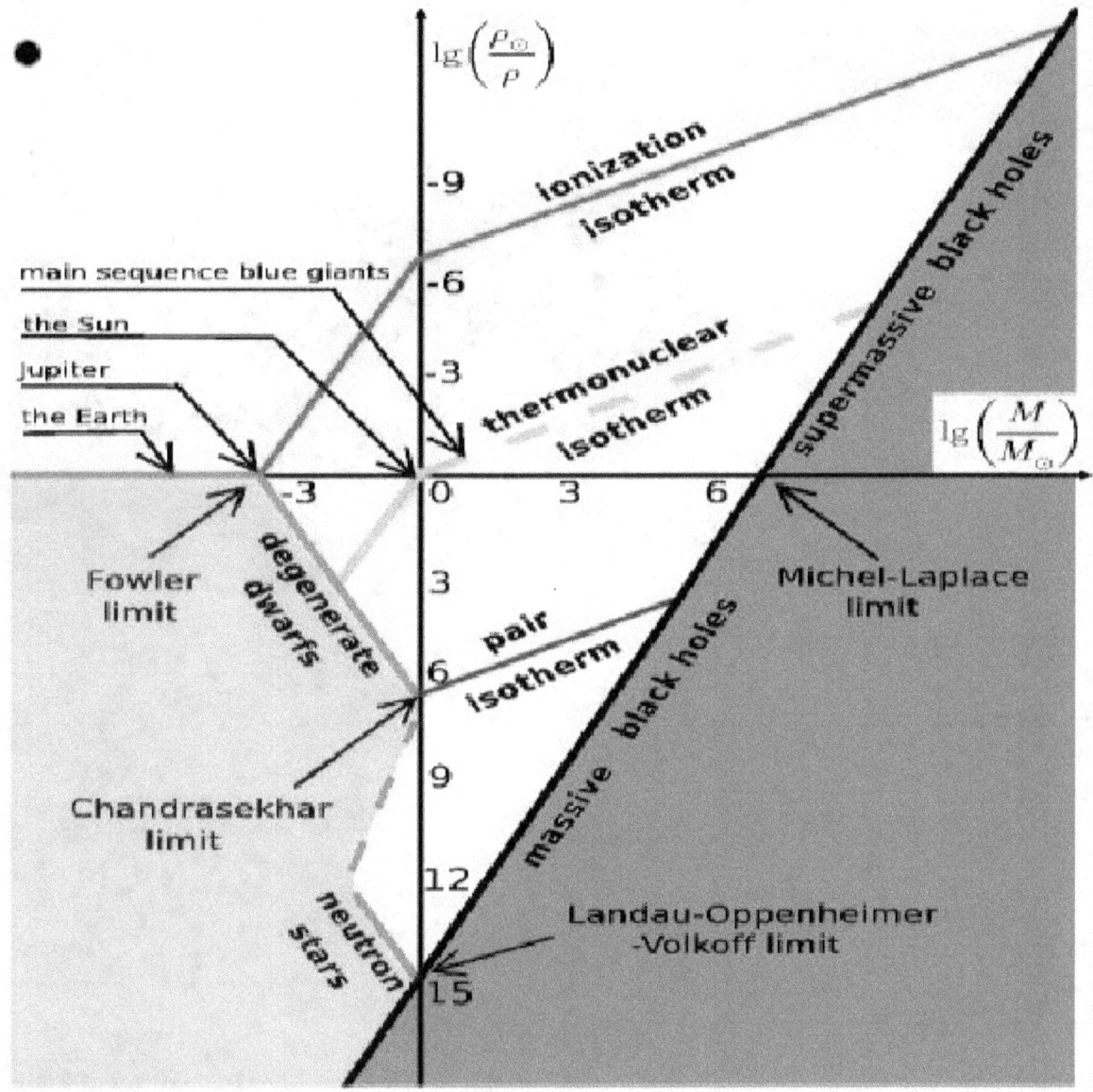

*Logarithmic plot of Mass against mean Density (with solar values as origin) showing possible kinds of stellar equilibrium state. For a configuration in the shaded region, beyond the black hole limit line, no equilibrium is possible, so runaway collapse will be inevitable.*

According to some theories, at a later stage, the collapsing object will reach the maximum possible energy density for a certain volume of space or the Planck density (as there is nothing that can stop it). This is when the known laws of gravity cease to be valid.[7] There are competing theories as to what occurs at this point, but it can no longer really be considered gravitational collapse at that stage.[8]

It might be thought that a sufficiently large neutron star could exist inside its Schwarzschild radius and appear like a black hole without having all the mass compressed to a singularity at the center; however, this is a misconception. Within the event horizon, matter would have to move outward faster than the speed of light in order to remain stable and avoid collapsing to the center. No physical force therefore can prevent the star from collapsing to a singularity (at least within the currently accepted framework of general relativity; this doesn't hold for the Einstein–Yang–Mills–Dirac system). A model for nonspherical collapse in general relativity with emission of matter and gravitational waves has been presented.[9]

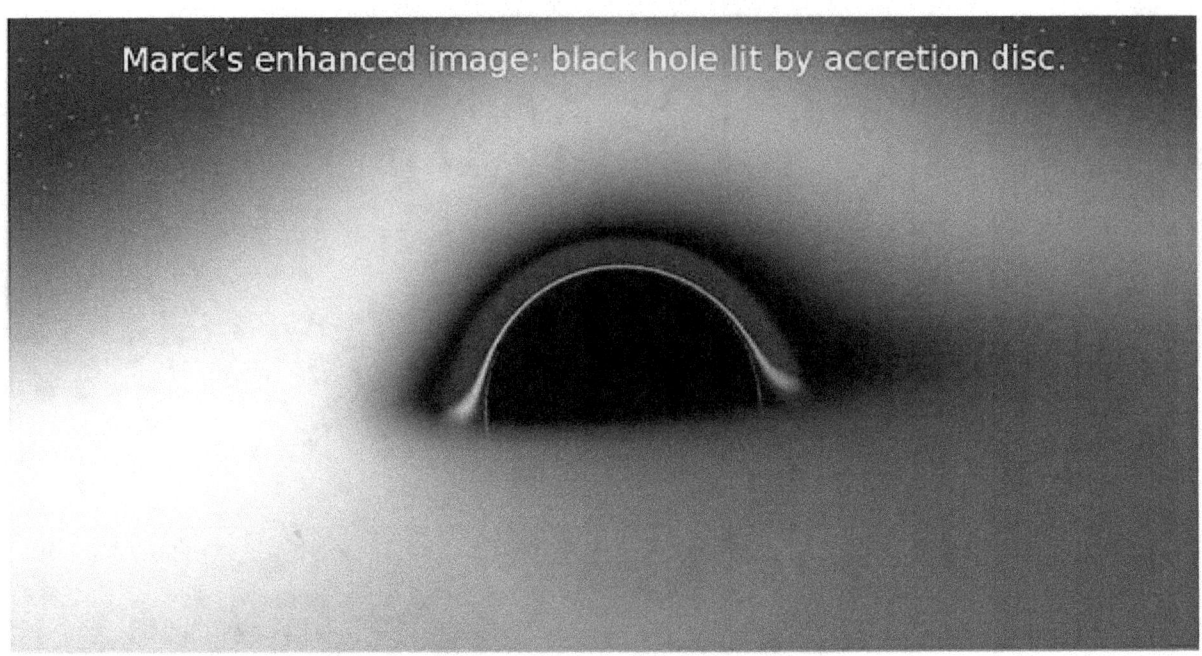

*Simulated view from outside black hole with thin accretion disc by J.A.Marck[4]*

## 7.3   See also

- Big Crunch

- Stellar evolution

- Thermal runaway

## 7.4   References

[1] Kwok, Sun (2006). *Physics and chemistry of the interstellar medium*. University Science Books. pp. 435–437. ISBN 1-891389-46-7.

[2] Prialnik, Dina (2000). *An Introduction to the Theory of Stellar Structure and Evolution*. Cambridge University Press. pp. 198–199. ISBN 0-521-65937-X.

[3] And theoretically Black dwarfs - but: *"...no black dwarfs are expected to exist in the universe yet"*

[4] Class. Quantum Grav. 13 (1996)p393

[5] "Gravitational collapse and space-time singularities", R. Penrose, Phys. Rev. Let. 14 (1965) p 57

[6] B. Carter, "Axisymmetric black hole has only two degrees of freedom", Phys. Rev. Let. 26 (1971) p331

[7] "Black Holes - Planck Unit? WIP". Physics Forums. Archived from the original on 2008-08-02.

[8] Brill, Dieter (19 January 2012). "Black Hole Horizons and How They Begin". *Astronomical Review*.

[9] Bedran, ML et al.(1996)."Model for nonspherical collapse and formation of black holes by emission of neutrinos, strings and gravitational waves", Phys. Rev. **D 54**(6),3826.

## 7.5 External links

- Gravitational collapse on arxiv.org

# Chapter 8

# White dwarf

For other uses, see White dwarf (disambiguation).

A **white dwarf**, also called a **degenerate dwarf**, is a stellar remnant composed mostly of electron-degenerate matter. A white dwarf is very dense: its mass is comparable to that of the Sun, and its volume is comparable to that of Earth. A white dwarf's faint luminosity comes from the emission of stored thermal energy.[1] The nearest known white dwarf is Sirius B, at 8.6 light years, the smaller component of the Sirius binary star. There are currently thought to be eight white dwarfs among the hundred star systems nearest the Sun.[2] The unusual faintness of white dwarfs was first recognized in 1910.[3] The name *white dwarf* was coined by Willem Luyten in 1922.[4]

White dwarfs are thought to be the final evolutionary state of stars (including our Sun) whose mass is not high enough to become a neutron star—over 97% of the stars in the Milky Way.[5, 81] After the hydrogen–fusing period of a main-sequence star of low or medium mass ends, a star will expand to a red giant during which it fuses helium to carbon and oxygen in its core by the triple-alpha process. If a red giant has insufficient mass to generate the core temperatures required to fuse carbon, around 1 billion K, an inert mass of carbon and oxygen will build up at its center. After shedding its outer layers to form a planetary nebula, it will leave behind this core, which forms the remnant white dwarf.[6] Usually, therefore, white dwarfs are composed of carbon and oxygen. If the mass of the progenitor is between 8 and 10.5 solar masses ($M\odot$), the core temperature is sufficient to fuse carbon but not neon, in which case an oxygen-neon–magnesium white dwarf may be formed.[7] Stars of very low mass will not be able to fuse helium, hence, a helium white dwarf[8][9] may be formed by mass loss in binary systems.

The material in a white dwarf no longer undergoes fusion reactions, so the star has no source of energy. As a result, it cannot support itself by the heat generated by fusion against gravitational collapse, but is supported only by electron degeneracy pressure, causing it to be extremely dense. The physics of degeneracy yields a maximum mass for a non-rotating white dwarf, the Chandrasekhar limit—approximately 1.4 $M\odot$—beyond which it cannot be supported by electron degeneracy pressure. A carbon-oxygen white dwarf that approaches this mass limit, typically by mass transfer from a companion star, may explode as a Type Ia supernova via a process known as carbon detonation.[1][6] (SN 1006 is thought to be a famous example.)

A white dwarf is very hot when it is formed, but since it has no source of energy, it will gradually radiate away its energy and cool. This means that its radiation, which initially has a high color temperature, will lessen and redden with time. Over a very long time, a white dwarf will cool to temperatures at which it will no longer emit significant heat or light, and it will become a cold *black dwarf*.[6] However, the length of time it takes for a white dwarf to reach this state is calculated to be longer than the current age of the universe (approximately 13.8 billion years),[10] and since no white dwarf can be older than the age of the universe, it is thought that no black dwarfs yet exist.[1][5] The oldest white dwarfs still radiate at temperatures of a few thousand kelvins.

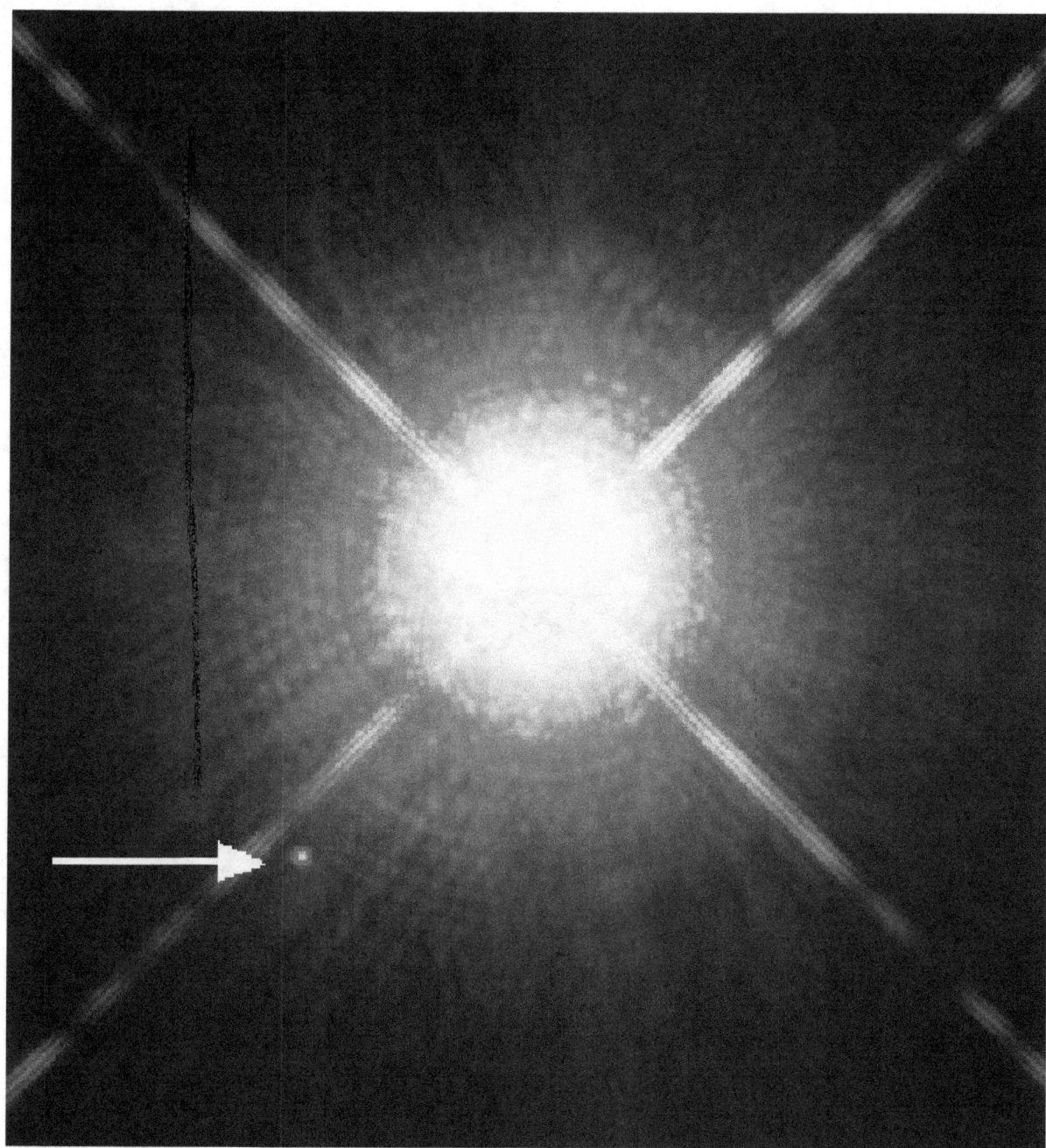

*Image of Sirius A and Sirius B taken by the Hubble Space Telescope. Sirius B, which is a white dwarf, can be seen as a faint pinprick of light to the lower left of the much brighter Sirius A.*

## 8.1 Discovery

See also: List of white dwarfs

The first white dwarf discovered was in the triple star system of 40 Eridani, which contains the relatively bright main sequence star 40 Eridani A, orbited at a distance by the closer binary system of the white dwarf 40 Eridani B and the main sequence red dwarf 40 Eridani C. The pair 40 Eridani B/C was discovered by William Herschel on 31 January 1783;[11], p. 73 it was again observed by Friedrich Georg Wilhelm Struve in 1825 and by Otto Wilhelm von Struve in

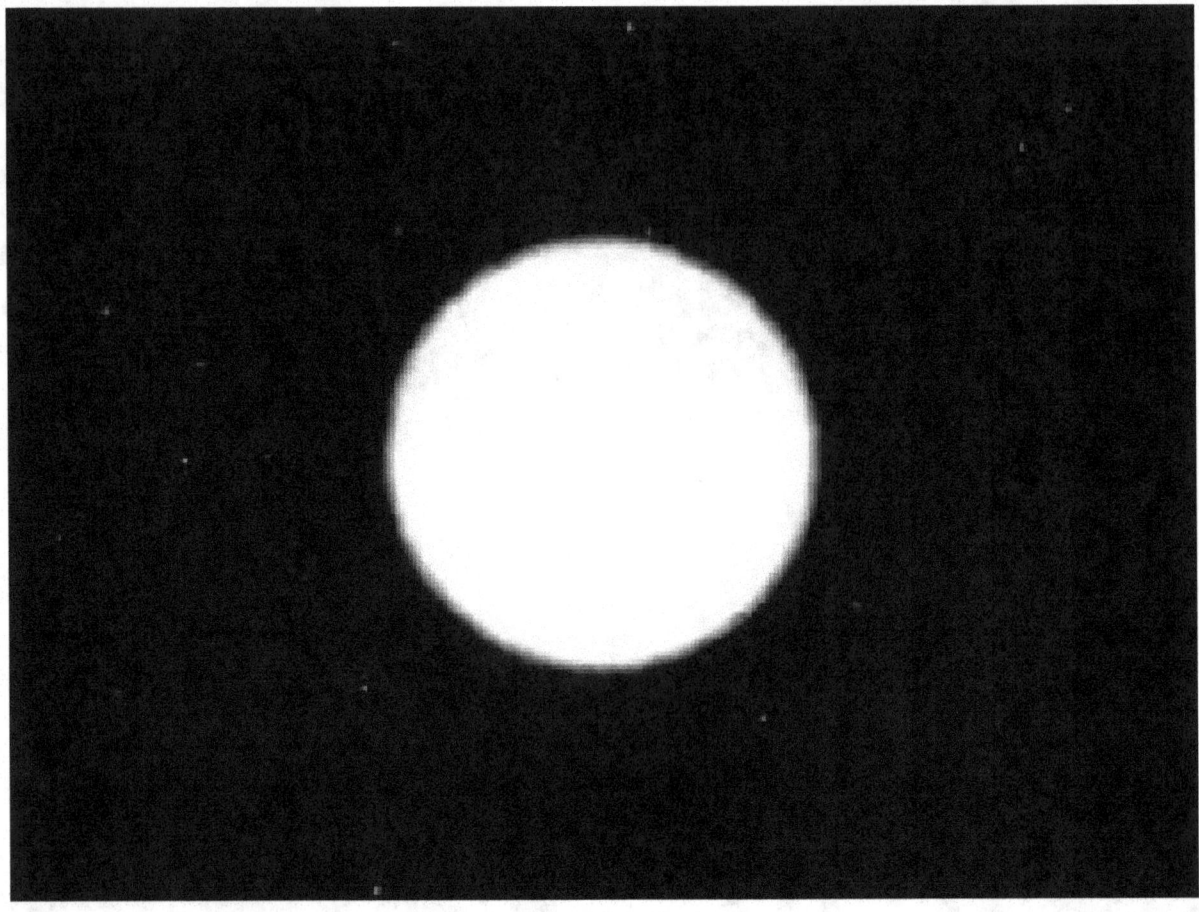

*Artist's concept of white dwarf aging.*

1851.[12][13] In 1910, Henry Norris Russell, Edward Charles Pickering and Williamina Fleming discovered that, despite being a dim star, 40 Eridani B was of spectral type A, or white.[4] In 1939, Russell looked back on the discovery:[3], p. 1

> I was visiting my friend and generous benefactor, Prof. Edward C. Pickering. With characteristic kindness, he had volunteered to have the spectra observed for all the stars—including comparison stars—which had been observed in the observations for stellar parallax which Hinks and I made at Cambridge, and I discussed. This piece of apparently routine work proved very fruitful—it led to the discovery that all the stars of very faint absolute magnitude were of spectral class M. In conversation on this subject (as I recall it), I asked Pickering about certain other faint stars, not on my list, mentioning in particular 40 Eridani B. Characteristically, he sent a note to the Observatory office and before long the answer came (I think from Mrs Fleming) that the spectrum of this star was A. I knew enough about it, even in these paleozoic days, to realize at once that there was an extreme inconsistency between what we would then have called "possible" values of the surface brightness and density. I must have shown that I was not only puzzled but crestfallen, at this exception to what looked like a very pretty rule of stellar characteristics; but Pickering smiled upon me, and said: "It is just these exceptions that lead to an advance in our knowledge", and so the white dwarfs entered the realm of study!

The spectral type of 40 Eridani B was officially described in 1914 by Walter Adams.[14]

The white dwarf companion of Sirius, Sirius B, was next to be discovered. During the nineteenth century, positional measurements of some stars became precise enough to measure small changes in their location. Friedrich Bessel used position measurements to determine that the stars Sirius ($\alpha$ Canis Majoris) and Procyon ($\alpha$ Canis Minoris) were changing their positions periodically. In 1844 he predicted that both stars had unseen companions:[15]

If we were to regard *Sirius* and *Procyon* as double stars, the change of their motions would not surprise us; we should acknowledge them as necessary, and have only to investigate their amount by observation. But light is no real property of mass. The existence of numberless visible stars can prove nothing against the existence of numberless invisible ones.

Bessel roughly estimated the period of the companion of Sirius to be about half a century:[15] C. A. F. Peters computed an orbit for it in 1851.[16] It was not until 31 January 1862 that Alvan Graham Clark observed a previously unseen star close to Sirius, later identified as the predicted companion.[16] Walter Adams announced in 1915 that he had found the spectrum of Sirius B to be similar to that of Sirius.[17]

In 1917, Adriaan van Maanen discovered Van Maanen's Star, an isolated white dwarf.[18] These three white dwarfs, the first discovered, are the so-called *classical white dwarfs*.[3], p. 2 Eventually, many faint white stars were found which had high proper motion, indicating that they could be suspected to be low-luminosity stars close to the Earth, and hence white dwarfs. Willem Luyten appears to have been the first to use the term *white dwarf* when he examined this class of stars in 1922;[4][19][20][21][22] the term was later popularized by Arthur Stanley Eddington.[4][23] Despite these suspicions, the first non-classical white dwarf was not definitely identified until the 1930s. 18 white dwarfs had been discovered by 1939.[3], p. 3 Luyten and others continued to search for white dwarfs in the 1940s. By 1950, over a hundred were known,[24] and by 1999, over 2,000 were known.[25] Since then the Sloan Digital Sky Survey has found over 9,000 white dwarfs, mostly new.[26]

## 8.2 Composition and structure

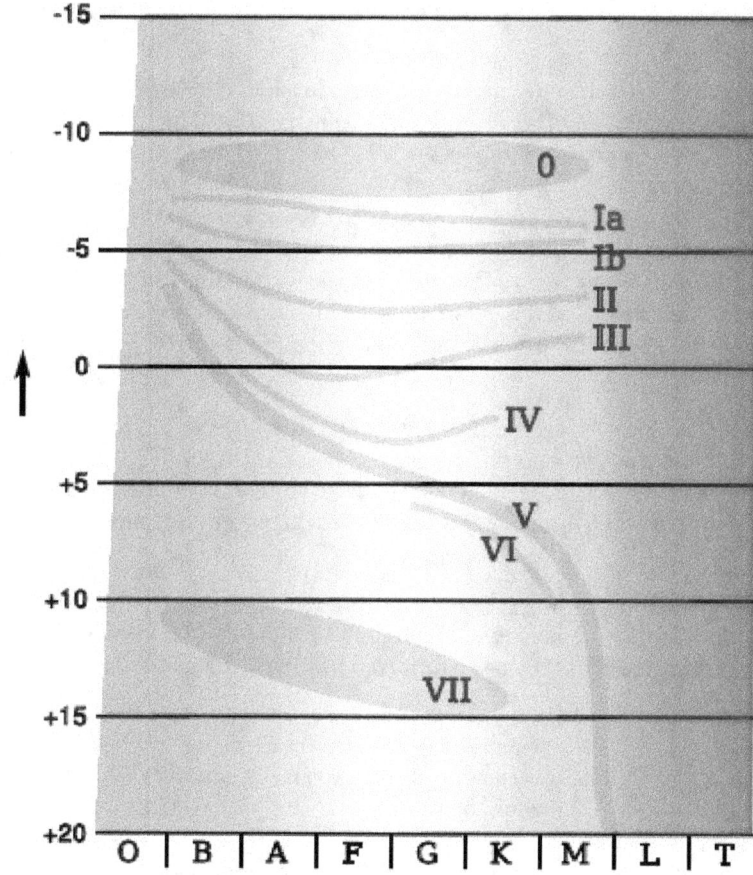

Hertzsprung–Russell diagram
Spectral type

Brown dwarfs
White dwarfs
Red dwarfs
Subdwarfs
Main sequence
("dwarfs")
Subgiants
Giants
Bright giants
Supergiants
Hypergiants
absolute
magni-
tude
(MV)

Although white dwarfs are known with estimated masses as low as 0.17 $M\odot$[27] and as high as 1.33 $M\odot$,[28] the mass distribution is strongly peaked at 0.6 $M\odot$, and the majority lie between 0.5 to 0.7 $M\odot$.[28] The estimated radii of observed white dwarfs, however, are typically 0.8–2 % the radius of the Sun;[29] this is comparable to the Earth's radius of approximately 0.9% solar radius. A white dwarf, then, packs mass comparable to the Sun's into a volume that is typically a million times smaller than the Sun's; the average density of matter in a white dwarf must therefore be, very roughly, 1,000,000 times greater than the average density of the Sun, or approximately $10^6$ g/cm$^3$, or 1 tonne per cubic centimetre.[1] A typical white dwarf has a density of between $10^7$ and $10^{11}$ kg per cubic meter. White dwarfs are composed of one of the densest forms of matter known, surpassed only by other compact stars such as neutron stars, black holes and, hypothetically, quark stars.[30]

White dwarfs were found to be extremely dense soon after their discovery. If a star is in a binary system, as is the case for Sirius B and 40 Eridani B, it is possible to estimate its mass from observations of the binary orbit. This was done for Sirius B by 1910,[31] yielding a mass estimate of 0.94 $M\odot$. (A more modern estimate is 1.00 $M\odot$.)[32] Since hotter bodies radiate more energy than colder ones, a star's surface brightness can be estimated from its effective surface temperature, and that from its spectrum. If the star's distance is known, its overall luminosity can also be estimated. From the luminosity and distance, the star's surface area and its radius can be calculated. Reasoning of this sort led to the realization, puzzling to astronomers at the time, that Sirius B and 40 Eridani B must be very dense. For example, when Ernst Öpik estimated the density of a number of visual binary stars in 1916, he found that 40 Eridani B had a density of over 25,000 times the Sun's, which was so high that he called it "impossible".[33] As Arthur Stanley Eddington put it later in 1927:[34], p. 50

> We learn about the stars by receiving and interpreting the messages which their light brings to us. The message of the Companion of Sirius when it was decoded ran: "I am composed of material 3,000 times denser than anything you have ever come across; a ton of my material would be a little nugget that you could put in a matchbox." What reply can one make to such a message? The reply which most of us made in 1914 was—"Shut up. Don't talk nonsense."

As Eddington pointed out in 1924, densities of this order implied that, according to the theory of general relativity, the light from Sirius B should be gravitationally redshifted.[23] This was confirmed when Adams measured this redshift in 1925.[35]

Such densities are possible because white dwarf material is not composed of atoms joined by chemical bonds, but rather consists of a plasma of unbound nuclei and electrons. There is therefore no obstacle to placing nuclei closer than normally allowed by electron orbitals limited by normal matter.[23] Eddington, however, wondered what would happen when this plasma cooled and the energy which kept the atoms ionized was no longer present.[38] This paradox was resolved by R. H. Fowler in 1926 by an application of the newly devised quantum mechanics. Since electrons obey the Pauli exclusion principle, no two electrons can occupy the same state, and they must obey Fermi–Dirac statistics, also introduced in 1926 to determine the statistical distribution of particles which satisfy the Pauli exclusion principle.[39] At zero temperature, therefore, electrons could not all occupy the lowest-energy, or *ground*, state; some of them had to occupy higher-energy states, forming a band of lowest-available energy states, the *Fermi sea*. This state of the electrons, called *degenerate*, meant that a white dwarf could cool to zero temperature and still possess high energy.[38][40]

Compression of a white dwarf will increase the number of electrons in a given volume. Applying the Pauli exclusion principle, we can see that this will increase the kinetic energy of the electrons, increasing the pressure.[38][41] This *electron degeneracy pressure* supports a white dwarf against gravitational collapse. The pressure depends only on density and not on temperature. Degenerate matter is relatively compressible; this means that the density of a high-mass white dwarf is much greater than that of a low-mass white dwarf and that the radius of a white dwarf decreases as its mass increases.[1]

The existence of a limiting mass that no white dwarf can exceed is another consequence of being supported by electron degeneracy pressure. These masses were first published in 1929 by Wilhelm Anderson[42] and in 1930 by Edmund C. Stoner.[43] The modern value of the limit was first published in 1931 by Subrahmanyan Chandrasekhar in his paper "The Maximum Mass of Ideal White Dwarfs".[44] For a non-rotating white dwarf, it is equal to approximately $5.7/\mu_e^2$ $M\odot$, where $\mu_e$ is the average molecular weight per electron of the star.[45], eq. (63) As the carbon-12 and oxygen-16 which predominantly compose a carbon-oxygen white dwarf both have atomic number equal to half their atomic weight, one should take $\mu_e$ equal to 2 for such a star,[40] leading to the commonly quoted value of 1.4 $M\odot$. (Near the beginning of the 20th century, there was reason to believe that stars were composed chiefly of heavy elements,[43], p. 955 so, in his 1931 paper, Chandrasekhar set the average molecular weight per electron, $\mu_e$, equal to 2.5, giving a limit of 0.91 $M\odot$.) Together with William Alfred Fowler, Chandrasekhar received the Nobel prize for this and other work in 1983.[46] The limiting mass is now called the *Chandrasekhar limit*.

If a white dwarf were to exceed the Chandrasekhar limit, and nuclear reactions did not take place, the pressure exerted by electrons would no longer be able to balance the force of gravity, and it would collapse into a denser object called a neutron star.[47] However, carbon-oxygen white dwarfs accreting mass from a neighboring star undergo a runaway nuclear fusion reaction, which leads to a Type Ia supernova explosion in which the white dwarf may be destroyed, before it reaches the limiting mass.[48]

New research indicates that many white dwarfs—at least in certain types of galaxies—may not approach that limit by way of accretion. It has been postulated that at least some of the white dwarfs that become supernovae attain the necessary mass by colliding with one another. It may be that in elliptical galaxies such collisions are the major source of supernovae. This hypothesis is based on the fact that less X-rays than expected are produced by the white dwarfs' accretion of matter. 30 to 50 times more X-rays would be expected to be produced by an amount of matter falling onto a population of accreting white dwarfs sufficient to produce supernovae at the observed rate. It has been concluded that no more than 5 percent of the supernovae in such galaxies could be created by the process of accretion onto white dwarfs. The significance of this finding is that there could be two types of supernovae, which could mean that the Chandrasekhar limit might not always apply in determining when a white dwarf goes supernova, given that two colliding white dwarfs could have a range of masses. This in turn would confuse efforts to use exploding white dwarfs as standard candles in determining distances.[49]

White dwarfs have low luminosity and therefore occupy a strip at the bottom of the Hertzsprung–Russell diagram, a graph of stellar luminosity versus color (or temperature). They should not be confused with low-luminosity objects at the low-mass end of the main sequence, such as the hydrogen-fusing red dwarfs, whose cores are supported in part by thermal pressure,[50] or the even lower-temperature brown dwarfs.[51]

## 8.2.1    Mass–radius relationship and mass limit

The relationship between the mass and radii of white dwarfs can be derived using an energy minimization argument . The energy of the white dwarf can be approximated by taking it to be the sum of its gravitational potential energy and kinetic energy. The gravitational potential energy of a unit mass piece of white dwarf, $E_g$, will be on the order of $-G M / R$, where G is the gravitational constant, $M$ is the mass of the white dwarf, and R is its radius.

$$E_g \approx \frac{-GM}{R}.$$

The kinetic energy of the unit mass, $E_k$, will primarily come from the motion of electrons, so it will be approximately $N$ $p^2 / 2m$, where p is the average electron momentum, m is the electron mass, and N is the number of electrons per unit mass. Since the electrons are degenerate, we can estimate p to be on the order of the uncertainty in momentum, $\Delta p$, given by the uncertainty principle, which says that $\Delta p \, \Delta x$ is on the order of the reduced Planck constant, $\hbar$. $\Delta x$ will be on the order of the average distance between electrons, which will be approximately $n^{-1/3}$, i.e., the reciprocal of the cube root of the number density, n, of electrons per unit volume. Since there are $N \cdot M$ electrons in the white dwarf, where $M$ is the star's mass and its volume is on the order of $R^3$, n will be on the order of $N \, M \, / \, R^3$.[40]

Solving for the kinetic energy per unit mass, $E_k$, we find that

$$E_k \approx \frac{N(\Delta p)^2}{2m} \approx \frac{N\hbar^2 n^{2/3}}{2m} \approx \frac{M^{2/3}N^{5/3}\hbar^2}{2mR^2}.$$

The white dwarf will be at equilibrium when its total energy, $E_g + E_k$, is minimized. At this point, the kinetic and gravitational potential energies should be comparable, so we may derive a rough mass-radius relationship by equating their magnitudes:

$$|E_g| \approx \frac{GM}{R} = E_k \approx \frac{M^{2/3}N^{5/3}\hbar^2}{2mR^2}.$$

Solving this for the radius, R, gives[40]

$$R \approx \frac{N^{5/3}\hbar^2}{2mGM^{1/3}}.$$

Dropping N, which depends only on the composition of the white dwarf, and the universal constants leaves us with a relationship between mass and radius:

$$R \sim M^{-1/3}$$

i.e., the radius of a white dwarf is inversely proportional to the cube root of its mass.

Since this analysis uses the non-relativistic formula $p^2 \, / \, 2m$ for the kinetic energy, it is non-relativistic. If we wish to analyze the situation where the electron velocity in a white dwarf is close to the speed of light, c, we should replace $p^2 \, / \, 2m$ by the extreme relativistic approximation $p \, c$ for the kinetic energy. With this substitution, we find

$$E_{k \text{ relativistic}} \approx \frac{M^{1/3}N^{4/3}\hbar c}{R}.$$

If we equate this to the magnitude of $E_g$, we find that R drops out and the mass, M, is forced to be[40]

$$M_{\text{limit}} \approx N^2 \left( \frac{\hbar c}{G} \right)^{3/2}.$$

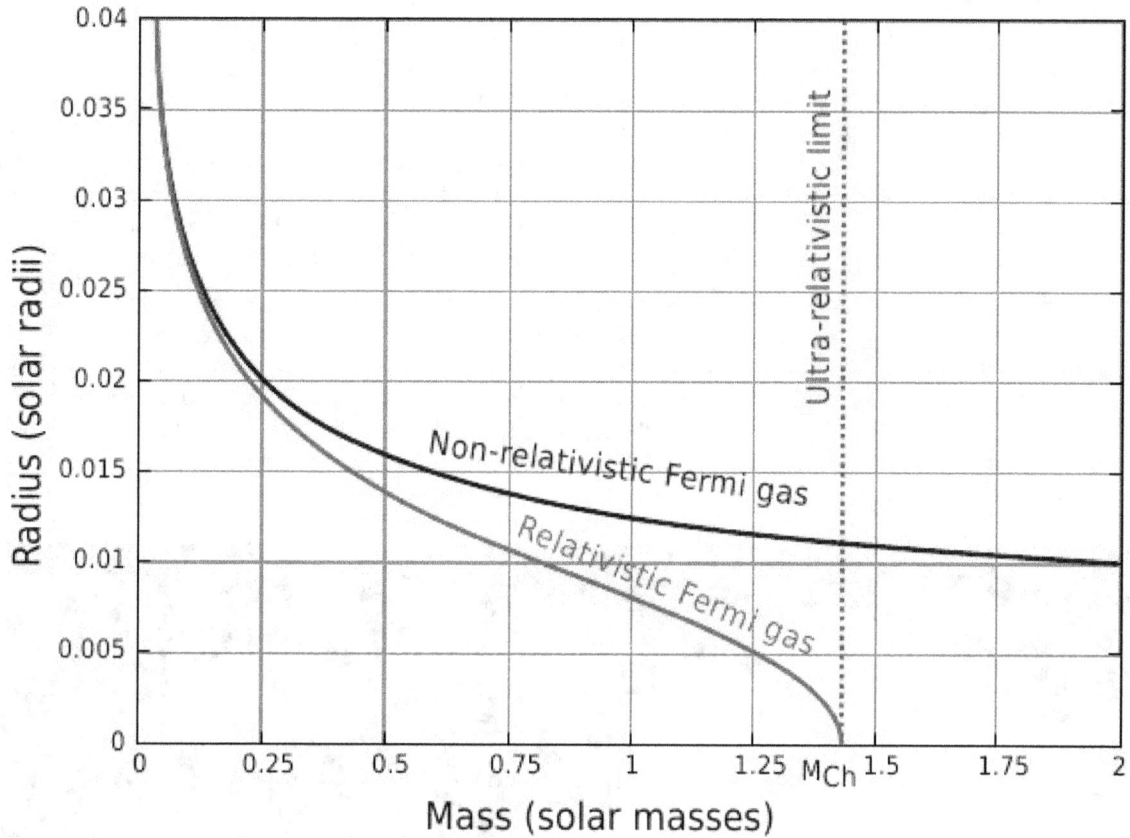

*Radius–mass relations for a model white dwarf. Mlimit is denoted as MCh*

To interpret this result, observe that as we add mass to a white dwarf, its radius will decrease, so, by the uncertainty principle, the momentum, and hence the velocity, of its electrons will increase. As this velocity approaches $c$, the extreme relativistic analysis becomes more exact, meaning that the mass M of the white dwarf must approach a limiting mass of $M_{\text{limit}}$. Therefore, no white dwarf can be heavier than the limiting mass $M_{\text{limit}}$, or 1.4 $M\odot$.

For a more accurate computation of the mass-radius relationship and limiting mass of a white dwarf, one must compute the equation of state which describes the relationship between density and pressure in the white dwarf material. If the density and pressure are both set equal to functions of the radius from the center of the star, the system of equations consisting of the hydrostatic equation together with the equation of state can then be solved to find the structure of the white dwarf at equilibrium. In the non-relativistic case, we will still find that the radius is inversely proportional to the cube root of the mass.[45], eq. (80) Relativistic corrections will alter the result so that the radius becomes zero at a finite value of the mass. This is the limiting value of the mass—called the *Chandrasekhar limit*—at which the white dwarf can no longer be supported by electron degeneracy pressure. The graph on the right shows the result of such a computation. It shows how radius varies with mass for non-relativistic (blue curve) and relativistic (green curve) models of a white dwarf. Both models treat the white dwarf as a cold Fermi gas in hydrostatic equilibrium. The average molecular weight per electron, $\mu_e$, has been set equal to 2. Radius is measured in standard solar radii and mass in standard solar masses.[45][52]

These computations all assume that the white dwarf is non-rotating. If the white dwarf is rotating, the equation of hydrostatic equilibrium must be modified to take into account the centrifugal pseudo-force arising from working in a rotating frame.[53] For a uniformly rotating white dwarf, the limiting mass increases only slightly. However, if the star is allowed to rotate nonuniformly, and viscosity is neglected, then, as was pointed out by Fred Hoyle in 1947,[54] there is no limit to the mass for which it is possible for a model white dwarf to be in static equilibrium. Not all of these model stars, however, will be dynamically stable.[55]

## 8.2.2  Radiation and cooling

The degenerate matter that makes up the bulk of a white dwarf has a very low opacity, because any absorption of a photon requires that an electron must transition to a higher empty state, which may not be possible as the energy of the photon may not be a match for the possible quantum states for that electron; it also has a high thermal conductivity. As a result, the interior of the white dwarf maintains a uniform temperature, approximately $10^7$ K. However, an outer shell of non-degenerate matter cools from approximately $10^7$ K to $10^4$ K. This matter radiates roughly as a black body. A white dwarf remains visible for a long time, as its tenuous outer atmosphere of normal matter begins to radiate at about $10^7$ K, upon formation, while its greater interior mass is at $10^7$ K but cannot radiate through its normal matter shell.[56]

The visible radiation emitted by white dwarfs varies over a wide color range, from the blue-white color of an O-type main sequence star to the red of an M-type red dwarf.[57] White dwarf effective surface temperatures extend from over 150,000 K[25] to barely under 4,000 K.[58][59] In accordance with the Stefan–Boltzmann law, luminosity increases with increasing surface temperature; this surface temperature range corresponds to a luminosity from over 100 times the Sun's to under 1/10,000 that of the Sun's.[59] Hot white dwarfs, with surface temperatures in excess of 30,000 K, have been observed to be sources of soft (i.e., lower-energy) X-rays. This enables the composition and structure of their atmospheres to be studied by soft X-ray and extreme ultraviolet observations.[60]

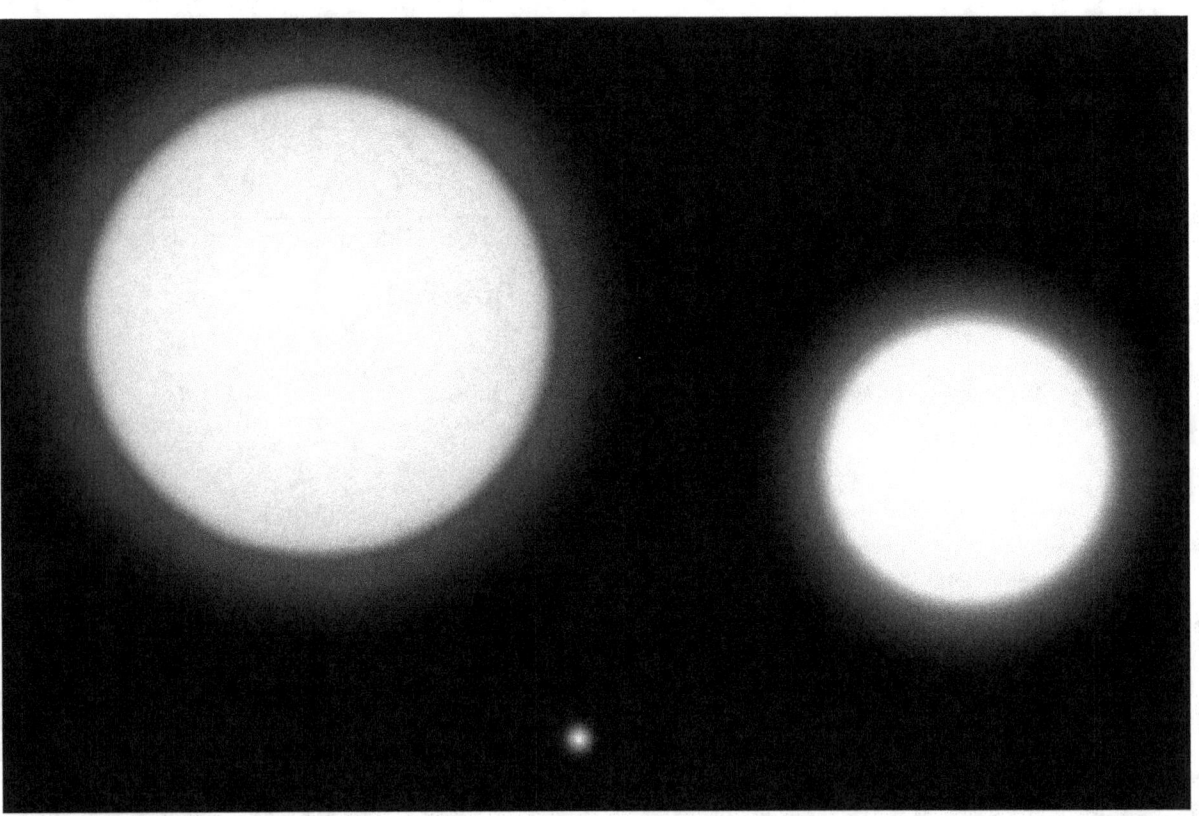

*A comparison between the white dwarf IK Pegasi B (center), its A-class companion IK Pegasi A (left) and the Sun (right). This white dwarf has a surface temperature of 35,500 K.*

As was explained by Leon Mestel in 1952, unless the white dwarf accretes matter from a companion star or other source, its radiation comes from its stored heat, which is not replenished.[61][62], §2.1. White dwarfs have an extremely small surface area to radiate this heat from, so they cool gradually, remaining hot for a long time.[6] As a white dwarf cools, its surface temperature decreases, the radiation which it emits reddens, and its luminosity decreases. Since the white dwarf has no energy sink other than radiation, it follows that its cooling slows with time. Pierre Bergeron, Maria Tereza Ruiz, and Sandy Leggett, for example, have estimated the rate of cooling for a carbon white dwarf of 0.59 $M\odot$ with a hydrogen atmosphere. After initially cooling to a surface temperature of 7,140 K, taking approximately 1.5 billion years, cooling approximately 500 more kelvins to 6,590 K takes around 0.3 billion years, but the next two steps of around 500 kelvins (to 6,030 K and 5,550 K) take first 0.4 and then 1.1 billion years.[63], Table 2.

Most observed white dwarfs have relatively high surface temperatures, between 8,000 K and 40,000 K.[26][64] A white dwarf, though, spends more of its lifetime at cooler temperatures than at hotter temperatures, so we should expect that there are more cool white dwarfs than hot white dwarfs. Once we adjust for the selection effect that hotter, more luminous white dwarfs are easier to observe, we do find that decreasing the temperature range examined results in finding more white dwarfs.[65] This trend stops when we reach extremely cool white dwarfs; few white dwarfs are observed with surface temperatures below 4,000 K,[66] and one of the coolest so far observed, WD 0346+246, has a surface temperature of approximately 3,900 K.[58] The reason for this is that the Universe's age is finite,[67][68] there has not been time for white dwarfs to cool below this temperature. The white dwarf luminosity function can therefore be used to find the time when stars started to form in a region; an estimate for the age of the Galactic disk found in this way is 8 billion years.[65] A white dwarf will eventually, in many trillion years, cool and become a non-radiating *black dwarf* in approximate thermal equilibrium with its surroundings and with the cosmic background radiation. However, no black dwarfs are thought to exist yet.[1]

Although white dwarf material is initially plasma—a fluid composed of nuclei and electrons—it was theoretically predicted in the 1960s that at a late stage of cooling, it should crystallize, starting at the center of the star.[69] The crystal structure is thought to be a body-centered cubic lattice.[51][70] In 1995 it was pointed out that asteroseismological observations of pulsating white dwarfs yielded a potential test of the crystallization theory,[71] and in 2004, Antonio Kanaan, Travis Metcalfe and a team of researchers with the Whole Earth Telescope estimated, on the basis of such observations, that approximately 90% of the mass of BPM 37093 had crystallized.[69][72][73] Other work gives a crystallized mass fraction of between 32% and 82%.[74] As a white dwarf core undergoes crystallization into a solid phase, latent heat is released which provides a source of thermal energy that delays the cooling of the star.[75]

Low-mass helium white dwarfs (with a mass < 0.20 Msun, often referred to as "extremely low-mass white dwarfs, ELM WDs") are formed in binary systems. As a result of their hydrogen-rich envelopes, residual hydrogen burning via the CNO cycle may keep these white dwarfs hot on a long timescale. In addition, they remain in a bloated proto-white dwarf stage for up to 2 Gyr before they reach the cooling track.[76]

### 8.2.3 Atmosphere and spectra

Although most white dwarfs are thought to be composed of carbon and oxygen, spectroscopy typically shows that their emitted light comes from an atmosphere which is observed to be either hydrogen-dominated or helium-dominated. The dominant element is usually at least 1,000 times more abundant than all other elements. As explained by Schatzman in the 1940s, the high surface gravity is thought to cause this purity by gravitationally separating the atmosphere so that heavy elements are below and the lighter above.[77][78], §5–6 This atmosphere, the only part of the white dwarf visible to us, is thought to be the top of an envelope which is a residue of the star's envelope in the AGB phase and may also contain material accreted from the interstellar medium. The envelope is believed to consist of a helium-rich layer with mass no more than 1/100 of the star's total mass, which, if the atmosphere is hydrogen-dominated, is overlain by a hydrogen-rich layer with mass approximately 1/10,000 of the stars total mass.[59][79], §4–5.

Although thin, these outer layers determine the thermal evolution of the white dwarf. The degenerate electrons in the bulk of a white dwarf conduct heat well. Most of a white dwarf's mass is therefore at almost the same temperature (isothermal), and it is also hot: a white dwarf with surface temperature between 8,000 K and 16,000 K will have a core temperature between approximately 5,000,000 K and 20,000,000 K. The white dwarf is kept from cooling very quickly only by its outer layers' opacity to radiation.[59]

The first attempt to classify white dwarf spectra appears to have been by G. P. Kuiper in 1941,[57][80] and various classification schemes have been proposed and used since then.[81][82] The system currently in use was introduced by Edward M. Sion, Jesse L. Greenstein and their coauthors in 1983 and has been subsequently revised several times. It classifies a spectrum by a symbol which consists of an initial D, a letter describing the primary feature of the spectrum followed by an optional sequence of letters describing secondary features of the spectrum (as shown in the table to the right), and a temperature index number, computed by dividing 50,400 K by the effective temperature. For example:

- A white dwarf with only He I lines in its spectrum and an effective temperature of 15,000 K could be given the classification of DB3, or, if warranted by the precision of the temperature measurement, DB3.5.

- A white dwarf with a polarized magnetic field, an effective temperature of 17,000 K, and a spectrum dominated by He I lines which also had hydrogen features could be given the classification of DBAP3.

The symbols ? and : may also be used if the correct classification is uncertain.[25][57]

White dwarfs whose primary spectral classification is DA have hydrogen-dominated atmospheres. They make up the majority (approximately 80%) of all observed white dwarfs.[59] The next class in number is of DBs (approximately 16%).[83] The hot (above 15,000 K) DQ class (roughly 0.1%) have carbon-dominated atmospheres.[84] Those classified as DB, DC, DO, DZ, and cool DQ have helium-dominated atmospheres. Assuming that carbon and metals are not present, which spectral classification is seen depends on the effective temperature. Between approximately 100,000 K to 45,000 K, the spectrum will be classified DO, dominated by singly ionized helium. From 30,000 K to 12,000 K, the spectrum will be DB, showing neutral helium lines, and below about 12,000 K, the spectrum will be featureless and classified DC.[79],§ 2.4,[59]

### Molecular hydrogen in white dwarf atmospheres

Main article: Molecules in stars

In 2013 S. Xu, M. Jura, D. Koster, B. Klein, and B. Zuckerman published a scientific paper in Astrophysical Journal Letters announcing the discovery of hydrogen($H_2$) in white dwarf stellar atmospheres [85]

## 8.2.4   Magnetic field

Magnetic fields in white dwarfs with a strength at the surface of ~1 million gauss (100 teslas) were predicted by P. M. S. Blackett in 1947 as a consequence of a physical law he had proposed which stated that an uncharged, rotating body should generate a magnetic field proportional to its angular momentum.[86] This putative law, sometimes called the *Blackett effect*, was never generally accepted, and by the 1950s even Blackett felt it had been refuted.[87], pp. 39–43 In the 1960s, it was proposed that white dwarfs might have magnetic fields because of conservation of total surface magnetic flux during the evolution of its non-degenerate progenitor star to a white dwarf. A surface magnetic field of ~100 gauss (0.01 T) in the progenitor star would thus become a surface magnetic field of ~100·100² = 1 million gauss (100 T) once the star's radius had shrunk by a factor of 100.[78], §8;[88], p. 484 The first magnetic white dwarf to be observed was GJ 742, which was detected to have a magnetic field in 1970 by its emission of circularly polarized light.[89] It is thought to have a surface field of approximately 300 million gauss (30 kT).[78], §8 Since then magnetic fields have been discovered in well over 100 white dwarfs, ranging from $2 \times 10^3$ to $10^9$ gauss (0.2 T to 100 kT). Only a small number of white dwarfs have been examined for fields, and it has been estimated that at least 10% of white dwarfs have fields in excess of 1 million gauss (100 T).[90][91]

### Chemical bonds

The magnetic fields in a white dwarf may allow for the existence of a new type of chemical bond, perpendicular paramagnetic bonding, in addition to ionic and covalent bonds, resulting in what has been initially described as "magnetized matter" in research published in 2012.[92]

## 8.3   Variability

Main article: Pulsating white dwarf
See also: Cataclysmic variables

Early calculations suggested that there might be white dwarfs whose luminosity varied with a period of around 10 seconds, but searches in the 1960s failed to observe this.[78], § 7.1.1;[95] The first variable white dwarf found was HL Tau 76; in 1965 and 1966, Arlo U. Landolt observed it to vary with a period of approximately 12.5 minutes.[96] The reason for this period being longer than predicted is that the variability of HL Tau 76, like that of the other pulsating variable white dwarfs known, arises from non-radial gravity wave pulsations.[78], § 7. Known types of pulsating white dwarf include the *DAV*, or *ZZ Ceti*, stars, including HL Tau 76, with hydrogen-dominated atmospheres and the spectral type DA;[78], pp. 891, 895 *DBV*, or *V777 Her*, stars, with helium-dominated atmospheres and the spectral type DB;[59], p. 3525 and *GW Vir stars* (sometimes subdivided into *DOV* and *PNNV* stars), with atmospheres dominated by helium, carbon, and oxygen.[94],§1.1, 1.2;[97],§1. GW Vir stars are not, strictly speaking, white dwarfs, but are stars which are in a position on the Hertzsprung-Russell diagram between the asymptotic giant branch and the white dwarf region. They may be called *pre-white dwarfs*.[94], § 1.1;[98] These variables all exhibit small (1%–30%) variations in light output, arising from a superposition of vibrational modes with periods of hundreds to thousands of seconds. Observation of these variations gives asteroseismological evidence about the interiors of white dwarfs.[99]

## 8.4 Formation

White dwarfs are thought to represent the end point of stellar evolution for main-sequence stars with masses from about 0.07 to 10 $M\odot$.[5][100] The composition of the white dwarf produced will differ depending on the initial mass of the star.

### 8.4.1 Stars with very low mass

If the mass of a main-sequence star is lower than approximately half a solar mass, it will never become hot enough to fuse helium at its core. It is thought that, over a lifespan that considerably exceeds the age (~13.8 billion years)[10] of the Universe, such a star will eventually burn all its hydrogen and end its evolution as a helium white dwarf composed chiefly of helium-4 nuclei.[101] Owing to the very long time this process takes, it is not thought to be the origin of the observed helium white dwarfs. Rather, they are thought to be the product of mass loss in binary systems[6][8][9][102][103][104] or mass loss due to a large planetary companion.[105][106]

### 8.4.2 Stars with low to medium mass

If the mass of a main-sequence star is between approximately 0.5 to 8 $M\odot$, its core will become sufficiently hot to fuse helium into carbon and oxygen via the triple-alpha process, but it will never become sufficiently hot to fuse carbon into neon. Near the end of the period in which it undergoes fusion reactions, such a star will have a carbon-oxygen core which does not undergo fusion reactions, surrounded by an inner helium-burning shell and an outer hydrogen-burning shell. On the Hertzsprung-Russell diagram, it will be found on the asymptotic giant branch. It will then expel most of its outer material, creating a planetary nebula, until only the carbon-oxygen core is left. This process is responsible for the carbon-oxygen white dwarfs which form the vast majority of observed white dwarfs.[102][107][108]

### 8.4.3 Stars with medium to high mass

If a star is massive enough, its core will eventually become sufficiently hot to fuse carbon to neon, and then to fuse neon to iron. Such a star will not become a white dwarf, because the mass of its central, non-fusing core, supported by electron degeneracy pressure, will eventually exceed the largest possible mass supportable by degeneracy pressure. At this point the core of the star will collapse and it will explode in a core-collapse supernova which will leave behind a remnant neutron star, black hole, or possibly a more exotic form of compact star.[100][109] Some main-sequence stars, of perhaps 8 to 10 $M\odot$, although sufficiently massive to fuse carbon to neon and magnesium, may be insufficiently massive to fuse neon. Such a star may leave a remnant white dwarf composed chiefly of oxygen, neon, and magnesium, provided that its core does not collapse, and provided that fusion does not proceed so violently as to blow apart the star in a supernova.[110][111] Although some isolated white dwarfs have been identified which may be of this type, most evidence for the existence of such stars comes from the novae called *ONeMg* or *neon* novae. The spectra of these novae exhibit abundances of neon, magnesium, and other intermediate-mass elements which appear to be only explicable by the accretion of material onto an oxygen-neon-magnesium white dwarf.[7][112][113]

### 8.4.4   Type Ia supernovae

Type Ia supernovae, that involve one or two previous white dwarfs, have been proposed to be a channel for transformation of this type of stellar remmant. In this scenario, the carbon detonation produced in a Type Ia supernova is too weak to destroy the white dwarf, expelling just a small part of its mass as ejecta and producing an asymmetric explosion that kicks the star at high speeds as a Hypervelocity star. The matter processed in the failed detonation is re-accreted back by the white dwarf with the heaviest elements such as iron falling to its core and accumulating there.[114]

These *iron-core* white dwarfs would be smaller than their carbon-oxygen kind of similar mass and would cool and crystallize faster than those.[115]

## 8.5   Fate

*Artist's impression of debris around a white dwarf*[116]

A white dwarf is stable once formed and will continue to cool almost indefinitely; eventually, it will become a black white dwarf, also called a black dwarf. Assuming that the Universe continues to expand, it is thought that in $10^{19}$ to $10^{20}$ years, the galaxies will evaporate as their stars escape into intergalactic space.[117], §IIIA. White dwarfs should generally survive this, although an occasional collision between white dwarfs may produce a new fusing star or a super-Chandrasekhar mass white dwarf which will explode in a Type Ia supernova.[117], §IIIC, IV. The subsequent lifetime of white dwarfs is thought to be on the order of the lifetime of the proton, known to be at least $10^{32}$ years. Some simple grand unified theories predict a proton lifetime of no more than $10^{49}$ years. If these theories are not valid, the proton may decay by more complicated nuclear processes, or by quantum gravitational processes involving a virtual black hole; in these cases, the lifetime is estimated to be no more than $10^{200}$ years. If protons do decay, the mass of a white dwarf will decrease very slowly with time as its nuclei decay, until it loses enough mass to become a nondegenerate lump of matter, and finally disappears completely.[117], §IV.

A white dwarf can also be cannibalized or evaporated by a companion star, losing so much mass that it becomes a planetary mass object. The resultant object, orbiting the former companion, now host star, could be a helium planet or diamond planet.[118][119]

## 8.6 Debris disks and planets

A white dwarf's stellar and planetary system is inherited from its progenitor star and may interact with the white dwarf in various ways. Infrared spectroscopic observations made by NASA's Spitzer Space Telescope of the central star of the Helix Nebula suggest the presence of a dust cloud, which may be caused by cometary collisions. It is possible that infalling material from this may cause X-ray emission from the central star.[120][121] Similarly, observations made in 2004 indicated the presence of a dust cloud around the young white dwarf G29-38 (estimated to have formed from its AGB progenitor about 500 million years ago), which may have been created by tidal disruption of a comet passing close to the white dwarf.[122] Some estimations based on the metal content of the atmospheres of the white dwarfs consider that at least a 15% of them may be orbited by planets and/or asteroids, or at least their debris.[123] Another suggested idea is that white dwarfs could be orbited by the stripped cores of rocky planets, that would have survived the red giant phase of their star but losing their outer layers and, given those planetary remnants would likely be made of metals, to attempt to detect them looking for the signatures of their interaction with the white dwarf's magnetic field.[124]

There is a planet in the white dwarf–pulsar binary system PSR B1620-26.

There are two circumbinary planets around the white dwarf–red dwarf binary NN Serpentis.

White dwarf WD 1145+017 is the first white dwarf observed with a disintegrating minor planet.[125][126]

## 8.7 Habitability

It has been proposed that white dwarfs with surface temperatures of less than 10,000 Kelvin could harbor a habitable zone at a distance between ~0.005 to 0.02 AU that would last upwards of 3 billion years. The goal is to search for transits of hypothetical Earth-like planets that could have migrated inward and/or formed there. As a white dwarf has a size similar to that of a planet, these kinds of transits would produce strong eclipses.[127] Newer research, however, casts some doubts on this idea, given that the close orbits of those hypothetical planets around their parent stars would subject them to strong tidal forces that could render them unhabitable by triggering a greenhouse effect.[128] Another suggested constraint to this idea is the origin of those planets. Leaving aside in-situ formation on an accretion disk surrounding the white dwarf, there are two ways a planet could end in a close orbit around stars of this kind: by surviving being engulfed by the star during its red giant phase, and then spiraling towards its core, or inward migration after the white dwarf has formed. The former case is implausible for low-mass bodies, as they are unlikely to survive being absorbed by their stars. In the latter case, the planets would have to expel so much orbital energy as heat, through tidal interactions with the white dwarf, that they would likely end as uninhabitable embers.[129] Some have suggested that Dyson spheres could be built around white dwarfs.[130]

## 8.8 Binary stars and novae

If a white dwarf is in a binary star system and is accreting matter from its companion, a variety of phenomena may occur, including novae and Type Ia supernovae. It may also be a super-soft x-ray source if it is able to take material from its companion fast enough to sustain fusion on its surface.[131] A close binary system of two white dwarfs can radiate energy in the form of gravitational waves, causing their mutual orbit to steadily shrink until the stars merge.[132][133]

### 8.8.1 Type Ia supernovae

Main article: Type Ia supernova

The mass of an isolated, nonrotating white dwarf cannot exceed the Chandrasekhar limit of ~1.4 $M\odot$. (This limit may increase if the white dwarf is rotating rapidly and nonuniformly.)[134] White dwarfs in binary systems, however, can accrete material from a companion star, increasing both their mass and their density. As their mass approaches the Chandrasekhar limit, this could theoretically lead to either the explosive ignition of fusion in the white dwarf or its collapse into a neutron star.[47]

Accretion provides the currently favored mechanism, the *single-degenerate model*, for Type Ia supernovae. In this model, a carbon–oxygen white dwarf accretes material from a companion star,[48]. p. 14. increasing its mass and compressing its core. It is believed that compressional heating of the core leads to ignition of carbon fusion as the mass approaches the Chandrasekhar limit.[48] Because the white dwarf is supported against gravity by quantum degeneracy pressure instead of by thermal pressure, adding heat to the star's interior increases its temperature but not its pressure, so the white dwarf does not expand and cool in response. Rather, the increased temperature accelerates the rate of the fusion reaction, in a runaway process that feeds on itself. The thermonuclear flame consumes much of the white dwarf in a few seconds, causing a Type Ia supernova explosion that obliterates the star.[1][48][135] In another possible mechanism for Type Ia supernovae, the *double-degenerate model*, two carbon-oxygen white dwarfs in a binary system merge, creating an object with mass greater than the Chandrasekhar limit in which carbon fusion is then ignited.[48]. p. 14.

Observations have failed to note signs of accretion leading up to Type Ia supernovae, and this is now thought to be because the star is first loaded up to above the Chandrasekhar limit while also being spun up to a very fast rate by the same process. Once the accretion stops the star gradually slows down until the spin is no longer fast enough to prevent the explosion.[136]

### 8.8.2 Cataclysmic variables

Main article: Cataclysmic variable star

Before accretion of material pushes a white dwarf close to the Chandrasekhar limit, accreted hydrogen-rich material on the surface may ignite in a less destructive type of thermonuclear explosion powered by hydrogen fusion. Since the white dwarf's core remains intact, these surface explosions can be repeated as long as accretion continues. This weaker kind of repetitive cataclysmic phenomenon is called a (classical) nova. Astronomers have also observed dwarf novae, which have smaller, more frequent luminosity peaks than classical novae. These are thought to be caused by the release of gravitational potential energy when part of the accretion disc collapses onto the star, rather than by fusion. In general, binary systems with a white dwarf accreting matter from a stellar companion are called cataclysmic variables. As well as novae and dwarf novae, several other classes of these variables are known.[1][48][137][138] Both fusion- and accretion-powered cataclysmic variables have been observed to be X-ray sources.[138]

## 8.9 Nearest

## 8.10 See also

- List of white dwarfs

- Planetary nebula

- PG 1159 star

- Stellar classification

- Timeline of white dwarfs, neutron stars, and supernovae

- Degenerate matter

- Black dwarf

- Robust Associations of Massive Baryonic Objects (RAMBOs)

- Neutron star

## 8.11   References

[1] Johnson, J. (2007). "Extreme Stars: White Dwarfs & Neutron Stars". *Lecture notes, Astronomy 162*. Ohio State University. Retrieved 17 October 2011.

[2] Henry, T. J. (1 January 2009). "The One Hundred Nearest Star Systems". Research Consortium On Nearby Stars. Retrieved 21 July 2010.

[3] *White Dwarfs*, E. Schatzman, Amsterdam: North-Holland, 1958.

[4] Holberg, J. B. (2005). "How Degenerate Stars Came to be Known as White Dwarfs". *American Astronomical Society Meeting 207* **207**: 1503. Bibcode:2005AAS...20720501H.

[5] Fontaine, G.; Brassard, P.; Bergeron, P. (2001). "The Potential of White Dwarf Cosmochronology". *Publications of the Astronomical Society of the Pacific* **113** (782): 409–435. Bibcode:2001PASP..113..409F. doi:10.1086/319535.

[6] Richmond, M. "Late stages of evolution for low-mass stars". *Lecture notes, Physics 230*. Rochester Institute of Technology. Retrieved 3 May 2007.

[7] Werner, K.; Hammer, N. J.; Nagel, T.; Rauch, T.; Dreizler, S. (2005). "On Possible Oxygen/Neon White Dwarfs: H1504+65 and the White Dwarf Donors in Ultracompact X-ray Binaries". *14th European Workshop on White Dwarfs* **334**: 165. arXiv:astro-ph/0410690. Bibcode:2005ASPC..334..165W.

[8] Liebert, J.; Bergeron, P.; Eisenstein, D.; Harris, H. C.; Kleinman, S. J.; Nitta, A.; Krzesinski, J. (2004). "A Helium White Dwarf of Extremely Low Mass". *The Astrophysical Journal* **606** (2): L147. arXiv:astro-ph/0404291. Bibcode:2004ApJ...606L.147L. doi:10.1086/421462.

[9] "Cosmic weight loss: The lowest mass white dwarf" (Press release). Harvard-Smithsonian Center for Astrophysics. 17 April 2007.

[10] Spergel, D. N.; Bean, R.; Doré, O.; Nolta, M. R.; Bennett, C. L.; Dunkley, J.; Hinshaw, G.; Jarosik, N.; et al. (2007). "Wilkinson Microwave Anisotropy Probe (WMAP) Three Year Results: Implications for Cosmology". *The Astrophysical Journal Supplement Series* **170** (2): 377–408. arXiv:astro-ph/0603449. Bibcode:2007ApJS..170..377S. doi:10.1086/513700.

[11] Herschel, W. (1785). "Catalogue of Double Stars. By William Herschel, Esq. F. R. S". *Philosophical Transactions of the Royal Society of London* **75**: 40–126. Bibcode:1785RSPT...75...40H. doi:10.1098/rstl.1785.0006. JSTOR 106749.

[12] Van Den Bos, W. H. (1926). "The orbit and the masses of 40 Eridani BC". *Bulletin of the Astronomical Institutes of the Netherlands* **3**: 128. Bibcode:1926BAN.....3..128V.

[13] Heintz, W. D. (1974). "Astrometric study of four visual binaries". *The Astronomical Journal* **79**: 819. Bibcode:1974AJ.....79..81 doi:10.1086/111614.

[14] Adams, W. S. (1914). "An A-Type Star of Very Low Luminosity". *Publications of the Astronomical Society of the Pacific* **26**: 198. Bibcode:1914PASP...26..198A. doi:10.1086/122337.

[15] "On the variations of the proper motions of Procyon and Sirius". *Monthly Notices of the Royal Astronomical Society* **6** (11): 136–141. 1844. Bibcode:1844MNRAS...6..136.. doi:10.1093/mnras/6.11.136a.

[16] Flammarion, Camille (1877). "The Companion of Sirius". *Astronomical register* **15**: 186. Bibcode:1877AReg...15..186F.

[17] Adams, W. S. (1915). "The Spectrum of the Companion of Sirius". *Publications of the Astronomical Society of the Pacific* **27**: 236. Bibcode:1915PASP...27..236A. doi:10.1086/122440.

[18] Van Maanen, A. (1917). "Two Faint Stars with Large Proper Motion". *Publications of the Astronomical Society of the Pacific* **29**: 258. Bibcode:1917PASP...29..258V. doi:10.1086/122654.

[19] Luyten, W. J. (1922). "The Mean Parallax of Early-Type Stars of Determined Proper Motion and Apparent Magnitude". *Publications of the Astronomical Society of the Pacific* **34**: 156. Bibcode:1922PASP...34..156L. doi:10.1086/123176.

[20] Luyten, W. J. (1922). "Note on Some Faint Early Type Stars with Large Proper Motions". *Publications of the Astronomical Society of the Pacific* **34**: 54. Bibcode:1922PASP...34...54L. doi:10.1086/123146.

[21] Luyten, W. J. (1922). "Additional Note on Faint Early-Type Stars with Large Proper-Motions". *Publications of the Astronomical Society of the Pacific* **34**: 132. Bibcode:1922PASP...34..132L. doi:10.1086/123168.

[22] Aitken, R. G. (1922). "Comet c 1922 (Baade)".*Publications of the Astronomical Society of the Pacific***34**: 353. Bibcode:1922PA doi:10.1086/123244.

[23] Eddington, A. S. (1924). "On the relation between the masses and luminosities of the stars". *Monthly Notices of the Royal Astronomical Society* **84**: 308. Bibcode:1924MNRAS..84..308E. doi:10.1093/mnras/84.5.308.

[24] Luyten, W. J. (1950). "The search for white dwarfs". *The Astronomical Journal* **55**: 86. Bibcode:1950AJ.....55...86L. doi:10.1086/106358.

[25] McCook, George P.; Sion, Edward M. (1999). "A Catalog of Spectroscopically Identified White Dwarfs". *The Astrophysical Journal Supplement Series* **121**: 1–130. Bibcode:1999ApJS..121....1M. doi:10.1086/313186.

[26] Eisenstein, Daniel J.; Liebert, James; Harris, Hugh C.; Kleinman, S. J.; Nitta, Atsuko; Silvestri, Nicole; Anderson, Scott A.; Barentine, J. C.; Brewington, Howard J.; Brinkmann, J.; Harvanek, Michael; Krzesiński, Jurek; Neilsen, Jr., Eric H.; Long, Dan; Schneider, Donald P.; Snedden, Stephanie A. (2006). "A Catalog of Spectroscopically Confirmed White Dwarfs from the Sloan Digital Sky Survey Data Release 4". *The Astrophysical Journal Supplement Series* **167**: 40–58. arXiv:astro-ph/0606700. Bibcode:2006ApJS..167...40E. doi:10.1086/507110.

[27] Kilic, M.; Allende Prieto, C.; Brown, Warren R.; Koester, D. (2007). "The Lowest Mass White Dwarf". *The Astrophysical Journal* **660** (2): 1451–1461. arXiv:astro-ph/0611498. Bibcode:2007ApJ...660.1451K. doi:10.1086/514327.

[28] Kepler, S. O.; Kleinman, S. J.; Nitta, A.; Koester, D.; Castanheira, B. G.; Giovannini, O.; Costa, A. F. M.; Althaus, L. (2007). "White dwarf mass distribution in the SDSS". *Monthly Notices of the Royal Astronomical Society* **375** (4): 1315–1324. arXiv:astro-ph/0612277. Bibcode:2007MNRAS.375.1315K. doi:10.1111/j.1365-2966.2006.11388.x.

[29] Shipman, H. L. (1979). "Masses and radii of white-dwarf stars. III – Results for 110 hydrogen-rich and 28 helium-rich stars". *The Astrophysical Journal* **228**: 240. Bibcode:1979ApJ...228..240S. doi:10.1086/156841.

[30] Sandin, F. (2005). "Exotic Phases of Matter in Compact Stars" (PDF). *Licentiate thesis*. Luleå University of Technology. Retrieved 20 August 2011.

[31] Boss, L. (1910). *Preliminary General Catalogue of 6188 stars for the epoch 1900*. Carnegie Institution of Washington. Bibcode:1910pgcs.book.....B. LCCN 10009645.

[32] Liebert, J.; Young, P. A.; Arnett, D.; Holberg, J. B.; Williams, K. A. (2005). "The Age and Progenitor Mass of Sirius B". *The Astrophysical Journal* **630**: L69. arXiv:astro-ph/0507523. Bibcode:2005ApJ...630L..69L. doi:10.1086/462419.

[33] Öpik, E. (1916). "The Densities of Visual Binary Stars". *The Astrophysical Journal* **44**: 292. Bibcode:1916ApJ....44..292O. doi:10.1086/142296.

[34] Eddington, A. S. (1927). *Stars and Atoms*. Clarendon Press. LCCN 27015694.

[35] Adams, W. S. (1925). "The Relativity Displacement of the Spectral Lines in the Companion of Sirius". *Proceedings of the National Academy of Sciences* **11** (7): 382–387. Bibcode:1925PNAS...11..382A. doi:10.1073/pnas.11.7.382.

[36] Nave, C. R. "Nuclear Size and Density". *HyperPhysics*. Georgia State University. Retrieved 26 June 2009.

[37] Adams, Steve (1997). *Relativity: an introduction to space-time physics*. CRC Press. p. 240. ISBN 0-7484-0621-2.

[38] Fowler, R. H. (1926). "On dense matter".*Monthly Notices of the Royal Astronomical Society***87**: 114. Bibcode:1926MNRAS..87 doi:10.1093/mnras/87.2.114.

[39] Hoddeson, L. H.; Baym, G. (1980). "The Development of the Quantum Mechanical Electron Theory of Metals: 1900–28". *Proceedings of the Royal Society of London* **371** (1744): 8–23. Bibcode:1980RSPSA.371....8H. doi:10.1098/rspa.1980.0051. JSTOR 2990270.

[40] "Estimating Stellar Parameters from Energy Equipartition". ScienceBits. Retrieved 9 May 2007.

[41] Bean, R. "Lecture 12 – Degeneracy pressure" (PDF). *Lecture notes, Astronomy 211*. Cornell University. Archived from the original (PDF) on 2007-09-25. Retrieved 21 September 2007.

[42] Anderson, W. (1929). "Über die Grenzdichte der Materie und der Energie". *Zeitschrift für Physik* **56** (11–12): 851–856. Bibcode:1929ZPhy...56..851A. doi:10.1007/BF01340146.

[43] Stoner, C. (1930). "The Equilibrium of Dense Stars". *Philosophical Magazine* **9**: 944.

[44] Chandrasekhar, S. (1931). "The Maximum Mass of Ideal White Dwarfs". *The Astrophysical Journal* **74**: 81. Bibcode:1931ApJ... doi:10.1086/143324.

[45] Chandrasekhar, S. (1935). "The highly collapsed configurations of a stellar mass (Second paper)". *Monthly Notices of the Royal Astronomical Society* **95**: 207. Bibcode:1935MNRAS..95..207C. doi:10.1093/mnras/95.3.207.

[46] "The Nobel Prize in Physics 1983". The Nobel Foundation. Retrieved 4 May 2007.

[47] Canal, R.; Gutierrez, J. (1997). "The Possible White Dwarf-Neutron Star Connection". arXiv:astro-ph/9701225 [astro-ph].

[48] Hillebrandt, W.; Niemeyer, J. C. (2000). "Type IA supernova explosion models". *Annual Review of Astronomy and Astrophysics* **38**: 191–230. arXiv:astro-ph/0006305. Bibcode:2000ARA&A..38..191H. doi:10.1146/annurev.astro.38.1.191.

[49] Overbye, D. (22 February 2010). "From the Clash of White Dwarfs, the Birth of a Supernova". *New York Times*. Retrieved 22 February 2010.

[50] Chabrier, G.; Baraffe, I. (2000). "Theory of low-Mass stars and substellar objects". *Annual Review of Astronomy and Astrophysics* **38**: 337–377. arXiv:astro-ph/0006383. Bibcode:2000ARA&A..38..337C. doi:10.1146/annurev.astro.38.1.337.

[51] Kaler, J. "The Hertzsprung-Russell (HR) diagram". Retrieved 5 May 2007.

[52] "Basic symbols". *Standards for Astronomical Catalogues, Version 2.0*. VizieR. Retrieved 12 January 2007.

[53] Tohline, J. E. "The Structure, Stability, and Dynamics of Self-Gravitating Systems". Retrieved 30 May 2007.

[54] Hoyle, F. (1947). "Stars, Distribution and Motions of, Note on equilibrium configurations for rotating white dwarfs". *Monthly Notices of the Royal Astronomical Society* **107**: 231–236. Bibcode:1947MNRAS.107..231H. doi:10.1093/mnras/107.2.231.

[55] Ostriker, J. P.; Bodenheimer, P. (1968). "Rapidly Rotating Stars. II. Massive White Dwarfs". *The Astrophysical Journal* **151**: 1089. Bibcode:1968ApJ...151.1089O. doi:10.1086/149507.

[56] Kutner, M. L. (2003). *Astronomy: A physical perspective*. Cambridge University Press. p. 189. ISBN 978-0-521-52927-3.

[57] Sion, E. M.; Greenstein, J. L.; Landstreet, J. D.; Liebert, J.; Shipman, H. L.; Wegner, G. A. (1983). "A proposed new white dwarf spectral classification system". *The Astrophysical Journal* **269**: 253. Bibcode:1983ApJ...269..253S. doi:10.1086/161036.

[58] Hambly, N. C.; Smartt, S. J.; Hodgkin, S. T. (1997). "WD 0346+246: A Very Low Luminosity, Cool Degenerate in Taurus". *The Astrophysical Journal* **489** (2): L157. Bibcode:1997ApJ...489L.157H. doi:10.1086/316797.

[59] Fontaine, G.; Wesemael, F. (2001). "White dwarfs". In Murdin, P. *Encyclopedia of Astronomy and Astrophysics*. IOP Publishing/Nature Publishing Group. ISBN 0-333-75088-8.

[60] Heise, J. (1985). "X-ray emission from isolated hot white dwarfs". *Space Science Reviews* **40**: 79–90. Bibcode:1985SSRv...40... doi:10.1007/BF00212870.

[61] Mestel, L. (1952). "On the theory of white dwarf stars. I. The energy sources of white dwarfs". *Monthly Notices of the Royal Astronomical Society* **112**: 583. Bibcode:1952MNRAS.112..583M. doi:10.1093/mnras/112.6.583.

[62] Kawaler, S. D. (1998). "White Dwarf Stars and the Hubble Deep Field". *The Hubble Deep Field : Proceedings of the Space Telescope Science Institute Symposium*. p. 252. arXiv:astro-ph/9802217. Bibcode:1998hdf..symp..252K. ISBN 0-521-63097-5.

[63]  Bergeron, P.; Ruiz, M. T.; Leggett, S. K. (1997). "The Chemical Evolution of Cool White Dwarfs and the Age of the Local Galactic Disk".*The Astrophysical Journal Supplement Series***108**: 339–387. Bibcode:1997ApJS..108..339B.doi:10.1086/3129

[64]  McCook, G. P.; Sion, E. M. "A Catalogue of Spectroscopically Identified White Dwarfs". *The Astrophysical Journal Supplement Series* **121** (1): 1–130. Bibcode:1999ApJS..121....1M. doi:10.1086/313186.

[65]  Leggett, S. K.; Ruiz, M. T.; Bergeron, P. (1998). "The Cool White Dwarf Luminosity Function and the Age of the Galactic Disk". *The Astrophysical Journal* **497**: 294–302. Bibcode:1998ApJ...497..294L. doi:10.1086/305463.

[66]  Gates, E.; Gyuk, G.; Harris, H. C.; Subbarao, M.; Anderson, S.; Kleinman, S. J.; Liebert, J.; Brewington, H.; et al. (2004). "Discovery of New Ultracool White Dwarfs in the Sloan Digital Sky Survey". *The Astrophysical Journal* **612** (2): L129. arXiv:astro-ph/0405566. Bibcode:2004ApJ...612L.129G. doi:10.1086/424568.

[67]  Winget, D. E.; Hansen, C. J.; Liebert, J.; Van Horn, H. M.; Fontaine, G.; Nather, R. E.; Kepler, S. O.; Lamb, D. Q. (1987). "An independent method for determining the age of the universe". *The Astrophysical Journal* **315**: L77. Bibcode:1987ApJ...315L..77W.doi:10.1086/184864.

[68]  Trefil, J. S. (2004). *The Moment of Creation: Big Bang Physics from Before the First Millisecond to the Present Universe*. Dover Publications. ISBN 0-486-43813-9.

[69]  Metcalfe, T. S.; Montgomery, M. H.; Kanaan, A. (2004). "Testing White Dwarf Crystallization Theory with Asteroseismology of the Massive Pulsating DA Star BPM 37093". *The Astrophysical Journal* **605** (2): L133. arXiv:astro-ph/0402046. Bibcode:2004ApJ...605L.133M. doi:10.1086/420884.

[70]  Barrat, J. L.; Hansen, J. P.; Mochkovitch, R. (1988). "Crystallization of carbon-oxygen mixtures in white dwarfs". *Astronomy and Astrophysics* **199**: L15. Bibcode:1988A&A...199L..15B.

[71]  Winget, D. E. (1995). "The Status of White Dwarf Asteroseismology and a Glimpse of the Road Ahead". *Baltic Astronomy* **4**: 129. Bibcode:1995BaltA...4..129W.

[72]  Diamond star thrills astronomers, David Whitehouse, BBC News, 16 February 2004. Accessed on line 6 January 2007.

[73]  Kanaan, A.; Nitta, A.; Winget, D. E.; Kepler, S. O.; Montgomery, M. H.; Metcalfe, T. S.; Oliveira, H.; Fraga, L.; et al. (2005). "Whole Earth Telescope observations of BPM 37093: A seismological test of crystallization theory in white dwarfs". *Astronomy and Astrophysics* **432**: 219–224. arXiv:astro-ph/0411199. Bibcode:2005A&A...432..219K. doi:10.1051/0004-6361:20041125.

[74]  Brassard, P.; Fontaine, G. (2005). "Asteroseismology of the Crystallized ZZ Ceti Star BPM 37093: A Different View". *The Astrophysical Journal* **622**: 572–576. Bibcode:2005ApJ...622..572B. doi:10.1086/428116.

[75]  B.M.S. Hansen, J. Liebert: Cool White Dwarfs, Annual Review of Astronomy and Astrophysics 41, 465 (2003)

[76]  Istrate et al. (2014), "The timescale of low-mass proto-helium white dwarf evolution" A&A Letter 571, L3

[77]  Schatzman, E. (1945). "Théorie du débit d'énergie des naines blanches".*Annales d'Astrophysique***8**: 143. Bibcode:1945AnAp..

[78]  Koester, D.; Chanmugam, G. (1990). "Physics of white dwarf stars". *Reports on Progress in Physics* **53** (7): 837–915. Bibcode:1990RPPh...53..837K. doi:10.1088/0034-4885/53/7/001.

[79]  Kawaler, S. D. (1997). "White Dwarf Stars". In Kawaler, S. D.; Novikov, I.; Srinivasan, G. *Stellar remnants*. 1997. ISBN 3-540-61520-2.

[80]  Kuiper, G. P. (1941). "List of Known White Dwarfs". *Publications of the Astronomical Society of the Pacific* **53**: 248. Bibcode:1941PASP...53..248K. doi:10.1086/125335.

[81]  Luyten, W. J. (1952). "The Spectra and Luminosities of White Dwarfs".*The Astrophysical Journal***116**: 283. Bibcode:1952Ap doi:10.1086/145612.

[82]  Greenstein, J. L. (1960). *Stellar atmospheres*. University of Chicago Press. Bibcode:1960stat.conf.....G. LCCN 61-9138.

[83]  Kepler, S. O.; Kleinman, S. J.; Nitta, A.; Koester, D.; Castanheira, B. G.; Giovannini, O.; Costa, A. F. M.; Althaus, L. (2007). "White dwarf mass distribution in the SDSS". *Monthly Notices of the Royal Astronomical Society* **375** (4): 1315–1324. arXiv:astro-ph/0612277. Bibcode:2007MNRAS.375.1315K. doi:10.1111/j.1365-2966.2006.11388.x.

[84] Dufour, P.; Liebert, J.; Fontaine, G.; Behara, N. (2007). "White dwarf stars with carbon atmospheres". *Nature* **450** (7169): 522–4. arXiv:0711.3227. Bibcode:2007Natur.450..522D. doi:10.1038/nature06318. PMID 18033290.

[85] Discovery of Molecular Hydrogen in White Dwarf Atmospheres - IOPscience

[86] Blackett, P. M. S. (1947). "The Magnetic Field of Massive Rotating Bodies".*Nature***159**(4046): 658–66. Bibcode:1947Natur. doi:10.1038/159658a0. PMID 20239729.

[87] Lovell, B. (1975). "Patrick Maynard Stuart Blackett, Baron Blackett, of Chelsea. 18 November 1897-13 July 1974". *Biographical Memoirs of Fellows of the Royal Society* **21**: 1–115. doi:10.1098/rsbm.1975.0001. JSTOR 769678.

[88] Ginzburg, V. L.; Zheleznyakov, V. V.; Zaitsev, V. V. (1969). "Coherent mechanisms of radio emission and magnetic models of pulsars". *Astrophysics and Space Science* **4** (4): 464–504. Bibcode:1969Ap&SS...4..464G. doi:10.1007/BF00651351.

[89] Kemp, J. C.; Swedlund, J. B.; Landstreet, J. D.; Angel, J. R. P. (1970). "Discovery of Circularly Polarized Light from a White Dwarf". *The Astrophysical Journal* **161**: L77. Bibcode:1970ApJ...161L..77K. doi:10.1086/180574.

[90] Jordan, S.; Aznar Cuadrado, R.; Napiwotzki, R.; Schmid, H. M.; Solanki, S. K. (2007). "The fraction of DA white dwarfs with kilo-Gauss magnetic fields". *Astronomy and Astrophysics* **462** (3): 1097–1101. arXiv:astro-ph/0610875. Bibcode:2007A&A.. .462.1097J.doi:10.1051/0004-6361:20066163.

[91] Liebert, James; Bergeron, P.; Holberg, J. B. (2003). "The True Incidence of Magnetism Among Field White Dwarfs". *The Astronomical Journal* **125**: 348–353. arXiv:astro-ph/0210319. Bibcode:2003AJ....125..348L. doi:10.1086/345573.

[92] Stars draw atoms closer together : Nature News & Comment

[93] ZZ Ceti variables, Association Française des Observateurs d'Etoiles Variables, web page at the Centre de Données astronomiques de Strasbourg. Accessed on line 6 June 2007.

[94] Quirion, P.-O.; Fontaine, G.; Brassard, P. (2007). "Mapping the Instability Domains of GW Vir Stars in the Effective Temperature–Surface Gravity Diagram". *The Astrophysical Journal Supplement Series* **171**: 219–248. Bibcode:2007ApJS..171.. 219Q.doi:10.1086/513870.

[95] Lawrence, G. M.; Ostriker, J. P.; Hesser, J. E. (1967). "Ultrashort-Period Stellar Oscillations. I. Results from White Dwarfs, Old Novae, Central Stars of Planetary Nebulae, 3c 273, and Scorpius XR-1". *The Astrophysical Journal* **148**: L161. Bibcode:1967ApJ ...148L.161L.doi:10.1086/180037.

[96] Landolt, A. U. (1968). "A New Short-Period Blue Variable". *The Astrophysical Journal* **153**: 151. Bibcode:1968ApJ...153..151L. doi:10.1086/149645.

[97] Nagel, T.; Werner, K. (2004). "Detection of non-radial g-mode pulsations in the newly discovered PG 1159 star HE 1429-1209". *Astronomy and Astrophysics* **426** (2): L45. arXiv:astro-ph/0409243. Bibcode:2004A&A...426L..45N. doi:10.1051/0004-6361:200400079.

[98] O'Brien, M. S. (2000). "The Extent and Cause of the Pre–White Dwarf Instability Strip". *The Astrophysical Journal* **532** (2): 1078–1088. arXiv:astro-ph/9910495. Bibcode:2000ApJ...532.1078O. doi:10.1086/308613.

[99] Winget, D. E. (1998). "Asteroseismology of white dwarf stars". *Journal of Physics: Condensed Matter* **10** (49): 11247–11261. Bibcode:1998JPCM...1011247W. doi:10.1088/0953-8984/10/49/014.

[100] Heger, A.; Fryer, C. L.; Woosley, S. E.; Langer, N.; Hartmann, D. H. (2003). "How Massive Single Stars End Their Life". *The Astrophysical Journal* **591**: 288–300. arXiv:astro-ph/0212469. Bibcode:2003ApJ...591..288H. doi:10.1086/375341.

[101] Laughlin, G.; Bodenheimer, P.; Adams, Fred C. (1997). "The End of the Main Sequence". *The Astrophysical Journal* **482**: 420–432. Bibcode:1997ApJ...482..420L. doi:10.1086/304125.

[102] Stars Beyond Maturity, Simon Jeffery, online article. Accessed on line 3 May 2007.

[103] Sarna, M.J.; Ergma, E.; Gerškevitš, J. (2001). "Helium core white dwarf evolution – including white dwarf companions to neutron stars". *Astronomische Nachrichten* **322** (5–6): 405–410. Bibcode:2001AN....322..405S. doi:10.1002/1521-3994(200112) 322:5/6<405::AID-ASNA405>3.0.CO;2-6.

[104] Benvenuto, O. G.; De Vito, M. A. (2005). "The formation of helium white dwarfs in close binary systems – II". *Monthly Notices of the Royal Astronomical Society* **362** (3): 891–905. Bibcode:2005MNRAS.362..891B. doi:10.1111/j.1365-2966.2005.09315.x.

[105] Nelemans, G.; Tauris, T. M. (1998). "Formation of undermassive single white dwarfs and the influence of planets on late stellar evolution". *Astronomy and Astrophysics* **335**: L85. arXiv:astro-ph/9806011. Bibcode:1998A&A...335L..85N.

[106] "Planet diet helps white dwarfs stay young and trim" (2639). NewScientist. 18 January 2008.

[107] the evolution of low-mass stars, Vik Dhillon, lecture notes, Physics 213, University of Sheffield. Accessed on line 3 May 2007.

[108] the evolution of high-mass stars, Vik Dhillon, lecture notes, Physics 213, University of Sheffield. Accessed on line 3 May 2007.

[109] Schaffner-Bielich, Jürgen (2005). "Strange quark matter in stars: A general overview". *Journal of Physics G: Nuclear and Particle Physics* **31** (6): S651. arXiv:astro-ph/0412215. Bibcode:2005JPhG...31S.651S. doi:10.1088/0954-3899/31/6/004.

[110] Nomoto, K. (1984). "Evolution of 8–10 solar mass stars toward electron capture supernovae. I – Formation of electron-degenerate O + NE + MG cores". *The Astrophysical Journal* **277**: 791. Bibcode:1984ApJ...277..791N. doi:10.1086/161749.

[111] Woosley, S. E.; Heger, A.; Weaver, T. A. (2002). "The evolution and explosion of massive stars". *Reviews of Modern Physics* **74** (4): 1015–1071. Bibcode:2002RvMP...74.1015W. doi:10.1103/RevModPhys.74.1015.

[112] Werner, K.; Rauch, T.; Barstow, M. A.; Kruk, J. W. (2004). "Chandra and FUSE spectroscopy of the hot bare stellar core H?1504+65". *Astronomy and Astrophysics* **421** (3): 1169–1183. arXiv:astro-ph/0404325. Bibcode:2004A&A...421.1169W. doi:10.1051/0004-6361:20047154.

[113] Livio, Mario; Truran, James W. (1994). "On the interpretation and implications of nova abundances: An abundance of riches or an overabundance of enrichments". *The Astrophysical Journal* **425**: 797. Bibcode:1994ApJ...425..797L. doi:10.1086/174024.

[114] Jordan, George C. IV.; Perets, Hagai B.; Fisher, Robert T.; van Rossum, Daniel R. (2012). "Failed-detonation Supernovae: Subluminous Low-velocity Ia Supernovae and their Kicked Remnant White Dwarfs with Iron-rich Cores". *The Astrophysical Journal Letters* **761** (2): L23. arXiv:1208.5069. Bibcode:2012ApJ...761L..23J. doi:10.1088/2041-8205/761/2/L23.

[115] Panei, J. A.; Althaus, L. G.; Benvenuto, O. G. (2000). "The evolution of iron-core white dwarfs". *Monthly Notices of the Royal Astronomical Society* **312** (3): 531–539. arXiv:astro-ph/9911371. Bibcode:2000MNRAS.312..531P. doi:10.1046/j.1365-8711.2000.03236.x.

[116] "Hubble finds dead stars "polluted" with planetary debris". *ESA/Hubble Press Release*. Retrieved 10 May 2013.

[117] Adams, Fred C.; Laughlin, Gregory (1997). "A dying universe: The long-term fate and evolutionof astrophysical objects". *Reviews of Modern Physics***69**(2): 337–372. arXiv:astro-ph/9701131. Bibcode:1997RvMP...69..337A.doi:10.1103/RevModPh

[118] S. Seager, M. Kuchner, C. Hier-Majumder, B. Militzer (19 July 2007). "Mass-Radius Relationships for Solid Exoplanets". *The Astrophysical Journal* (November 2007) **669** (2): 1279–1297. arXiv:0707.2895. Bibcode:2007ApJ...669.1279S. doi:10.1086/521346.

[119] Michael Lemonick (26 August 2011). "Scientists Discover a Diamond as Big as a Planet". *Time Magazine*.

[120] Comet clash kicks up dusty haze, BBC News, 13 February 2007. Accessed on line 20 September 2007.

[121] Su, K. Y. L.; Chu, Y.-H.; Rieke, G. H.; Huggins, P. J.; Gruendl, R.; Napiwotzki, R.; Rauch, T.; Latter, W. B.; Volk, K. (2007). "A Debris Disk around the Central Star of the Helix Nebula?". *The Astrophysical Journal* **657**: L41. arXiv:astro-ph/0702296. Bibcode:2007ApJ...657L..41S. doi:10.1086/513018.

[122] Reach, William T.; Kuchner, Marc J.; Von Hippel, Ted; Burrows, Adam; Mullally, Fergal; Kilic, Mukremin; Winget, D. E. (2005). "The Dust Cloud around the White Dwarf G29-38". *The Astrophysical Journal* **635** (2): L161. arXiv:astro-ph/0511358. Bibcode:2005ApJ...635L.161R. doi:10.1086/499561.

[123] Sion, Edward M.; Holberg, J. B.; Oswalt, Terry D.; McCook, George P.; Wasatonic, Richard (2009). "The White Dwarfs Within 20 Parsecs of the Sun: Kinematics and Statistics". *The Astronomical Journal* **138** (6): 1681–1689. arXiv:0910.1288. Bibcode:2009AJ....138.1681S. doi:10.1088/0004-6256/138/6/1681.

[124] Li, Jianke; Ferrario, Lilia; Wickramasinghe, Dayal (1998). "Planets around White Dwarfs". *Astrophysical Journal Letters* **503** (1): L151. Bibcode:1998ApJ...503L.151L. doi:10.1086/311546. p. L51.

[125] Lemonick, Michael D.; 21, National Geographic PUBLISHED October. "Zombie Star Caught Feasting On Asteroids". *National Geographic News*. Retrieved 2015-10-22.

[126] Vanderburg, Andrew; Johnson, John Asher; Rappaport, Saul; Bieryla, Allyson; Irwin, Jonathan; Lewis, John Arban; Kipping, David; Brown, Warren R.; Dufour, Patrick (2015-10-22). "A disintegrating minor planet transiting a white dwarf". *Nature* **526** (7574): 546–549. doi:10.1038/nature15527. ISSN 0028-0836.

[127] Agol, Eric (2011). "Transit Surveys for Earths in the Habitable Zones of White Dwarfs" (PDF). *The Astrophysical Journal Letters* **635** (2): L31. arXiv:1103.2791. Bibcode:2011ApJ...731L..31A. doi:10.1088/2041-8205/731/2/L31.

[128] Barnes, Rory; Heller, René (2011). "Habitable Planets Around White and Brown Dwarfs: The Perils of a Cooling Primary". *Astrobiology* **13** (3): 279–291. arXiv:1211.6467. Bibcode:2013AsBio..13..279B. doi:10.1089/ast.2012.0867. PMC 3612282. PMID 23537137.

[129] Nordhaus, J.; Spiegel, D.S. (2013). "On the orbits of low-mass companions to white dwarfs and the fates of the known exoplanets". *Monthly Notices of the Royal Astronomical Society* **432** (1): 500–505. arXiv:1211.1013. Bibcode:2013MNRAS.432..500N. doi:10.1093/mnras/stt569.

[130] İbrahim Semiz, Salim Oğur. "Dyson Spheres around White Dwarfs".

[131] Di Stefano, R.; Nelson, L. A.; Lee, W.; Wood, T. H.; Rappaport, S. (1997). P. Ruiz-Lapuente; R. Canal; J. Isern, eds. *Luminous Supersoft X-ray Sources as Type Ia Progenitors*. NATO ASI series: Mathematical and physical sciences (486). Springer. pp. 148–149. ISBN 0-7923-4359-X.

[132] Aguilar, David A.; Pulliam, Christine (16 November 2010). "Astronomers Discover Merging Star Systems that Might Explode". Harvard-Smithsonian Center for Astrophysics. Retrieved 16 February 2011.

[133] Aguilar, David A.; Pulliam, Christine (13 July 2011). "Evolved Stars Locked in Fatalistic Dance". Harvard-Smithsonian Center for Astrophysics. Retrieved 17 July 2011.

[134] Yoon, S.-C.; Langer, N. (2004). "Presupernova evolution of accreting white dwarfs with rotation". *Astronomy and Astrophysics* **419** (2): 623–644. arXiv:astro-ph/0402287. Bibcode:2004A&A...419..623Y. doi:10.1051/0004-6361:20035822.

[135] Blinnikov, S. I.; Röpke, F. K.; Sorokina, E. I.; Gieseler, M.; Reinecke, M.; Travaglio, C.; Hillebrandt, W.; Stritzinger, M. (2006). "Theoretical light curves for deflagration models of type Ia supernova". *Astronomy and Astrophysics* **453**: 229–240. arXiv:astro-ph/0603036. Bibcode:2006A&A...453..229B. doi:10.1051/0004-6361:20054594.

[136] O'Neill, Ian. "Don't Slow Down White Dwarf, You Might Explode." *Discovery Communications, LLC* 6 September 2011.

[137] Imagine the Universe! Cataclysmic Variables, fact sheet at NASA Goddard. Accessed on line 4 May 2007.

[138] Introduction to Cataclysmic Variables (CVs), fact sheet at NASA Goddard. Accessed on line 4 May 2007.

[139] Giammichele, N.; Bergeron, P.; Dufour, P. (April 2012). "Know Your Neighborhood: A Detailed Model Atmosphere Analysis of Nearby White Dwarfs". *The Astrophysical Journal Supplement* **199** (2): 35. Bibcode:2012ApJS..199...29G. doi:10.1088/0067-0049/199/2/29, 29.

## 8.12 External links and further reading

### 8.12.1 General

- Kawaler, S. D. (1997). "White Dwarf Stars". In Kawaler, S. D.; Novikov, I.; Srinivasan, G. *Stellar remnants*. 1997. ISBN 3-540-61520-2.

### 8.12.2 Physics

- *Black holes, white dwarfs, and neutron stars: the physics of compact objects*, Stuart L. Shapiro and Saul A. Teukolsky, New York: Wiley, 1983. ISBN 0-471-87317-9.

- Koester, D; Chanmugam, G (1990). "Physics of white dwarf stars". *Reports on Progress in Physics* **53** (7): 837–915. Bibcode:1990RPPh...53..837K. doi:10.1088/0034-4885/53/7/001.

- *White dwarf stars and the Chandrasekhar limit*, Dave Gentile, Master's thesis, DePaul University, 1995.

- Estimating Stellar Parameters from Energy Equipartition, sciencebits.com. Discusses how to find mass-radius relations and mass limits for white dwarfs using simple energy arguments.

### 8.12.3   Variability

- Winget, D E (1998). "Asteroseismology of white dwarf stars". *Journal of Physics: Condensed Matter* **10** (49): 11247–11261. Bibcode:1998JPCM...1011247W. doi:10.1088/0953-8984/10/49/014.

### 8.12.4   Magnetic field

- Wickramasinghe, D. T.; Ferrario, Lilia (2000). "Magnetism in Isolated and Binary White Dwarfs". *Publications of the Astronomical Society of the Pacific* **112** (773): 873–924. Bibcode:2000PASP..112..873W. doi:10.1086/316593.

### 8.12.5   Frequency

- Gibson, B. K.; Flynn, C (2001). "White Dwarfs and Dark Matter".*Science***292**(5525): 2211a. doi:10.1126/science PMID 11423620.

### 8.12.6   Observational

- Provencal, J. L.; Shipman, H. L.; Hog, Erik; Thejll, P. (1998). "Testing the White Dwarf Mass-Radius Relation withHipparcos". *The Astrophysical Journal* **494** (2): 759–767. Bibcode:1998ApJ...494..759P. doi:10.1086/305238.

- Gates, Evalyn; Gyuk, Geza; Harris, Hugh C.; Subbarao, Mark; Anderson, Scott; Kleinman, S. J.; Liebert, James; Brewington, Howard; et al. (2004). "Discovery of New Ultracool White Dwarfs in the Sloan Digital Sky Survey". *The Astrophysical Journal***612**(2): L129. arXiv:astro-ph/0405566. Bibcode:2004ApJ...612L.129G.doi:10.1086/

- Villanova University White Dwarf Catalogue WD, G. P. McCook and E. M. Sion.

- Dufour, P.; Liebert, J.; Fontaine, G.; Behara, N. (2007). "White dwarf stars with carbon atmospheres". *Nature* **450** (7169): 522–4. arXiv:0711.3227. Bibcode:2007Natur.450..522D. doi:10.1038/nature06318. PMID 18033290.

### 8.12.7   Images

- Astronomy Picture of the Day

  - NGC 2440: Cocoon of a New White Dwarf 2010 February 21
  - Dust and the Helix Nebula 2009 December 31
  - The Helix Nebula from La Silla Observatory 2009 March 3
  - IC 4406: A Seemingly Square Nebula 2008 July 27
  - A Nearby Supernova in Spiral Galaxy M100 2006 March 7
  - Astronomy Picture of the Day: White Dwarf Star Spiral 2005 June 1

The merger process of two co-orbiting white dwarfs produces gravitational waves

# Chapter 9

# Stellar evolution

| Mass (solar masses) | Time (years) | Spectral type |
| --- | --- | --- |
| 60 | 3 million | O3 |
| 30 | 11 million | O7 |
| 10 | 32 million | B4 |
| 3 | 370 million | A5 |
| 1.5 | 3 billion | F5 |
| 1 | 10 billion | G2 (Sun) |
| 0.1 | 1000s billions | M7 |

*Representative lifetimes of stars as a function of their masses*

**Stellar evolution** is the process by which a star changes during its lifetime. Depending on the mass of the star, this lifetime ranges from a few million years for the most massive to trillions of years for the least massive, which is considerably longer than the age of the universe. The table shows the lifetimes of stars as a function of their masses.[1] All stars are born from collapsing clouds of gas and dust, often called nebulae or molecular clouds. Over the course of millions of years, these protostars settle down into a state of equilibrium, becoming what is known as a main-sequence star.

Nuclear fusion powers a star for most of its life. Initially the energy is generated by the fusion of hydrogen atoms at the core of the main-sequence star. Later, as the preponderance of atoms at the core becomes helium, stars like the Sun begin to fuse hydrogen along a spherical shell surrounding the core. This process causes the star to gradually grow in size, passing through the subgiant stage until it reaches the red giant phase. Stars with at least half the mass of the Sun can also begin to generate energy through the fusion of helium at their core, whereas more-massive stars can fuse heavier elements along a series of concentric shells. Once a star like the Sun has exhausted its nuclear fuel, its core collapses into a dense white dwarf and the outer layers are expelled as a planetary nebula. Stars with around ten or more times the mass of the Sun can explode in a supernova as their inert iron cores collapse into an extremely dense neutron star or black hole. Although the universe is not old enough for any of the smallest red dwarfs to have reached the end of their lives, stellar models suggest they will slowly become brighter and hotter before running out of hydrogen fuel and becoming low-mass white dwarfs.[2]

Stellar evolution is not studied by observing the life of a single star, as most stellar changes occur too slowly to be detected, even over many centuries. Instead, astrophysicists come to understand how stars evolve by observing numerous stars at

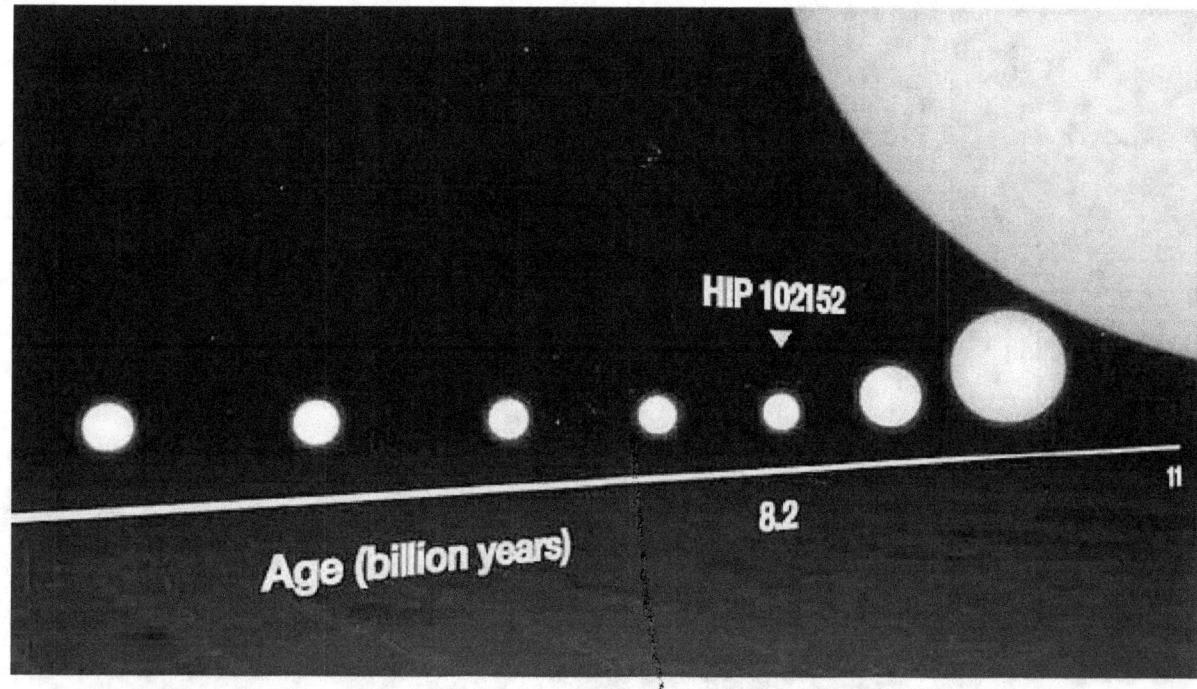

*The life cycle of a Sun-like star.*

various points in their lifetime, and by simulating stellar structure using computer models.

In June 2015, astronomers reported evidence for Population III stars in the Cosmos Redshift 7 galaxy at $z = 6.60$. Such stars are likely to have existed in the very early universe (i.e., at high redshift), and may have started the production of chemical elements heavier than hydrogen that are needed for the later formation of planets and life as we know it.[3][4]

## 9.1   Birth of a star

Main article: Star formation

### 9.1.1   Protostar

Stellar evolution starts with the gravitational collapse of a giant molecular cloud. Typical giant molecular clouds are roughly 100 light-years ($9.5 \times 10^{14}$ km) across and contain up to 6,000,000 solar masses ($1.2 \times 10^{37}$ kg). As it collapses, a giant molecular cloud breaks into smaller and smaller pieces. In each of these fragments, the collapsing gas releases gravitational potential energy as heat. As its temperature and pressure increase, a fragment condenses into a rotating sphere of superhot gas known as a protostar.[5]

A protostar continues to grow by accretion of gas and dust from the molecular cloud, becoming a pre-main-sequence star as it reaches its final mass. Further development is determined by its mass. (Mass is compared to the mass of the Sun: 1.0 M☉ ($2.0 \times 10^{30}$ kg) means 1 solar mass.)

Protostars are encompassed in dust, and are thus more readily visible at infrared wavelengths. Observations from the Wide-field Infrared Survey Explorer (WISE) have been especially important for unveiling numerous Galactic protostars and their parent star clusters.[6][7]

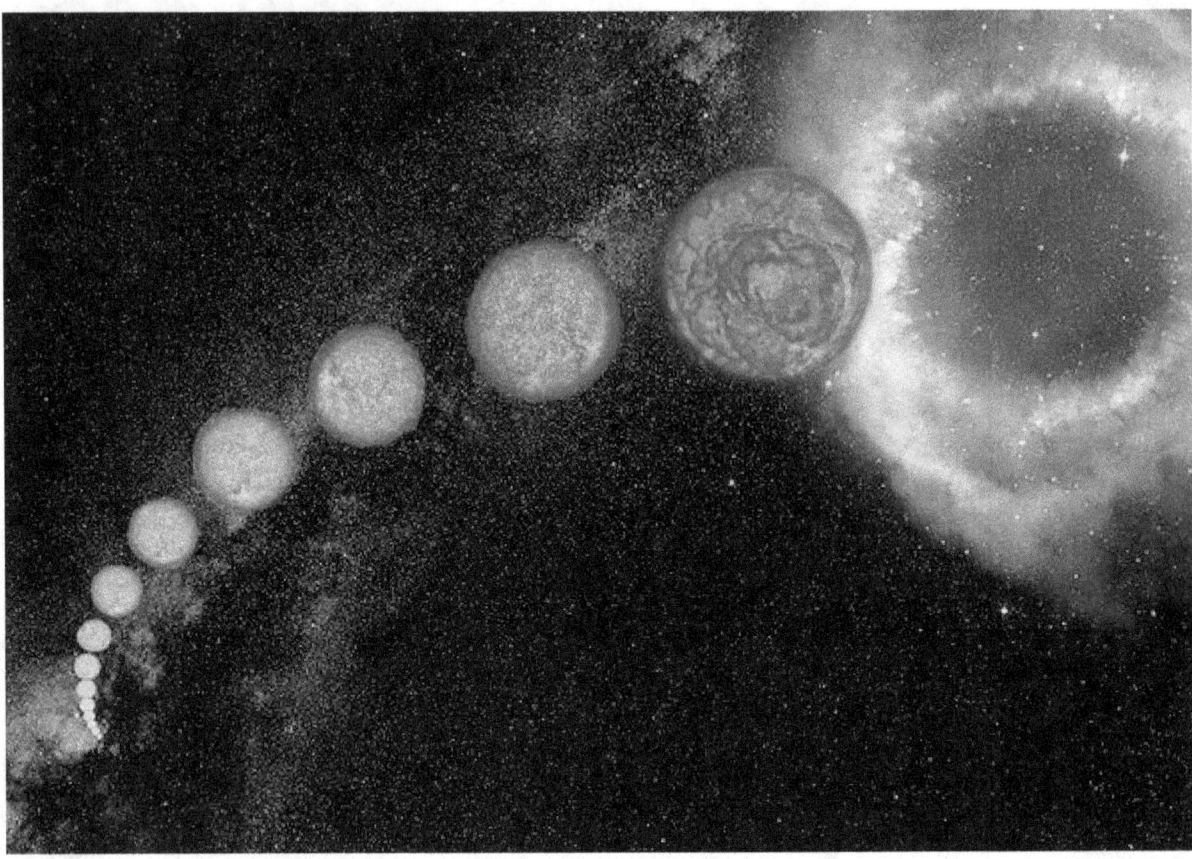

*Artist's depiction of the life cycle of a Sun-like star, starting as a main-sequence star at lower left then expanding through the subgiant and giant phases, until its outer envelope is expelled to form a planetary nebula at upper right.*

### 9.1.2   Brown dwarfs and sub-stellar objects

Protostars with masses less than roughly 0.08 M⊙ ($1.6\times10^{29}$ kg) never reach temperatures high enough for nuclear fusion of hydrogen to begin. These are known as brown dwarfs. The International Astronomical Union defines brown dwarfs as stars massive enough to fuse deuterium at some point in their lives (13 Jupiter masses ($MJ$), $2.5 \times 10^{28}$ kg, or 0.0125 M⊙). Objects smaller than 13 $MJ$ are classified as sub-brown dwarfs (but if they orbit around another stellar object they are classified as planets).[8] Both types, deuterium-burning and not, shine dimly and die away slowly, cooling gradually over hundreds of millions of years.

### 9.1.3   Hydrogen fusion

For a more-massive protostar, the core temperature will eventually reach 10 million kelvin, initiating the proton–proton chain reaction and allowing hydrogen to fuse, first to deuterium and then to helium. In stars of slightly over 1 M⊙ ($2.0\times10^{30}$ kg), the carbon–nitrogen–oxygen fusion reaction (CNO cycle) contributes a large portion of the energy generation. The onset of nuclear fusion leads relatively quickly to a hydrostatic equilibrium in which energy released by the core exerts a "radiation pressure" balancing the weight of the star's matter, preventing further gravitational collapse. The star thus evolves rapidly to a stable state, beginning the main-sequence phase of its evolution.

A new star will sit at a specific point on the main sequence of the Hertzsprung–Russell diagram, with the main-sequence spectral type depending upon the mass of the star. Small, relatively cold, low-mass red dwarfs fuse hydrogen slowly and will remain on the main sequence for hundreds of billions of years or longer, whereas massive, hot O-type stars will leave the main sequence after just a few million years. A mid-sized yellow dwarf star, like the Sun, will remain on the main sequence for about 10 billion years. The Sun is thought to be in the middle of its main sequence lifespan.

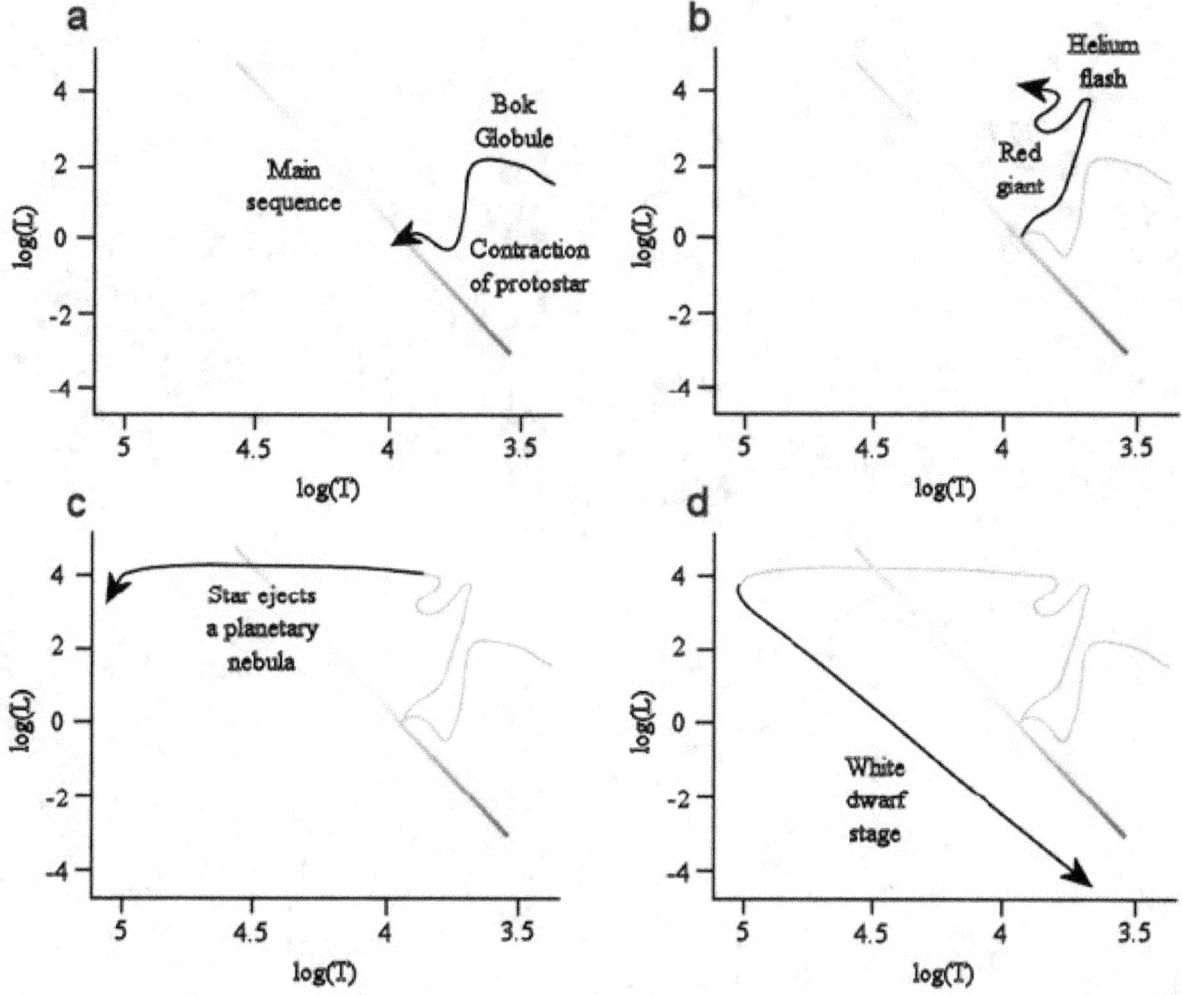

*Schematic of stellar evolution.*

WR
LBV
YHG
BSG
RSG
AGB
RG

The evolutionary tracks of stars with different initial masses on the Hertzsprung–Russell diagram. The tracks start once the star has evolved to the main sequence and stop when fusion stops (for massive stars) and at the end of the red giant branch (for stars 1 $M_\odot$ and less).[9]

A yellow track is shown for the Sun, which will become a red giant after its main-sequence phase ends before expanding further along the asymptotic giant branch, which will be the last phase in which the Sun undergoes fusion.

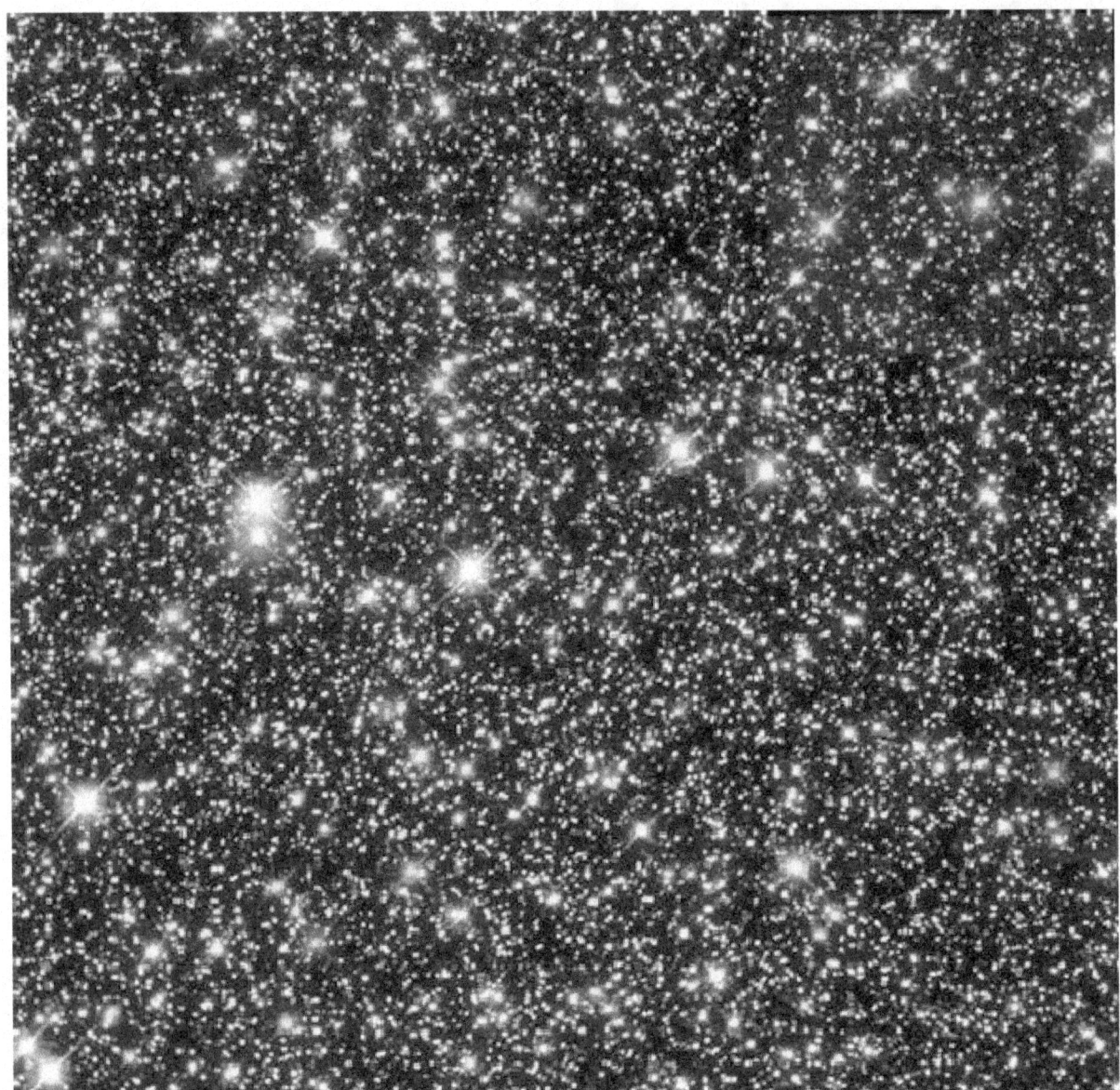

*A dense starfield in Sagittarius*

## 9.2 Mature stars

Eventually the core exhausts its supply of hydrogen and the star begins to evolve off of the main sequence. Without the outward pressure generated by the fusion of hydrogen to counteract the force of gravity the core contracts until either electron degeneracy pressure becomes sufficient to oppose gravity or the core becomes hot enough (around 100 MK) for helium fusion to begin. Which of these happens first depends upon the star's mass.

### 9.2.1 Low-mass stars

What happens after a low-mass star ceases to produce energy through fusion has not been directly observed; the universe is around 13.8 billion years old, which is less time (by several orders of magnitude, in some cases) than it takes for fusion to cease in such stars.

Recent astrophysical models suggest that red dwarfs of 0.1 $M\odot$ may stay on the main sequence for some six to twelve trillion years, gradually increasing in both temperature and luminosity, and take several hundred billion more to collapse, slowly, into a white dwarf.[10][11] Such stars will not become red giants as they are fully convective and will not develop a degenerate helium core with a shell burning hydrogen. Instead, hydrogen fusion will proceed until almost the whole star is helium.

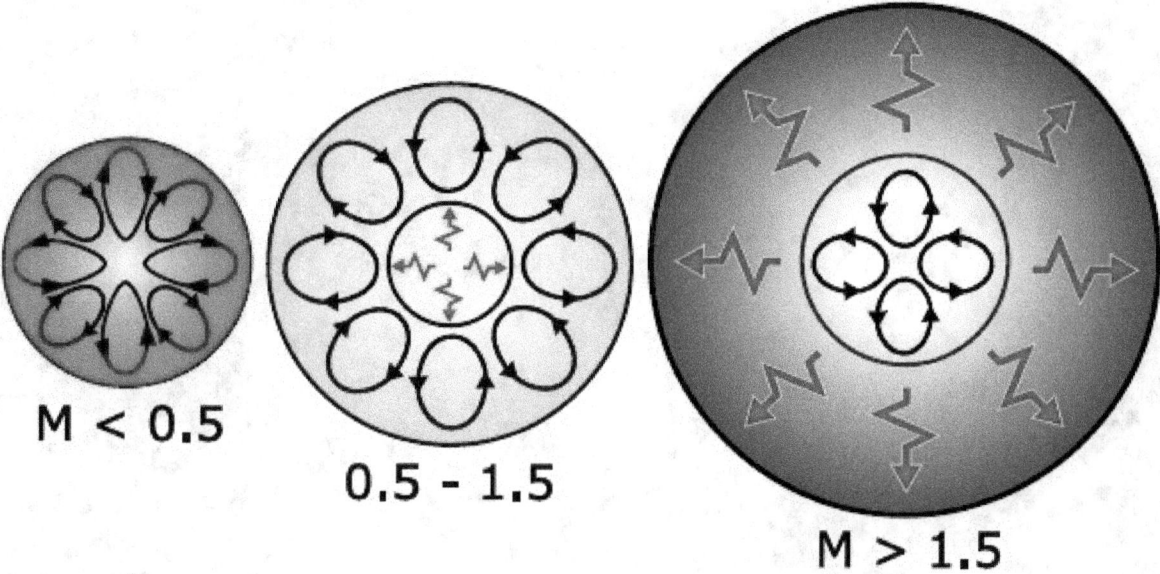

*Internal structures of main-sequence stars, convection zones with arrowed cycles and radiative zones with red flashes. To the left a* **low-mass** *red dwarf, in the center a* **mid-sized** *yellow dwarf and at the right a* **massive** *blue-white main-sequence star.*

Slightly more massive stars do expand into red giants, but their helium cores are not massive enough to reach the temperatures required for helium fusion so they never reach the tip of the red giant branch. When hydrogen shell burning finishes, these stars move directly off the red giant branch like a post AGB star, but at lower luminosity, to become a white dwarf.[2] A star of about 0.5 $M\odot$ will be able to reach temperatures high enough to fuse helium, and these "mid-sized" stars go on to further stages of evolution beyond the red giant branch.[12]

### 9.2.2 Mid-sized stars

Stars of roughly 0.5–10 $M\odot$ become red giants, which are large non-main-sequence stars of stellar classification K or M. Red giants lie along the right edge of the Hertzsprung–Russell diagram due to their red color and large luminosity. Examples include Aldebaran in the constellation Taurus and Arcturus in the constellation of Boötes. Red giants all have inert cores with hydrogen-burning shells: concentric layers atop the core that are still fusing hydrogen into helium.

Mid-sized stars are red giants during two different phases of their post-main-sequence evolution: red-giant-branch stars, whose inert cores are made of helium, and asymptotic-giant-branch stars, whose inert cores are made of carbon. Asymptotic-giant-branch stars have helium-burning shells inside the hydrogen-burning shells, whereas red-giant-branch stars have hydrogen-burning shells only.[13] In either case, the accelerated fusion in the hydrogen-containing layer immediately over the core causes the star to expand. This lifts the outer layers away from the core, reducing the gravitational pull on them, and they expand faster than the energy production increases. This causes the outer layers of the star to cool, which causes the star to become redder than it was on the main sequence.

**Red-giant-branch phase**

The red-giant-branch phase of a star's life follows the main sequence. Initially, the cores of red-giant-branch stars collapse, as the internal pressure of the core is insufficient to balance gravity. This gravitational collapse releases energy, heating

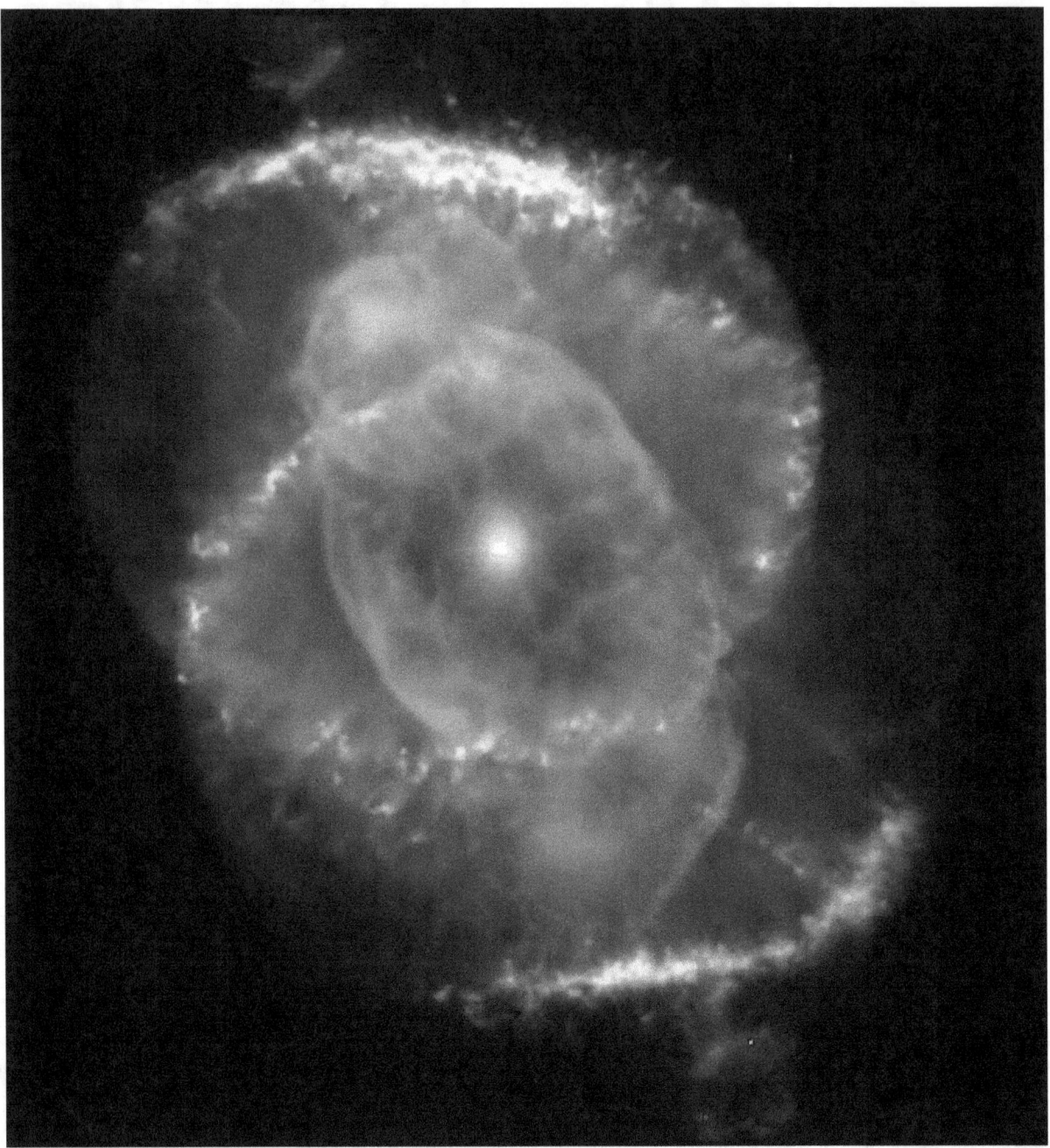

*The Cat's Eye Nebula, a planetary nebula formed by the death of a star with about the same mass as the Sun*

concentric shells immediately outside the inert helium core so that hydrogen fusion continues in these shells. The core of a red-giant-branch star of up to a few solar masses stops collapsing when it is dense enough to be supported by electron degeneracy pressure. Once this occurs, the core reaches hydrostatic equilibrium: the electron degeneracy pressure is sufficient to balance gravitational pressure.[14] The core's gravity compresses the hydrogen in the layer immediately above it, causing it to fuse faster than hydrogen would fuse in a main-sequence star of the same mass. This in turn causes the star to become more luminous (from 1,000–10,000 times brighter) and expand; the degree of expansion outstrips the increase in luminosity, causing the effective temperature to decrease.

The expanding outer layers of the star are convective, with the material being mixed by turbulence from near the fusing regions up to the surface of the star. For all but the lowest-mass stars, the fused material has remained deep in the stellar

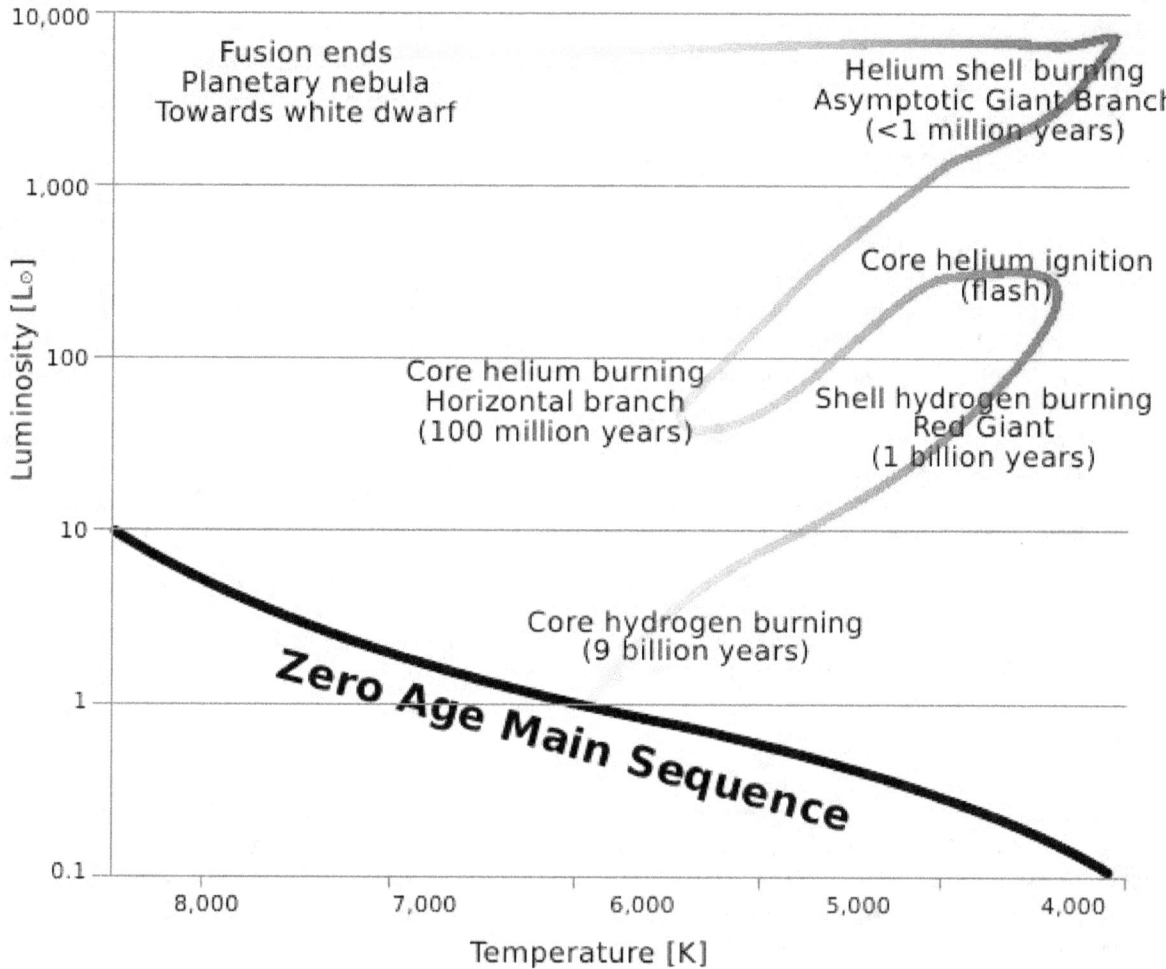

*Evolution of a Sun-like star*

interior prior to this point, so the convecting envelope makes fusion products visible at the star's surface for the first time. At this stage of evolution, the results are subtle, with the largest effects, alterations to the isotopes of hydrogen and helium, being unobservable. The effects of the CNO cycle appear at the surface, with lower $^{12}C/^{13}C$ ratios and altered proportions of carbon and nitrogen. These are detectable with spectroscopy and have been measured for many evolved stars.

As the hydrogen in the shell around the core is consumed, the helium core grows. Eventually, the electrons in the helium core of stars less than about 2.5 $M\odot$ become degenerate, preventing the helium core from contracting further.[15] Later in the red giant phase, the cores of stars more massive than 0.5 $M\odot$ get hot enough to start helium fusion by the triple-alpha process. In stars of approximately one solar mass, it can take a billion years or more for the core to reach helium ignition temperatures.[16]

If the core is largely supported by electron degeneracy pressure, helium fusion will ignite on a timescale of days in a helium flash. In more massive stars, the ignition of helium fusion occurs relatively slowly with no flash.[17] The nuclear power released during the helium flash is very large, on the order of $10^8$ times the luminosity of the Sun for a few days[15] and $10^{11}$ times the luminosity of the Sun (roughly the luminosity of the Milky Way Galaxy) for a few seconds.[18] However, the energy is absorbed by the stellar envelope and thus cannot be seen from outside the star.[15][18] The energy released by helium fusion causes the core to expand, so that hydrogen fusion in the overlying layers slows and total energy generation decreases. The star contracts, although not all the way to the main sequence, and it migrates to the horizontal branch on the Hertzsprung–Russell diagram, gradually shrinking in radius and increasing its surface temperature. Core helium flash stars evolve to the red end of the horizontal branch but do not migrate to higher temperatures before they gain a degenerate carbon-oxygen core and start helium shell burning. These stars are often observed as a red clump of stars in the colour-magnitude diagram of a cluster, hotter and less luminous than the red giants. Higher-mass stars with larger helium cores move along the horizontal branch to higher temperatures, some becoming unstable pulsating stars in the yellow instability strip (RR Lyrae variables), whereas some become even hotter and can form a blue tail or blue hook to the horizontal branch. The exact morphology of the horizontal branch depends on parameters such as metallicity, age, and helium content, but the exact details are still being modelled.[19]

**Asymptotic-giant-branch phase**

After a star has consumed the helium at the core, hydrogen and helium fusion continues in shells around a hot core of carbon and oxygen. The star follows the asymptotic giant branch on the Hertzsprung–Russell diagram, paralleling the original red giant evolution, but with even faster energy generation (which lasts for a shorter time).[20] Although helium is being burnt in a shell, the majority of the energy is produced by hydrogen burning in a shell further from the core of the star. Helium from these hydrogen burning shells drops towards the center of the star and periodically the energy output from the helium shell increases dramatically. This is known as a thermal pulse and they occur towards the end of the asymptotic-giant-branch phase, sometimes even into the post-asymptotic-giant-branch phase. Depending on mass and composition, there may be several to hundreds of thermal pulses.

There is a phase on the ascent of the asymptotic-giant-branch where a deep convective zone forms and can bring carbon from the core to the surface. This is known as the second dredge up, and in some stars there may even be a third dredge up. In this way a carbon star is formed, very cool and strongly reddened stars showing strong carbon lines in their spectra. A process known as hot bottom burning may convert carbon into oxygen and nitrogen before it can be dredged to the surface, and the interaction between these processes determines the observed luminosities and spectra of carbon stars in particular clusters.[21]

Another well known class of asymptotic-giant-branch stars are the Mira variables, which pulsate with well-defined periods of tens to hundreds of days and large amplitudes up to about 10 magnitudes (in the visual, total luminosity changes by a much smaller amount). In more-massive stars the stars become more luminous and the pulsation period is longer, leading to enhanced mass loss, and the stars become heavily obscured at visual wavelengths. These stars can be observed as OH/IR stars, pulsating in the infra-red and showing OH maser activity. These stars are clearly oxygen rich, in contrast to the carbon stars, but both must be produced by dredge ups.

These mid-range stars ultimately reach the tip of the asymptotic-giant-branch and run out of fuel for shell burning. They are not sufficiently massive to start full-scale carbon fusion, so they contract again, going through a period of post-asymptotic-giant-branch superwind to produce a planetary nebula with an extremely hot central star. The central star then cools to a white dwarf. The expelled gas is relatively rich in heavy elements created within the star and may be particularly oxygen or carbon enriched, depending on the type of the star. The gas builds up in an expanding shell called a circumstellar envelope and cools as it moves away from the star, allowing dust particles and molecules to form. With the high infrared energy input from the central star, ideal conditions are formed in these circumstellar envelopes for maser excitation.

It is possible for thermal pulses to be produced once post-asymptotic-giant-branch evolution has begun, producing a variety of unusual and poorly understood stars known as born-again asymptotic-giant-branch stars.[22] These may result in extreme horizontal-branch stars (subdwarf B stars), hydrogen deficient post-asymptotic-giant-branch stars, variable planetary nebula central stars, and R Coronae Borealis variables.

### 9.2.3   Massive stars

In massive stars, the core is already large enough at the onset of the hydrogen burning shell that helium ignition will occur before electron degeneracy pressure has a chance to become prevalent. Thus, when these stars expand and cool, they do not brighten as much as lower-mass stars; however, they were much brighter than lower-mass stars to begin with, and are thus still brighter than the red giants formed from less-massive stars. These stars are unlikely to survive as red supergiants;

*The Crab Nebula, the shattered remnants of a star which exploded as a supernova, the light of which reached Earth in 1054 AD*

instead they will destroy themselves as type II supernovas.

Extremely massive stars (more than approximately 40 $M\odot$), which are very luminous and thus have very rapid stellar winds, lose mass so rapidly due to radiation pressure that they tend to strip off their own envelopes before they can expand to become red supergiants, and thus retain extremely high surface temperatures (and blue-white color) from their main-sequence time onwards. The largest stars of the current generation are about 100-150 $M\odot$ because the outer layers would be expelled by the extreme radiation. Although lower-mass stars normally do not burn off their outer layers so rapidly, they can likewise avoid becoming red giants or red supergiants if they are in binary systems close enough so that the companion star strips off the envelope as it expands, or if they rotate rapidly enough so that convection extends all the way from the core to the surface, resulting in the absence of a separate core and envelope due to thorough mixing.[23]

The core grows hotter and denser as it gains material from fusion of hydrogen at the base of the envelope. In all massive stars, electron degeneracy pressure is insufficient to halt collapse by itself, so as each major element is consumed in the center, progressively heavier elements ignite, temporarily halting collapse. If the core of the star is not too massive (less than approximately 1.4 $M\odot$, taking into account mass loss that has occurred by this time), it may then form a white dwarf (possibly surrounded by a planetary nebula) as described above for less-massive stars, with the difference that the white dwarf is composed chiefly of oxygen, neon, and magnesium.

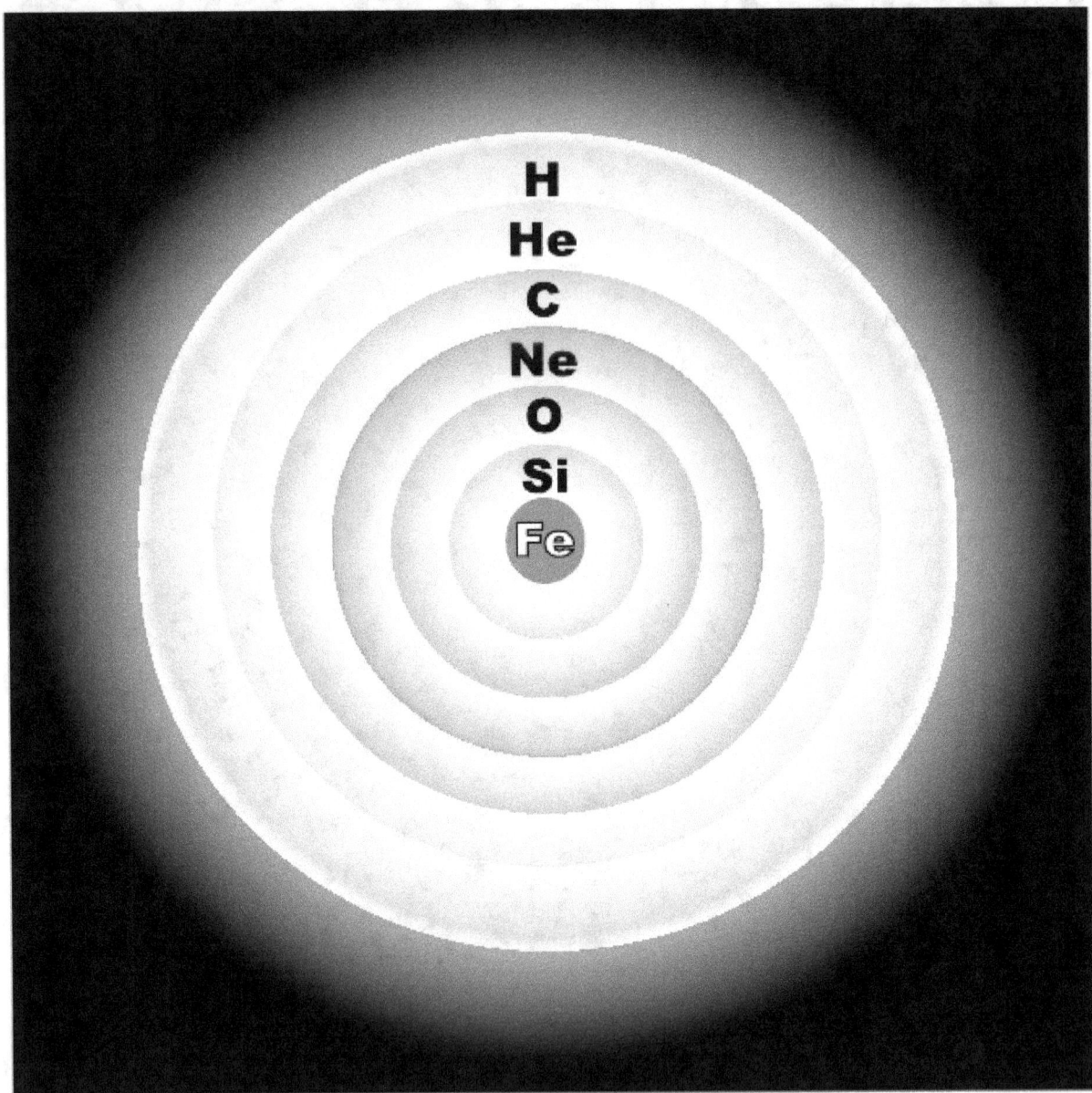

*The onion-like layers of a massive, evolved star just before core collapse. (Not to scale.)*

Above a certain mass (estimated at approximately 2.5 $M\odot$ and whose star's progenitor was around 10 $M\odot$), the core will reach the temperature (approximately 1.1 gigakelvins) at which neon partially breaks down to form oxygen and helium, the latter of which immediately fuses with some of the remaining neon to form magnesium; then oxygen fuses to form sulfur, silicon, and smaller amounts of other elements. Finally, the temperature gets high enough that any nucleus can be partially broken down, most commonly releasing an alpha particle (helium nucleus) which immediately fuses with another nucleus, so that several nuclei are effectively rearranged into a smaller number of heavier nuclei, with net release of energy because the addition of fragments to nuclei exceeds the energy required to break them off the parent nuclei.

A star with a core mass too great to form a white dwarf but insufficient to achieve sustained conversion of neon to oxygen and magnesium, will undergo core collapse (due to electron capture) before achieving fusion of the heavier elements.[24] Both heating and cooling caused by electron capture onto minor constituent elements (such as aluminum and sodium) prior to collapse may have a significant impact on total energy generation within the star shortly before collapse.[25] This may produce a noticeable effect on the abundance of elements and isotopes ejected in the subsequent supernova.

### Supernova

Main article: Supernova

Once the nucleosynthesis process arrives at iron-56, the continuation of this process consumes energy (the addition of fragments to nuclei releases less energy than required to break them off the parent nuclei). If the mass of the core exceeds the Chandrasekhar limit, electron degeneracy pressure will be unable to support its weight against the force of gravity, and the core will undergo sudden, catastrophic collapse to form a neutron star or (in the case of cores that exceed the Tolman-Oppenheimer-Volkoff limit), a black hole. Through a process that is not completely understood, some of the gravitational potential energy released by this core collapse is converted into a Type Ib, Type Ic, or Type II supernova. It is known that the core collapse produces a massive surge of neutrinos, as observed with supernova SN 1987A. The extremely energetic neutrinos fragment some nuclei; some of their energy is consumed in releasing nucleons, including neutrons, and some of their energy is transformed into heat and kinetic energy, thus augmenting the shock wave started by rebound of some of the infalling material from the collapse of the core. Electron capture in very dense parts of the infalling matter may produce additional neutrons. Because some of the rebounding matter is bombarded by the neutrons, some of its nuclei capture them, creating a spectrum of heavier-than-iron material including the radioactive elements up to (and likely beyond) uranium.[26] Although non-exploding red giants can produce significant quantities of elements heavier than iron using neutrons released in side reactions of earlier nuclear reactions, the abundance of elements heavier than iron (and in particular, of certain isotopes of elements that have multiple stable or long-lived isotopes) produced in such reactions is quite different from that produced in a supernova. Neither abundance alone matches that found in the Solar System, so both supernovae and ejection of elements from red giants are required to explain the observed abundance of heavy elements and isotopes thereof.

The energy transferred from collapse of the core to rebounding material not only generates heavy elements, but provides for their acceleration well beyond escape velocity, thus causing a Type Ib, Type Ic, or Type II supernova. Note that current understanding of this energy transfer is still not satisfactory; although current computer models of Type Ib, Type Ic, and Type II supernovae account for part of the energy transfer, they are not able to account for enough energy transfer to produce the observed ejection of material.[27]

Some evidence gained from analysis of the mass and orbital parameters of binary neutron stars (which require two such supernovae) hints that the collapse of an oxygen-neon-magnesium core may produce a supernova that differs observably (in ways other than size) from a supernova produced by the collapse of an iron core.[28]

The most-massive stars that exist today may be completely destroyed by a supernova with an energy greatly exceeding its gravitational binding energy. This rare event, caused by pair-instability, leaves behind no black hole remnant.[29] In the past history of the universe, some stars were even larger than the largest that exists today, and they would immediately collapse into a black hole at the end of their lives, due to photodisintegration.

## 9.3   Stellar remnants

After a star has burned out its fuel supply, its remnants can take one of three forms, depending on the mass during its lifetime.

### 9.3.1   White and black dwarfs

Main articles: White dwarf and Black dwarf

For a star of 1 $M\odot$, the resulting white dwarf is of about 0.6 $M\odot$, compressed into approximately the volume of the Earth. White dwarfs are stable because the inward pull of gravity is balanced by the degeneracy pressure of the star's electrons,

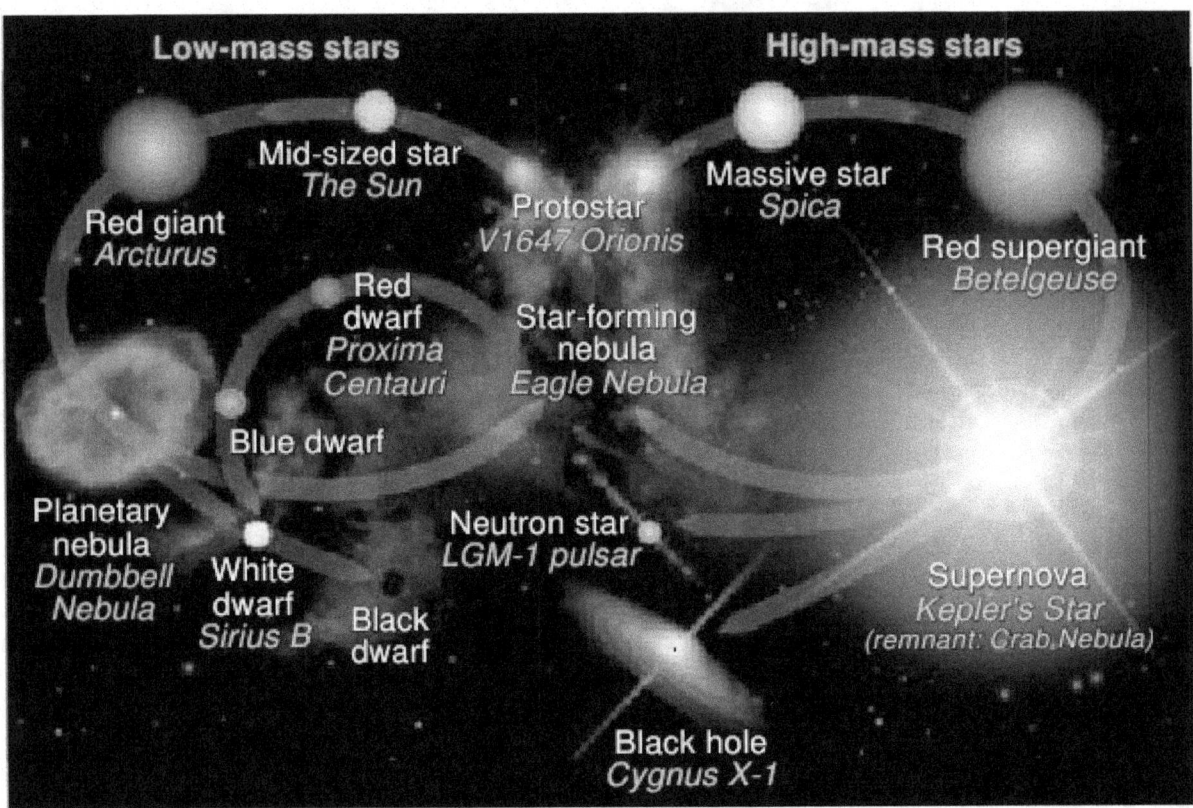

*Stellar evolution of low-mass (left cycle) and high-mass (right cycle) stars, with examples in italics*

a consequence of the Pauli exclusion principle. Electron degeneracy pressure provides a rather soft limit against further compression; therefore, for a given chemical composition, white dwarfs of higher mass have a smaller volume. With no fuel left to burn, the star radiates its remaining heat into space for billions of years.

A white dwarf is very hot when it first forms, more than 100,000 K at the surface and even hotter in its interior. It is so hot that a lot of its energy is lost in the form of neutrinos for the first 10 million years of its existence, but will have lost most of its energy after a billion years.[30]

The chemical composition of the white dwarf depends upon its mass. A star of a few solar masses will ignite carbon fusion to form magnesium, neon, and smaller amounts of other elements, resulting in a white dwarf composed chiefly of oxygen, neon, and magnesium, provided that it can lose enough mass to get below the Chandrasekhar limit (see below), and provided that the ignition of carbon is not so violent as to blow the star apart in a supernova.[31] A star of mass on the order of magnitude of the Sun will be unable to ignite carbon fusion, and will produce a white dwarf composed chiefly of carbon and oxygen, and of mass too low to collapse unless matter is added to it later (see below). A star of less than about half the mass of the Sun will be unable to ignite helium fusion (as noted earlier), and will produce a white dwarf composed chiefly of helium.

In the end, all that remains is a cold dark mass sometimes called a black dwarf. However, the universe is not old enough for any black dwarfs to exist yet.

If the white dwarf's mass increases above the Chandrasekhar limit, which is 1.4 $M_\odot$ for a white dwarf composed chiefly of carbon, oxygen, neon, and/or magnesium, then electron degeneracy pressure fails due to electron capture and the star collapses. Depending upon the chemical composition and pre-collapse temperature in the center, this will lead either to collapse into a neutron star or runaway ignition of carbon and oxygen. Heavier elements favor continued core collapse, because they require a higher temperature to ignite, because electron capture onto these elements and their fusion products is easier; higher core temperatures favor runaway nuclear reaction, which halts core collapse and leads to a Type Ia supernova.[32] These supernovae may be many times brighter than the Type II supernova marking the death of a massive star, even though the latter has the greater total energy release. This inability to collapse means that no white dwarf more massive than approximately 1.4 $M\odot$ can exist (with a possible minor exception for very rapidly spinning white dwarfs, whose centrifugal force due to rotation partially counteracts the weight of their matter). Mass transfer in a binary system may cause an initially stable white dwarf to surpass the Chandrasekhar limit.

If a white dwarf forms a close binary system with another star, hydrogen from the larger companion may accrete around and onto a white dwarf until it gets hot enough to fuse in a runaway reaction at its surface, although the white dwarf remains below the Chandrasekhar limit. Such an explosion is termed a nova.

### 9.3.2  Neutron stars

Main article: Neutron star

When a stellar core collapses, the pressure causes electron capture, thus converting the great majority of the protons into neutrons. The electromagnetic forces keeping separate nuclei apart are gone (proportionally, if nuclei were the size of dust mites, atoms would be as large as football stadiums), and most of the core of the star becomes a dense ball of contiguous neutrons (in some ways like a giant atomic nucleus), with a thin overlying layer of degenerate matter (chiefly iron unless matter of different composition is added later). The neutrons resist further compression by the Pauli Exclusion Principle, in a way analogous to electron degeneracy pressure, but stronger.

These stars, known as neutron stars, are extremely small—on the order of radius 10 km, no bigger than the size of a large city—and are phenomenally dense. Their period of rotation shortens dramatically as the stars shrink (due to conservation of angular momentum); observed rotational periods of neutron stars range from about 1.5 milliseconds (over 600 revolutions per second) to several seconds.[33] When these rapidly rotating stars' magnetic poles are aligned with the Earth, we detect a pulse of radiation each revolution. Such neutron stars are called pulsars, and were the first neutron stars to be discovered. Though electromagnetic radiation detected from pulsars is most often in the form of radio waves, pulsars have also been detected at visible, X-ray, and gamma ray wavelengths.[34]

### 9.3.3  Black holes

Main article: Black hole

If the mass of the stellar remnant is high enough, the neutron degeneracy pressure will be insufficient to prevent collapse below the Schwarzschild radius. The stellar remnant thus becomes a black hole. The mass at which this occurs is not known with certainty, but is currently estimated at between 2 and 3 $M\odot$.

Black holes are predicted by the theory of general relativity. According to classical general relativity, no matter or information can flow from the interior of a black hole to an outside observer, although quantum effects may allow deviations from this strict rule. The existence of black holes in the universe is well supported, both theoretically and by astronomical observation.

Because the core-collapse supernova mechanism itself is imperfectly understood, it is still not known whether it is possible for a star to collapse directly to a black hole without producing a visible supernova, or whether some supernovae initially form unstable neutron stars which then collapse into black holes; the exact relation between the initial mass of the star and the final remnant is also not completely certain. Resolution of these uncertainties requires the analysis of more supernovae and supernova remnants.

## 9.4  Models

A stellar evolutionary model is a mathematical model that can be used to compute the evolutionary phases of a star from its formation until it becomes a remnant. The mass and chemical composition of the star are used as the inputs, and the luminosity and surface temperature are the only constraints. The model formulae are based upon the physical understanding of the star, usually under the assumption of hydrostatic equilibrium. Extensive computer calculations are then run to determine the changing state of the star over time, yielding a table of data that can be used to determine the evolutionary track of the star across the Hertzsprung–Russell diagram, along with other evolving properties.[35] Accurate

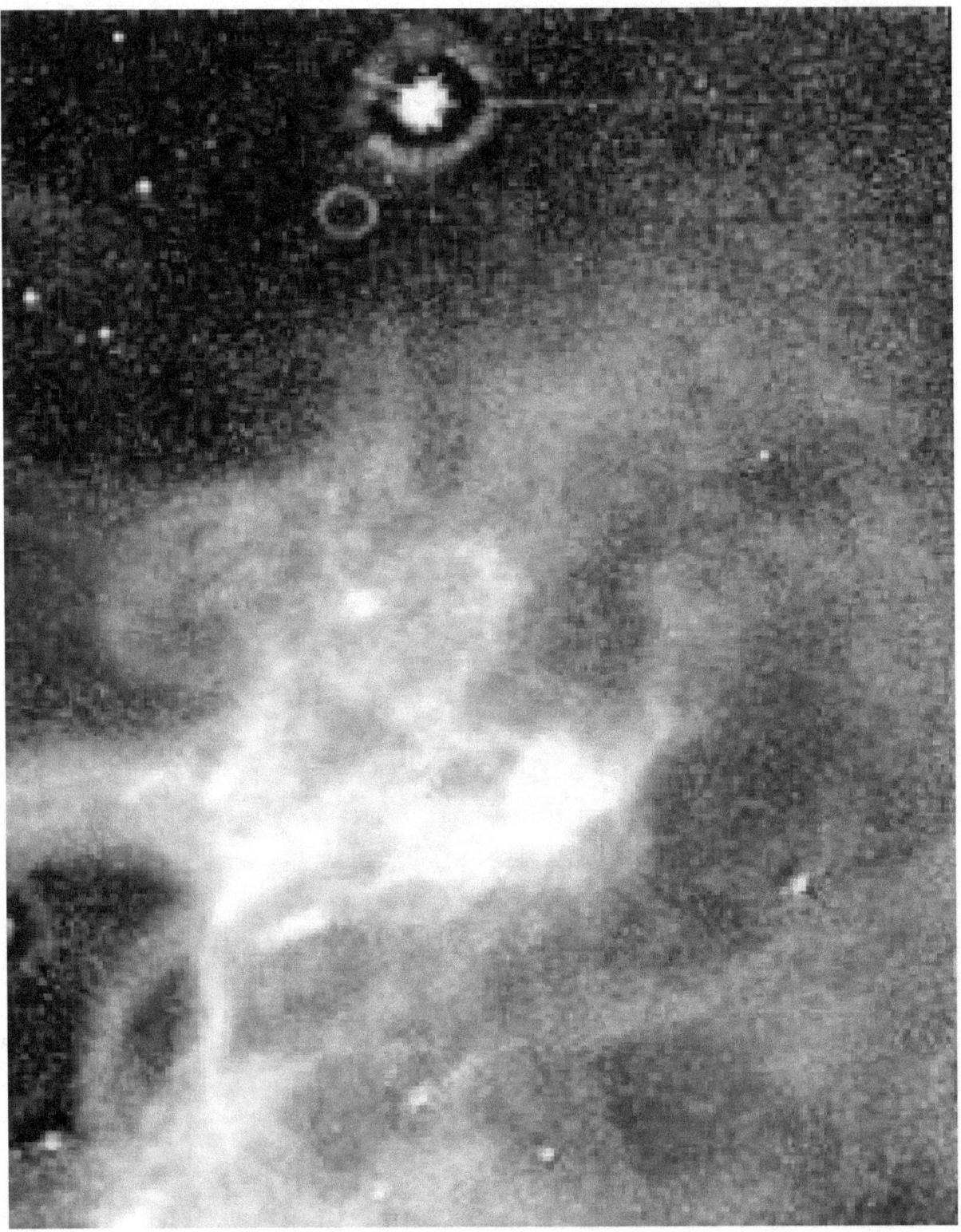

*Bubble-like shock wave still expanding from a supernova explosion 15,000 years ago.*

models can be used to estimate the current age of a star by comparing its physical properties with those of stars along a matching evolutionary track.[36]

## 9.5 See also

- Galaxy formation and evolution

- Nucleosynthesis

- Standard Solar Model

- Stellar populations (metallicity)

- Stellar rotation#After formation - Rotations slow as stars age

- Timeline of stellar astronomy

## 9.6 Further reading

- Astronomy 606 (Stellar Structure and Evolution) lecture notes, Cole Miller, Department of Astronomy, University of Maryland

- Astronomy 162, Unit 2 (The Structure & Evolution of Stars) lecture notes, Richard W. Pogge, Department of Astronomy, Ohio State University

- Hansen, Carl J.; Kawaler, Steven D.; Trimble, Virginia (2004). *Stellar interiors: physical principles, structure, and evolution* (2nd ed.). Springer-Verlag. ISBN 0-387-20089-4

- Prialnik, Dina (2000). *An Introduction to the Theory of Stellar Structure and Evolution*. Cambridge University Press. ISBN 0-521-65065-8.

- Ryan, Sean G.; Norton, Andrew J. (2010). *Stellar Evolution and Nucleosynthesis*. Cambridge University Press. p. 125. ISBN 0521133203.

## 9.7 External links

- Stellar evolution simulator

- Pisa Stellar Models

## 9.8 References

[1] Bertulani, Carlos A. (2013). *Nuclei in the Cosmos*. World Scientific. ISBN 978-981-4417-66-2.

[2] Laughlin, Gregory; Bodenheimer, Peter; Adams, Fred C. (1997). "The End of the Main Sequence". *The Astrophysical Journal* **482**: 420–432. Bibcode:1997ApJ...482..420L. doi:10.1086/304125.

[3] Sobral, David; Matthee, Jorryt; Darvish, Behnam; Schaerer, Daniel; Mobasher, Bahram; Röttgering, Huub J. A.; Santos, Sérgio; Hemmati, Shoubaneh (4 June 2015). "Evidence For POPIII-Like Stellar Populations In The Most Luminous LYMAN-α Emitters At The Epoch Of Re-Ionisation: Spectroscopic Confirmation" (PDF). *The Astrophysical Journal*. doi:10.1088/0004-637x/808/2/139. Retrieved 17 June 2015.

[4] Overbye, Dennis (17 June 2015). "Astronomers Report Finding Earliest Stars That Enriched Cosmos". *New York Times*. Retrieved 17 June 2015.

[5] Prialnik (2000, Chapter 10)

[6] "Wide-field Infrared Survey Explorer Mission". NASA.

[7]  Majaess, D. (2013). *Discovering protostars and their host clusters via WISE*. ApSS, 344, 1 (*VizieR catalog*)

[8]  "Working Group on Extrasolar Planets: Definition of a "Planet"". *IAU position statement*. 2003-02-28. Archived from the original on February 4, 2012. Retrieved 2012-05-30.

[9]  Prialnik (2000, Fig. 8.19, p. 174)

[10]  "Why the Smallest Stars Stay Small". *Sky & Telescope* (22). November 1997.

[11]  Adams, F. C.; P. Bodenheimer; G. Laughlin (2005). "M dwarfs: planet formation and long term evolution". *Astronomische Nachrichten* **326** (10): 913–919. Bibcode:2005AN....326..913A. doi:10.1002/asna.200510440.

[12]  Ryan & Norton (2010, p. 114)

[13]  Hansen, Kawaler & Trimble (2004, pp. 55–56)

[14]  Hansen, Kawaler & Trimble (2004, pp. 62–63)

[15]  Ryan & Norton (2010, p. 115)

[16]  Prialnik (2000, p. 155, Table 8.4)

[17]  Ryan & Norton (2010, p. 125)

[18]  Prialnik (2000, p. 151)

[19]  Gratton, R. G.; Carretta, E.; Bragaglia, A.; Lucatello, S.; d'Orazi, V. (2010). "The second and third parameters of the horizontal branch in globular clusters". *Astronomy and Astrophysics* **517**: A81. arXiv:1004.3862. Bibcode:2010A&A...517A..81G. doi:10.1051/0004-6361/200912572.

[20]  Sackmann, I. -J.; Boothroyd, A. I.; Kraemer, K. E. (1993). "Our Sun. III. Present and Future". *The Astrophysical Journal* **418**: 457. Bibcode:1993ApJ...418..457S. doi:10.1086/173407.

[21]  van Loon; Zijlstra; Whitelock; Peter te Lintel Hekkert; Chapman; Cecile Loup; Groenewegen; Waters; Trams (1998). "Obscured Asymptotic Giant Branch stars in the Magellanic Clouds IV. Carbon stars and OH/IR stars". *Astronomy and Astrophysics* **329**: 169–85. arXiv:astro-ph/9709119v1. CiteSeerX: 10.1.1.389.3269.

[22]  Bibcode: 1991IAUS..145..363H

[23]  D. Vanbeveren; De Loore, C.; Van Rensbergen, W. (1998). "Massive stars" (PDF). *The Astronomy and Astrophysics Review* **9** (1–2): 63–152. Bibcode:1998A&ARv...9...63V. doi:10.1007/s001590050015. Archived from the original (PDF) on 2009-03-27.

[24]  Ken'ichi Nomoto (1987). "Evolution of 8–10 $M\odot$ stars toward electron capture supernovae. II – Collapse of an O + Ne + Mg core". *Astrophysical Journal*. 322. Part 1: 206–214. Bibcode:1987ApJ...322..206N. doi:10.1086/165716.

[25]  Claudio Ritossa; et al. (1999). "On the Evolution of Stars that Form Electron-degenerate Cores Processed by Carbon Burning. V. Shell Convection Sustained by Helium Burning, Transient Neon Burning, Dredge-out, URCA Cooling, and Other Properties of an 11 M_solar Population I Model Star". *The Astrophysical Journal* **515** (1): 381–397. Bibcode:1999ApJ...515..381R. doi:10.1086/307017.

[26]  How do Massive Stars Explode?

[27]  Robert Buras; et al. (June 2003). "Supernova Simulations Still Defy Explosions". *Research Highlights*. Max-Planck-Institut für Astrophysik.

[28]  E. P. J. van den Heuvel (2004). "X-Ray Binaries and Their Descendants: Binary Radio Pulsars; Evidence for Three Classes of Neutron Stars?". *Proceedings of the 5th INTEGRAL Workshop on the INTEGRAL Universe (ESA SP-552)* **552**: 185–194. arXiv:astro-ph/0407451. Bibcode:2004inun.conf..185V.

[29]  Pair Instability Supernovae and Hypernovae., Nicolay J. Hammer, (2003), accessed May 7, 2007. Archived June 8, 2012 at the Wayback Machine

[30]  Fossil Stars (1): White Dwarfs

[31] Ken'ichi Nomoto (1984). "Evolution of 8–10 $M\odot$ stars toward electron capture supernovae. I – Formation of electron-degenerate O + Ne + Mg cores". *Astrophysical Journal*. Part I **277**: 791–805. Bibcode:1984ApJ...277..791N. doi:10.1086/161749.

[32] Ken'ichi Nomoto & Yoji Kondo (1991). "Conditions for accretion-induced collapse of white dwarfs". *Astrophysical Journal*. 367. Part 2: L19–L22. Bibcode:1991ApJ...367L..19N. doi:10.1086/185922.

[33] D'Amico, N.; Stappers, B. W.; Bailes, M.; Martin, C. E.; Bell, J. F.; Lyne, A. G.; Manchester, R. N (1998). "The Parkes Southern Pulsar Survey - III. Timing of long-period pulsars". *Monthly Notices of the Royal Astronomical Society* **297**: 28–40. Bibcode:1998MNRAS.297...28D. doi:10.1046/j.1365-8711.1998.01397.x.

[34] Courtland, Rachel (17 October 2008). "Pulsar Detected by Gamma Waves Only". *New Scientist*. Archived from the original on April 2, 2013.

[35] Demarque, P.; Guenther, D. B.; Li, L. H.; Mazumdar, A.; Straka, C. W. (August 2008). "YREC: the Yale rotating stellar evolution code". *Astrophysics and Space Science* **316** (1–4): 31–41. arXiv:0710.4003. Bibcode:2008Ap&SS.316...31D. doi:10.1007/s10509-007-9698-y. ISBN 9781402094408.

[36] Ryan, Seán; Norton, Andrew J. (2010). "Assigning ages from hydrogen-burning timescales". *Stellar Evolution and Nucleosynthesis*. Cambridge University Press. p. 79. ISBN 0-521-13320-3.

# Chapter 10

# Gravitational energy

**Gravitational energy** is potential energy associated with the gravitational field. This phrase is found frequently in scientific writings about quasars (quasi-stellar objects) and other active galaxies. Quasars generate and emit their energy from a very small region. The emission of large amounts of power from a small region requires a power source far more efficient than the nuclear fusion that powers stars. The release of gravitational energy[1] by matter falling towards a massive black hole is the only process known that can produce such high power continuously. Stellar explosions – supernovas and gamma-ray bursts can do so, but only for a few weeks.[1]

## 10.1  Newtonian mechanics

According to classical mechanics, between two or more masses (or other forms of energy–momentum) a gravitational potential energy exists. Conservation of energy requires that this gravitational field energy is always negative.[2]

Particularly, between any two point masses $m$ and $M$ (this works for the spherical bodies also), there always exists a gravitational force of $F = GmM/r^2$ where r is the distance between their centers. Increasing the distance from $r = r_0$ to $r = r_1$ reduces the force, but, since forces in Newton mechanics indicate how much potential energy is lost over space, $F = -\frac{dU}{dx}$ , this separation requires $\int_{r_0}^{r_1} \frac{mMG}{r^2} dr = \frac{mMG}{r}\big|_{r_1}^{r_0} = \frac{mMG}{r_0} - \frac{mMG}{r_1} = E$ of energy. Performing positive work equal to E units of energy, we can recede objects from $r_0$ to $r_1$ special units apart. By performing positive work equal to $E = mMG/r_0$ , the second term vanishes and objects are infinitely separated ( $r_1 = \infty$ ). Because gravitational force stops pulling objects together at that distance, $E = mMG/r_0$ is known as gravitational binding energy, which is infinite at $r_0 = 0$ since the gravitational force is infinite there.

## 10.2  General relativity

Main article: Mass in general relativity

In general relativity gravitational energy is extremely complex, and there is no single agreed upon definition of the concept. It is sometimes modeled via the Landau–Lifshitz pseudotensor[3] which allows for the energy-momentum conservation laws of classical mechanics to be retained. Addition of the matter stress–energy–momentum tensor to the Landau–Lifshitz pseudotensor results in a combined matter plus gravitational energy pseudotensor which has a vanishing 4-divergence in all frames; the vanishing divergence ensures the conservation law. Some people object to this derivation on the grounds that pseudotensors are inappropriate in general relativity, but the divergence of the combined matter plus gravitational energy pseudotensor is a tensor.

## 10.3 See also

- Gravitational binding energy

- Gravitational potential

- Standard gravitational parameter

- Gravitational wave

## 10.4 References

[1] Lambourne, Robert J. A. (2010). *Relativity, Gravitation and Cosmology* (Illustrated ed.). Cambridge University Press. p. 222. ISBN 0521131383. Retrieved 2012-11-20.

[2] Alan Guth *The Inflationary Universe: The Quest for a New Theory of Cosmic Origins* (1997), Random House , ISBN 0-224-04448-6 Appendix A: *Gravitational Energy* demonstrates the negativity of gravitational energy.

[3] Lev Davidovich Landau & Evgeny Mikhailovich Lifshitz, *The Classical Theory of Fields*, (1951), Pergamon Press, ISBN 7-5062-4256-7

# Chapter 11

# History of supernova observation

The known **history of supernova observation** goes back to 185 CE, when, supernova SN 185 appeared, the oldest appearance of a supernova recorded by humankind. Several additional supernovae within the Milky Way galaxy have been recorded since that time, with SN 1604 being the most recent supernova to be observed in this galaxy.[1]

Since the development of the telescope, the field of supernova discovery has expanded to other galaxies. These occurrences provide important information on the distances of galaxies. Successful models of supernova behavior have also been developed, and the role of supernovae in the star formation process is now increasingly understood.

## 11.1   Early history

The supernova explosion that formed the Vela Supernova Remnant most likely occurred 10,000–20,000 years ago. In 1976, NASA astronomers suggested that inhabitants of the southern hemisphere may have witnessed this explosion and recorded it symbolically. A year later, archaeologist George Michanowsky recalled some incomprehensible ancient markings in Bolivia that were left by Native Americans. The carvings showed four small circles flanked by two larger circles. The smaller circles resemble stellar groupings in the constellations Vela and Carina. One of the larger circles may represent the star Capella. Another circle is located near the position of the supernova remnant, George Michanowsky suggested this may represent the supernova explosion as witnessed by the indigenous residents.[2]

In 185 CE, Chinese astronomers recorded the appearance of a bright star in the sky, and observed that it took about eight months to fade from the sky. It was observed to sparkle like a star and did not move across the heavens like a comet. These observations are consistent with the appearance of a supernova, and this is believed to be the oldest confirmed record of a supernova event by humankind. SN 185 may have also possibly been recorded in Roman literature, though no records have survived.[3] The gaseous shell RCW 86 is suspected as being the remnant of this event, and recent X-ray studies show a good match for the expected age.[4]

In 393 CE, the Chinese recorded the appearance of another "guest star", SN 393, in the modern constellation of Scorpius.[5] Additional unconfirmed supernovae events may have been observed in 369 CE, 386 CE, 437 CE, 827 CE and 902 CE.[1] However these have not yet been associated with a supernova remnant, and so they remain only candidates. Over a span of about 2,000 years, Chinese astronomers recorded a total of twenty such candidate events, including later explosions noted by Islamic, European, and possibly Indian and other observers.[1][6]

The supernova SN 1006 appeared in the southern constellation of Lupus during the year 1006 CE. This was the brightest recorded star ever to appear in the night sky, and its presence was noted in China, Egypt, Iraq, Italy, Japan and Switzerland. It may also have been noted in France, Syria, and North America. Egyptian physician, astronomer and astrologer Ali ibn Ridwan gave the brightness of this star as one-quarter the brightness of the Moon. Modern astronomers have discovered the faint remnant of this explosion and determined that it was only 7,100 light-years from the Earth.[7]

Supernova SN 1054 was another widely-observed event, with Arab, Chinese, and Japanese astronomers recording the star's appearance in 1054 CE. It may also have been recorded by the Anasazi as a petroglyph.[8] This explosion appeared

*The Crab Nebula is a pulsar wind nebula associated with the 1054 supernova.*

in the constellation of Taurus, where it produced the Crab Nebula remnant. At its peak, the luminosity of SN 1054 may have been four times as bright as Venus, and it remained visible in daylight for 23 days and was visible in the night sky for 653 days.[9][10]

There are fewer records of supernova SN 1181, which occurred in the constellation Cassiopeia just over a century after SN 1054. It was noted by Chinese and Japanese astronomers, however. The pulsar 3C58 may be the stellar relic from this event.[11]

The Danish astronomer Tycho Brahe was noted for his careful observations of the night sky from his observatory on the island of Hven. In 1572 he noted the appearance of a new star, also in the constellation Cassiopeia. Later called SN 1572, this supernova was associated with a remnant during the 1960s.[12]

A common belief in Europe during this period was the Aristotelian idea that the world beyond the Moon and planets was immutable. So observers argued that the phenomenon was something in the Earth's atmosphere. However Tycho noted that the object remained stationary from night to night—never changing its parallax—so it must lie far away.[13][14] He

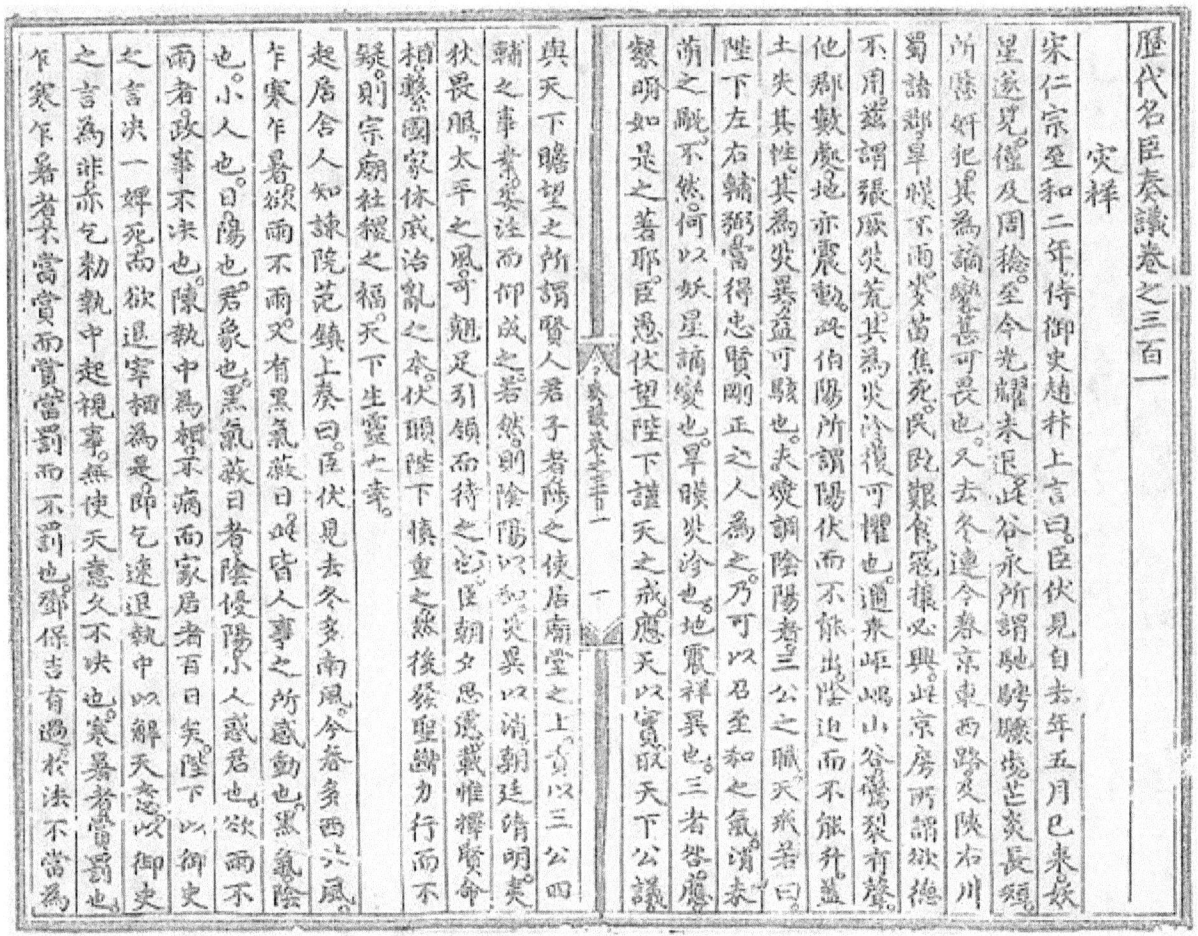

*The guest star reported by Chinese astronomers in 1054 is identified as SN 1054. The highlighted passages refer to the supernova.*

published his observations in the small book *De nova et nullius aevi memoria prius visa stella* (Latin for "Concerning the new and previously unseen star") in 1573. It is from the title of this book that the modern word *nova* for cataclysmic variable stars is derived.[15]

The most recent supernova to be seen in the Milky Way galaxy was SN 1604, which was observed October 9, 1604. Several people noted the sudden appearance of this star, but it was Johannes Kepler who became noted for his systematic study of the object. He published his observations in the work *De Stella nova in pede Serpentarii*.[16]

Galileo, like Tycho before him, tried in vain to measure the parallax of this new star, and then argued against the Aristotelian view of an immutable heavens.[17] The remnant of this supernova was identified in 1941 at the Mount Wilson Observatory.[18]

## 11.2   Telescope observation

The true nature of the supernova remained obscure for some time. Observers slowly came to recognize a class of stars that undergo long-term periodic fluctuations in luminosity. Both John Russell Hind in 1848 and Norman Pogson in 1863 had charted stars that underwent sudden changes in brightness. However these received little attention from the astronomical community. Finally, in 1866, English astronomer William Huggins made the first spectroscopic observations of a nova, discovering lines of hydrogen in the unusual spectrum of the recurrent nova T Coronae Borealis.[19] Huggins proposed a cataclysmic explosion as the underlying mechanism, and his efforts drew interest from other astronomers.[20]

In 1885, a nova-like outburst was observed in the direction of the Andromeda galaxy by Ernst Hartwig in Estonia. S

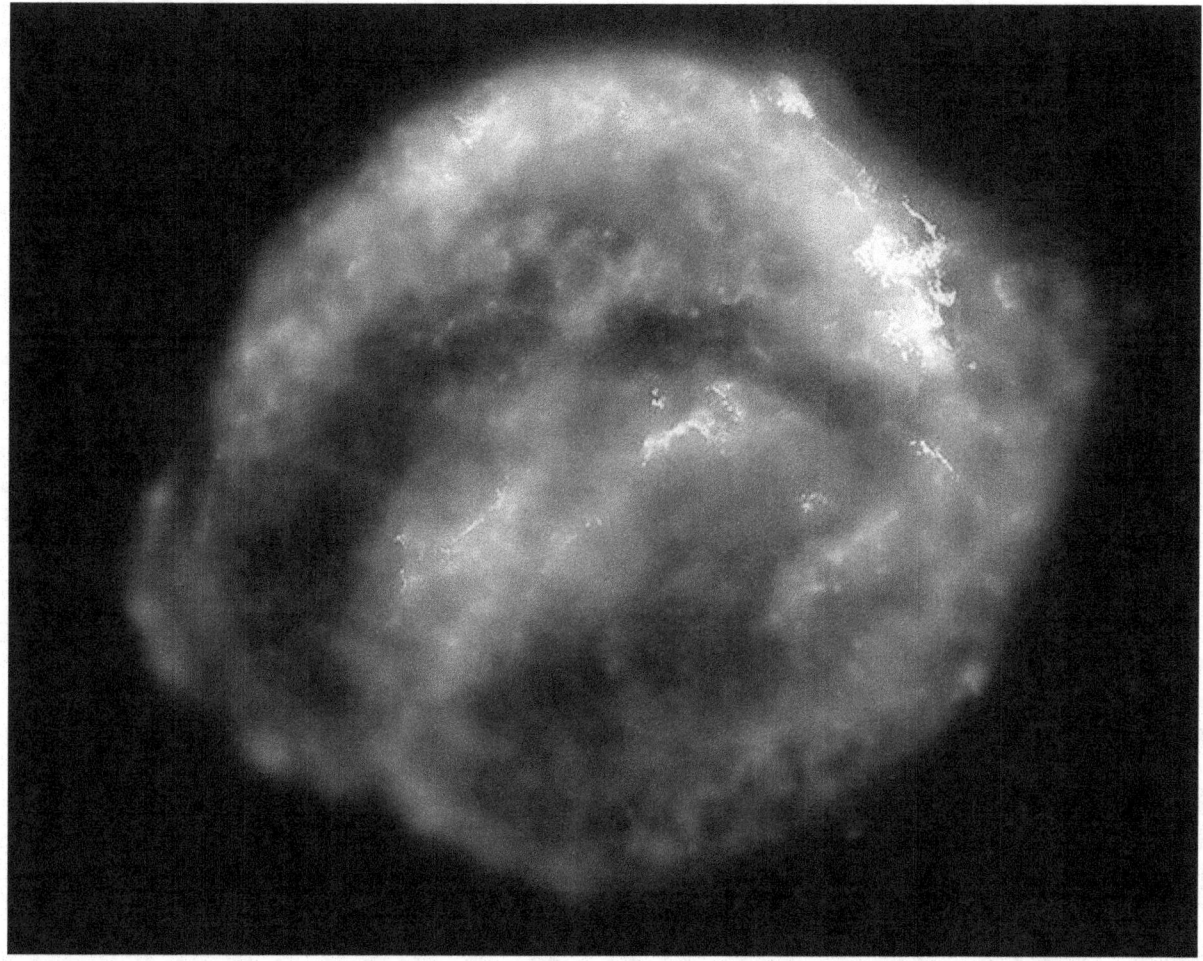

*Multiwavelength X-ray image of the remnant of Kepler's Supernova, SN 1604. (Chandra X-ray Observatory)*

Andromedae increased to 6th magnitude, outshining the entire nucleus of the galaxy, then faded in a manner much like a nova. In 1917, George W. Ritchey measured the distance to the Andromeda galaxy and discovered it lay much farther than had previously been thought. This meant that S Andromedae, which did not just lie along the line of sight to the galaxy but had actually resided in the nucleus, released a much greater amount of energy than was typical for a nova.[21]

Early work on this new category of nova was performed during the 1930s by Walter Baade and Fritz Zwicky at Mount Wilson Observatory.[22] They identified S Andromedae, what they considered a typical supernova, as an explosive event that released radiation approximately equal to the Sun's total energy output for $10^7$ years. They decided to call this new class of cataclysmic variables super-novae, and postulated that the energy was generated by the gravitational collapse of ordinary stars into neutron stars.[23] The name *super-novae* was first used in a 1931 lecture at Caltech by Zwicky, then used publicly in 1933 at a meeting of the American Physical Society. By 1938, the hyphen had been lost and the modern name was in use.[24]

Although supernovae are relatively rare events, occurring on average about once every 50 years in the Milky Way,[25] observations of distant galaxies allowed supernovae to be discovered and examined more frequently. The first supernova detection patrol was begun by Zwicky in 1933. He was joined by Josef J. Johnson from Caltech in 1936. Using a 45-cm Schmidt telescope at Palomar observatory, they discovered twelve new supernovae within three years by comparing new photographic plates to reference images of extragalactic regions.[26]

In 1938, Walter Baade became the first astronomer to identify a nebula as a supernova remnant when he suggested that the Crab Nebula was the remains of SN 1054. He noted that, while it had the appearance of a planetary nebula, the measured velocity of expansion was much too large to belong to that classification.[27] During the same year, Baade first

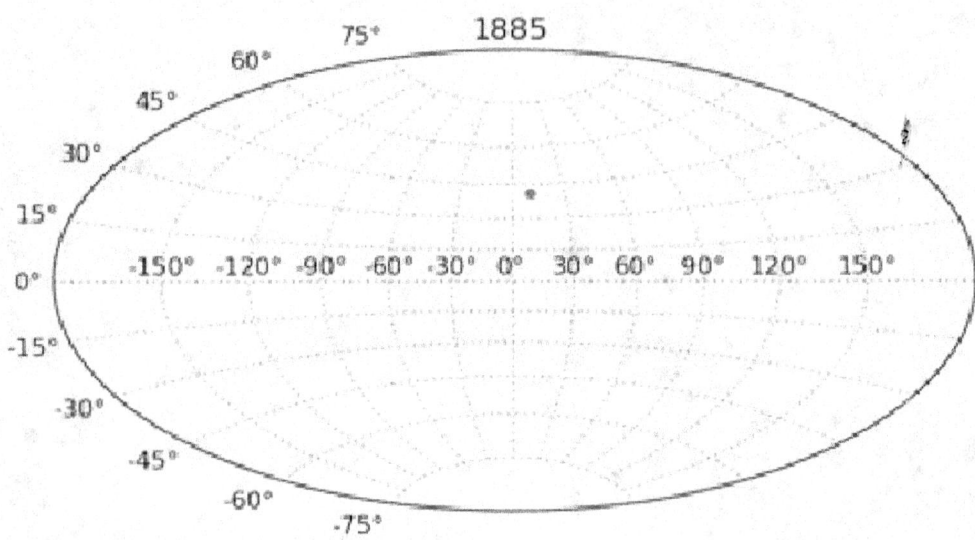

*Animation showing R.A. and Dec. of supernovae discovered since 1885. Some recent survey contributions are highlighted in color.*

proposed the use of the Type Ia supernova as a secondary distance indicator. Later, the work of Allan Sandage and Gustav Tammann helped refine the process so that Type Ia supernovae became a type of standard candle for measuring large distances across the cosmos.[28][29]

The first spectral classification of these distant supernovae was performed by Rudolph Minkowski in 1941. He categorized them into two types, based on whether or not lines of the element hydrogen appeared in the supernova spectrum.[30] Zwicky later proposed additional types III, IV, and V, although these are no longer used and now appear to be associated with single peculiar supernova types. Further sub-division of the spectra categories resulted in the modern supernova classification scheme.[31]

In the aftermath of the Second World War, Fred Hoyle worked on the problem of how the various observed elements in the universe were produced. In 1946 he proposed that a massive star could generate the necessary thermonuclear reactions, and the nuclear reactions of heavy elements were responsible for the removal of energy necessary for a gravitational collapse to occur. The collapsing star became rotationally unstable, and produced an explosive expulsion of elements that were distributed into interstellar space.[32] The concept that rapid nuclear fusion was the source of energy for a supernova explosion was developed by Hoyle and William Fowler during the 1960s.[33]

The first computer-controlled search for supernovae was begun in the 1960s at Northwestern University. They built a 24-inch telescope at Corralitos Observatory in New Mexico that could be repositioned under computer control. The telescope displayed a new galaxy each minute, with observers checking the view on a television screen. By this means, they discovered 14 supernovae over a period of two years.[34]

## 11.3   1970–1999

The modern standard model for Type Ia supernovae explosions is founded on a proposal by Whelan and Iben in 1973, and is based upon a mass-transfer scenario to a degenerate companion star.[35] In particular, the light curve of SN1972e in NGC 5253, which was observed for more than a year, was followed long enough to discover that after its broad "hump" in brightness, the supernova faded at a nearly constant rate of about 0.01 magnitudes per day. Translated to another system of units, this is nearly the same as the decay rate of cobalt$-$56 ($^{56}$Co), whose half-life is 77 days. The degenerate explosion model predicts the production of about a solar mass of nickel$-$56 ($^{56}$Ni) by the exploding star. The $^{56}$Ni decays with a half-life of 6.8 days to $^{56}$Co, and the decay of the nickel and cobalt provides the energy radiated away by the supernova late in its history. The agreement in both total energy production and the fade rate between the theoretical models and the observations of 1972e led to rapid acceptance of the degenerate-explosion model.[36]

Through observation of the light curves of many Type Ia supernovae, it was discovered that they appear to have a common peak luminosity.[37] By measuring the luminosity of these events, the distance to their host galaxy can be estimated with good accuracy. Thus this category of supernovae has become highly useful as a standard candle for measuring cosmic distances. In 1998, the High-Z Supernova Search and the Supernova Cosmology Project discovered that the most distant Type Ia supernovae appeared dimmer than expected. This has provided evidence that the expansion of the universe may be accelerating.[38][39]

Although no supernova has been observed in the Milky Way since 1604, it appears that a supernova exploded in the constellation Cassiopeia about 300 years ago, around the year 1667 or 1680. The remnant of this explosion, Cassiopeia A—is heavily obscured by interstellar dust, which is possibly why it did not make a notable appearance. However it can be observed in other parts of the spectrum, and it is currently the brightest radio source beyond our solar system.[40]

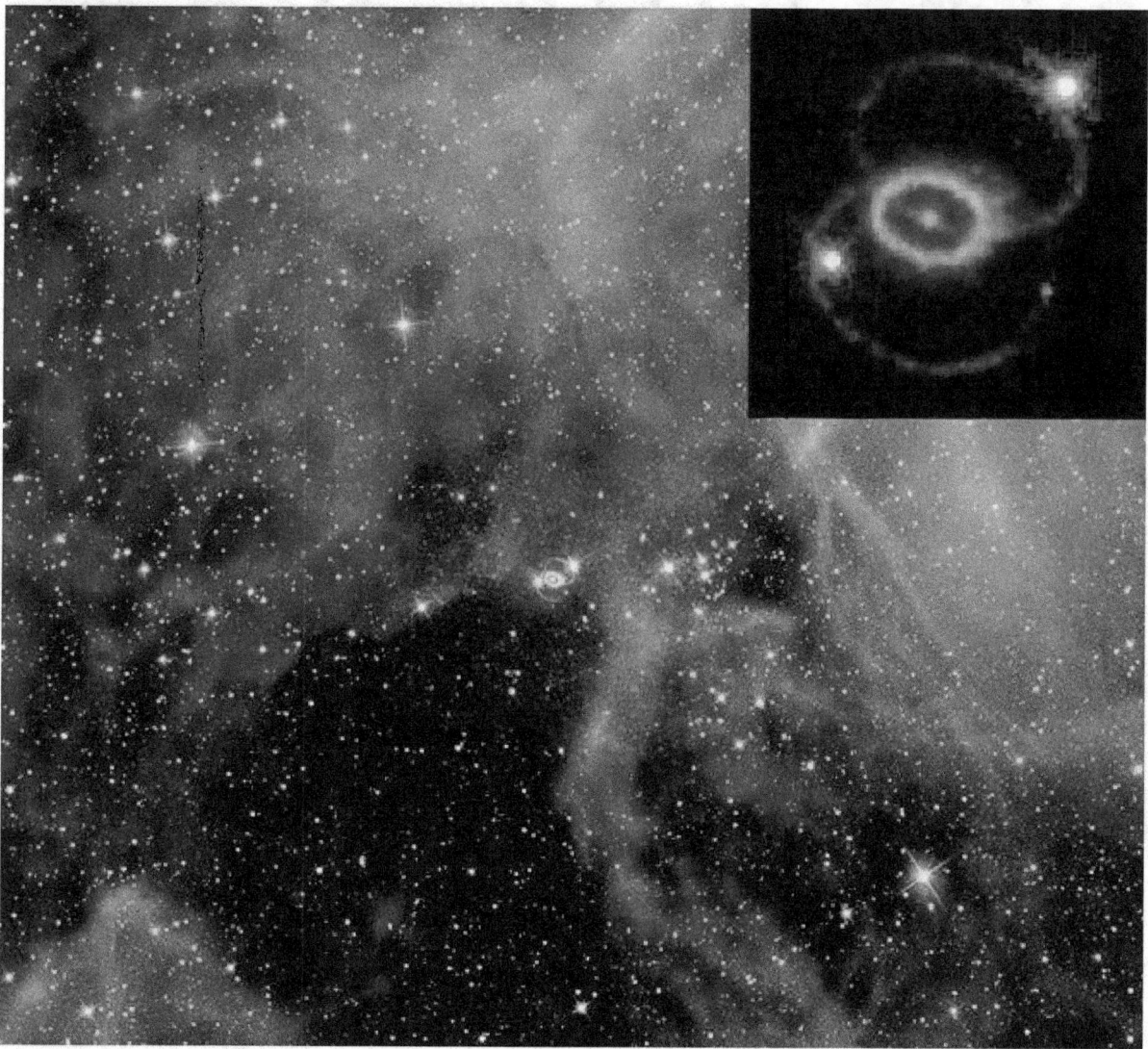

*Supernova 1987A remnant near the center*

In 1987, Supernova 1987A in the Large Magellanic Cloud was observed within hours of its start. It was the first supernova to be detected through its neutrino emission and the first to be observed across every band of the electromagnetic spectrum. The relative proximity of this supernova has allowed detailed observation, and it provided the first opportunity for modern theories of supernova formation to be tested against observations.[41][42]

The rate of supernova discovery steadily increased throughout the twentieth century.[43] In the 1990s, several automated supernova search programs were initiated. The Leuschner Observatory Supernova Search program was begun in 1992 at Leuschner Observatory. It was joined the same year by the Berkeley Automated Imaging Telescope program. These were succeeded in 1996 by the Katzman Automatic Imaging Telescope at Lick Observatory, which was primarily used for the Lick Observatory Supernova Search (LOSS). By 2000, the Lick program resulted in the discovery of 96 supernovae, making it the world's most successful Supernova search program.[44]

In the late 1990s it was proposed that recent supernova remnants could be found by looking for gamma rays from the decay of titanium-44. This has a half-life of 90 years and the gamma rays can traverse the galaxy easily, so it permits us to see any remnants from the last millennium or so. Two sources were found, the previously discovered Cassiopeia A remnant, and the RX J0852.0-4622 remnant, which had just been discovered overlapping the Vela Supernova Remnant[45]

*In 1999 a star within IC 755 was seen to explode as a supernova and named SN 1999an.*

This remnant (RX J0852.0-4622) had been found in front (apparently) of the larger Vela Supernova Remnant.[46] The gamma rays from the decay of titanium-44 showed that it must have exploded fairly recently (perhaps around 1200 AD), but there is no historical record of it. The flux of gamma rays and x-rays indicates that the supernova was relatively close to us (perhaps 200 parsecs or 600 ly). If so, this is a surprising event because supernovae less than 200 parsecs away are estimated to occur less than once per 100,000 years.[47]

## 11.4   2000 to present

The "SN 2003fg" was discovered in a forming galaxy in 2003. The appearance of this supernova was studied in "real-time", and it has posed several major physical questions as it seems more massive than the Chandrasekhar limit would allow.[49]

First observed in September 2006, the supernova SN 2006gy, which occurred in a galaxy called NGC 1260 (240 million light-years away), is the largest and, until confirmation of luminosity of SN 2005ap in October 2007, the most luminous supernova ever observed. The explosion was at least 100 times more luminous than any previously observed supernova,[50][51] with the progenitor star being estimated 150 times more massive than the Sun.[52] Although this had some characteristics of a Type Ia supernova, Hydrogen was found in the spectrum.[53] It is thought that SN 2006gy is a likely candidate for a pair-instability supernova. SN 2005ap, which was discovered by Robert Quimby who also discovered SN 2006gy, was about twice as bright as SN 2006gy and about 300 times as bright as a normal type II supernova.[54]

On May 21, 2008, astronomers announced that they had for the first time caught a supernova on camera just as it was exploding. By chance, a burst of X-rays was noticed while looking at galaxy NGC 2770, 88 million light-years from

*Cosmic lens MACS J1720+35 helps Hubble to find a distant supernova.*[48]

Earth, and a variety of telescopes were aimed in that direction just in time to capture what has been named SN 2008D. "This eventually confirmed that the big X-ray blast marked the birth of a supernova," said Alicia Soderberg of Princeton University.[56]

One of the many amateur astronomers looking for supernovae, Caroline Moore, a member of the Puckett Observatory Supernova Search Team, found supernova SN 2008ha late November 2008. At the age of 14 she has now been declared the youngest person ever to find a supernova.[57][58] However, in January 2011, 10-year-old Kathryn Aurora Gray from Canada was reported to have discovered a supernova, making her the youngest ever to find a supernova.[59] Ms. Gray, her father, and a friend spotted SN 2010lt, a magnitude 17 supernova in galaxy UGC 3378 in the constellation Camelopardalis, about 240 million light years away.

In 2009, researchers have found nitrates in ice cores from Antarctica at depths corresponding to the known supernovae of 1006 and 1054 AD, as well as from around 1060 AD. The nitrates were apparently formed from nitrogen oxides created by gamma rays from the supernovae. This technique should be able to detect supernovae going back several thousand years.[61]

On November 15, 2010, astronomers using NASA's Chandra X-ray Observatory announced that, while viewing the remnant of SN 1979C in the galaxy Messier 100, they have discovered an object which could be a young, 30-year-old black hole. NASA also noted the possibility this object could be a spinning neutron star producing a wind of high energy particles.[62]

On August 24, 2011, the Palomar Transient Factory automated survey discovered a new Type Ia supernova (SN 2011fe) in the Pinwheel Galaxy (M101) shortly after it burst into existence. Being only 21 million lightyears away and detected so early after the event started, it will allow scientists to learn more about the early developments of these types of supernovae.[63]

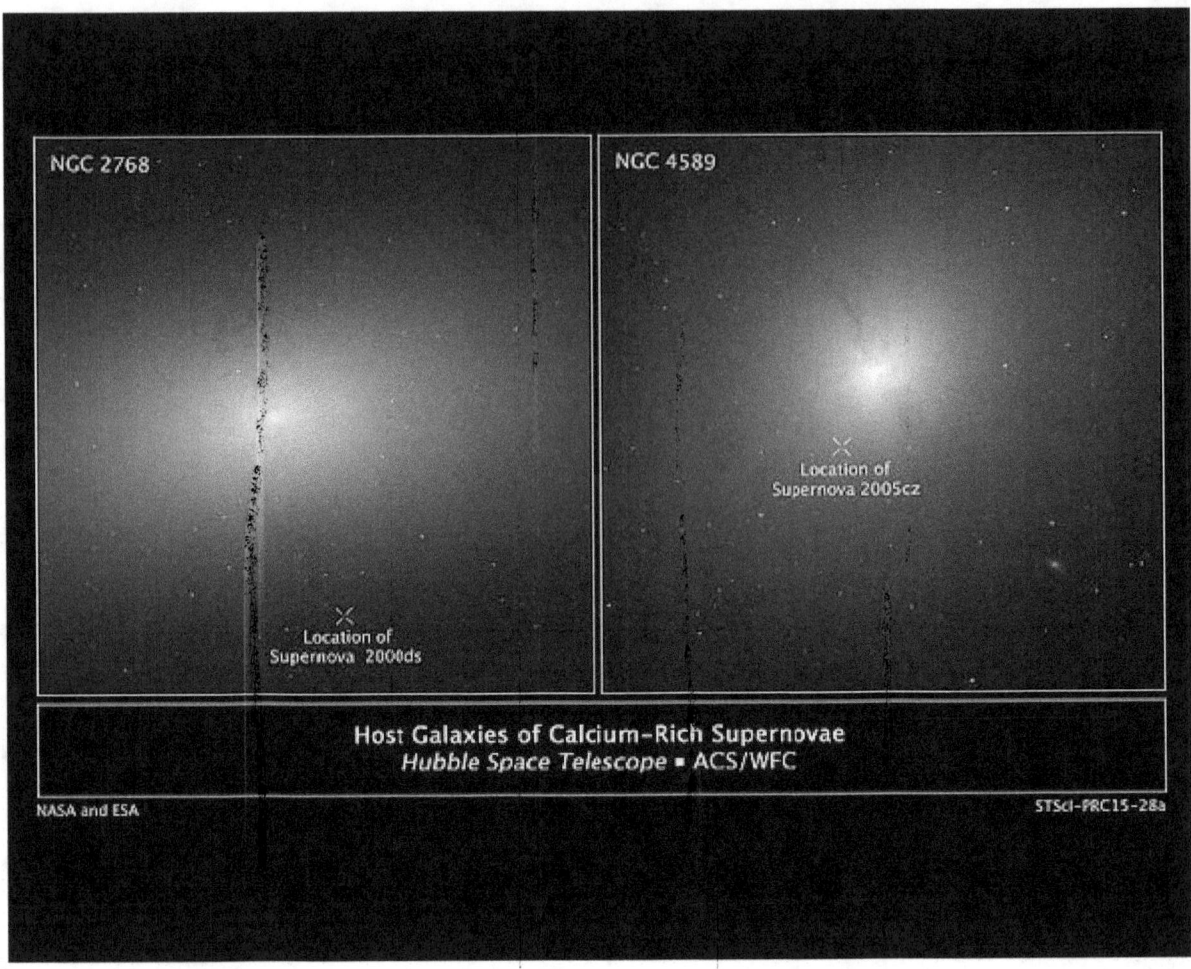

*Host Galaxies of Calcium-Rich Supernovae.*[55]

On 16 March 2012, a Type II supernova, designated as SN 2012aw, was discovered in M95.[64][65][66]

On January 22, 2014, students at the University of London Observatory spotted an exploding star SN 2014J in the nearby galaxy M82 (the Cigar Galaxy). At a distance of around 12 million light years, the supernova is one of the nearest to be observed in recent decades.[67]

## 11.5   Future

The estimated rate of supernova production in a galaxy the size of the Milky Way is about one every 50 years. This is much higher than the actual observed rate, implying that a portion of these events have been obscured from the Earth by interstellar dust. The deployment of new instruments that can observe across a wide range of the electromagnetic spectrum, along with neutrino detectors, means that the next such event will almost certainly be detected.[25]

## 11.6   See also

- History of astronomy

*Supernova SN 2012cg in spiral galaxy NGC 4424.[60]*

## 11.7 References

[1] Clark, D. H.; Stephenson, F. R. (June 29, 1981). "The Historical Supernovae". *Supernovae: A survey of current research; Proceedings of the Advanced Study Institute*. Cambridge, England: Dordrecht, D. Reidel Publishing Co. pp. 355–370. Bibcode:1982sscr.conf..355C.

[2] "Science: Homage to a Star". *Time Magazine*. October 22, 1973. Retrieved 2011-01-28.

[3] Stothers, Richard (1977). "Is the Supernova of CE 185 Recorded in Ancient Roman Literature". *Isis* **68** (3): 443447. doi:10.1086/351822.

[4] "New evidence links stellar remains to oldest recorded supernova". ESA News. September 18, 2006. Retrieved 2006-05-24.

[5] Wang, Z.-R.; Qu, Q. Y.; Chen, Y. (1998). "The AD 393 Guest Star: the SNR RX 51713.7-3946". *Proceedings of IAU Symposium #188*. Dordrecht: Kluwer Academic. p. 262. Bibcode:1998IAUS..188..262W.

[6] Hartmut Frommert, Christine Kronberg. "Supernovae observed in the Milky Way: Historical Supernovae". SEDS. Retrieved 2007-01-03.

[7] "Astronomers Peg Brightness of History's Brightest Star". NAOA News. March 5, 2003. Retrieved 2006-06-08.

[8] Greening, Dan (1995). "1054 Supernova Petrograph". Pomona College Astronomy Program. Retrieved 2006-09-25.

[9] Collins II, G. W.; Claspy, W. P.; Martin, J. C. (1999). "A Reinterpretation of Historical References to the Supernova of A.D. 1054". *Publications of the Astronomical Society of the Pacific* **111** (761): 871–880. arXiv:astro-ph/9904285. Bibcode:1999 doi:10.1086/316401.

[10] Brecher, K.; Fesen; Maran; Brandt (1983). "Ancient records and the Crab Nebula supernova". *The Observatory* **103**: 106–113. Bibcode:1983Obs...103..106B.

[11] "3C58: Pulsar Gives Insight on Ultra Dense Matter and Magnetic Fields". Harvard-Smithsonian Center for Astrophysics. December 14, 2004. Retrieved 2006-09-26.

[12] Villard, R.; Sanders, R. (July 24, 1991). "Stellar survivor from 1572 CE explosion supports supernova theory". UCBerkeley News. Retrieved 2006-09-25.

[13] Cowen, R. (1999). "Danish astronomer argues for a changing cosmos". *Science News* **156** (25 & 26).

[14] Nardo, Don (2007). *Tycho Brahe: Pioneer of Astronomy*. Compass Point Books. ISBN 0-7565-3309-0.

[15] Stacey, Blake. "Supernovas: Making Astronomical History". SNEWS: Supernova Early Warning System. Retrieved 2006-09-25.

[16] "Johannes Kepler: De Stella Nova". New York Society Library. Archived from the original on 2007-09-28. Retrieved 2009-07-17.

[17] Wilson, Fred L. (July 7, 1996). "History of Science: Galileo and the Rise of Mechanism". Rochester Institute of Technology. Archived from the original on 2007-06-17. Retrieved 2009-07-17.

[18] Blair, Bill. "Bill Blair's Kepler's Supernova Remnant Page". NASA and Johns Hopkins University. Retrieved 2006-09-20.

[19] Higgins, William (1866). "On a New Star". *Monthly Notices of the Royal Astronomical Society* **26**: 275. Bibcode:1866MNRAS.

[20] Becker, Barbara J. (1993). "Eclecticism, Opportunism, and the Evolution of a New Research Agenda: William and Margaret Huggins and the Origins of Astrophysics". University of California—Irvine. Retrieved 2006-09-27.

[21] van Zyl, Jan Eben (2003). "Variable Stars VI". Astronomical Society of Southern Africa. Archived from the original on 2006-09-23. Retrieved 2009-07-17.

[22] Baade, W.; Zwicky, F. (1934). "On Super-Novae". *Proceedings of the National Academy of Sciences of the United States of America* **20** (5): 254–259. Bibcode:1934PNAS...20..254B. doi:10.1073/pnas.20.5.254. PMC 1076395. PMID 16587881. Retrieved 2008-06-04.

[23] Osterbrock, D. E. (1999). "Who Really Coined the Word Supernova? Who First Predicted Neutron Stars?". *Bulletin of the American Astronomical Society* **33**: 1330. Bibcode:2001AAS...199.1501O.

[24] Murdin, Paul; Murdin, Lesley (1985). *Supernovae* (2nd ed.). Cambridge University Press. p. 42. ISBN 0-521-30038-X.

[25] Türler, Marc (2006). "INTEGRAL reveals Milky Ways' supernova rate". *CERN Courier* **46** (1). Retrieved 2008-06-04.

[26] Heilbron, John Lewis (2005). *The Oxford guide to the history of physics and astronomy* **10**. Oxford University Press US. p. 315. ISBN 0-19-517198-5.

[27] Baade, W. (October 1938). "The Absolute Photographic Magnitude of Supernovae". *Astrophysical Journal* **88**: 285–304. Bibcode:1938ApJ....88..285B. doi:10.1086/143983.

[28] Lynden-Bell, Donald (December 24, 2010). "Allan Sandage (1926–2010)". *Science* **330**(6012): 1763. Bibcode:2010Sci...330. doi:10.1126/science.1201221. PMID 21205661.

[29] Perlmutter, Saul (April 2003). "Supernovae, Dark Energy, and the Accelerating Universe". *Physics Today* **56** (4): 53–62. Bibcode:2003PhT....56d..53P. doi:10.1063/1.1580050.

[30] Rudolph, Minkowski (1941). "Spectra of Supernovae". *Publications of the Astronomical Society of the Pacific* **53** (314): 224. Bibcode:1941PASP...53..224M. doi:10.1086/125315.

[31] da Silva, L. A. L. (1993). "The Classification of Supernovae". *Astrophysics and Space Science* **202**(2): 215–236. Bibcode:1993 doi:10.1007/BF00626878.

[32] Hoyle, Fred (1946). "The Synthesis of the Elements from Hydrogen". *Monthly Notices of the Royal Astronomical Society* **106**: 343–383. Bibcode:1946MNRAS.106..343H. doi:10.1093/mnras/106.5.343.

[33] Woosley, S. E. (1999). "Hoyle & Fowler's Nucleosynthesis in Supernovae". *Astrophysical Journal* **525C**: 924. Bibcode:1999Ap

[34] Marschall, Laurence A. (1994). *The supernova story*. Princeton science library. Princeton University Press. pp. 112–113. ISBN 0-691-03633-0.

[35] Whelan, J.; Iben Jr., I. (1973). "Binaries and Supernovae of Type I". *Astrophysical Journal* **186**: 1007–1014. Bibcode:1973ApJ doi:10.1086/152565.

[36] Trimble, V. (1982). "Supernovae. Part I: the events". *Reviews of Modern Physics* **54**(4): 1183–1224. Bibcode:1982RvMP...54 doi:10.1103/RevModPhys.54.1183.

[37] Kowal, C. T. (1968). "Absolute magnitudes of supernovae". *Astronomical Journal* **73**: 1021–1024. Bibcode:1968AJ.....73.102 doi:10.1086/110763.

[38] Leibundgut, B.; Sollerman, J. (2001). "A cosmological surprise: the universe accelerates". *Europhysics News* **32** (4): 121–125. Bibcode:2001ENews..32..121L. doi:10.1051/epn:2001401. Retrieved 2008-06-04.

[39] "Confirmation of the accelerated expansion of the Universe". Centre National de la Recherche Scientifique. September 19, 2003. Retrieved 2006-11-03.

[40] "Cassiopeia A - SNR". Caltech/NASA Infrared Processing and Analysis Center. Retrieved 2006-10-02.

[41] McCray, Richard (1993). "Supernova 1987A revisited". *Annual review of astronomy and astrophysics* **31** (1): 175–216. Bibcode:1993ARA&A..31..175M. doi:10.1146/annurev.aa.31.090193.001135.

[42] Comins, Neil F.; Kaufmann, William J. (2008). *Discovering the Universe: From the Stars to the Planets*. Macmillan. p. 230. ISBN 1-4292-3042-8.

[43] Kowal, C. T.; Sargent, W. L. W. (November 1971). "Supernovae discovered since 1885". *Astronomical Journal* **76**: 756–764. Bibcode:1971AJ.....76..756K. doi:10.1086/111193.

[44] Filippenko, Alexei V.; Li, W. D.; Treffers, R. R.; Modjaz, Maryam (2001). "The Lick Observatory Supernova Search with the Katzman Automatic Imaging Telescope". In Bohdan Paczynski, Wen-Ping Chen, Claudia Lemme. *Small Telescope Astronomy on Global Scales, IAU Colloquium 183*. ASP Conference Series **246**. San Francisco. Bibcode:2001ASPC..246..121F. ISBN 1-58381-084-6.

[45] Iyudin, A. F.; et al. (November 1998). "Emission from $^{44}$Ti associated with a previously unknown Galactic supernova". *Nature* **396** (6707): 142–144. Bibcode:1998Natur.396..142I. doi:10.1038/24106.

[46] Aschenbach, Bernd (1998-11-12). "Discovery of a young nearby supernova remnant". *Letters to Nature* **396** (6707): 141–142. Bibcode:1998Natur.396..141A. doi:10.1038/24103.

[47] Fields, B. D.; Ellis, J. (1999). "On Deep-Ocean Fe-60 as a Fossil of a Near-Earth Supernova". *New Astronomy* **4** (6): 419–430. arXiv:astro-ph/9811457. Bibcode:1999NewA....4..419F. doi:10.1016/S1384-1076(99)00034-2.

[48] "Hubble astronomers check the prescription of a cosmic lens". *ESA/Hubble Press Release*. Retrieved 2 May 2014.

[49] Howell, D. A.; et al. (2006). "Snls-03d3bb: An Overluminous, Low Velocity Type Ia Supernova Discovered At Z=0.244". *American Astronomical Society Meeting 208*. Bibcode:2006AAS...208.0203H.

[50] Berardelli, Phil (May 7, 2007). "Star Goes Out Big Time". Science Magazine ScienceNOW Daily News. Retrieved 2008-06-04.

[51] Grey Hautaluoma, Grey Hautaluoma and Megan Watzke (May 7, 2007). "NASA's Chandra Sees Brightest Supernova Ever". NASA. Retrieved 2008-06-04.

[52] Dunham, Will (May 8, 2007). "Brightest supernova ever seen". News in Science, Space and Astronomy.

[53] Shiga, David (January 3, 2007). "Brightest supernova discovery hints at stellar collision". New Scientist. Retrieved 2009-07-17.

[54] Than, Ker (October 11, 2007). "Supernova blazed like 100 billion suns". MSNBC. Retrieved 2007-10-17.

[55] "Host Galaxies of Calcium-Rich Supernovae". Retrieved 17 August 2015.

[56] Anonymous (May 21, 2008). "Supernova caught exploding on camera". Reuters UK. Retrieved 2009-07-17.

[57] Moore, Robert E. (2008-11-13). *The story about SN2008ha* "Rare supernova found by 14-year-old amateur astronomer" Check |url= scheme (help). Deer Pond Observatory. Retrieved 2008-12-19.

[58] Bishop, David (2008-12-19). "Supernova 2008ha in UGC 12682". Rochester Academy of Sciences. Retrieved 2008-12-19.

[59] Cohen, Tobi (January 3, 2011). "N.B. girl youngest ever to discover a supernova". The Vancouver Sun. Retrieved 2011-01-04.

[60] "A galactic cloak for an exploding star". *ESA/Hubble Picture of the Week*. ESA/Hubble. Retrieved 26 February 2015.

[61] "Ancient supernovae found written into the Antarctic ice". *New Scientist* (2698). 2009-03-04. Retrieved 2009-03-09. Refers to .

[62] Perrotto, Trent; Anderson, Janet; Watzke, Megan (November 15, 2010). "NASA'S Chandra Finds Youngest Nearby Black Hole". NASA. Retrieved 2010-11-19.

[63] Beatty, Kelly (25 August 2011). "Supernova Erupts in Pinwheel Galaxy". *Sky & Telescope*. Retrieved 26 August 2011

[64] "Deep Sky Videos". Retrieved 19 March 2012.

[65] "Supernova 2012aw: the pictures!". Retrieved 19 March 2012.

[66] "List of Recent Supernovae". Retrieved 8 April 2012.

[67] "UCL students discover a supernova".

## 11.8   External links

- Hecht, Jeff (June 19, 2006). "Enigmatic object baffles supernova team". NewScientist.com. Retrieved 2009-07-17.

# Chapter 12

# Type Ia supernova

*This artist's impression video shows the central part of the planetary nebula Henize 2-428. The core of this unique object consists of two white dwarf stars, each with a mass a little less than that of the Sun. They are expected to slowly draw closer to each other and merge in around 700 million years. This event will create a dazzling supernova of Type Ia and destroy both stars.*

**Type Ia supernovae** occur in binary systems (two stars orbiting one another) in which one of the stars is a white dwarf while the other can vary from a giant star to an even smaller white dwarf.[1] A white dwarf is the remnant of a star that has completed its normal life cycle and has ceased nuclear fusion. However, white dwarfs of the common carbon-oxygen variety are capable of further fusion reactions that release a great deal of energy if their temperatures rise high enough.

Physically, carbon-oxygen white dwarfs with a low rate of rotation are limited to below 1.38 solar masses ($M\odot$).[2][3] Beyond this, they re-ignite and in some cases trigger a supernova explosion. Somewhat confusingly, this limit is often referred to as the Chandrasekhar mass, despite being marginally different from the absolute Chandrasekhar limit where electron degeneracy pressure is unable to prevent catastrophic collapse. If a white dwarf gradually accretes mass from a binary companion, the general hypothesis is that its core will reach the ignition temperature for carbon fusion as it approaches the limit. If the white dwarf merges with another white dwarf (a very rare event), it will momentarily exceed the limit and begin to collapse, again raising its temperature past the nuclear fusion ignition point. Within a few seconds of initiation of nuclear fusion, a substantial fraction of the matter in the white dwarf undergoes a runaway reaction, releasing enough energy ($1$–$2 \times 10^{44}$ J)[4] to unbind the star in a supernova explosion.[5]

This category of supernovae produces consistent peak luminosity because of the uniform mass of white dwarfs that explode via the accretion mechanism. The stability of this value allows these explosions to be used as standard candles to measure the distance to their host galaxies because the visual magnitude of the supernovae depends primarily on the distance.

In May 2015, NASA reported that the *Kepler* space observatory observed KSN 2011b, a Type Ia supernova in the process of exploding. Details of the pre-nova moments may help scientists better understand dark energy.[6]

## 12.1   Consensus model

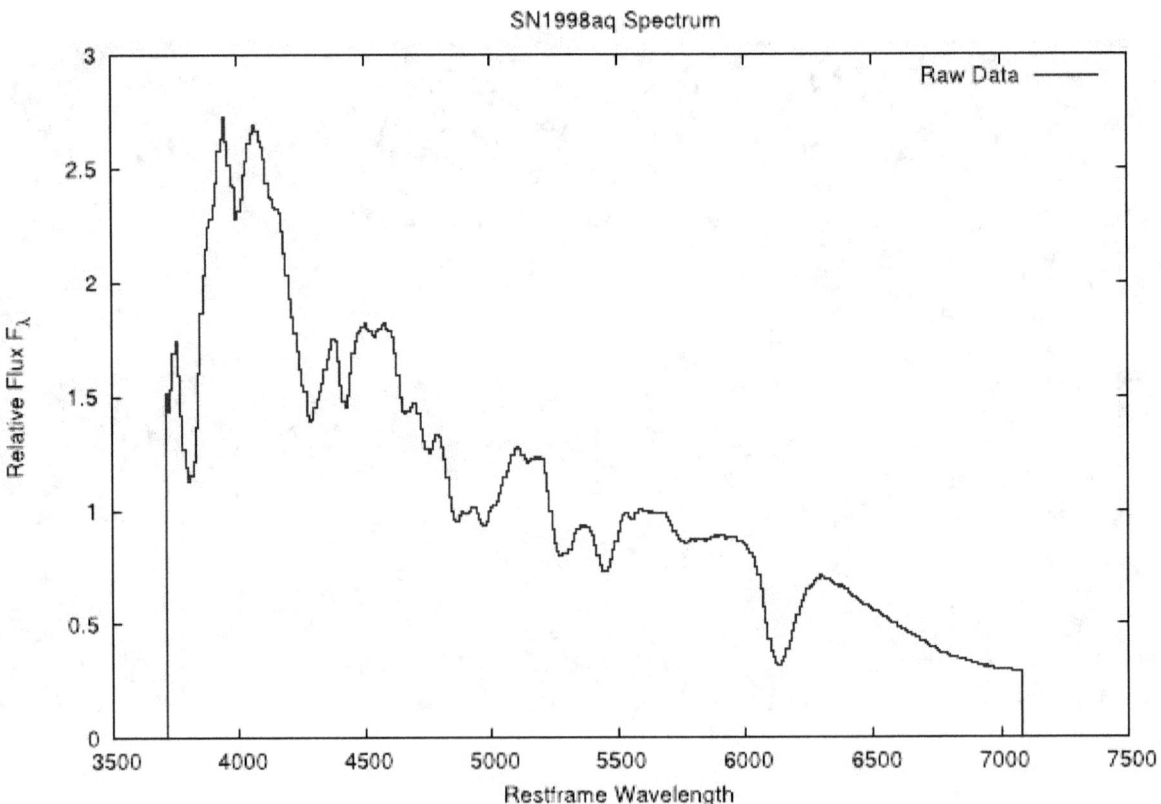

*Spectrum of SN1998aq, a Type Ia supernova, one day after maximum light in the B band*[7]

The Type Ia supernova is a sub-category in the Minkowski-Zwicky supernova classification scheme, which was devised by American astronomer Rudolph Minkowski and Swiss astronomer Fritz Zwicky.[8] There are several means by which a supernova of this type can form, but they share a common underlying mechanism. Theoretical astronomers long believed the progenitor star for this type of supernova is a white dwarf and empirical evidence for this was found in 2014 when a Type Ia supernova was observed in the galaxy, Messier 82.[9] When a slowly-rotating[2] carbon-oxygen white dwarf accretes matter from a companion, it can exceed the Chandrasekhar limit of about 1.44 $M\odot$, beyond which it can no longer support its weight with electron degeneracy pressure.[10] In the absence of a countervailing process, the white dwarf would collapse to form a neutron star,[11] as normally occurs in the case of a white dwarf that is primarily composed of magnesium, neon, and oxygen.[12]

The current view among astronomers who model Type Ia supernova explosions, however, is that this limit is never actually attained and collapse is never initiated. Instead, the increase in pressure and density due to the increasing weight raises the temperature of the core,[3] and as the white dwarf approaches about 99% of the limit,[13] a period of convection ensues, lasting approximately 1,000 years.[14] At some point in this simmering phase, a deflagration flame front is born, powered by carbon fusion. The details of the ignition are still unknown, including the location and number of points where the flame begins.[15] Oxygen fusion is initiated shortly thereafter, but this fuel is not consumed as completely as carbon.[16]

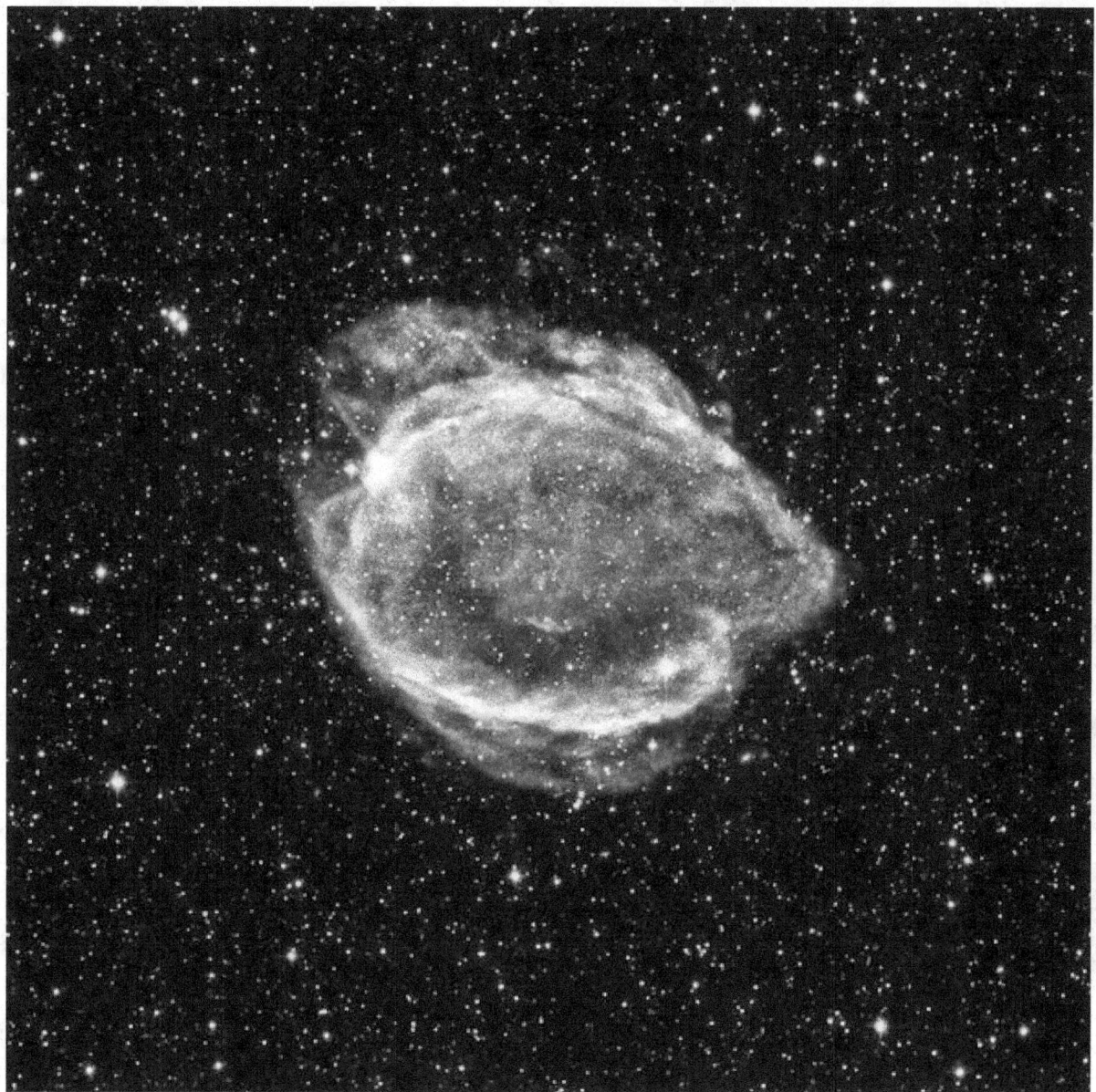

*G299 Type Ia supernova remnant.*

Once fusion has begun, the temperature of the white dwarf starts to rise. A main sequence star supported by thermal pressure would expand and cool which automatically counterbalances an increase in thermal energy. However, degeneracy pressure is independent of temperature; the white dwarf is unable to regulate the fusion process in the manner of normal stars, so it is vulnerable to a runaway fusion reaction. The flame accelerates dramatically, in part due to the Rayleigh–Taylor instability and interactions with turbulence. It is still a matter of considerable debate whether this flame transforms into a supersonic detonation from a subsonic deflagration.[14][17]

Regardless of the exact details of this nuclear fusion, it is generally accepted that a substantial fraction of the carbon and oxygen in the white dwarf are converted into heavier elements within a period of only a few seconds,[16] raising the internal temperature to billions of degrees. This energy release from thermonuclear fusion ($1–2\times10^{44}$ J[14]) is more than enough to unbind the star; that is, the individual particles making up the white dwarf gain enough kinetic energy to fly apart from each other. The star explodes violently and releases a shock wave in which matter is typically ejected at speeds on the order of 5,000–20000 km/s, roughly 6% of the speed of light. The energy released in the explosion also causes an extreme increase in luminosity. The typical visual absolute magnitude of Type Ia supernovae is $M_v = -19.3$ (about 5 billion times brighter than the Sun), with little variation.[14]

The theory of this type of supernovae is similar to that of novae, in which a white dwarf accretes matter more slowly and does not approach the Chandrasekhar limit. In the case of a nova, the in-falling matter causes a hydrogen fusion surface explosion that does not disrupt the star.[14] This type of supernova differs from a core-collapse supernova, which is caused by the cataclysmic explosion of the outer layers of a massive star as its core implodes.[18]

## 12.2    Formation

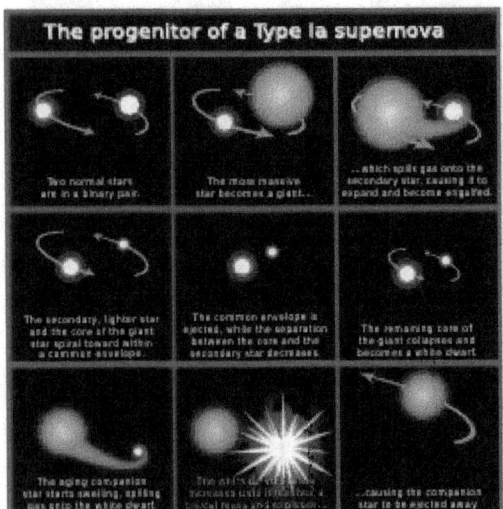

Formation process

Gas is being stripped from a giant star to form an accretion disc around a compact companion (such as a white dwarf star). *NASA image*

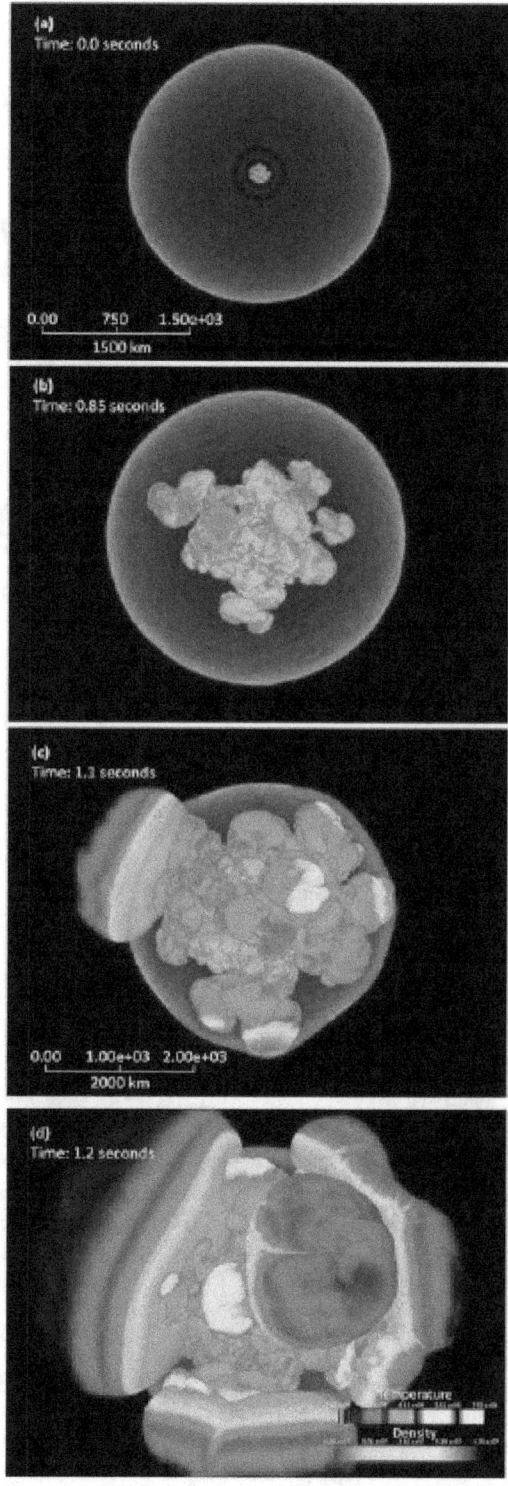

Simulation of the explosion phase of the deflagration-to-detonation model of supernovae formation, run on scientific supercomputer.

### 12.2.1   Single degenerate progenitors

One model for the formation of this category of supernova is a close binary star system. The progenitor binary system consists of main sequence stars, with the primary possessing more mass than the secondary. Being greater in mass, the primary is the first of the pair to evolve onto the asymptotic giant branch, where the star's envelope expands considerably. If the two stars share a common envelope then the system can lose significant amounts of mass, reducing the angular momentum, orbital radius and period. After the primary has degenerated into a white dwarf, the secondary star later evolves into a red giant and the stage is set for mass accretion onto the primary. During this final shared-envelope phase, the two stars spiral in closer together as angular momentum is lost. The resulting orbit can have a period as brief as a few hours.[19][20] If the accretion continues long enough, the white dwarf may eventually approach the Chandrasekhar limit.

The white dwarf companion could also accrete matter from other types of companions, including a subgiant or (if the orbit is sufficiently close) even a main sequence star. The actual evolutionary process during this accretion stage remains uncertain, as it can depend both on the rate of accretion and the transfer of angular momentum to the white dwarf companion.[21]

It has been estimated that single degenerate progenitors account for no more than 20% of all Type Ia supernovae.[22]

### 12.2.2   Double degenerate progenitors

A second possible mechanism for triggering a Type Ia supernova is the merger of two white dwarfs whose combined mass exceeds the Chandrasekhar limit. The resulting merger is called a super-Chandrasekhar mass white dwarf.[23][24] In such a case, the total mass would not be constrained by the Chandrasekhar limit.

Collisions of solitary stars within the Milky Way occur only once every $10^7 - 10^{13}$ years: far less frequently than the appearance of novae.[25] Collisions occur with greater frequency in the dense core regions of globular clusters.[26] (Cf. blue stragglers) A likely scenario is a collision with a binary star system, or between two binary systems containing white dwarfs. This collision can leave behind a close binary system of two white dwarfs. Their orbit decays and they merge through their shared envelope.[27] However, a study based on SDSS spectra found 15 double systems of the 4,000 white dwarfs tested, implying a double white dwarf merger every 100 years in the Milky Way. Conveniently, this rate matches the number of Type Ia supernovae detected in our neighborhood.[28]

A double degenerate scenario is one of several explanations proposed for the anomalously massive ($2\ M\odot$) progenitor of the SN 2003fg.[29][30] It is the only possible explanation for SNR 0509-67.5, as all possible models with only one white dwarf have been ruled out.[31] It has also been strongly suggested for SN 1006, given that no companion star remnant has been found there.[22] Observations made with NASA's Swift space telescope ruled out existing supergiant or giant companion stars of every Type Ia supernovae studied. The supergiant companion's blown out outer shell should emit X-rays, but this glow wasn't detected by Swift's *XRT (X-Ray telescope)* in the 53 closest supernova remnants. For 12 Type Ia supernovae observed within 10 days of the explosion, the satellite's *UVOT (Ultraviolet/Optical Telescope)* showed no ultraviolet radiation originating from the heated companion star's surface hit by the supernova shock wave, meaning there were no red giants or larger stars orbiting those supernova progenitors. In the case of SN 2011fe, the companion star must have been smaller than the Sun, if it existed.[32] The Chandra X-ray Observatory revealed that the X-ray radiation of five elliptical galaxies and the bulge of the Andromeda galaxy is 30-50 times fainter than expected. X-ray radiation should be emitted by the accretion discs of Type Ia supernova progenitors. The missing radiation indicates that few white dwarfs possess accretion discs, ruling out the common, accretion-based model of Ia supernovae.[33] Inward spiraling white dwarf pairs must be strong sources of gravitational waves, but this can't be detected as of 2012.

Double degenerate scenarios raise questions about the applicability of Type Ia supernovae as standard candles, since total mass of the two merging white dwarfs varies significantly, meaning luminosity also varies.

### 12.2.3   Type Iax

It has been proposed that a group of sub-luminous supernovae that occur when helium accretes onto a white dwarf should be classified as **Type Iax**.[34][35] This type of supernova may not always completely destroy the white dwarf progenitor.[36]

## 12.3 Observation

Unlike the other types of supernovae, Type Ia supernovae generally occur in all types of galaxies, including ellipticals. They show no preference for regions of current stellar formation.[37] As white dwarf stars form at the end of a star's main sequence evolutionary period, such a long-lived star system may have wandered far from the region where it originally formed. Thereafter a close binary system may spend another million years in the mass transfer stage (possibly forming persistent nova outbursts) before the conditions are ripe for a Type Ia supernova to occur.[38]

A long-standing problem in astronomy has been the identification of supernova progenitors. Direct observation of a progenitor would provide useful constraints on supernova models. As of 2006, the search for such a progenitor had been ongoing for longer than a century.[39] Observation of the supernova SN 2011fe has provided useful constraints. Previous observations with the Hubble Space Telescope did not show a star at the position of the event, thereby excluding a red giant as the source. The expanding plasma from the explosion was found to contain carbon and oxygen, making it likely the progenitor was a white dwarf primarily composed of these elements.[40] Similarly, observations of the nearby SN PTF 11kx,[41] discovered January 16, 2011 (UT) by the Palomar Transient Factory (PTF), lead to the conclusion that this explosion arises from single-degenerate progenitor, with a red giant companion, thus suggesting there is no single progenitor path to SN Ia. Direct observations of the progenitor of PTF11kx were reported in the August 24 edition of Science and support this conclusion, and also show that the progenitor star experienced periodic nova eruptions before the supernova - another surprising discovery. [42][43] However, later analysis revealed that the CSM is too massive for the single-degenerate scenario, and fits better the core-degenerate scenario.[44]

### 12.3.1   Light curve

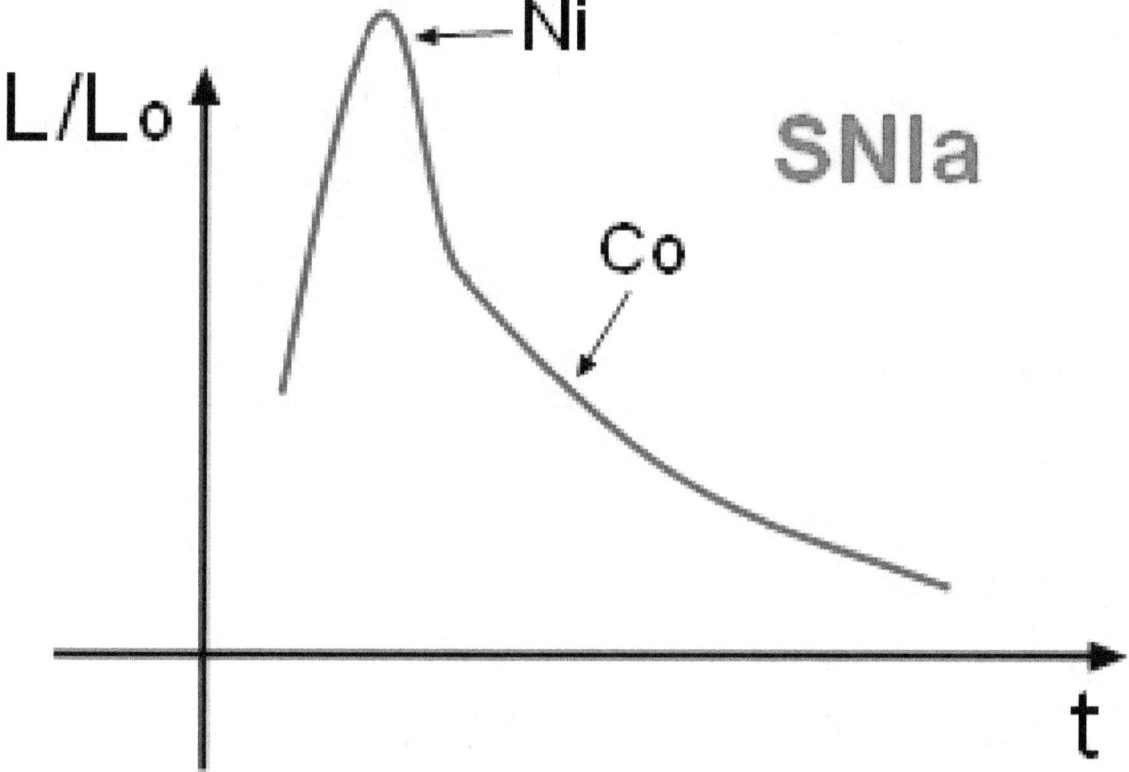

*This plot of luminosity (relative to the Sun, $L_{\odot}$) versus time shows the characteristic light curve for a Type Ia supernova. The peak is primarily due to the decay of Nickel (Ni), while the later stage is powered by Cobalt (Co).*

Type Ia supernovae have a characteristic light curve, their graph of luminosity as a function of time after the explosion. Near the time of maximum luminosity, the spectrum contains lines of intermediate-mass elements from oxygen to calcium; these are the main constituents of the outer layers of the star. Months after the explosion, when the outer layers have expanded to the point of transparency, the spectrum is dominated by light emitted by material near the core of the star, heavy elements synthesized during the explosion; most prominently isotopes close to the mass of iron (or iron peak elements). The radioactive decay of nickel−56 through cobalt−56 to iron−56 produces high-energy photons which dominate the energy output of the ejecta at intermediate to late times.[14]

The use of Type Ia supernovae to measure precise distances was pioneered by a collaboration of Chilean and US astronomers, the Calán/Tololo Supernova Survey.[45] In a series of papers in the 1990s the survey showed that while Type Ia supernovae do not all reach the same peak luminosity, a single parameter measured from the light curve can be used to correct unreddened Type Ia supernovae to standard candle values. The original correction to standard candle value is known as the Phillips relationship [46] and was shown by this group to be able to measure relative distances to 7% accuracy.[47] The cause of this uniformity in peak brightness is related to the amount of nickel-56 produced in white dwarfs presumably exploding near the Chandrasekhar limit.[48]

The similarity in the absolute luminosity profiles of nearly all known Type Ia supernovae has led to their use as a secondary standard candle in extragalactic astronomy. [49] Improved calibrations of the Cepheid variable distance scale [50] and direct geometric distance measurements to NGC 4258 from the dynamics of maser emission [51] when combined with the Hubble diagram of the Type Ia supernova distances have led to an improved value of the Hubble constant.

In 1998, observations of distant Type Ia supernovae indicated the unexpected result that the Universe seems to undergo an accelerating expansion.[52][53] Three members from two teams were subsequently awarded Nobel Prizes for this discovery.[54]

## 12.4   Two distinct types

Recently it has been discovered that type Ia supernovae which were considered the same are in fact different, moreover a form of the type Ia supernova which is relatively infrequent today was far more common earlier in the history of the universe. This could have far reaching cosmological significance and could lead to revision of estimation of the rate of expansion of the universe and the prevalence of dark energy. More research is needed.[55][56]

## 12.5   See also

- Carbon detonation

- History of supernova observation

- Supernova remnant

- Extragalactic Distance Scale

## 12.6   References

[1] HubbleSite - Dark Energy - Type Ia Supernovae

[2] Yoon, S.-C.; Langer, L. (2004). "Presupernova Evolution of Accreting White Dwarfs with Rotation". *Astronomy and Astrophysics* **419** (2): 623–644. arXiv:astro-ph/0402287. Bibcode:2004A&A...419..623Y. doi:10.1051/0004-6361:20035822. Retrieved 2007-05-30.

[3] Mazzali, P. A.; Röpke, F. K.; Benetti, S.; Hillebrandt, W. (2007). "A Common Explosion Mechanism for Type Ia Supernovae". *Science* **315** (5813): 825–828. arXiv:astro-ph/0702351. Bibcode:2007Sci...315..825M. doi:10.1126/science.1136259. PMID 17289993.

[4]  Khokhlov, A.; Müller, E.; Höflich, P. (1993). "Light curves of Type IA supernova models with different explosion mechanisms". *Astronomy and Astrophysics* **270** (1–2): 223–248. Bibcode:1993A&A...270..223K.

[5]  Staff (2006-09-07). "Introduction to Supernova Remnants". NASA Goddard/SAO. Retrieved 2007-05-01.

[6]  Johnson, Michele; Chandler, Lynn (May 20, 2015). "NASA Spacecraft Capture Rare, Early Moments of Baby Supernovae". *NASA*. Retrieved May 21, 2015.

[7]  Matheson, Thomas; Kirshner, Robert; Challis, Pete; Jha, Saurabh; et al. (2008). "Optical Spectroscopy of Type Ia Supernovae". *Astronomical Journal* **135** (4): 1598–1615. arXiv:0803.1705. Bibcode:2008AJ....135.1598M. doi:10.1088/0004-6256/135/4/1598.

[8]  da Silva, L. A. L. (1993). "The Classification of Supernovae". *Astrophysics and Space Science* **202** (2): 215–236. Bibcode:1 doi:10.1007/BF00626878.

[9]  Type Ia Supernovae: Why Our Standard Candle Isn't Really Standard

[10]  Lieb, E. H.; Yau, H.-T. (1987). "A rigorous examination of the Chandrasekhar theory of stellar collapse". *Astrophysical Journal* **323** (1): 140–144. Bibcode:1987ApJ...323..140L. doi:10.1086/165813.

[11]  Canal, R.; Gutiérrez, J. (1997). "The possible white dwarf-neutron star connection". *Astrophysics and Space Science Library*. Astrophysics and Space Science Library **214**: 49–55. arXiv:astro-ph/9701225. Bibcode:1997astro.ph..1225C. doi:10.1007/978-94-011-5542-7_7. ISBN 978-0-7923-4585-5.

[12]  Fryer, C. L.; New, K. C. B. (2006-01-24). "2.1 Collapse scenario". *Gravitational Waves from Gravitational Collapse*. Max-Planck-Gesellschaft. Retrieved 2007-06-07.

[13]  Wheeler, J. Craig (2000-01-15). *Cosmic Catastrophes: Supernovae, Gamma-Ray Bursts, and Adventures in Hyperspace*. Cambridge, UK: Cambridge University Press. p. 96. ISBN 0-521-65195-6.

[14]  Hillebrandt, W.; Niemeyer, J. C. (2000). "Type IA Supernova Explosion Models". *Annual Review of Astronomy and Astrophysics* **38** (1): 191–230. arXiv:astro-ph/0006305. Bibcode:2000ARA&A..38..191H. doi:10.1146/annurev.astro.38.1.191.

[15]  "Science Summary". ASC / Alliances Center for Astrophysical Thermonuclear Flashes. 2001. Retrieved 2006-11-27.

[16]  Röpke, F. K.; Hillebrandt, W. (2004). "The case against the progenitor's carbon-to-oxygen ratio as a source of peak luminosity variations in Type Ia supernovae". *Astronomy and Astrophysics* **420** (1): L1–L4. arXiv:astro-ph/0403509. Bibcode:2004A&A...

[17]  Gamezo, V. N.; Khokhlov, A. M.; Oran, E. S.; Chtchelkanova, A. Y.; Rosenberg, R. O. (2003-01-03). "Thermonuclear Supernovae: Simulations of the Deflagration Stage and Their Implications". *Science* **299** (5603): 77–81. doi:10.1126/science.1078129. PMID 12446871. Retrieved 2006-11-28.

[18]  Gilmore, Gerry (2004). "The Short Spectacular Life of a Superstar". *Science* **304** (5697): 1915–1916. doi:10.1126/science.110 PMID 15218132. Retrieved 2007-05-01.

[19]  Paczynski, B. (July 28 – August 1, 1975). "Common Envelope Binaries". *Structure and Evolution of Close Binary Systems*. Cambridge, England: Dordrecht, D. Reidel Publishing Co. pp. 75–80. Bibcode:1976IAUS...73...75P.

[20]  Postnov, K. A.; Yungelson, L. R. (2006). "The Evolution of Compact Binary Star Systems". Living Reviews in Relativity. Retrieved 2007-01-08.

[21]  Langer, N.; Yoon, S.-C.; Wellstein, S.; Scheithauer, S. (2002). "On the evolution of interacting binaries which contain a white dwarf". In Gänsicke, B. T.; Beuermann, K.; Rein, K. *The Physics of Cataclysmic Variables and Related Objects, ASP Conference Proceedings*. San Francisco, California: Astronomical Society of the Pacific. p. 252. Bibcode:2002ASPC..261..252L.

[22]  González Hernández, J. I.; Ruiz-Lapuente, P.; Tabernero, H. M.; Montes, D.; Canal, R.; Méndez, J.; Bedin, L. R. (2012). "No surviving evolved companions of the progenitor of SN 1006". *Nature* **489** (7417): 533–536. doi:10.1038/nature11447. PMID 23018963. See also lay reference: John Matson (December 2012). "No Star Left Behind". *Scientific American* **307** (6). p. 16

[23]  Staff. "Type Ia Supernova Progenitors". Swinburne University. Retrieved 2007-05-20.

[24]  "Brightest supernova discovery hints at stellar collision". New Scientist. 2007-01-03. Retrieved 2007-01-06.

[25] Whipple, Fred L. (1939). "Supernovae and Stellar Collisions". *Proceedings of the National Academy of Sciences of the United States of America* **25** (3): 118–125. Bibcode:1939PNAS...25..118W. doi:10.1073/pnas.25.3.118.

[26] Rubin, V. C.; Ford, W. K. J. (1999). "A Thousand Blazing Suns: The Inner Life of Globular Clusters". *Mercury* **28**: 26. Bibcode:1999Mercu..28d..26M. Retrieved 2006-06-02.

[27] Middleditch, J. (2004). "A White Dwarf Merger Paradigm for Supernovae and Gamma-Ray Bursts". *The Astrophysical Journal* **601** (2): L167–L170. arXiv:astro-ph/0311484. Bibcode:2003astro.ph.11484M. doi:10.1086/382074.

[28] "Important Clue Uncovered for the Origins of a Type of Supernovae Explosion, Thanks to a Research Team at the University of Pittsburgh". University of Pittsburgh. Retrieved 23 March 2012.

[29] "The Weirdest Type Ia Supernova Yet". Lawrence Berkeley National Laboratory. 2006-09-20. Retrieved 2006-11-02.

[30] "Bizarre Supernova Breaks All The Rules". New Scientist. 2006-09-20. Retrieved 2007-01-08.

[31] Schaefer, Bradley E.; Pagnotta, Ashley (2012). "An absence of ex-companion stars in the type Ia supernova remnant SNR 0509-67.5". *Nature* **481** (7380): 164–166. Bibcode:2012Natur.481..164S. doi:10.1038/nature10692. PMID 22237107.

[32] "NASA'S Swift Narrows Down Origin of Important Supernova Class". NASA. Retrieved 24 March 2012.

[33] "NASA's Chandra Reveals Origin of Key Cosmic Explosions". Chandra X-Ray Observatory website. Retrieved 28 March 2012.

[34] Bo Wang; Stephen Justham; Zhanwen Han (2013). "Double-detonation explosions as progenitors of Type Iax supernovae". arXiv:1301.1047v1 [astro-ph.SR].

[35] Ryan J. Foley; P. J. Challis; R. Chornock; M. Ganeshalingam; W. Li; G. H. Marion; N. I. Morrell; G. Pignata; M. D. Stritzinger; J. M. Silverman; X. Wang; J. P. Anderson; A. V. Filippenko; W. L. Freedman; M. Hamuy; S. W. Jha; R. P. Kirshner; C. McCully; S. E. Persson; M. M. Phillips; D. E. Reichart; A. M. Soderberg (2012). "Type Iax Supernovae: A New Class of Stellar Explosion". arXiv:1212.2209v2 [astro-ph.SR].

[36] "Hubble finds supernova star system linked to potential 'zombie star'". SpaceDaily. 6 August 2014.

[37] van Dyk, Schuyler D. (1992). "Association of supernovae with recent star formation regions in late type galaxies". *Astronomical Journal* **103** (6): 1788–1803. Bibcode:1992AJ....103.1788V. doi:10.1086/116195.

[38] Hoeflich, N.; Deutschmann, A.; Wellstein, S.; Höflich, P. (1999). "The evolution of main sequence star + white dwarf binary systems towards Type Ia supernovae".*Astronomy and Astrophysics***362**: 1046–1064. arXiv:astro-ph/0008444. Bibcode:2000A&A

[39] Kotak, R. (December 2008). "Progenitors of Type Ia Supernovae". Written at Keele University, Keele, United Kingdom. In Evans, A.; Bode, M. F.; O'Brien, T. J.; Darnley, M. J. *RS Ophiuchi (2006) and the Recurrent Nova Phenomenon, proceedings of the conference held 12–14 June 2007*. ASP Conference Series **401**. San Francisco: Astronomical Society of the Pacific, 2008. p. 150. Bibcode:2008ASPC..401..150K.

[40] Nugent, Peter E.; Sullivan, Mark; Cenko, S. Bradley; Thomas, Rollin C.; Kasen, Daniel; Howell, D. Andrew; Bersier, David; Bloom, Joshua S.; Kulkarni, S. R.; Kandrashoff, Michael T.; Filippenko, Alexei V.; Silverman, Jeffrey M.; Marcy, Geoffrey W.; Howard, Andrew W.; Isaacson, Howard T.; Maguire, Kate; Suzuki, Nao; Tarlton, James E.; Pan, Yen-Chen; Bildsten, Lars; Fulton, Benjamin J.; Parrent, Jerod T.; Sand, David; Podsiadlowski, Philipp; Bianco, Federica B.; Dilday, Benjamin; Graham, Melissa L.; Lyman, Joe; James, Phil; et al. (December 2011). "Supernova 2011fe from an Exploding Carbon-Oxygen White Dwarf Star". *Nature* **480** (7377): 344–347. arXiv:1110.6201. Bibcode:2011Natur.480..344N. doi:10.1038/nature10644. PMID 22170680

[41] Dilday, B.; Howell, DA; Cenko, SB; Silverman, JM; Nugent, PE; Sullivan, M; Ben-Ami, S; Bildsten, L; Bolte, M; Endl, M; Filippenko, A. V.; Gnat, O; Horesh, A; Hsiao, E; Kasliwal, MM; Kirkman, D; Maguire, K; Marcy, GW; Moore, K; Pan, Y; Parrent, J. T.; Podsiadlowski, P; Quimby, RM; Sternberg, A; Suzuki, N; Tytler, DR; Xu, D; Bloom, JS; Gal-Yam, A; et al. (2012). "PTF11kx: A Type-Ia Supernova with a Symbiotic Nova Progenitor". *Science* **337** (6097): 942–5. arXiv:1207.1306. Bibcode:2012Sci...337..942D. doi:10.1126/science.1219164. PMID 22923575.

[42] Dilday, B.; Howell, DA; Cenko, SB; Silverman, JM; Nugent, PE; Sullivan, M; Ben-Ami, S; Bildsten, L; Bolte, M; Endl, M; Filippenko, A. V.; Gnat, O; Horesh, A; Hsiao, E; Kasliwal, MM; Kirkman, D; Maguire, K; Marcy, GW; Moore, K; Pan, Y; Parrent, J. T.; Podsiadlowski, P; Quimby, RM; Sternberg, A; Suzuki, N; Tytler, DR; Xu, D; Bloom, JS; Gal-Yam, A; et al. (24 August 2012). "PTF 11kx: A Type Ia Supernova with a Symbiotic Nova Progenitor". *Science* **337** (6097): 942–945. arXiv:1207.1306. Bibcode:2012Sci...337..942D. doi:10.1126/science.1219164. PMID 22923575.

[43] "The First-Ever Direct Observations of a Type Ia Supernova Progenitor System". *Scitech daily.* - popular account of the discovery

[44] Soker, Noam; Kashi, Amit; García-Berro, Enrique; Torres, Santiago; Camacho, Judit (2013). "Explaining the Type Ia supernova PTF 11kx with a violent prompt merger scenario". *Monthly Notices of the Royal Astronomical Society* **431**: 1541. Bibcode:2013MNRAS.431.1541S. doi:10.1093/mnras/stt271.

[45] Hamuy, M.; et al. (1993). "The 1990 Calan/Tololo Supernova Search".*Astronomical Journal***106**(6): 2392. Bibcode:1993AJ. doi:10.1086/116811.

[46] Phillips, M. M. (1993). "The absolute magnitudes of Type IA supernovae". *Astrophysical Journal Letters* **413** (2): L105. Bibcode:1993ApJ...413L.105P. doi:10.1086/186970.

[47] Hamuy, M.; et al. (1996). "The Absolute Luminosities of the Calan/Tololo Type IA Supernovae". *Astronomical Journal* **112**: 2391. arXiv:astro-ph/9609059. Bibcode:1996AJ....112.2391H. doi:10.1086/118190.

[48] Colgate, S. A. (1979). "Supernovae as a standard candle for cosmology".*Astrophysical Journal***232**(1): 404–408. Bibcode:1979 doi:10.1086/157300.

[49] Hamuy, M.; et al. (1996). "A Hubble diagram of distant type IA supernovae".*Astronomical Journal***109**: 1. Bibcode:1995AJ. doi:10.1086/117251.

[50] Freedman, W.; et al. "Final Results from the Hubble Space Telescope Key Project to Measure the Hubble Constant". *Astrophysical Journal* **553** (1): 47–72. arXiv:astro-ph/0012376. Bibcode:2001ApJ...553...47F. doi:10.1086/320638.

[51] Macri, L. M.; Stanek, K. Z.; Bersier, D.; Greenhill, L. J.; Reid, M. J. (2006). "A New Cepheid Distance to the Maser-Host Galaxy NGC 4258 and Its Implications for the Hubble Constant". *Astrophysical Journal* **652** (2): 1133–1149. arXiv:astro-ph/0608211. Bibcode:2006ApJ...652.1133M. doi:10.1086/508530.

[52] Perlmutter S, Supernova Cosmology Project, Goldhaber G, Knop RA, Nugent P, Castro PG, Deustua S, Fabbro S, Goobar A, Groom DE, Hook IM, Kim AG, Kim MY, Lee JC, Nunes NJ, Pain R, Pennypacker CR, Quimby R, Lidman C, Ellis RS, Irwin M, McMahon RG, Ruiz-Lapuente P, Walton N, Schaefer B, Boyle BJ, Filippenko AV, Matheson T, Fruchter AS, et al. (1999). "Measurements of Omega and Lambda from 42 high redshift supernovae". *Astrophysical Journal* **517** (2): 565–86. arXiv:astro-ph/9812133. Bibcode:1999ApJ...517..565P. doi:10.1086/307221.

[53] Riess AG, et al. (1998). "Observational evidence from supernovae for an accelerating Universe and a cosmological constant". *Astronomical Journal* **116** (3): 1009–38. arXiv:astro-ph/9805201. Bibcode:1998AJ....116.1009R. doi:10.1086/300499.

[54] *Cosmology*, Steven Weinberg, Oxford University Press, 2008

[55] Accelerating universe? Not so fast

[56] Nielsen, J. T.; Guffanti, A.; Sarkar, S. (2015). "Marginal evidence for cosmic acceleration from Type Ia supernovae". arXiv:150

## 12.7 External links

- Falck, Bridget (2006). "Type Ia Supernova Cosmology with ADEPT". Johns Hopkins University. Retrieved 2007-05-20.

- Staff (February 27, 2007). "Sloan Supernova Survey". Sloan Digital Sky Survey. Retrieved 2007-05-25.

- "Novae and Supernovae". peripatus.gen.nz. Retrieved 2007-05-25.

- "Source for major type of supernova". Pole Star Publications Ltd. August 6, 2003. Retrieved 2007-11-25. (A Type Ia progenitor found)

- "Novae and Supernovae explosions found". peripatus.gen.nz. Retrieved 2007-05-25.

- SNFactory Shows Type Ia 'Standard Candles' Have Many Masses (March 4, 2014)

# Chapter 13

# Type Ib and Ic supernovae

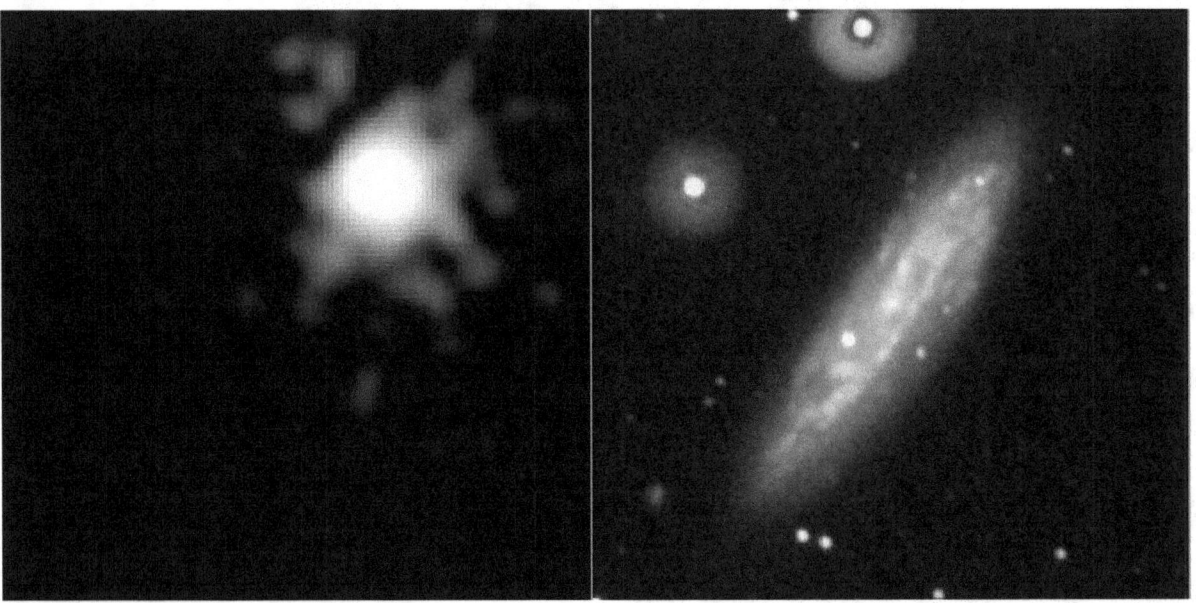

*The Type Ib supernova Supernova 2008D[1][2] in galaxy NGC 2770, shown in X-ray (left) and visible light (right), at the corresponding positions of the images. NASA image.[3]*

**Types Ib and Ic supernovae** are categories of stellar explosions that are caused by the core collapse of massive stars. These stars have shed (or been stripped of) their outer envelope of hydrogen, and, when compared to the spectrum of Type Ia supernovae, they lack the absorption line of silicon. Compared to Type Ib, Type Ic supernovae are hypothesized to have lost more of their initial envelope, including most of their helium. The two types are usually referred to as **stripped core-collapse supernovae**.

## 13.1   Spectra

When a supernova is observed, it can be categorized in the Minkowski–Zwicky supernova classification scheme based upon the absorption lines that appear in its spectrum.[4] A supernova is first categorized as either a Type I or Type II, then sub-categorized based on more specific traits. Supernovae belonging to the general category Type I lack hydrogen lines in their spectra; in contrast to Type II supernovae which do display lines of hydrogen. The Type I category is sub-divided into Type Ia, Type Ib and Type Ic supernovae.[5]

Type Ib/Ic supernovae are distinguished from Type Ia by the lack of an absorption line of singly ionized silicon at a wavelength of 635.5 nanometres.[6] As Type Ib/Ic supernovae age, they also display lines from elements such as oxygen, calcium and magnesium. In contrast, Type Ia spectra become dominated by lines of iron.[7] Type Ic supernovae are distinguished from Type Ib in that the former also lack lines of helium at 587.6 nm.[7]

## 13.2 Formation

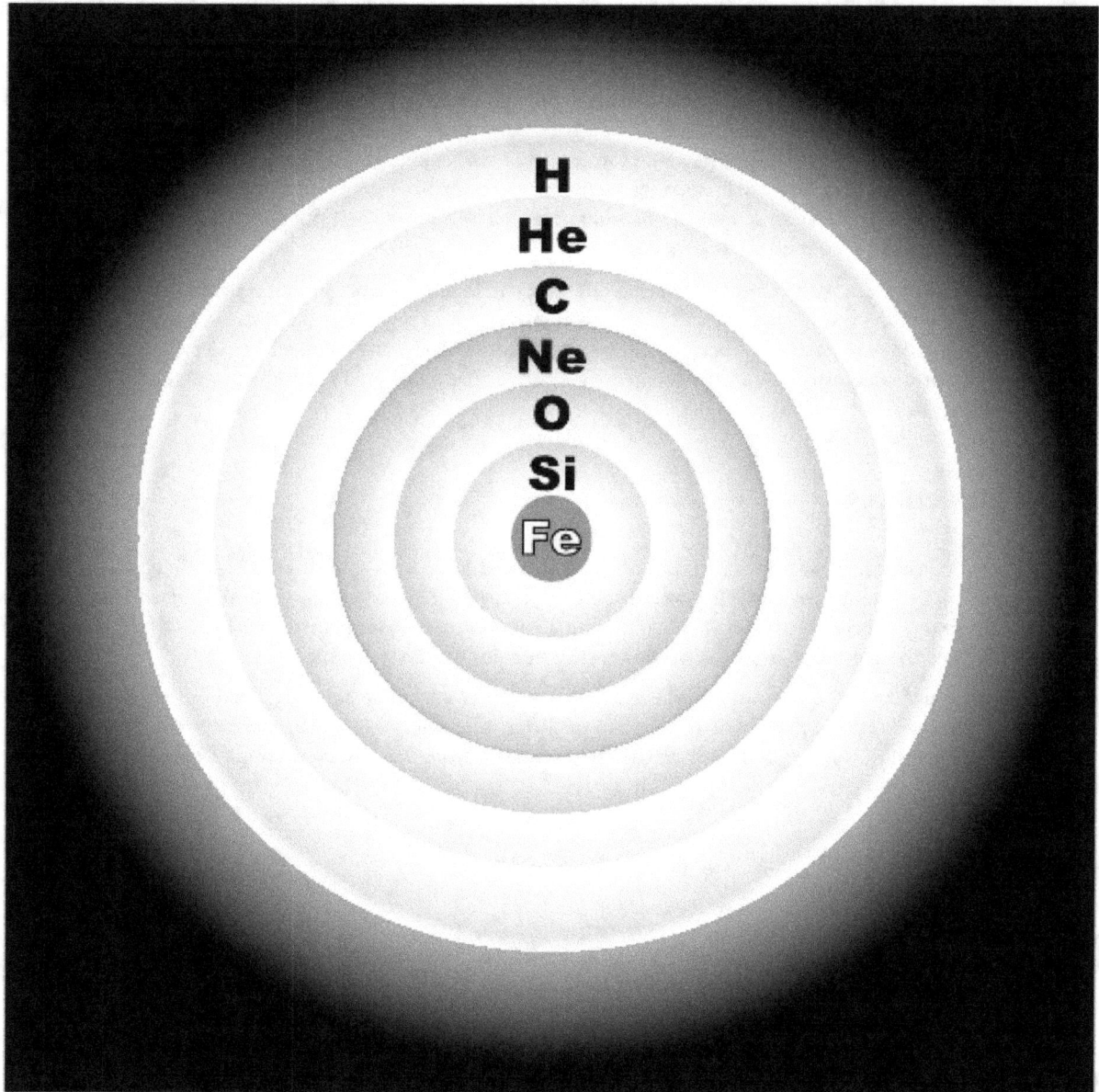

*The onion-like layers of an evolved, massive star (not to scale).*

Prior to becoming a supernova, an evolved massive star is organized in the manner of an onion, with layers of different elements undergoing fusion. The outermost layer consists of hydrogen, followed by helium, carbon, oxygen, and so forth. Thus when the outer envelope of hydrogen is shed, this exposes the next layer that consists primarily of helium (mixed with other elements). This can occur when a very hot, massive star reaches a point in its evolution when significant mass loss is occurring from its stellar wind. Highly massive stars (with 25 or more times the mass of the Sun) can lose up to

$10^{-5}$ solar masses ($M\odot$) each year—the equivalent of 1 $M\odot$ every 100,000 years.[8]

Type Ib and Ic supernovae are hypothesized to have been produced by core collapse of massive stars that have lost their outer layer of hydrogen and helium, either via winds or mass transfer to a companion.[6] The progenitors of Types Ib and Ic have lost most of their outer envelopes due to strong stellar winds or else from interaction with a close companion of about 3–4 $M\odot$.[9][10] Rapid mass loss can occur in the case of a Wolf-Rayet star, and these massive objects show a spectrum that is lacking in hydrogen. Type Ib progenitors have ejected most of the hydrogen in their outer atmospheres, while Type Ic progenitors have lost both the hydrogen and helium shells; in other words, Type Ic have lost more of their envelope (i.e., much of the helium layer) than the progenitors of Type Ib.[6] In other respects, however, the underlying mechanism behind Type Ib and Ic supernovae is similar to that of a Type II supernova, thus placing Type Ib/c between Type Ia and Type II.[6] Because of their similarity, Type Ib and Ic supernovae are sometimes collectively called Type Ibc supernovae.[11]

There is some evidence that a small percent of the Type Ic supernovae may be the progenitors of gamma ray bursts (GRB); in particular, type Ic supernovae that have broad spectral lines corresponding to high-velocity outflows are thought to be strongly associated with gamma ray bursts (GRB). However, it is also hypothesized that any hydrogen-stripped Type Ib or Ic supernova could be a GRB, dependent upon the geometry of the explosion.[12] In any case, astronomers believe that most Type Ib, and probably Type Ic as well, result from core collapse in stripped, massive stars, rather than from the thermonuclear runaway of white dwarfs.[6]

As they are formed from rare, very massive stars, the rate of Type Ib and Ic supernovae occurrence is much lower than the corresponding rate for Type II supernovae.[13] They normally occur in regions of new star formation, and have never been observed in an elliptical galaxy.[10] Because they share a similar operating mechanism, Type Ib/c and the various Type II supernovae are collectively called core-collapse supernovae. In particular, Type Ib/c may be referred to as *stripped core-collapse supernovae*.[6]

## 13.3   Light curves

The light curves (a plot of luminosity versus time) of Type Ib supernovae vary in form, but in some cases can be nearly identical to those of Type Ia supernovae. However, Type Ib light curves may peak at lower luminosity and may be redder. In the infrared portion of the spectrum, the light curve of a Type Ib supernova is similar to a Type II-L light curve. (See Supernova.)[14] Type Ib supernovae usually have slower decline rates for the spectral curves than Ic.[6]

Type Ia supernovae light curves are useful for measuring distances on a cosmological scale. That is, they serve as standard candles. However, due to the similarity of the spectra of Type Ib and Ic supernovae, the latter can form a source of contamination of supernova surveys and must be carefully removed from the observed samples before making distance estimates.[15]

## 13.4   See also

- Supernova

- Type Ia supernova

## 13.5   References

[1] Malesani, D.; et al. (2008). "Early spectroscopic identification of SN 2008D". *Astrophys. J.* **692** (2): L84–L87. arXiv:0805.1188. Bibcode:2009ApJ...692L..84M. doi:10.1088/0004-637X/692/2/L84.

[2] Soderberg, A.M.; et al. (2008). "An extremely luminous X-ray outburst at the birth of a supernova". *Nature* **453** (7194): 469–74. arXiv:0802.1712. Bibcode:2008Natur.453..469S. doi:10.1038/nature06997. PMID 18497815.

[3] Naeye, R.; Gutro, R. (21 May 2008). "NASA's Swift Satellite Catches First Supernova in the Act of Exploding". NASA/GSFC. Retrieved 2008-05-22.

[4] da Silva, L.A.L. (1993). "The Classification of Supernovae". *Astrophysics and Space Science* **202**(2): 215–236. Bibcode:1993 doi:10.1007/BF00626878.

[5] Montes, M. (12 February 2002). "Supernova Taxonomy". Naval Research Laboratory. Retrieved 2006-11-09.

[6] Filippenko, A.V. (2004). "Supernovae and Their Massive Star Progenitors". arXiv:astro-ph/0412029 [astro-ph].

[7] "Type Ib Supernova Spectra". *COSMOS - The SAO Encyclopedia of Astronomy*. Swinburne University of Technology. Retrieved 2010-05-05.

[8] Dray, L.M.; Tout, C.A.; Karaks, A.I.; Lattanzio, J.C. (2003). "Chemical enrichment by Wolf-Rayet and asymptotic giant branch stars". *Monthly Notices of the Royal Astronomical Society* **338** (4): 973–989. Bibcode:2003MNRAS.338..973D. doi:10.1046/j.1365-8711.2003.06142.x. .

[9] Pols, O. (26 October – 1 November 1995). "Close Binary Progenitors of Type Ib/Ic and IIb/II-L Supernovae". *Proceedings of The Third Pacific Rim Conference on Recent Development on Binary Star Research*. Chiang Mai, Thailand. pp. 153–158. Bibcode:1997rdbs.conf..153P.

[10] Woosley, S. E.; Eastman, R.G. (June 20–30, 1995). "Type Ib and Ic Supernovae: Models and Spectra". *Proceedings of the NATO Advanced Study Institute*. Begur, Girona, Spain: Kluwer Academic Publishers. p. 821. Bibcode:1997thsu.conf..821W.

[11] Williams, A.J. "Initial Statistics from the Perth Automated Supernova Search". *Publications of the Astronomical Society of Australia* **14** (2): 208–13. Bibcode:1997PASA...14..208W. doi:10.1071/AS97208.

[12] Ryder, S.D.; et al. (2004). "Modulations in the radio light curve of the Type IIb supernova 2001ig: evidence for a Wolf-Rayet binary progenitor?". *Monthly Notices of the Royal Astronomical Society* **349** (3): 1093–1100. arXiv:astro-ph/0401135. Bibcode:2004MNRAS.349.1093R. doi:10.1111/j.1365-2966.2004.07589.x.

[13] Sadler, E.M.; Campbell, D. (1997). "A first estimate of the radio supernova rate". Astronomical Society of Australia. Retrieved 2007-02-08.

[14] Tsvetkov, D.Yu. (1987). "Light curves of type Ib supernova: SN 1984l in NGC 991". *Soviet Astronomy Letters* **13**: 376–378. Bibcode:1987SvAL...13..376T.

[15] Homeier, N.L. (2005). "The Effect of Type Ibc Contamination in Cosmological Supernova Samples". *The Astrophysical Journal* **620** (1): 12–20. arXiv:astro-ph/0410593. Bibcode:2005ApJ...620...12H. doi:10.1086/427060.

# Chapter 14

# Type II supernova

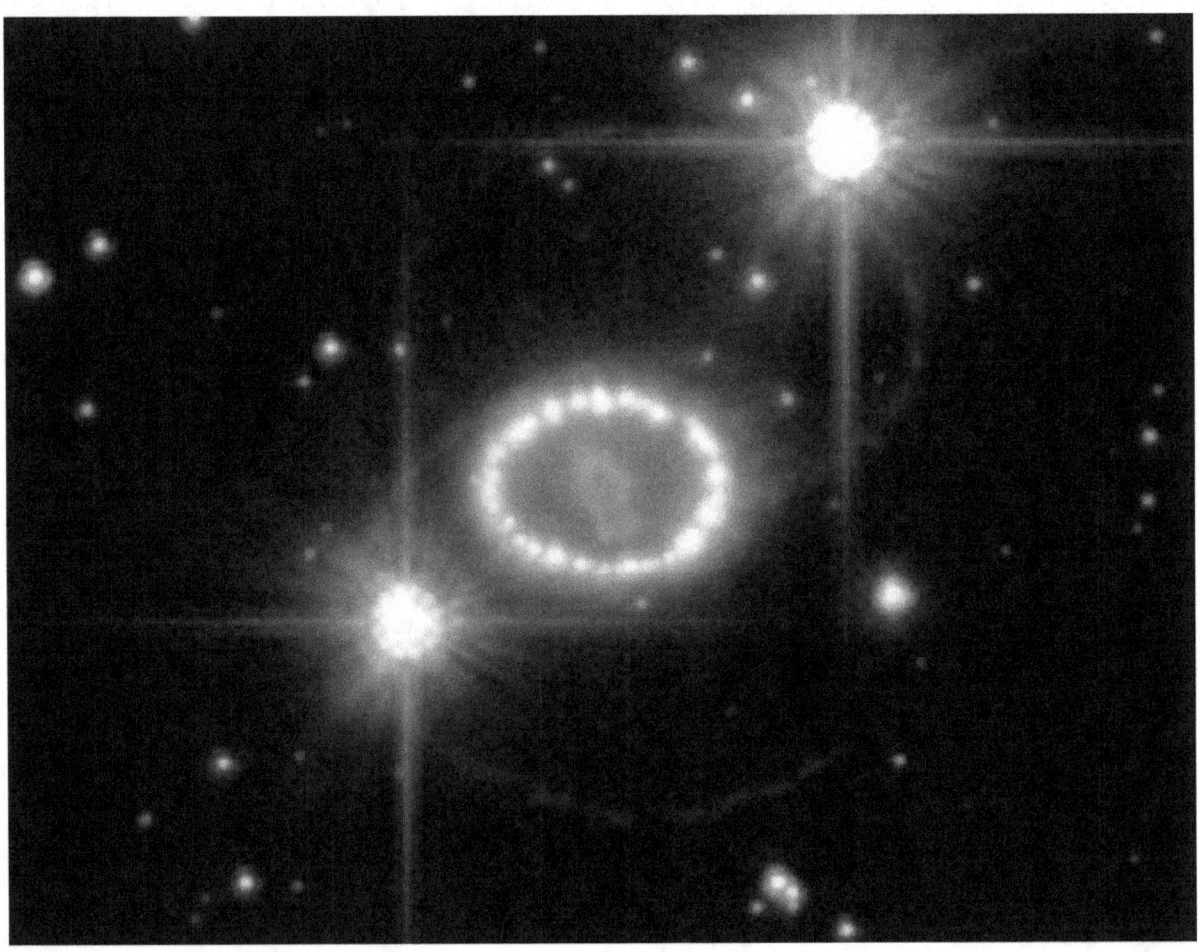

*The expanding remnant of SN 1987A, a Type II-P supernova in the Large Magellanic Cloud.* NASA image.

A **Type II supernova** (plural: *supernovae* or *supernovas*) results from the rapid collapse and violent explosion of a massive star. A star must have at least 8 times, and no more than 40–50 times, the mass of the Sun ($M\odot$) for this type of explosion.[1] It is distinguished from other types of supernovae by the presence of hydrogen in its spectrum. Type II supernovae are mainly observed in the spiral arms of galaxies and in H II regions, but not in elliptical galaxies.

Stars generate energy by the nuclear fusion of elements. Unlike the Sun, massive stars possess the mass needed to fuse elements that have an atomic mass greater than hydrogen and helium, albeit at increasingly higher temperatures and pressures, causing increasingly shorter stellar life spans. The degeneracy pressure of electrons and the energy generated by these fusion reactions are sufficient to counter the force of gravity and prevent the star from collapsing, maintaining stellar equilibrium. The star fuses increasingly higher mass elements, starting with hydrogen and then helium, progressing up through the periodic table until a core of iron and nickel is produced. Fusion of iron or nickel produces no net energy output, so no further fusion can take place, leaving the nickel-iron core inert. Due to the lack of energy output allowing outward pressure, equilibrium is broken.

When the mass of the inert core exceeds the Chandrasekhar limit of about 1.4 $M\odot$, electron degeneracy alone is no longer sufficient to counter gravity and maintain stellar equilibrium. A cataclysmic implosion takes place within seconds, in which the outer core reaches an inward velocity of up to 23% of the speed of light and the inner core reaches temperatures of up to 100 billion kelvin. Neutrons and neutrinos are formed via reversed beta-decay, releasing about $10^{46}$ joules (100 foe) in a ten-second burst. The collapse is halted by neutron degeneracy, causing the implosion to rebound and bounce outward. The energy of this expanding shock wave is sufficient to accelerate the surrounding stellar material to escape velocity, forming a supernova explosion, while the shock wave and extremely high temperature and pressure briefly allow for the production of elements heavier than iron.[2] Depending on initial size of the star, the remnants of the core form a neutron star or a black hole. Because of the underlying mechanism, the resulting nova is also described as a core-collapse supernova.

There exist several categories of Type II supernova explosions, which are categorized based on the resulting light curve—a graph of luminosity versus time—following the explosion. Type II-L supernovae show a steady (linear) decline of the light curve following the explosion, whereas Type II-P display a period of slower decline (a plateau) in their light curve followed by a normal decay. Type Ib and Ic supernovae are a type of core-collapse supernova for a massive star that has shed its outer envelope of hydrogen and (for Type Ic) helium. As a result, they appear to be lacking in these elements.

## 14.1 Formation

Stars far more massive than the sun evolve in more complex ways. In the core of the star, hydrogen is fused into helium, releasing thermal energy that heats the sun's core and provides outward pressure that supports the sun's layers against collapse in a process known as stellar or hydrostatic equilibrium. The helium produced in the core accumulates there since temperatures in the core are not yet high enough to cause it to fuse. Eventually, as the hydrogen at the core is exhausted, fusion starts to slow down, and gravity causes the core to contract. This contraction raises the temperature high enough to initiate a shorter phase of helium fusion, which accounts for less than 10% of the star's total lifetime. In stars with fewer than eight solar masses, the carbon produced by helium fusion does not fuse, and the star gradually cools to become a white dwarf.[3][4] White dwarf stars, if they have a near companion, may then become Type Ia supernovae.

A much larger star, however, is massive enough to create temperatures and pressures needed to cause the carbon in the core to begin to fuse once the star contracts at the end of the helium-burning stage. The cores of these massive stars become layered like onions as progressively heavier atomic nuclei build up at the center, with an outermost layer of hydrogen gas, surrounding a layer of hydrogen fusing into helium, surrounding a layer of helium fusing into carbon via the triple-alpha process, surrounding layers that fuse to progressively heavier elements. As a star this massive evolves, it undergoes repeated stages where fusion in the core stops, and the core collapses until the pressure and temperature are sufficient to begin the next stage of fusion, reigniting to halt collapse.[3][4]

## 14.2 Core collapse

The factor limiting this process is the amount of energy that is released through fusion, which is dependent on the binding energy that holds together these atomic nuclei. Each additional step produces progressively heavier nuclei, which release progressively less energy when fusing. In addition, from carbon-burning onwards, energy loss via neutrino production becomes significant, leading to a higher rate of reaction than would otherwise take place.[6] This continues until nickel-56 is produced, which decays radioactively into cobalt-56 and then iron-56 over the course of a few months. As iron and

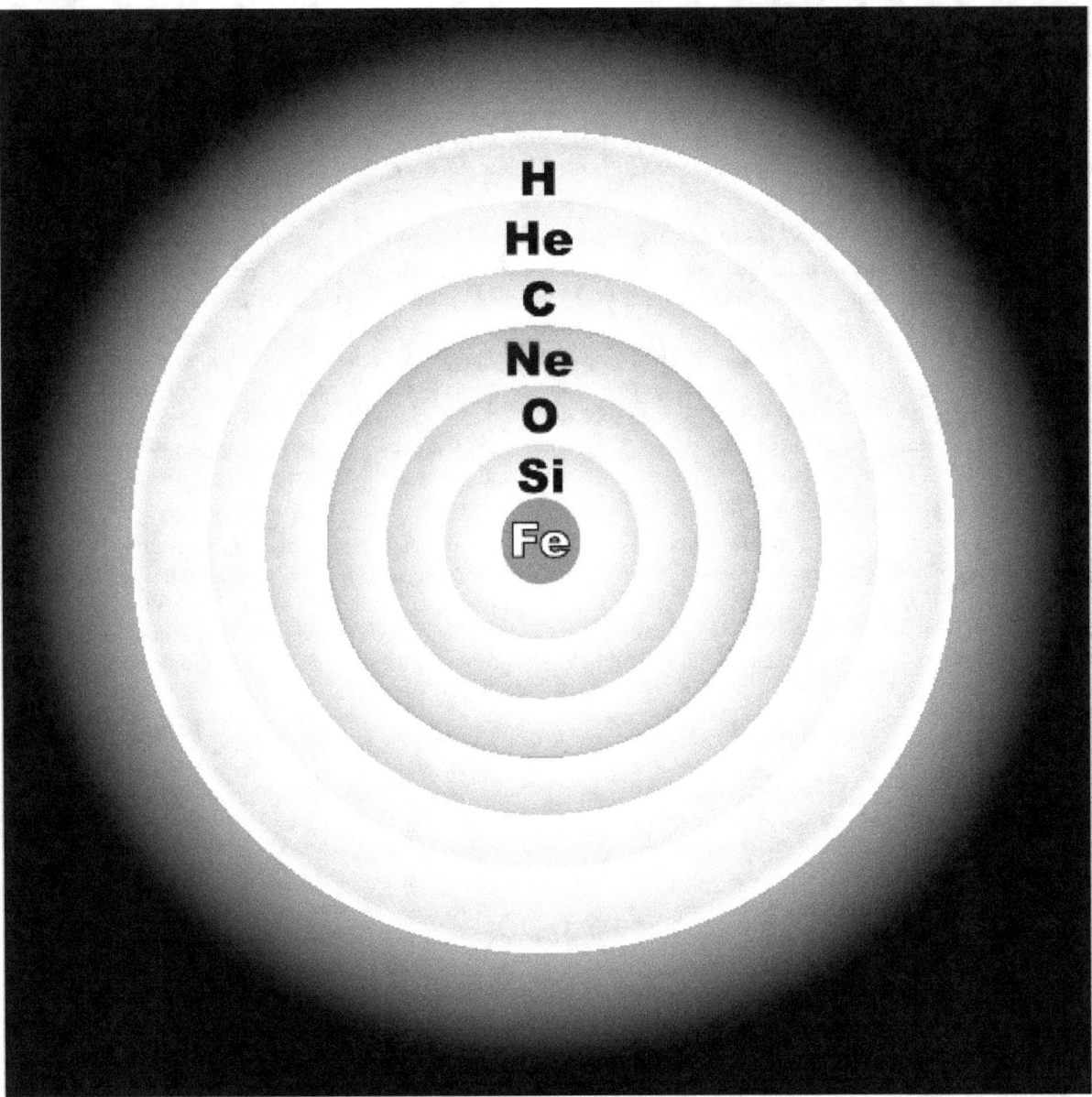

*The onion-like layers of a massive, evolved star just before core collapse. (Not to scale.)*

nickel have the highest binding energy per nucleon of all the elements,[7] energy cannot be produced at the core by fusion, and a nickel-iron core grows.[4][8] This core is under huge gravitational pressure. As there is no fusion to further raise the star's temperature to support it against collapse, it is supported only by degeneracy pressure of electrons. In this state, matter is so dense that further compaction would require electrons to occupy the same energy states. However, this is forbidden for identical fermion particles, such as the electron – a phenomenon called the Pauli exclusion principle.

When the core's mass exceeds the Chandrasekhar limit of about 1.4 $M\odot$, degeneracy pressure can no longer support it, and catastrophic collapse ensues.[9] The outer part of the core reaches velocities of up to 70,000 km/s (23% of the speed of light) as it collapses toward the center of the star.[10] The rapidly shrinking core heats up, producing high-energy gamma rays that decompose iron nuclei into helium nuclei and free neutrons via photodisintegration. As the core's density increases, it becomes energetically favorable for electrons and protons to merge via inverse beta decay, producing neutrons and elementary particles called neutrinos. Because neutrinos rarely interact with normal matter, they can escape from the core, carrying away energy and further accelerating the collapse, which proceeds over a timescale of milliseconds. As the core detaches from the outer layers of the star, some of these neutrinos are absorbed by the star's outer layers, beginning the supernova explosion.[11]

For Type II supernovae, the collapse is eventually halted by short-range repulsive neutron-neutron interactions, mediated by the strong force, as well as by degeneracy pressure of neutrons, at a density comparable to that of an atomic nucleus. Once collapse stops, the infalling matter rebounds, producing a shock wave that propagates outward. The energy from this shock dissociates heavy elements within the core. This reduces the energy of the shock, which can stall the explosion within the outer core.[12]

The core collapse phase is so dense and energetic that only neutrinos are able to escape. As the protons and electrons combine to form neutrons by means of electron capture, an electron neutrino is produced. In a typical Type II supernova, the newly formed neutron core has an initial temperature of about 100 billion kelvin, $10^4$ times the temperature of the sun's core. Much of this thermal energy must be shed for a stable neutron star to form, otherwise the neutrons would "boil away". This is accomplished by a further release of neutrinos.[13] These 'thermal' neutrinos form as neutrino-antineutrino pairs of all flavors, and total several times the number of electron-capture neutrinos.[14] The two neutrino production mechanisms convert the gravitational potential energy of the collapse into a ten-second neutrino burst, releasing about $10^{46}$ joules (100 foe).[15]

Through a process that is not clearly understood, about $10^{44}$ joules (1 foe) is reabsorbed by the stalled shock, producing an explosion.[a][12] The neutrinos generated by a supernova were actually observed in the case of Supernova 1987A, leading astronomers to conclude that the core collapse picture is basically correct. The water-based Kamiokande II and IMB instruments detected antineutrinos of thermal origin,[13] while the gallium−71-based Baksan instrument detected neutrinos (lepton number = 1) of either thermal or electron-capture origin.

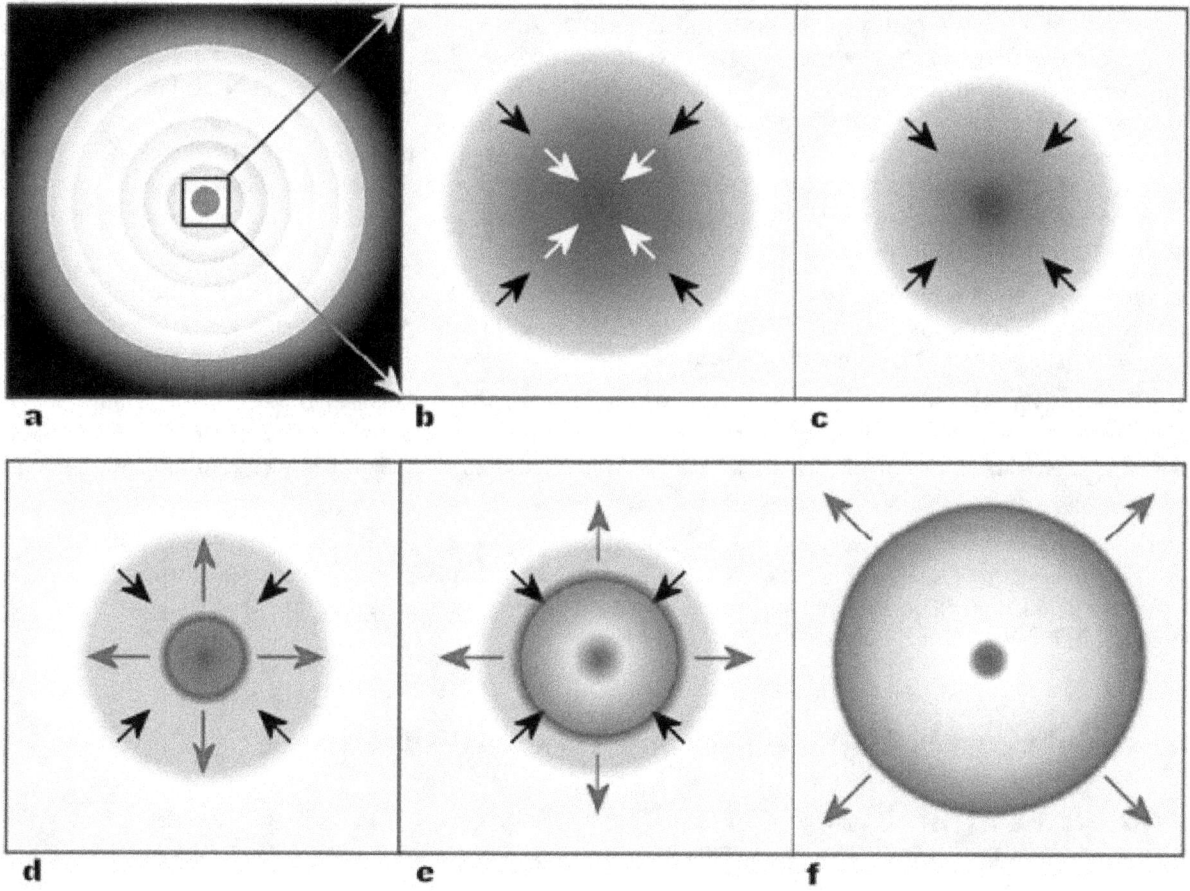

*Within a massive, evolved star (a) the onion-layered shells of elements undergo fusion, forming a nickel-iron core (b) that reaches Chandrasekhar-mass and starts to collapse. The inner part of the core is compressed into neutrons (c), causing infalling material to bounce (d) and form an outward-propagating shock front (red). The shock starts to stall (e), but it is re-invigorated by neutrino interaction. The surrounding material is blasted away (f), leaving only a degenerate remnant.*

When the progenitor star is below about 20 $M\odot$ – depending on the strength of the explosion and the amount of material that falls back – the degenerate remnant of a core collapse is a neutron star.[10] Above this mass, the remnant collapses to form a black hole.[4][16] The theoretical limiting mass for this type of core collapse scenario is about 40–50 $M\odot$. Above that mass, a star is believed to collapse directly into a black hole without forming a supernova explosion,[17] although uncertainties in models of supernova collapse make calculation of these limits uncertain.

## 14.3   Theoretical models

The Standard Model of particle physics is a theory which describes three of the four known fundamental interactions between the elementary particles that make up all matter. This theory allows predictions to be made about how particles will interact under many conditions. The energy per particle in a supernova is typically one to one hundred and fifty picojoules (tens to hundreds of MeV).[18] The per-particle energy involved in a supernova is small enough that the predictions gained from the Standard Model of particle physics are likely to be basically correct. But the high densities may require corrections to the Standard Model.[19] In particular, Earth-based particle accelerators can produce particle interactions which are of much higher energy than are found in supernovae,[20] but these experiments involve individual particles interacting with individual particles, and it is likely that the high densities within the supernova will produce novel effects. The interactions between neutrinos and the other particles in the supernova take place with the weak nuclear force, which is believed to be well understood. However, the interactions between the protons and neutrons involve the strong nuclear force, which is much less well understood.[21]

The major unsolved problem with Type II supernovae is that it is not understood how the burst of neutrinos transfers its energy to the rest of the star producing the shock wave which causes the star to explode. From the above discussion, only one percent of the energy needs to be transferred to produce an explosion, but explaining how that one percent of transfer occurs has proven very difficult, even though the particle interactions involved are believed to be well understood. In the 1990s, one model for doing this involved convective overturn, which suggests that convection, either from neutrinos from below, or infalling matter from above, completes the process of destroying the progenitor star. Heavier elements than iron are formed during this explosion by neutron capture, and from the pressure of the neutrinos pressing into the boundary of the "neutrinosphere", seeding the surrounding space with a cloud of gas and dust which is richer in heavy elements than the material from which the star originally formed.[22]

Neutrino physics, which is modeled by the Standard Model, is crucial to the understanding of this process.[19] The other crucial area of investigation is the hydrodynamics of the plasma that makes up the dying star; how it behaves during the core collapse determines when and how the "shock wave" forms and when and how it "stalls" and is reenergized.[23]

In fact, some theoretical models incorporate a hydrodynamical instability in the stalled shock known as the "Standing Accretion Shock Instability" (SASI). This instability comes about as a consequence of non-spherical perturbations oscillating the stalled shock thereby deforming it. The SASI is often used in tandem with neutrino theories in computer simulations for re-energizing the stalled shock.[24]

Computer models have been very successful at calculating the behavior of Type II supernovae once the shock has been formed. By ignoring the first second of the explosion, and assuming that an explosion is started, astrophysicists have been able to make detailed predictions about the elements produced by the supernova and of the expected light curve from the supernova.[25][26][27]

## 14.4   Light curves for Type II-L and Type II-P supernovae

When the spectrum of a Type II supernova is examined, it normally displays Balmer absorption lines – reduced flux at the characteristic frequencies where hydrogen atoms absorb energy. The presence of these lines is used to distinguish this category of supernova from a Type I supernova.

When the luminosity of a Type II supernova is plotted over a period of time, it shows a characteristic rise to a peak brightness followed by a decline. These light curves have an average decay rate of 0.008 magnitudes per day; much lower than the decay rate for Type Ia supernovae. Type II are sub-divided into two classes, depending on the shape of the light curve. The light curve for a Type II-L supernova shows a steady (linear) decline following the peak brightness.

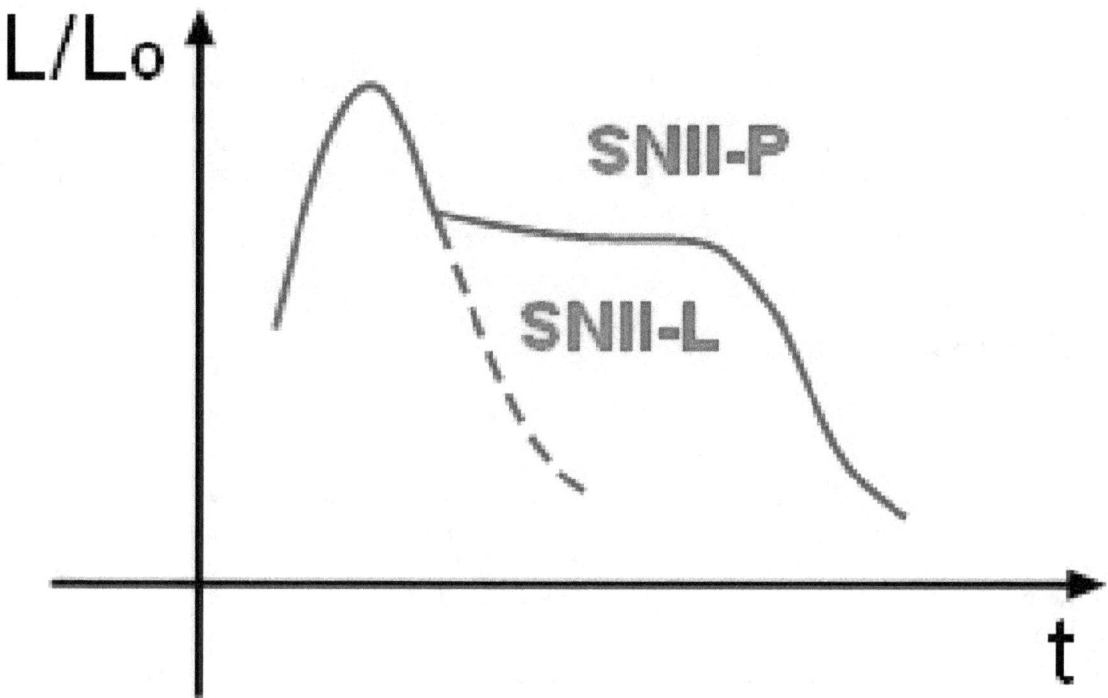

*This graph of the luminosity as a function of time shows the characteristic shapes of the light curves for a Type II-L and II-P supernova.*

By contrast, the light curve of a Type II-P supernova has a distinctive flat stretch (called a plateau) during the decline; representing a period where the luminosity decays at a slower rate. The net luminosity decay rate is lower, at 0.0075 magnitudes per day for Type II-P, compared to 0.012 magnitudes per day for Type II-L.[28]

The difference in the shape of the light curves is believed to be caused, in the case of Type II-L supernovae, by the expulsion of most of the hydrogen envelope of the progenitor star.[28] The plateau phase in Type II-P supernovae is due to a change in the opacity of the exterior layer. The shock wave ionizes the hydrogen in the outer envelope – stripping the electron from the hydrogen atom – resulting in a significant increase in the opacity. This prevents photons from the inner parts of the explosion from escaping. Once the hydrogen cools sufficiently to recombine, the outer layer becomes transparent.[29]

## 14.5   Type IIn supernovae

The "n" denotes narrow, which indicates the presence of intermediate or very narrow width H emission lines in the spectra. In the intermediate width case, the ejecta from the explosion may be interacting strongly with gas around the star – the circumstellar medium. [30][31] The estimated circumstellar density required to explain the observational properties is much higher than that expected from the standard stellar evolution theory.[32] It is generally assumed that the high circumstellar density is due to the high mass-loss rates of the SN IIn progenitors. The estimated mass-loss rates are typically higher than $10^{-3}$ M$\odot$ yr$^{-1}$. There are indications that they originate as stars similar to Luminous blue variables with large mass losses before exploding.[33] SN 1998S and SN 2005gl are famous examples of Type IIn; SN 2006gy, an extremely energetic supernova, may be another example.[34]

## 14.6   Type IIb supernovae

A *Type IIb supernova* has a weak hydrogen line in its initial spectrum, which is why it is classified as a Type II. However, later on the H emission becomes undetectable, and there is also a second peak in the light curve that has a spectrum which more closely resembles a Type Ib supernova. The progenitor could have been a giant star which lost most of its hydrogen envelope due to interactions with a companion in a binary system, leaving behind the core that consisted almost entirely of helium.[35] As the ejecta of a Type IIb expands, the hydrogen layer quickly becomes more transparent and reveals the deeper layers.[35] The classic example of a Type IIb supernova is Supernova 1993J,[36][37] while another example is Cassiopeia A.[38] The IIb class was first introduced (as a theoretical concept) by Ensman & Woosley 1987.

## 14.7   Hypernovae (collapsars)

Main article: Hypernova

Hypernovae are a rare type of supernova substantially more luminous and energetic than standard supernovae. Examples are 1997ef (type Ic) and 1997cy (type IIn). Hypernovae are produced by more than one type of event: relativistic jets during formation of a black hole from fallback of material onto the neutron star core, the collapsar model; interaction with a dense envelope of circumstellar material, the CSM model; the highest mass pair instability supernovae; possibly others such as binary and quark star model.

Stars with initial masses between about 25 and 90 times the sun develop cores large enough that after a supernova explosion, some material will fall back onto the neutron star core and create a black hole. In many cases this reduces the luminosity of the supernova, and above 90 $M\odot$ the star collapses directly into a black hole without a supernova explosion. However, if the progenitor is spinning quickly enough the infalling material generates relativistic jets that emit more energy than the original explosion.[39] They may also be seen directly if beamed towards us, giving the impression of an even more luminous object. In some cases these can produce gamma-ray bursts, although not all gamma-ray bursts are from supernovae.[40]

In some cases a type II supernova occurs when the star is surrounded by a very dense cloud of material, most likely expelled during luminous blue variable eruptions. This material is shocked by the explosion and becomes more luminous than a standard supernova. It is likely that there is a range of luminosities for these type IIn supernovae with only the brightest qualifying as a hypernova.

Pair instability supernovae occur when an oxygen core in an extremely massive star becomes hot enough that gamma rays spontaneously produce electron-positron pairs.[41] This causes the core to collapse, but where the collapse of an iron core causes endothermic fusion to heavier elements, the collapse of an oxygen core creates runaway exothermic fusion which completely unbinds the star. The total energy emitted depends on the initial mass, with much of the core being converted to $^{56}$Ni and ejected which then powers the supernova for many months. At the lower end stars of about 140 $M\odot$ produce supernovae that are long-lived but otherwise typical, while the highest mass stars of around 250 $M\odot$ produce supernovae that are extremely luminous and also very long lived; hypernovae. More massive stars die by photodisintegration. Only population III stars, with very low metallicity, can reach this stage. Stars with more heavy elements are more opaque and blow away their outer layers until they are small enough to explode as a normal type Ib/c supernova. It is thought that even in our own galaxy, mergers of old low metallicity stars may form massive stars capable of creating a pair instability supernova.

## 14.8   See also

- History of supernova observation

- Supernova nucleosynthesis

- Supernova remnant

# 14.9 References

[1] Gilmore, Gerry (2004). "The Short Spectacular Life of a Superstar". *Science* **304** (5697): 1915–1916. doi:10.1126/science. PMID 15218132.

[2] Staff (2006-09-07). "Introduction to Supernova Remnants". NASA Goddard/SAO. Retrieved 2007-05-01.

[3] Richmond, Michael. "Late stages of evolution for low-mass stars". Rochester Institute of Technology. Retrieved 2006-08-04.

[4] Hinshaw, Gary (2006-08-23). "The Life and Death of Stars". NASA Wilkinson Microwave Anisotropy Probe (WMAP) Mission. Retrieved 2006-09-01.

[5] Woosley, S.; Janka, H.-T. (December 2005). "The Physics of Core-Collapse Supernovae". *Nature Physics* **1** (3): 147–154. arXiv:astro-ph/0601261. Bibcode:2005NatPh...1..147W. doi:10.1038/nphys172.

[6] Clayton, Donald (1983). *Principles of Stellar Evolution and Nucleosynthesis*. University of Chicago Press. ISBN 978-0-226-10953-4.

[7] Fewell, M. P. (1995). "The atomic nuclide with the highest mean binding energy". *American Journal of Physics* **63** (7): 653–658. Bibcode:1995AmJPh..63..653F. doi:10.1119/1.17828.

[8] Fleurot, Fabrice. "Evolution of Massive Stars". Laurentian University. Retrieved 2007-08-13.

[9] Lieb, E. H.; Yau, H.-T. (1987). "A rigorous examination of the Chandrasekhar theory of stellar collapse". *Astrophysical Journal* **323** (1): 140–144. Bibcode:1987ApJ...323..140L. doi:10.1086/165813.

[10] Fryer, C. L.; New, K. C. B. (2006-01-24). "Gravitational Waves from Gravitational Collapse". Max Planck Institute for Gravitational Physics. Retrieved 2006-12-14.

[11] Hayakawa, T.; Iwamoto, N.; Kajino, T.; Shizuma, T.; Umeda, H.; Nomoto, K. (2006). "Principle of Universality of Gamma-Process Nucleosynthesis in Core-Collapse Supernova Explosions". *The Astrophysical Journal* **648** (1): L47–L50. Bibcode:2006 doi:10.1086/507703.

[12] Fryer, C. L.; New, K. B. C. (2006-01-24). "Gravitational Waves from Gravitational Collapse, section 3.1". Los Alamos National Laboratory. Retrieved 2006-12-09.

[13] Mann, Alfred K. (1997). *Shadow of a star: The neutrino story of Supernova 1987A*. New York: W. H. Freeman. p. 122. ISBN 0-7167-3097-9.

[14] Gribbin, John R.; Gribbin, Mary (2000). *Stardust: Supernovae and Life – The Cosmic Connection*. New Haven: Yale University Press. p. 173. ISBN 978-0-300-09097-0.

[15] Barwick, S.; Beacom, J.; et al. (2004-10-29). "APS Neutrino Study: Report of the Neutrino Astrophysics and Cosmology Working Group" (PDF). American Physical Society. Retrieved 2006-12-12.

[16] Fryer, Chris L. (2003). "Black Hole Formation from Stellar Collapse". *Classical and Quantum Gravity* **20** (10): S73–S80. Bibcode:2003CQGra..20S..73F. doi:10.1088/0264-9381/20/10/309.

[17] Fryer, Chris L. (1999). "Mass Limits For Black Hole Formation". *The Astrophysical Journal* **522** (1): 413–418. arXiv:astro-ph/9902315. Bibcode:1999ApJ...522..413F. doi:10.1086/307647.

[18] Izzard, R. G.; Ramirez-Ruiz, E.; Tout, C. A. (2004). "Formation rates of core-collapse supernovae and gamma-ray bursts". *Monthly Notices of the Royal Astronomical Society* **348** (4): 1215. arXiv:astro-ph/0311463. Bibcode:2004MNRAS.348.1215I. doi:10.1111/j.1365-2966.2004.07436.x.

[19] Rampp, M.; Buras, R.; Janka, H.-Th.; Raffelt, G. (February 11–16, 2002). "Core-collapse supernova simulations: Variations of the input physics". *Proceedings of the 11th Workshop on "Nuclear Astrophysics"*. Ringberg Castle, Tegernsee, Germany. pp. 119–125. Bibcode:2002nuas.conf..119R.

[20] The OPAL Collaboration; Ackerstaff, K.; et al. (1998). "Tests of the Standard Model and Constraints on New Physics from Measurements of Fermion-pair Production at 189 GeV at LEP". *Submitted to The European Physical Journal C* **2** (3): 441–472. doi:10.1007/s100529800851. Retrieved 2007-03-18.

[21] Staff (2004-10-05). "The Nobel Prize in Physics 2004". Nobel Foundation. Retrieved 2007-05-30.

[22] Stover, Dawn (2006). "Life In A Bubble". *Popular Science* **269** (6): 16.

[23] Janka, H.-Th.; Langanke, K.; Marek, A.; Martinez-Pinedo, G.; Mueller, B. (2006). "Theory of Core-Collapse Supernovae". *Bethe Centennial Volume of Physics Reports (submitted)* **142** (1–4): 229. arXiv:astro-ph/0612072. Bibcode:1993JHyd..142..229H. doi:10.1016/0022-1694(93)90012-X.

[24] Wakana Iwakami; Kei Kotake; Naofumi Ohnishi; Shoichi Yamada; Keisuke Sawada (March 10–15, 2008). "3D Simulations of Standing Accretion Shock Instability in Core-Collapse Supernovae" (PDF). *3D Simulations of Standing Accretion Shock Instability in Core-Collapse Supernovae*. 14th Workshop on "Nuclear Astrophysics". Retrieved 30 January 2013.

[25] Blinnikov, S.I.; Röpke, F. K.; Sorokina, E. I.; Gieseler, M.; Reinecke, M.; Travaglio, C.; Hillebrandt, W.; Stritzinger, M. (2006). "Theoretical light curves for deflagration models of type Ia supernova". *Astronomy and Astrophysics* **453** (1): 229–240. arXiv:astro-ph/0603036. Bibcode:2006A&A...453..229B. doi:10.1051/0004-6361:20054594.

[26] Young, Timothy R. (2004). "A Parameter Study of Type II Supernova Light Curves Using 6 M He Cores". *The Astrophysical Journal* **617** (2): 1233–1250. arXiv:astro-ph/0409284. Bibcode:2004ApJ...617.1233Y. doi:10.1086/425675.

[27] Rauscher, T.; Heger, A.; Hoffman, R. D.; Woosley, S. E. (2002). "Nucleosynthesis in Massive Stars With Improved Nuclear and Stellar Physics". *The Astrophysical Journal* **576** (1): 323–348. arXiv:astro-ph/0112478. Bibcode:2002ApJ...576..323R. doi:10.1086/341728.

[28] Doggett, J. B.; Branch, D. (1985). "A Comparative Study of Supernova Light Curves". *Astronomical Journal* **90**: 2303–2311. Bibcode:1985AJ.....90.2303D. doi:10.1086/113934.

[29] "Type II Supernova Light Curves". Swinburne University of Technology. Retrieved 2007-03-17.

[30] Filippenko, A. V. (1997). "Optical Spectra of Supernovae". *Annual Review of Astronomy and Astrophysics* **35**: 309–330. Bibcode:1997ARA&A..35..309F. doi:10.1146/annurev.astro.35.1.309.

[31] Pastorello, A.; Turatto, M.; Benetti, S.; Cappellaro, E.; Danziger, I. J.; Mazzali, P. A.; Patat, F.; Filippenko, A. V.; Schlegel, D. J.; Matheson, T. (2002). "The type IIn supernova 1995G: interaction with the circumstellar medium". *Monthly Notices of the Royal Astronomical Society* **333** (1): 27–38. arXiv:astro-ph/0201483. Bibcode:2002MNRAS.333...27P. doi:10.1046/j.1365-8711.2002.05366.x.

[32] Langer, N. (22 September 2012). "Presupernova Evolution of Massive Single and Binary Stars". *Annual Review of Astronomy and Astrophysics* **50** (1): 107–164. arXiv:1206.5443. Bibcode:2012ARA&A..50..107L. doi:10.1146/annurev-astro-081811-125534.

[33] Michael Kiewe; Avishay Gal-Yam; Iair Arcavi; Leonard; Emilio Enriquez; Bradley Cenko; Fox; Dae-Sik Moon; Sand; Soderberg, Alicia M.; Cccp, The (2010). "Caltech Core-Collapse Project (CCCP) observations of type IIn supernovae: typical properties and implications for their progenitor stars". *ApJ* **744** (10): 10. arXiv:1010.2689. Bibcode:2012ApJ...744...10K. doi:10.1088/0004-637X/744/1/10.

[34] Smith, N.; Chornock, R.; Silverman, J. M.; Filippenko, A. V.; Foley, R. J. (2010). "Spectral Evolution of the Extraordinary Type IIn Supernova 2006gy" (pdf). *The Astrophysical Journal* **709** (2): 856–883. arXiv:0906.2200. Bibcode:2010ApJ...709..856S. doi:10.1088/0004-637X/709/2/856.

[35] Utrobin, V. P. (1996). "Nonthermal ionization and excitation in Type IIb supernova 1993J". *Astronomy and Astrophysics* **306** (5940): 219–231. Bibcode:1996A&A...306..219U.

[36] Nomoto, K.; Suzuki, T.; Shigeyama, T.; Kumagai, S.; Yamaoka, H.; Saio, H. (1993). "A type IIb model for supernova 1993J". *Nature* **364** (6437): 507. Bibcode:1993Natur.364..507N. doi:10.1038/364507a0.

[37] Chevalier, R. A.; Soderberg, A. M. (2010). "Type IIb Supernovae with Compact and Extended Progenitors". *The Astrophysical Journal* **711**: L40. arXiv:0911.3408. Bibcode:2010ApJ...711L..40C. doi:10.1088/2041-8205/711/1/L40.

[38] Krause, O.; Birkmann, S.; Usuda, T.; Hattori, T.; Goto, M.; Rieke, G.; Misselt, K. (2008). "The Cassiopeia A supernova was of type IIb". *Science* **320** (5880): 1195–1197. arXiv:0805.4557. Bibcode:2008Sci...320.1195K. doi:10.1126/science.1155788. PMID 18511684.

[39] Nomoto, K. I.; Tanaka, M.; Tominaga, N.; Maeda, K. (2010). "Hypernovae, gamma-ray bursts, and first stars". *New Astronomy Reviews* **54** (3–6): 191. Bibcode:2010NewAR..54..191N. doi:10.1016/j.newar.2010.09.022.

[40] "Cosmological Gamma-Ray Bursts and Hypernovae Conclusively Linked". European Organisation for Astronomical Research in the Southern Hemisphere (ESO). 2003-06-18. Retrieved 2006-10-30.

[41] Kasen, D.; Woosley, S. E.; Heger, A. (2011). "Pair Instability Supernovae: Light Curves, Spectra, and Shock Breakout" (pdf). *The Astrophysical Journal* **734** (2): 102. arXiv:1101.3336. Bibcode:2011ApJ...734..102K. doi:10.1088/0004-637X/734/2/102.

## 14.10    External links

- Merrifield, Michael. "Type II Supernova". *Sixty Symbols*. Brady Haran for the University of Nottingham.

# Chapter 15

# Light curve

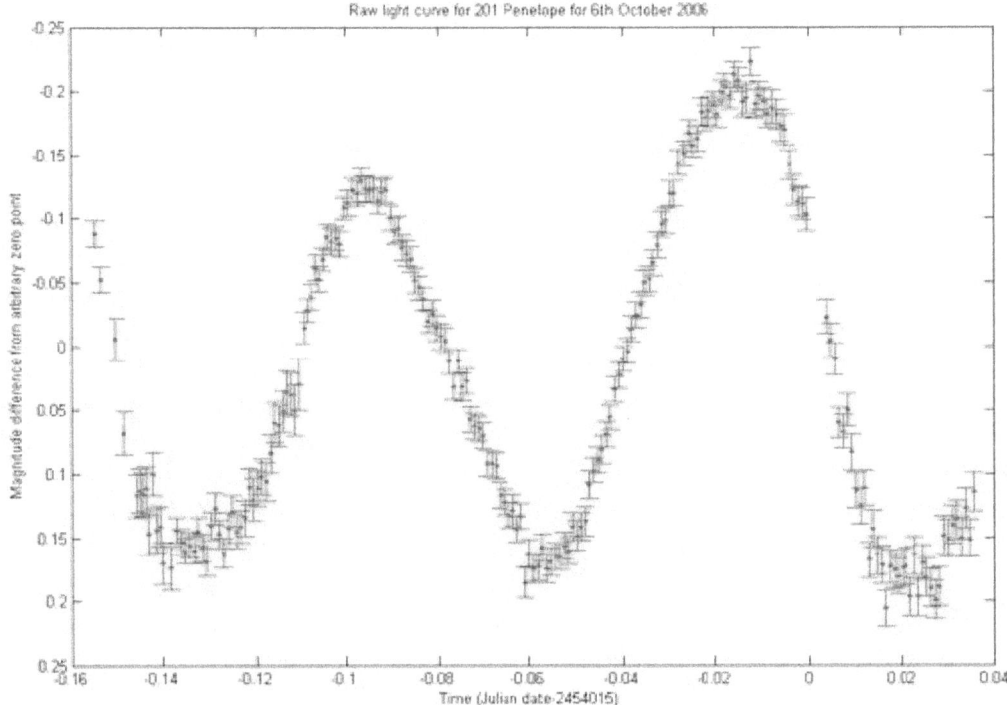

*Light curve of the asteroid 201 Penelope based on images taken on 6 October 2006 at Mount John University Observatory. Shows just over one full rotation, which lasts 3.7474 hours.*

In astronomy, a **light curve** is a graph of light intensity of a celestial object or region, as a function of time. The light is usually in a particular frequency interval or band. Light curves can be periodic, as in the case of eclipsing binaries, Cepheid variables, other periodic variables, and transiting extrasolar planets, or aperiodic, like the light curve of a nova, a cataclysmic variable star, a supernova or a microlensing event. The study of the light curve, together with other observations, can yield considerable information about the physical process that produces it or constrain the physical theories about it.[1] Light waves can also be used in botany to determine a plant's reactions to light intensities.

## 15.1 Astronomy

Main article: Supernova

In astronomy, light curves from a supernova are used to determine what type of supernova it is. If the supernova's light curve has a sharp maximum and slopes down gradually, then it is a type I supernova. If the supernova's light curve has a less sharp maximum, slopes down quickly, and then levels off, it is a type II supernova.[2]

## 15.2 Planetology

In planetology, a light curve can be used to estimate the rotation period of a minor planet, moon, or comet nucleus. From the Earth there is often no way to resolve a small object in our Solar System, even in the most powerful of telescopes, since the apparent angular size of the object is smaller than one pixel in the detector. Thus, astronomers measure the amount of light produced by an object as a function of time (the light curve). The time separation of peaks in the light curve gives an estimate of the rotational period of the object. The difference between the maximum and minimum brightnesses (the amplitude of the light curve) can be due to the shape of the object, or to bright and dark areas on its surface. For example, an asymmetrical asteroid's light curve generally has more pronounced peaks, while a more spherical object's light curve will be flatter.[3]

## 15.3 Botany

In botany, a light curve shows the photosynthetic response of leaf tissue or algal communities to varying light intensities. The shape of the curve illustrates the principle of limiting factors; in low light levels, the rate of photosynthesis is limited by the concentration of chlorophyll and the efficiency of the light-dependent reactions, but in higher light levels it is limited by the efficiency of RuBisCo and the availability of carbon dioxide. The point on the curve where these two differing slopes meet is called the light saturation point and is where the light-dependent reactions are producing more ATP and NADPH than can be utilized by the light-independent reactions. Since photosynthesis is also limited by ambient carbon dioxide levels, light curves are often repeated at several different constant carbon dioxide concentrations.[4]

## 15.4 References

[1] S. V. H. Haugan *Separating intrinsic and microlensing variability using parallax measurements* (astro-ph/9508112. August 1995)

[2] "Supernova". *Georgia State University – Hyperphysics – Carl Rod Nave*. 1998.

[3] Harris, A. W.; Warner, B. D.; Pravec, P. (Eds.) (2006). "Asteroid Lightcurve Derived Data. EAR-A-5-DDR-DERIVED-LIGHTCURVE-V8.0.". NASA Planetary Data System. Archived from the original on January 28, 2007. Retrieved 2007-03-15.

[4] Smith, E.L. (August 1936). "Photosynthesis in Relation to Light and Carbon Dioxide".*PNAS***22**(8): 504–511. Bibcode:1936P doi:10.1073/pnas.22.8.504. JSTOR 86299. PMC 1079215. PMID 16577734.

## 15.5 External links

- The AAVSO online light curve generator can plot light curves for thousands of variable stars

- Lightcurves: An Introduction by NASA's Imagine the Universe

# Chapter 16

# Apparent magnitude

For a more detailed discussion of the history of the magnitude system, see Magnitude (astronomy).

The **apparent magnitude** (*m*) of a celestial object is a measure of its brightness as seen by an observer on Earth, adjusted

*Asteroid 65 Cybele and two stars, with their magnitudes labeled*

to the value it would have in the absence of the atmosphere. The brighter an object appears, the lower its magnitude value (i.e. inverse relation). In addition, the magnitude scale is logarithmic: a difference of one in magnitude corresponds to a change in brightness by a factor of about 2.5.

Generally, the visible spectrum (vmag) is used as a basis for the apparent magnitude. However, other spectra are also used (e.g. the near-infrared J-band). In the visible spectrum, Sirius is the brightest star after the Sun. In the near-infrared J-band, Betelgeuse is the brightest. The apparent magnitude of stars is measured with a bolometer.

## 16.1 History

The scale used to indicate magnitude originates in the Hellenistic practice of dividing stars visible to the naked eye into six *magnitudes*. The brightest stars in the night sky were said to be of first magnitude ($m = 1$), whereas the faintest were of sixth magnitude ($m = 6$), the limit of human visual perception (without the aid of a telescope). Each grade of magnitude was considered twice the brightness of the following grade (a logarithmic scale), although that ratio was subjective as no photodetectors existed. This rather crude scale for the brightness of stars was popularized by Ptolemy in his *Almagest*, and is generally believed to have originated with Hipparchus.

In 1856, Norman Robert Pogson formalized the system by defining a first magnitude star as a star that is 100 times as bright as a sixth-magnitude star, thereby establishing the logarithmic scale still in use today. This implies that a star of magnitude $m$ is 2.512 times as bright as a star of magnitude $m+1$. This figure, the fifth root of 100 became known as *Pogson's Ratio*.[4] The zero point of Pogson's scale was originally defined by assigning Polaris a magnitude of exactly 2. Astronomers later discovered that Polaris is slightly variable, so they switched to Vega as the standard reference star, assigning the brightness of Vega as the definition of zero magnitude at any specified wavelength.

Apart from small corrections, the brightness of Vega still serves as the definition of zero magnitude for visible and near infrared wavelengths, where its spectral energy distribution (SED) closely approximates that of a black body for a temperature of 11,000 K. However, with the advent of infrared astronomy it was revealed that Vega's radiation includes an Infrared excess presumably due to a circumstellar disk consisting of dust at warm temperatures (but much cooler than the star's surface). At shorter (e.g. visible) wavelengths, there is negligible emission from dust at these temperatures. However, in order to properly extend the magnitude scale further into the infrared, this peculiarity of Vega should not affect the definition of the magnitude scale. Therefore, the magnitude scale was extrapolated to *all* wavelengths on the basis of the black body radiation curve for an ideal stellar surface at 11,000 K uncontaminated by circumstellar radiation. On this basis the spectral irradiance (usually expressed in janskys) for the zero magnitude point, as a function of wavelength can be computed (see ). Small deviations are specified between systems using measurement appartuses developed independently so that data obtained by different astronomers can be properly compared; of greater practical importance is the definition of magnitude not at a single wavelength but applying to the response of standard spectral filters used in photometry over various wavelength bands.

With the modern magnitude systems, brightness over a very wide range is specified according to the logarithmic definition detailed below, using this zero reference. In practice such apparent magnitudes do not exceed 30 (for detectable measurements). The brightness of Vega is exceeded by four stars in the night sky at visible wavelengths (and more at infrared wavelengths) as well as bright planets such as Venus, Mars, and Jupiter, and these must be described by *negative* magnitudes. For example, Sirius, the brightest star of the celestial sphere, has an apparent magnitude of −1.4 in the visible; negative magnitudes for other very bright astronomical objects can be found in the table below.

## 16.2 Calculations

As the amount of light received actually depends on the thickness of the Earth's atmosphere in the line of sight to the object, the apparent magnitudes are adjusted to the value they would have in the absence of the atmosphere. The dimmer an object appears, the higher the numerical value given to its apparent magnitude. Note that brightness varies with distance; an extremely bright object may appear quite dim, if it is far away. More exactly, brightness varies inversely with the square of the distance. The absolute magnitude, $M$, of a celestial body (outside the Solar System) is the apparent magnitude it would have if it were at 10 parsecs (~32.6 light years) and that of a planet (or other Solar System body) is the apparent magnitude it would have if it were 1 astronomical unit from both the Sun and Earth. The absolute magnitude of the Sun is 4.83 in the V band (yellow) and 5.48 in the B band (blue).[5]

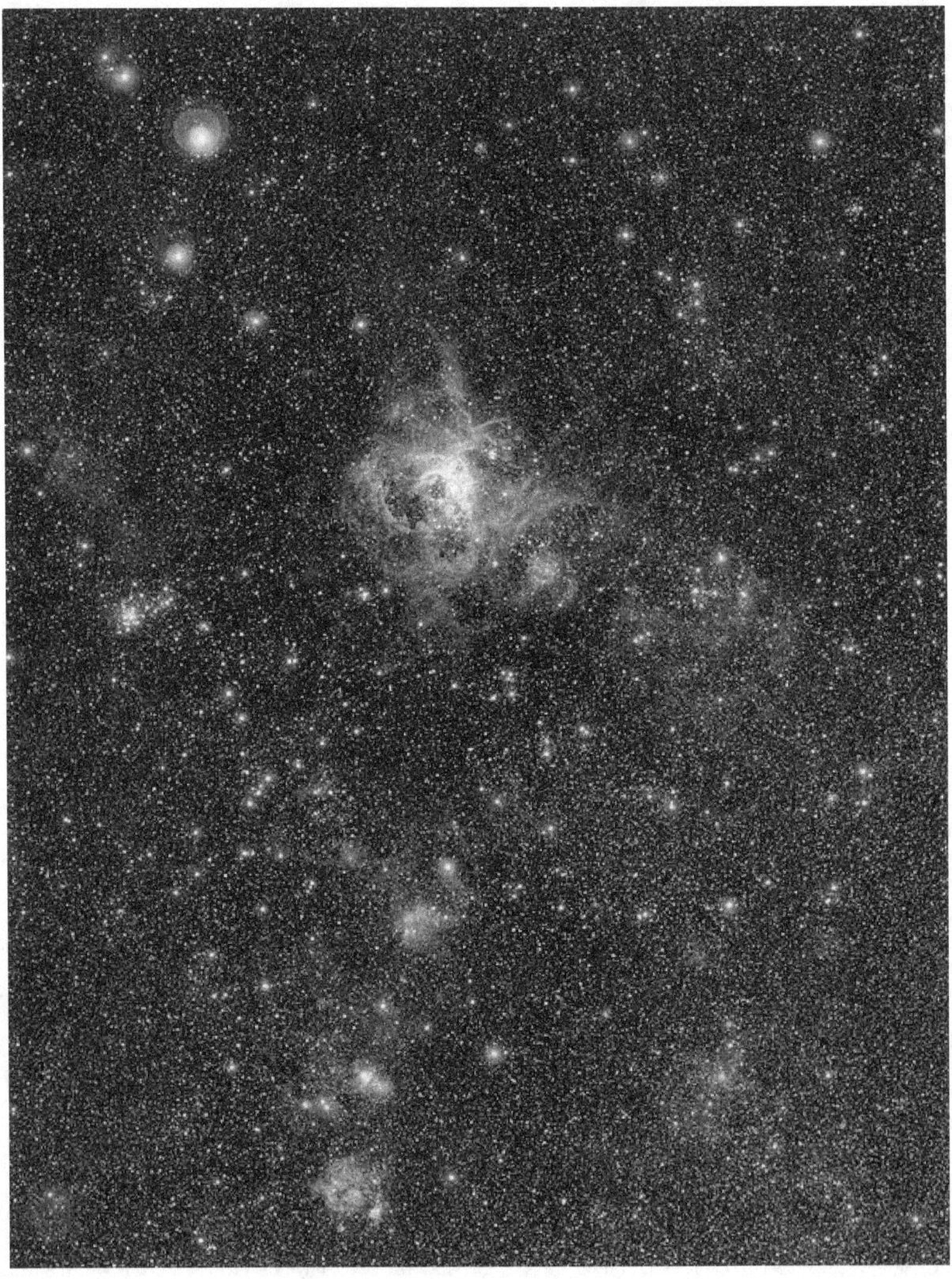

*30 Doradus image taken by ESO's VISTA. This nebula has an apparent magnitude of 8.*

The apparent magnitude, $m$, in the band, $x$, can be defined as,

$$m_x - m_{x,0} = -2.5 \log_{10} \left( \frac{F_x}{F_{x,0}} \right)$$

where $F_x$ is the observed flux in the band x, and $m_{x,0}$ and $F_{x,0}$ are a reference magnitude, and reference flux in the same band x, such as that of Vega. An increase of 1 in the magnitude scale corresponds to a decrease in brightness by a factor of $\approx 2.512$. Based on the properties of logarithms, a difference in magnitudes, $m_1 - m_2 = \Delta m$, can be converted to a variation in brightness as $F_2/F_1 \approx 2.512^{\Delta m}$.

### 16.2.1 Example: Sun and Moon

*What is the ratio in brightness between the Sun and the full moon?*

The apparent magnitude of the Sun is −26.74 (brighter), and the mean apparent magnitude of the full moon is −12.74 (dimmer).

**Difference in magnitude** :

$$x = m_1 - m_2 = (-12.74) - (-26.74) = 14.00$$

**Variation in Brightness** :

$$v_b = 2.512^x = 2.512^{14.00} \approx 400,000$$

The Sun appears about 400,000 times brighter than the full moon.

### 16.2.2 Magnitude addition

Sometimes, it might be useful to add magnitudes. For example, to determine the combined magnitude of a double star when the magnitudes of the individual components are known. This can be done by setting an equation using the brightness (in linear units) of each magnitude.[6]

$$2.512^{-m_f} = 2.512^{-m_1} + 2.512^{-m_2}$$

Solving for $m_f$ yields

$$m_f = -\log_{2.512} \left( 2.512^{-m_1} + 2.512^{-m_2} \right)$$

where $m_f$ is the resulting magnitude after adding $m_1$ and $m_2$. Note that the negative of each magnitude is used because greater intensities equate to lower magnitudes.

## 16.3 Standard reference values

It is important to note that the scale is logarithmic: the relative brightness of two objects is determined by the difference of their magnitudes. For example, a difference of 3.2 means that one object is about 19 times as bright as the other, because Pogson's Ratio raised to the power 3.2 is approximately 19.05. A common misconception is that the logarithmic nature of the scale is because the human eye itself has a logarithmic response. In Pogson's time this was thought to be true (see Weber-Fechner law), but it is now believed that the response is a power law (see Stevens' power law).[8]

Magnitude is complicated by the fact that light is not monochromatic. The sensitivity of a light detector varies according to the wavelength of the light, and the way it varies depends on the type of light detector. For this reason, it is necessary to specify how the magnitude is measured for the value to be meaningful. For this purpose the UBV system is widely used, in which the magnitude is measured in three different wavelength bands: U (centred at about 350 nm, in the near ultraviolet), B (about 435 nm, in the blue region) and V (about 555 nm, in the middle of the human visual range in daylight). The V band was chosen for spectral purposes and gives magnitudes closely corresponding to those seen by the light-adapted human eye, and when an apparent magnitude is given without any further qualification, it is usually the V magnitude that is meant, more or less the same as **visual magnitude**.

Because cooler stars, such as red giants and red dwarfs, emit little energy in the blue and UV regions of the spectrum their power is often under-represented by the UBV scale. Indeed, some L and T class stars have an estimated magnitude of well over 100, because they emit extremely little visible light, but are strongest in infrared.

Measures of magnitude need cautious treatment and it is extremely important to measure like with like. On early 20th century and older orthochromatic (blue-sensitive) photographic film, the relative brightnesses of the blue supergiant Rigel and the red supergiant Betelgeuse irregular variable star (at maximum) are reversed compared to what human eyes perceive, because this archaic film is more sensitive to blue light than it is to red light. Magnitudes obtained from this method are known as photographic magnitudes, and are now considered obsolete.

For objects within the Milky Way with a given absolute magnitude, 5 is added to the apparent magnitude for every tenfold increase in the distance to the object. This relationship does not apply for objects at very great distances (far beyond the Milky Way), because a correction for general relativity must then be taken into account due to the non-Euclidean nature of space.

For planets and other Solar System bodies the apparent magnitude is derived from its phase curve and the distances to the Sun and observer.

## 16.4   Table of notable celestial objects

Some of the above magnitudes are only approximate. Telescope sensitivity also depends on observing time, optical bandpass, and interfering light from scattering and airglow.

## 16.5   See also

- Absolute magnitude

- Magnitude (astronomy)

- Photographic magnitude

- Luminosity in astronomy

- List of brightest stars

- List of nearest bright stars

- List of nearest stars

- Lux

- Surface brightness

- Distance modulus

# 16.6 References

[1] "Vmag<6.5". SIMBAD Astronomical Database. Retrieved 2010-06-25.

[2] "Magnitude". National Solar Observatory—Sacramento Peak. Archived from the original on 2008-02-06. Retrieved 2006-08-23.

[3] Bright Star Catalogue

[4] Magnitudes of Thirty-six of the Minor Planets for the first day of each month of the year 1857, N. Pogson, MNRAS Vol. 17, p. 12 (1856)

[5] Prof. Aaron Evans. "Some Useful Astronomical Definitions" (PDF). Stony Brook Astronomy Program. Retrieved 2009-07-12.

[6] "Magnitude Arithmetic". *Weekly Topic*. Caglow. Retrieved 30 January 2012.

[7] Prof. Gregory D. Wirth. "Astronomical Magnitude Systems". Department of Physics and Astronomy, University of Toronto. Retrieved 2012-08-15.

[8] E. Schulman and C. V. Cox (1997). "Misconceptions About Astronomical Magnitudes". *American Journal of Physics* **65**: 1003. Bibcode:1997AmJPh..65.1003S. doi:10.1119/1.18714.

[9] Williams, Dr. David R. (2004-09-01). "Sun Fact Sheet". NASA (National Space Science Data Center). Archived from the original on 15 July 2010. Retrieved 2010-07-03.

[10] Introduction to Astrophysics: The Stars – Jean Dufay, page 3

[11] Ian S. McLean, *Electronic imaging in astronomy: detectors and instrumentation* Springer, 2008, ISBN 3-540-76582-4 page 529

[12] Williams, Dr. David R. (2010-02-02). "Moon Fact Sheet". NASA (National Space Science Data Center). Archived from the original on 23 March 2010. Retrieved 2010-04-09.

[13] "Brightest comets seen since 1935". International Comet Quarterly. Retrieved 18 December 2011.

[14] Winkler, P. Frank; Gupta, Gaurav; Long, Knox S. (2003). "The SN 1006 Remnant: Optical Proper Motions, Deep Imaging, Distance, and Brightness at Maximum". *The Astrophysical Journal* **585**: 324–335. arXiv:astro-ph/0208415. Bibcode:2003ApJ. .doi:10.1086/345985.

[15] Supernova 1054 – Creation of the Crab Nebula

[16] "ISS Information - Heavens-above.com". Heavens-above. Retrieved 2007-12-22.

[17] "HORIZONS Web-Interface for Venus (Major Body=299)" (2006 February 27 (GEOPHYSICAL DATA)). JPL Horizons On-Line Ephemeris System. Retrieved 2010-11-28. (Using JPL Horizons you can see that on 2013-Dec-08 Venus will have an apmag of −4.89)

[18] Williams, David R. (2007-11-02). "Jupiter Fact Sheet". *National Space Science Data Center*. NASA. Retrieved 2010-06-25.

[19] Williams, David R. (2007-11-29). "Mars Fact Sheet". *National Space Science Data Center*. NASA. Archived from the original on 12 June 2010. Retrieved 2010-06-25.

[20] "Sirius". SIMBAD Astronomical Database. Retrieved 2010-06-26.

[21] "Canopus". SIMBAD Astronomical Database. Retrieved 2010-06-26.

[22] "Arcturus". SIMBAD Astronomical Database. Retrieved 2010-06-26.

[23] "Vega". SIMBAD Astronomical Database. Retrieved 2010-04-14.

[24] "SIMBAD-M31". SIMBAD Astronomical Database. Retrieved 2009-11-29.

[25] Yeomans and Chamberlin. "Horizon Online Ephemeris System for Ganymede (Major Body 503)". California Institute of Technology, Jet Propulsion Laboratory. Retrieved 2010-04-14. (4.38 on 1951-Oct-03)

[26] "M41 possibly recorded by Aristotle". SEDS (Students for the Exploration and Development of Space). 2006-07-28. Retrieved 2009-11-29.

[27] Williams, David R. (2005-01-31). "Uranus Fact Sheet". *National Space Science Data Center*. NASA. Archived from the original on 29 June 2010. Retrieved 2010-06-25.

[28] "SIMBAD-M33". SIMBAD Astronomical Database. Retrieved 2009-11-28.

[29] Lodriguss, Jerry (1993). "M33 (Triangulum Galaxy)". Retrieved 2009-11-27. (shows b mag not v mag)

[30] "Messier 81". SEDS (Students for the Exploration and Development of Space). 2007-09-02. Retrieved 2009-11-28.

[31] John E. Bortle (February 2001). "The Bortle Dark-Sky Scale". Sky & Telescope. Retrieved 2009-11-18.

[32] Williams, David R. (2007-11-29). "Neptune Fact Sheet". *National Space Science Data Center*. NASA. Archived from the original on 1 July 2010. Retrieved 2010-06-25.

[33] Yeomans and Chamberlin. "Horizon Online Ephemeris System for Titan (Major Body 606)". California Institute of Technology, Jet Propulsion Laboratory. Retrieved 2010-06-28. (8.10 on 2003-Dec-30)

[34] "Classic Satellites of the Solar System". Observatorio ARVAL. Archived from the original on 31 July 2010. Retrieved 2010-06-25.

[35] "Planetary Satellite Physical Parameters". JPL (Solar System Dynamics). 2009-04-03. Archived from the original on 23 July 2009. Retrieved 2009-07-25.

[36] "AstDys (10) Hygiea Ephemerides". Department of Mathematics, University of Pisa, Italy. Retrieved 2010-06-26.

[37] Ed Zarenski (2004). "Limiting Magnitude in Binoculars" (PDF). Cloudy Nights. Retrieved 2011-05-06.

[38] Williams, David R. (2006-09-07). "Pluto Fact Sheet". *National Space Science Data Center*. NASA. Archived from the original on 1 July 2010. Retrieved 2010-06-26.

[39] "AstDys (2060) Chiron Ephemerides". Department of Mathematics, University of Pisa, Italy. Retrieved 2010-06-26.

[40] "AstDys (136472) Makemake Ephemerides". Department of Mathematics, University of Pisa, Italy. Retrieved 2010-06-26.

[41] "AstDys (136108) Haumea Ephemerides". Department of Mathematics, University of Pisa, Italy. Retrieved 2010-06-26.

[42] Steve Cullen (sgcullen) (2009-10-05). "17 New Asteroids Found by LightBuckets". LightBuckets. Retrieved 2009-11-15.

[43] Cooperation with Ken Crawford

[44] "CRedshift 6 Quasar (CFHQS J1641 +3755)".

[45] Scott S. Sheppard. "Saturn's Known Satellites". Carnegie Institution (Department of Terrestrial Magnetism). Retrieved 2010-06-28.

[46] Magnitude difference is $2.512*\log_{10}[(5000/5)^2 \times (4999/4)^2] \approx 30.6$, so Jupiter is 30.6 mag fainter at 5000 au

[47] "New Image of Comet Halley in the Cold". ESO. 2003-09-01. Archived from the original on 1 March 2009. Retrieved 2009-02-22.

[48] The HST eXtreme Deep Field XDF: Combining all ACS and WFC3/IR Data on the HUDF Region into the Deepest Field Ever

## 16.7 External links

- The astronomical magnitude scale (International Comet Quarterly)

# Chapter 17

# Chandrasekhar limit

The **Chandrasekhar limit** (/tʃʌndrə'ʃeɪkɑr/) is the maximum mass of a stable white dwarf star. The limit was first indicated in papers published by Wilhelm Anderson and E. C. Stoner, and was named after Subrahmanyan Chandrasekhar, the Indian astrophysicist who independently discovered and improved upon the accuracy of the calculation in 1930, at the age of 19, in India. This limit was initially ignored by the community of scientists because such a limit would logically require the existence of black holes, which were considered a scientific impossibility at the time. White dwarfs resist gravitational collapse primarily through electron degeneracy pressure. (By comparison, main sequence stars resist collapse through thermal pressure.) The Chandrasekhar limit is the mass above which electron degeneracy pressure in the star's core is insufficient to balance the star's own gravitational self-attraction. Consequently, white dwarfs with masses greater than the limit would be subject to further gravitational collapse, evolving into a different type of stellar remnant, such as a neutron star or black hole. (However, white dwarfs generally avoid this fate by exploding before they undergo collapse.) Those with masses under the limit remain stable as white dwarfs.[1]

The currently accepted value of the limit is about 1.39 $M_\odot$ ( $2.765 \times 10^{30}$ kg).[2][3][4]

## 17.1 Physics

Electron degeneracy pressure is a quantum-mechanical effect arising from the Pauli exclusion principle. Since electrons are fermions, no two electrons can be in the same state, so not all electrons can be in the minimum-energy level. Rather, electrons must occupy a band of energy levels. Compression of the electron gas increases the number of electrons in a given volume and raises the maximum energy level in the occupied band. Therefore, the energy of the electrons will increase upon compression, so pressure must be exerted on the electron gas to compress it, producing electron degeneracy pressure. With sufficient compression, electrons are forced into nuclei in the process of electron capture, relieving the pressure.

In the nonrelativistic case, electron degeneracy pressure gives rise to an equation of state of the form $P = K_1 \rho^{\frac{5}{3}}$, where $P$ is the pressure, $\rho$ is the mass density, and $K_1$ is a constant. Solving the hydrostatic equation then leads to a model white dwarf which is a polytrope of index 3/2 and therefore has radius inversely proportional to the cube root of its mass, and volume inversely proportional to its mass.[5]

As the mass of a model white dwarf increases, the typical energies to which degeneracy pressure forces the electrons are no longer negligible relative to their rest masses. The velocities of the electrons approach the speed of light, and special relativity must be taken into account. In the strongly relativistic limit, the equation of state takes the form $P = K_2 \rho^{\frac{4}{3}}$. This will yield a polytrope of index 3, which will have a total mass, $M_{\text{limit}}$ say, depending only on $K_2$.[6]

For a fully relativistic treatment, the equation of state used will interpolate between the equations $P = K_1 \rho^{\frac{5}{3}}$ for small $\rho$ and $P = K_2 \rho^{\frac{4}{3}}$ for large $\rho$. When this is done, the model radius still decreases with mass, but becomes zero at $M_{\text{limit}}$. This is the Chandrasekhar limit.[7] The curves of radius against mass for the non-relativistic and relativistic models are shown in the graph. They are colored blue and green, respectively. $\mu_e$ has been set equal to 2. Radius is measured in standard solar radii[8] or kilometers, and mass in standard solar masses.

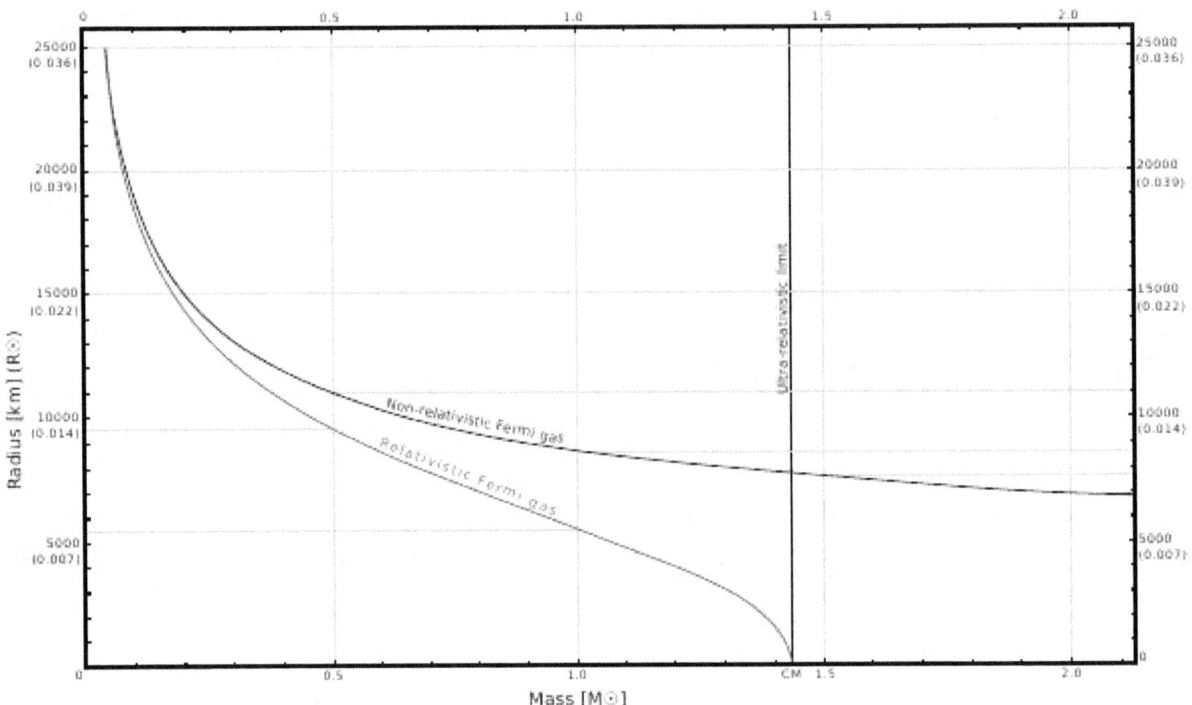

*Radius–mass relations for a model white dwarf. The green curve uses the general pressure law for an ideal Fermi gas, while the blue curve is for a non-relativistic ideal Fermi gas. The black line marks the ultrarelativistic limit.*

Calculated values for the limit will vary depending on the nuclear composition of the mass.[9] Chandrasekhar[10]. gives the following expression, based on the equation of state for an ideal Fermi gas:

$$M_{\text{limit}} = \frac{\omega_3^0 \sqrt{3\pi}}{2} \left( \frac{\hbar c}{G} \right)^{3/2} \frac{1}{(\mu_e m_H)^2}.$$

where:

- $\hbar$ is the reduced Planck constant

- $c$ is the speed of light

- $G$ is the gravitational constant

- $\mu_e$ is the average molecular weight per electron, which depends upon the chemical composition of the star.

- $mH$ is the mass of the hydrogen atom.

- $\omega_3^0 \approx 2.018236$ is a constant connected with the solution to the Lane-Emden equation.

As $\sqrt{\hbar c/G}$ is the Planck mass, the limit is of the order of

$$\frac{M_{Pl}^3}{m_H^2}.$$

A more accurate value of the limit than that given by this simple model requires adjusting for various factors, including electrostatic interactions between the electrons and nuclei and effects caused by nonzero temperature.[9] Lieb and Yau[12] have given a rigorous derivation of the limit from a relativistic many-particle Schrödinger equation.

## 17.2 History

In 1926, the British physicist Ralph H. Fowler observed that the relationship among the density, energy and temperature of white dwarfs could be explained by viewing them as a gas of nonrelativistic, non-interacting electrons and nuclei which obeyed Fermi–Dirac statistics.[13] This Fermi gas model was then used by the British physicist E. C. Stoner in 1929 to calculate the relationship among the mass, radius, and density of white dwarfs, assuming them to be homogeneous spheres.[14] Wilhelm Anderson applied a relativistic correction to this model, giving rise to a maximum possible mass of approximately $1.37 \times 10^{30}$ kg.[15] In 1930, Stoner derived the internal energy–density equation of state for a Fermi gas, and was then able to treat the mass–radius relationship in a fully relativistic manner, giving a limiting mass of approximately (for $\mu_e=2.5$) $2.19 \cdot 10^{30}$ kg.[16] Stoner went on to derive the pressure–density equation of state, which he published in 1932.[17] These equations of state were also previously published by the Soviet physicist Yakov Frenkel in 1928, together with some other remarks on the physics of degenerate matter.[18] Frenkel's work, however, was ignored by the astronomical and astrophysical community.[19]

A series of papers published between 1931 and 1935 had its beginning on a trip from India to England in 1930, where the Indian physicist Subrahmanyan Chandrasekhar worked on the calculation of the statistics of a degenerate Fermi gas.[20] In these papers, Chandrasekhar solved the hydrostatic equation together with the nonrelativistic Fermi gas equation of state,[5] and also treated the case of a relativistic Fermi gas, giving rise to the value of the limit shown above.[6][7][10][21] Chandrasekhar reviews this work in his Nobel Prize lecture.[11] This value was also computed in 1932 by the Soviet physicist Lev Davidovich Landau,[22] who, however, did not apply it to white dwarfs.

Chandrasekhar's work on the limit aroused controversy, owing to the opposition of the British astrophysicist Arthur Stanley Eddington. Eddington was aware that the existence of black holes was theoretically possible, and also realized that the existence of the limit made their formation possible. However, he was unwilling to accept that this could happen. After a talk by Chandrasekhar on the limit in 1935, he replied:

> The star has to go on radiating and radiating and contracting and contracting until, I suppose, it gets down to a few km radius, when gravity becomes strong enough to hold in the radiation, and the star can at last find peace. ... I think there should be a law of Nature to prevent a star from behaving in this absurd way!
> — [23]

Eddington's proposed solution to the perceived problem was to modify relativistic mechanics so as to make the law $P=K_1\rho^{5/3}$ universally applicable, even for large $\rho$.[24] Although Bohr, Fowler, Pauli, and other physicists agreed with Chandrasekhar's analysis, at the time, owing to Eddington's status, they were unwilling to publicly support Chandrasekhar.[25]. Through the rest of his life, Eddington held to his position in his writings,[26][27][28][29][30]including his work on his fundamental theory.[31] The drama associated with this disagreement is one of the main themes of *Empire of the Stars*, Arthur I. Miller's biography of Chandrasekhar.[25] In Miller's view:

> Chandra's discovery might well have transformed and accelerated developments in both physics and astrophysics in the 1930s. Instead, Eddington's heavy-handed intervention lent weighty support to the conservative community astrophysicists, who steadfastly refused even to consider the idea that stars might collapse to nothing. As a result, Chandra's work was almost forgotten.
> — p. 150. [25]

## 17.3 Applications

The core of a star is kept from collapsing by the heat generated by the fusion of nuclei of lighter elements into heavier ones. At various stages of stellar evolution, the nuclei required for this process will be exhausted, and the core will collapse, causing it to become denser and hotter. A critical situation arises when iron accumulates in the core, since iron nuclei are incapable of generating further energy through fusion. If the core becomes sufficiently dense, electron degeneracy pressure will play a significant part in stabilizing it against gravitational collapse.[32]

If a main-sequence star is not too massive (less than approximately 8 solar masses), it will eventually shed enough mass to form a white dwarf having mass below the Chandrasekhar limit, which will consist of the former core of the star. For more-massive stars, electron degeneracy pressure will not keep the iron core from collapsing to very great density, leading to formation of a neutron star, black hole, or, speculatively, a quark star. (For very massive, low-metallicity stars, it is also possible that instabilities will destroy the star completely.)[33][34][35][36] During the collapse, neutrons are formed by the capture of electrons by protons in the process of electron capture, leading to the emission of neutrinos.[32]. pp. 1046–1047. The decrease in gravitational potential energy of the collapsing core releases a large amount of energy which is on the order of $10^{46}$ joules (100 foes). Most of this energy is carried away by the emitted neutrinos.[37] This process is believed to be responsible for supernovae of types Ib, Ic, and II.[32]

Type Ia supernovae derive their energy from runaway fusion of the nuclei in the interior of a white dwarf. This fate may befall carbon–oxygen white dwarfs that accrete matter from a companion giant star, leading to a steadily increasing mass. As the white dwarf's mass approaches the Chandrasekhar limit, its central density increases, and, as a result of compressional heating, its temperature also increases. This eventually ignites nuclear fusion reactions, leading to an immediate carbon detonation which disrupts the star and causes the supernova.[38]. §5.1.2

A strong indication of the reliability of Chandrasekhar's formula is that the absolute magnitudes of supernovae of Type Ia are all approximately the same; at maximum luminosity, MV is approximately $-19.3$, with a standard deviation of no more than 0.3.[38], (1) A 1-sigma interval therefore represents a factor of less than 2 in luminosity. This seems to indicate that all type Ia supernovae convert approximately the same amount of mass to energy.

## 17.4   Super-Chandrasekhar mass supernovae

Main article: Champagne Supernova

In April 2003, the Supernova Legacy Survey observed a type Ia supernova, designated SNLS-03D3bb, in a galaxy approximately 4 billion light years away. According to a group of astronomers at the University of Toronto and elsewhere, the observations of this supernova are best explained by assuming that it arose from a white dwarf which grew to twice the mass of the Sun before exploding. They believe that the star, dubbed the "Champagne Supernova" by University of Oklahoma astronomer David R. Branch, may have been spinning so fast that centrifugal force allowed it to exceed the limit. Alternatively, the supernova may have resulted from the merger of two white dwarfs, so that the limit was only violated momentarily. Nevertheless, they point out that this observation poses a challenge to the use of type Ia supernovae as standard candles.[39][40][41]

Since the observation of the Champagne Supernova in 2003, more very bright type Ia supernovae have been observed that are thought to have originated from white dwarfs whose masses exceeded the Chandrasekhar limit. These include SN 2006gz, SN 2007if and SN 2009dc.[42] The super-Chandrasekhar mass white dwarfs that gave rise to these supernovae are believed to have had masses up to 2.4–2.8 solar masses.[42] One way to potentially explain the problem of the Champagne Supernova was considering it the result of an aspherical explosion of a white dwarf. However, spectropolarimetric observations of SN 2009dc showed it had a polarization smaller than 0.3, making the large asphericity theory unlikely.[42]

## 17.5   Tolman–Oppenheimer–Volkoff limit

After a supernova explosion, a neutron star may be left behind. Like white dwarfs these objects are extremely compact and are supported by degeneracy pressure, but a neutron star is so massive and compressed that electrons and protons have combined to form neutrons, and the star is thus supported by neutron degeneracy pressure instead of electron degeneracy pressure. The limit of neutron degeneracy pressure, analogous to the Chandrasekhar limit, is known as the Tolman–Oppenheimer–Volkoff limit.

# 17.6 References

[1] Sean Carroll, Ph.D., Cal Tech, 2007, The Teaching Company, *Dark Matter, Dark Energy: The Dark Side of the Universe*, Guidebook Part 2 page 44, Accessed Oct. 7, 2013, "...Chandrasekhar limit: The maximum mass of a white dwarf star, about 1.4 times the mass of the Sun. Above this mass, the gravitational pull becomes too great, and the star must collapse to a neutron star or black hole..."

[2] Israel, edited by S.W. Hawking, W. (1989). *Three hundred years of gravitation* (1st pbk. ed., with corrections. ed.). Cambridge [Cambridgeshire]: Cambridge University Press. ISBN 0-521-37976-8.

[3] p. 55, How A Supernova Explodes, Hans A. Bethe and Gerald Brown, pp. 51–62 in *Formation And Evolution of Black Holes in the Galaxy: Selected Papers with Commentary*, Hans Albrecht Bethe, Gerald Edward Brown, and Chang-Hwan Lee, River Edge, New Jersey: World Scientific: 2003. ISBN 981-238-250-X.

[4] Mazzali, P. A.; Röpke, F. K.; Benetti, S.; Hillebrandt, W. (2007). "A Common Explosion Mechanism for Type Ia Supernovae". *Science* (PDF) **315** (5813): 825–828. arXiv:astro-ph/0702351v1. Bibcode:2007Sci...315..825M. doi:10.1126/science.1136259. PMID 17289993.

[5] The Density of White Dwarf Stars, S. Chandrasekhar, *Philosophical Magazine* (7th series) **11** (1931), pp. 592–596.

[6] The Maximum Mass of Ideal White Dwarfs, S. Chandrasekhar, *Astrophysical Journal* **74** (1931), pp. 81–82.

[7] The Highly Collapsed Configurations of a Stellar Mass (second paper), S. Chandrasekhar, *Monthly Notices of the Royal Astronomical Society*, **95** (1935), pp. 207--225.

[8] *Standards for Astronomical Catalogues, Version 2.0*, section 3.2.2, web page, accessed 12-1-2007.

[9] The Neutron Star and Black Hole Initial Mass Function, F. X. Timmes, S. E. Woosley, and Thomas A. Weaver, *Astrophysical Journal* **457** (February 1, 1996), pp. 834–843.

[10] The Highly Collapsed Configurations of a Stellar Mass, S. Chandrasekhar, *Monthly Notices of the Royal Astronomical Society* **91** (1931), 456–466.

[11] *On Stars, Their Evolution and Their Stability*, Nobel Prize lecture, Subrahmanyan Chandrasekhar, December 8, 1983.

[12] A rigorous examination of the Chandrasekhar theory of stellar collapse, Elliott H. Lieb and Horng-Tzer Yau, *Astrophysical Journal* **323** (1987), pp. 140–144.

[13] On Dense Matter, R. H. Fowler, *Monthly Notices of the Royal Astronomical Society* **87** (1926), pp. 114–122.

[14] The Limiting Density of White Dwarf Stars, Edmund C. Stoner, *Philosophical Magazine* (7th series) **7** (1929), pp. 63–70.

[15] Über die Grenzdichte der Materie und der Energie, Wilhelm Anderson, *Zeitschrift für Physik* **56**, #11–12 (November 1929), pp. 851–856. DOI 10.1007/BF01340146.

[16] The Equilibrium of Dense Stars, Edmund C. Stoner, *Philosophical Magazine* (7th series) **9** (1930), pp. 944–963.

[17] The minimum pressure of a degenerate electron gas, E. C. Stoner, *Monthly Notices of the Royal Astronomical Society* **92** (May 1932), pp. 651–661.

[18] Anwendung der Pauli-Fermischen Elektronengastheorie auf das Problem der Kohäsionskräfte, J. Frenkel, *Zeitschrift für Physik* **50**, #3–4 (March 1928), pp. 234–248. DOI 10.1007/BF01328867.

[19] The article by Ya I Frenkel' on 'binding forces' and the theory of white dwarfs, D. G. Yakovlev, *Physics Uspekhi* **37**, #6 (1994), pp. 609–612.

[20] Chandrasekhar's biographical memoir at the National Academy of Sciences, web page, accessed 12-1-2007.

[21] Stellar Configurations with degenerate Cores, S. Chandrasekhar, *The Observatory* **57** (1934), pp. 373–377.

[22] On the Theory of Stars, in *Collected Papers of L. D. Landau*, ed. and with an introduction by D. ter Haar, New York: Gordon and Breach, 1965; originally published in *Phys. Z. Sowjet.* **1** (1932), 285.

[23] Meeting of the Royal Astronomical Society, Friday, 1935 January 11, *The Observatory* **58** (February 1935), pp. 33–41.

[24]  On "Relativistic Degeneracy", Sir A. S. Eddington, *Monthly Notices of the Royal Astronomical Society* **95** (1935), 194–206.

[25]  *Empire of the Stars: Obsession, Friendship, and Betrayal in the Quest for Black Holes*, Arthur I. Miller, Boston, New York: Houghton Mifflin, 2005, ISBN 0-618-34151-X; reviewed at *The Guardian*: The battle of black holes.

[26]  The International Astronomical Union meeting in Paris, 1935, *The Observatory* **58** (September 1935), pp. 257–265, at p. 259.

[27]  Note on "Relativistic Degeneracy", Sir A. S. Eddington, *Monthly Notices of the Royal Astronomical Society* **96** (November 1935), 20–21.

[28]  The Pressure of a Degenerate Electron Gas and Related Problems, Arthur Eddington, *Proceedings of the Royal Society of London. Series A, Mathematical and Physical Sciences* **152** (November 1, 1935), pp. 253–272.

[29]  *Relativity Theory of Protons and Electrons*, Sir Arthur Eddington, Cambridge: Cambridge University Press, 1936, chapter 13.

[30]  The physics of white dwarf matter, Sir A. S. Eddington, *Monthly Notices of the Royal Astronomical Society* **100** (June 1940), pp. 582–594.

[31]  *Fundamental Theory*, Sir A. S. Eddington, Cambridge: Cambridge University Press, 1946, §43–45.

[32]  The evolution and explosion of massive stars, S. E. Woosley, A. Heger, and T. A. Weaver, *Reviews of Modern Physics* **74**, #4 (October 2002), pp. 1015–1071.

[33]  White dwarfs in open clusters. VIII. NGC 2516: a test for the mass-radius and initial-final mass relations, D. Koester and D. Reimers, *Astronomy and Astrophysics* **313** (1996), pp. 810–814.

[34]  An Empirical Initial-Final Mass Relation from Hot, Massive White Dwarfs in NGC 2168 (M35), Kurtis A. Williams, M. Bolte, and Detlev Koester, *Astrophysical Journal* **615**, #1 (2004), pp. L49–L52; also arXiv astro-ph/0409447.

[35]  How Massive Single Stars End Their Life, A. Heger, C. L. Fryer, S. E. Woosley, N. Langer, and D. H. Hartmann, *Astrophysical Journal* **591**, #1 (2003), pp. 288–300.

[36]  Strange quark matter in stars: a general overview, Jürgen Schaffner-Bielich, *Journal of Physics G: Nuclear and Particle Physics* **31**, #6 (2005), pp. S651–S657; also arXiv astro-ph/0412215.

[37]  The Physics of Neutron Stars, by J. M. Lattimer and M. Prakash, *Science* **304**, #5670 (2004), pp. 536–542; also arXiv astro-ph/0405262.

[38]  Type IA Supernova Explosion Models, Wolfgang Hillebrandt and Jens C. Niemeyer, *Annual Review of Astronomy and Astrophysics* **38** (2000), pp. 191–230.

[39]  The weirdest Type Ia supernova yet, LBL press release, web page accessed 13-1-2007.

[40]  Champagne Supernova Challenges Ideas about How Supernovae Work, web page, spacedaily.com, accessed 13-1-2007.

[41]  The type Ia supernova SNLS-03D3bb from a super-Chandrasekhar-mass white dwarf star, D. Andrew Howell et al., *Nature* **443** (September 21, 2006), pp. 308–311; also, arXiv:astro-ph/0609616.

[42]  Hachisu, Izumi; Kato, M.; et al. (2012). "A single degenerate progenitor model for type Ia supernovae highly exceeding the Chandrasekhar mass limit". *The Astrophysical Journal* **744** (1): 76–79 (Article ID 69). arXiv:1106.3510. Bibcode:2012ApJ...744 doi:10.1088/0004-637X/744/1/69.

## 17.7   Further reading

- *On Stars, Their Evolution and Their Stability*, Nobel Prize lecture, Subrahmanyan Chandrasekhar, December 8, 1983.

- *White dwarf stars and the Chandrasekhar limit*, Masters' thesis, Dave Gentile, DePaul University, 1995.

- Estimating Stellar Parameters from Energy Equipartition, sciencebits.com. Discusses how to find mass-radius relations and mass limits for white dwarfs using simple energy arguments.

# Chapter 18

# Electron degeneracy pressure

**Electron degeneracy pressure** is a particular manifestation of the more general phenomenon of quantum degeneracy pressure. The Pauli exclusion principle disallows two identical half-integer spin particles (electrons and all other fermions) from simultaneously occupying the same quantum state. The result is an emergent pressure against compression of matter into smaller volumes of space. Electron degeneracy pressure results from the same underlying mechanism that defines the electron orbital structure of elemental matter. Freeman Dyson showed that the imperviousness of solid matter is due to quantum degeneracy pressure rather than electrostatic repulsion as had been previously assumed.[1][2][3] Furthermore, electron degeneracy creates a barrier to the gravitational collapse of dying stars and is responsible for the formation of white dwarfs.

When electrons are squeezed together too closely, the exclusion principle requires them to have different energy levels. To add another electron to a given volume requires raising an electron's energy level to make room, and this requirement for energy to compress the material manifests as a pressure.

Electron degeneracy pressure in a material can be computed as[4]

$$P = \frac{2}{3}\frac{E_{tot}}{V} = \frac{2}{3}\frac{\hbar^2 k_F^5}{10\pi^2 m_e} = \frac{(3\pi^2)^{2/3}\hbar^2}{5m_e}\rho_N^{5/3},$$

where $\hbar$ is the reduced Planck constant, $m_e$ is the mass of the electron, and $\rho_N$ is the free electron density (the number of free electrons per unit volume). When particle energies reach relativistic levels, a modified formula is required.

This is derived from the energy of each electron with wave number $k = \frac{2\pi}{\lambda}$, having $E = \frac{p^2}{2m} = \frac{\hbar^2 k^2}{2m}$, and every possible momentum state of an electron within this volume up to the Fermi energy being occupied.

This degeneracy pressure is omnipresent and is in addition to the normal gas pressure $P = NkT/V$. At commonly encountered densities, this pressure is so low that it can be neglected. Matter is electron degenerate when the density (proportional to $n/V$) is high enough, and the temperature low enough, that the sum is dominated by the degeneracy pressure.

Perhaps useful in appreciating electron degeneracy pressure is the Heisenberg uncertainty principle, which states that

$$\Delta x \Delta p \geq \frac{\hbar}{2}$$

where $\Delta x$ is the uncertainty of the position measurements and $\Delta p$ is the uncertainty (standard deviation) of the momentum measurements. A material subjected to ever-increasing pressure will compact more, and, for electrons within it, their delocalization, $\Delta x$, will decrease. Thus, as dictated by the uncertainty principle, the spread in the momenta of the electrons, $\Delta p$, will grow. Thus, no matter how low the temperature drops, the electrons must be traveling at this "Heisenberg speed", contributing to the pressure. When the pressure due to this "Heisenberg motion" exceeds that of the pressure from the thermal motions of the electrons, the electrons are referred to as degenerate, and the material is termed degenerate matter.

Electron degeneracy pressure will halt the gravitational collapse of a star if its mass is below the Chandrasekhar limit (1.44 solar masses[5]). This is the pressure that prevents a white dwarf star from collapsing. A star exceeding this limit and without significant thermally generated pressure will continue to collapse to form either a neutron star or black hole, because the degeneracy pressure provided by the electrons is weaker than the inward pull of gravity.

Main article: Degenerate matter
See also: Exchange interaction

# 18.1 References

[1] Dyson, F. J.; Lenard, A. (March 1967). "Stability of Matter I". *J. Math. Phys.* **8** (3): 423–434. Bibcode:1967JMP.....8..423D. doi:10.1063/1.1705209.

[2] Lenard, A; Dyson, F. J. (May 1968). "Stability of Matter II". *J. Math. Phys.* **9** (5): 698–711. Bibcode:1968JMP.....9..698L. doi:10.1063/1.1664631.

[3] Dyson, F. J. (August 1967). "Ground-State Energy of a Finite System of Charged Particles". *J. Math. Phys.* **8** (8): 1538–1545. Bibcode:1967JMP.....8.1538D. doi:10.1063/1.1705389.

[4] Griffiths (1994). *Introduction to Quantum Mechanics*. London, UK: Prentice Hall. ISBN 0131244051.Equation 5.46

[5] Mazzali, P. A.; K. Röpke, F. K.; Benetti, S.; Hillebrandt, W. (2007). "A Common Explosion Mechanism for Type Ia Supernovae". *Science* **315** (5813): 825–828. arXiv:astro-ph/0702351. Bibcode:2007Sci...315..825M. doi:10.1126/science.1136259. PMID 17289993.

# Chapter 19

# Carbon-burning process

"Carbon burning" redirects here. For combustion of carbon containing compounds, see combustion.

The **carbon-burning process** or **carbon fusion** is a set of nuclear fusion reactions that take place in massive stars (at least 8 $M_\odot$ at birth) that have used up the lighter elements in their cores. It requires high temperatures ($> 5 \times 10^8$ K or 50 keV) and densities ($> 3 \times 10^9$ kg/m$^3$).[1]

These figures for temperature and density are only a guide. More massive stars burn their nuclear fuel more quickly, since they have to offset greater gravitational forces to stay in (approximate) hydrostatic equilibrium. That generally means higher temperatures, although lower densities, than for less massive stars.[2] To get the right figures for a particular mass, and a particular stage of evolution, it is necessary to use a numerical stellar model computed with computer algorithms.[3] Such models are continually being refined based on particle physics experiments (which measure nuclear reaction rates) and astronomical observations (which include direct observation of mass loss, detection of nuclear products from spectrum observations after convection zones develop from the surface to fusion-burning regions – known as 'dredge-up' events – and so bring nuclear products to the surface, and many other observations relevant to models).[4]

## 19.1   Fusion reactions

The principal reactions are:[5]

## 19.2   Reaction products

This sequence of reactions can be understood by thinking of the two interacting carbon nuclei as coming together to form an excited state of the Mg-24 nucleus, which then decays in one of the five ways listed above.[6] The first two reactions are strongly exothermic, as indicated by the large positive energies released, and are the most frequent results of the interaction. The third reaction is strongly endothermic, as indicated by the large negative energy indicating that energy is absorbed rather than emitted. This makes it much less likely, yet still possible in the high-energy environment of carbon burning.[5] But the production of a few neutrons by this reaction is important, since these neutrons can combine with heavy nuclei, present in tiny amounts in most stars, to form even heavier isotopes in the s-process.[7]

The fourth reaction might be expected to be the most common from its large energy release, but in fact it is extremely improbable because it proceeds via the electromagnetic interaction,[5] as it produces a gamma ray photon, rather than utilising the strong force between nucleons as do the first two reactions. Nucleons look a lot bigger to each other than they do to photons of this energy. However, the Mg-24 produced in this reaction is the only magnesium left in the core when the carbon-burning process ends, as Mg-23 is radioactive.

The last reaction is also very unlikely since it involves three reaction products,[5] as well as being endothermic—think of the reaction proceeding in reverse, it would require the three products all to converge at the same time, which is less likely than two-body interactions.

The protons produced by the second reaction can take part in the proton-proton chain reaction, or the CNO cycle, but they can also be captured by Na-23 to form Ne-20 plus a He-4 nucleus.[5] In fact, a significant fraction of the Na-23 produced by the second reaction gets used up this way.[6] The oxygen (O-16) already produced by helium fusion in the previous stage of stellar evolution manages to survive the carbon-burning process pretty well, despite some of it being used up by capturing He-4 nuclei, in stars between 9 and 11 solar masses.[11][8] So the end result of carbon burning is a mixture mainly of oxygen, neon, sodium and magnesium.[3][5]

The fact that the mass-energy sum of the two carbon nuclei is similar to that of an excited state of the magnesium nucleus is known as 'resonance'. Without this resonance, carbon burning would only occur at temperatures one hundred times higher. The experimental and theoretical investigation of such resonances is still a subject of research.[9] A similar resonance increases the probability of the triple-alpha process, which is responsible for the original production of carbon.

## 19.3   Neutrino losses

Neutrino losses start to become a major factor in the fusion processes in stars at the temperatures and densities of carbon burning. Though the main reactions don't involve neutrinos, the side reactions such as the proton-proton chain reaction do. But the main source of neutrinos at these high temperatures involves a process in quantum theory known as pair production. A high energy gamma ray which has a greater energy than the rest mass of two electrons (mass-energy equivalence) can interact with electromagnetic fields of the atomic nuclei in the star, and become a particle and anti-particle pair of an electron and positron.

Normally, the positron quickly annihilates with another electron, producing two photons, and this process can be safely ignored at lower temperatures. But around 1 in $10^{19}$ pair productions[2] end with a weak interaction of the electron and positron, which replaces them with a neutrino and anti-neutrino pair. Since they move at virtually the speed of light and interact very weakly with matter, these neutrino particles usually escape the star without interacting, carrying away their mass-energy. This energy loss is comparable to the energy output from the carbon fusion.

Neutrino losses, by this and similar processes, play an increasingly important part in the evolution of the most massive stars. They force the star to burn its fuel at a higher temperature to offset them.[2] Fusion processes are very sensitive to temperature so the star can produce more energy to retain hydrostatic equilibrium, at the cost of burning through successive nuclear fuels ever more rapidly. Fusion produces less energy per unit mass as the fuel nuclei get heavier, and the core of the star contracts and heats up when switching from one fuel to the next, so both these processes also significantly reduce the lifetime of each successive fusion-burning fuel.

Up to helium burning, the neutrino losses are negligible, but from carbon burning the reduction in lifetime due to them roughly matches that due to fuel change and core contraction. In successive fuel changes in the most massive stars, the reduction in lifetime is dominated by the neutrino losses. For example, a star of 25 solar masses burns hydrogen in the core for $10^7$ years, helium for $10^6$ years and carbon for only $10^3$ years.[10]

## 19.4   Stellar evolution

Main article: Stellar evolution

During helium fusion, stars build up an inert core rich in carbon and oxygen. The inert core eventually reaches sufficient mass to collapse due to gravitation, whilst the helium burning moves gradually outward. This decrease in the inert core volume raises the temperature to the carbon ignition temperature. This will raise the temperature around the core and allow helium to burn in a shell around the core.[11] Outside this is another shell burning hydrogen. The resulting carbon burning provides energy from the core to restore the star's mechanical equilibrium. However, the balance is only short-lived; in a star of 25 solar masses, the process will use up most of the carbon in the core in only 600 years. The duration of this process varies significantly depending on the mass of the star.[12]

Stars with below 8–9 Solar masses never reach high enough core temperature to burn carbon, instead ending their lives as carbon-oxygen white dwarfs after shell helium flashes gently expel the outer envelope in a planetary nebula.[3][13]

In stars with masses between 8 and 11 solar masses, the carbon-oxygen core is under degenerate conditions and carbon ignition takes place in a *carbon flash*, that lasts just milliseconds and disrupts the stellar core.[14] In the late stages of this nuclear burning they develop a massive stellar wind, which quickly ejects the outer envelope in a planetary nebula leaving behind an O-Ne-Na-Mg white dwarf core of about 1.1 solar masses.[3] The core never reaches high enough temperature for further fusion burning of heavier elements than carbon.[13]

Stars with more than 11 solar masses start carbon burning in a non-degenerate core,[14] and after carbon exhaustion proceed with the neon-burning process once contraction of the inert (O, Ne, Na, Mg) core raises the temperature sufficiently.[13]

## 19.5 See also

- Proton–proton chain reaction
- CNO process
- Triple alpha process
- Alpha process
- Carbon detonation
- Neon burning

## 19.6 References

[1] Ryan, Sean G.; Norton, Andrew J. (2010). *Stellar Evolution and Nucleosynthesis*. Cambridge University Press. p. 135. ISBN 978-0-521-13320-3.

[2] Clayton, Donald (1983). *Principles of Stellar Evolution and Nucleosynthesis*. University of Chicago Press. ISBN 978-0-226-10953-4.

[3] Siess L. (2007). "Evolution of massive AGB stars. I. Carbon burning phase". *Astronomy and Astrophysics* **476** (2): 893–909. Bibcode:2006A&A...448..717S. doi:10.1051/0004-6361:20053043.

[4] Hernandez, G.; et al. (Dec 2006). "Rubidium-Rich Asymptotic Giant Branch Stars". *Science* **314** (5806): 1751–1754. arXiv:astro-ph/0611319. Bibcode:2006Sci...314.1751G. doi:10.1126/science.1133706. PMID 17095658.

[5] Camiel, W. H.; de Loore; C. Doom (1992). "Structure and evolution of single and binary stars". In Camiel W. H. de Loore. *Volume 179 of Astrophysics and space science library*. Springer. pp. 95–97. ISBN 978-0-7923-1768-5.

[6] Rose., William K. (1998). *Advanced Stellar Astrophysics*. Cambridge University Press. pp. 227–229. ISBN 978-0-521-58833-1.

[7] Rose (1998), pp. 229–234

[8] Camiel (1992), pp.97–98

[9] Strandberg, E.; et al. (May 2008). "Mg24($\alpha,\gamma$)Si28 resonance parameters at low $\alpha$-particle energies". *Physical Review C* **77** (5): 055801–+. Bibcode:2008PhRvC..77e5801S. doi:10.1103/PhysRevC.77.055801.

[10] Woosley, S.; Janka, H.-T. (2006-01-12). "The Physics of Core-Collapse Supernovae". *Nature Physics* **1** (3): 147–154. arXiv:astro-ph/0601261. Bibcode:2005NatPh...1..147W. doi:10.1038/nphys172.

[11] Ostlie, Dale A. and Carrol, Bradley W., *An introduction to Modern Stellar Astrophysics*, Addison-Wesley (2007)

[12] Anderson, Scott R., *Open Course: Astronomy: Lecture 19: Death of High-Mass Stars*, GEM (2001)

[13]  Ryan (2010), pp.147–148

[14]  *The Carbon Flash*

# Chapter 20

# Main sequence

For the racehorse, see Main Sequence (horse).

In astronomy, the **main sequence** is a continuous and distinctive band of stars that appears on plots of stellar color versus brightness. These color-magnitude plots are known as Hertzsprung–Russell diagrams after their co-developers, Ejnar Hertzsprung and Henry Norris Russell. Stars on this band are known as **main-sequence stars** or "dwarf" stars.[1][2]

After a star has formed, it generates thermal energy in the dense core region through the nuclear fusion of hydrogen atoms into helium. During this stage of the star's lifetime, it is located along the main sequence at a position determined primarily by its mass, but also based upon its chemical composition and other factors. All main-sequence stars are in hydrostatic equilibrium, where outward thermal pressure from the hot core is balanced by the inward pressure of gravitational collapse from the overlying layers. The strong dependence of the rate of energy generation in the core on the temperature and pressure helps to sustain this balance. Energy generated at the core makes its way to the surface and is radiated away at the photosphere. The energy is carried by either radiation or convection, with the latter occurring in regions with steeper temperature gradients, higher opacity or both.

The main sequence is sometimes divided into upper and lower parts, based on the dominant process that a star uses to generate energy. Stars below about 1.5 times the mass of the Sun (or 1.5 solar masses ($M\odot$)) primarily fuse hydrogen atoms together in a series of stages to form helium, a sequence called the proton–proton chain. Above this mass, in the upper main sequence, the nuclear fusion process mainly uses atoms of carbon, nitrogen and oxygen as intermediaries in the CNO cycle that produces helium from hydrogen atoms. Main-sequence stars with more than two solar masses undergo convection in their core regions, which acts to stir up the newly created helium and maintain the proportion of fuel needed for fusion to occur. Below this mass, stars have cores that are entirely radiative with convective zones near the surface. With decreasing stellar mass, the proportion of the star forming a convective envelope steadily increases, whereas main-sequence stars below 0.4 $M\odot$ undergo convection throughout their mass. When core convection does not occur, a helium-rich core develops surrounded by an outer layer of hydrogen.

In general, the more massive a star is, the shorter its lifespan on the main sequence. After the hydrogen fuel at the core has been consumed, the star evolves away from the main sequence on the HR diagram. The behavior of a star now depends on its mass, with stars below 0.23 $M\odot$ becoming white dwarfs directly, whereas stars with up to ten solar masses pass through a red giant stage.[3] More massive stars can explode as a supernova,[4] or collapse directly into a black hole.

## 20.1 History

In the early part of the 20th century, information about the types and distances of stars became more readily available. The spectra of stars were shown to have distinctive features, which allowed them to be categorized. Annie Jump Cannon and Edward C. Pickering at Harvard College Observatory developed a method of categorization that became known as the Harvard Classification Scheme, published in the *Harvard Annals* in 1901.[6]

In Potsdam in 1906, the Danish astronomer Ejnar Hertzsprung noticed that the reddest stars—classified as K and M in the Harvard scheme—could be divided into two distinct groups. These stars are either much brighter than the Sun, or

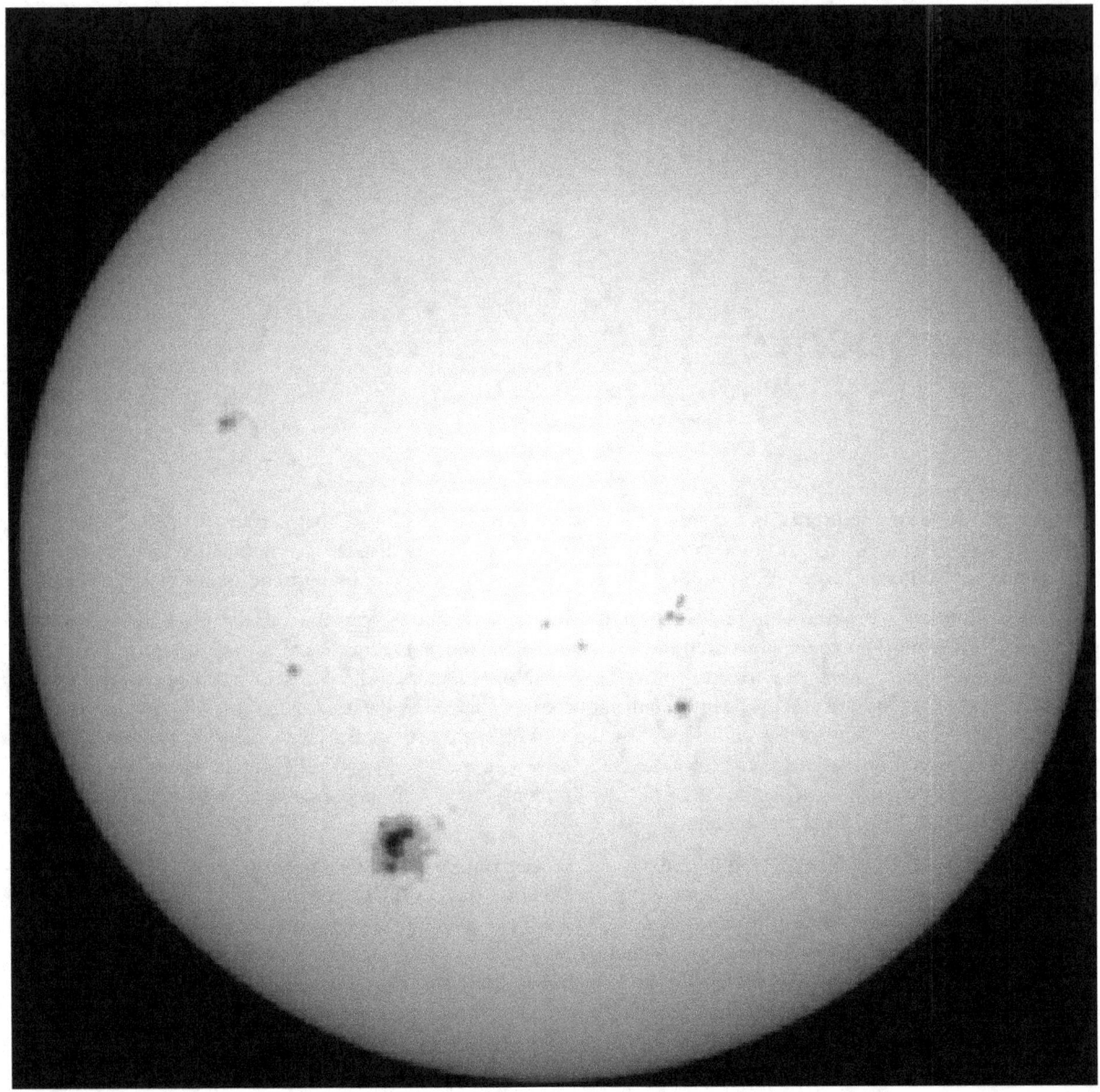

*The Sun is the most familiar example of a main-sequence star*

much fainter. To distinguish these groups, he called them "giant" and "dwarf" stars. The following year he began studying star clusters; large groupings of stars that are co-located at approximately the same distance. He published the first plots of color versus luminosity for these stars. These plots showed a prominent and continuous sequence of stars, which he named the Main Sequence.[7]

At Princeton University, Henry Norris Russell was following a similar course of research. He was studying the relationship between the spectral classification of stars and their actual brightness as corrected for distance—their absolute magnitude. For this purpose he used a set of stars that had reliable parallaxes and many of which had been categorized at Harvard. When he plotted the spectral types of these stars against their absolute magnitude, he found that dwarf stars followed a distinct relationship. This allowed the real brightness of a dwarf star to be predicted with reasonable accuracy.[8]

Of the red stars observed by Hertzsprung, the dwarf stars also followed the spectra-luminosity relationship discovered by Russell. However, the giant stars are much brighter than dwarfs and so, do not follow the same relationship. Russell proposed that the "giant stars must have low density or great surface-brightness, and the reverse is true of dwarf stars". The same curve also showed that there were very few faint white stars.[8]

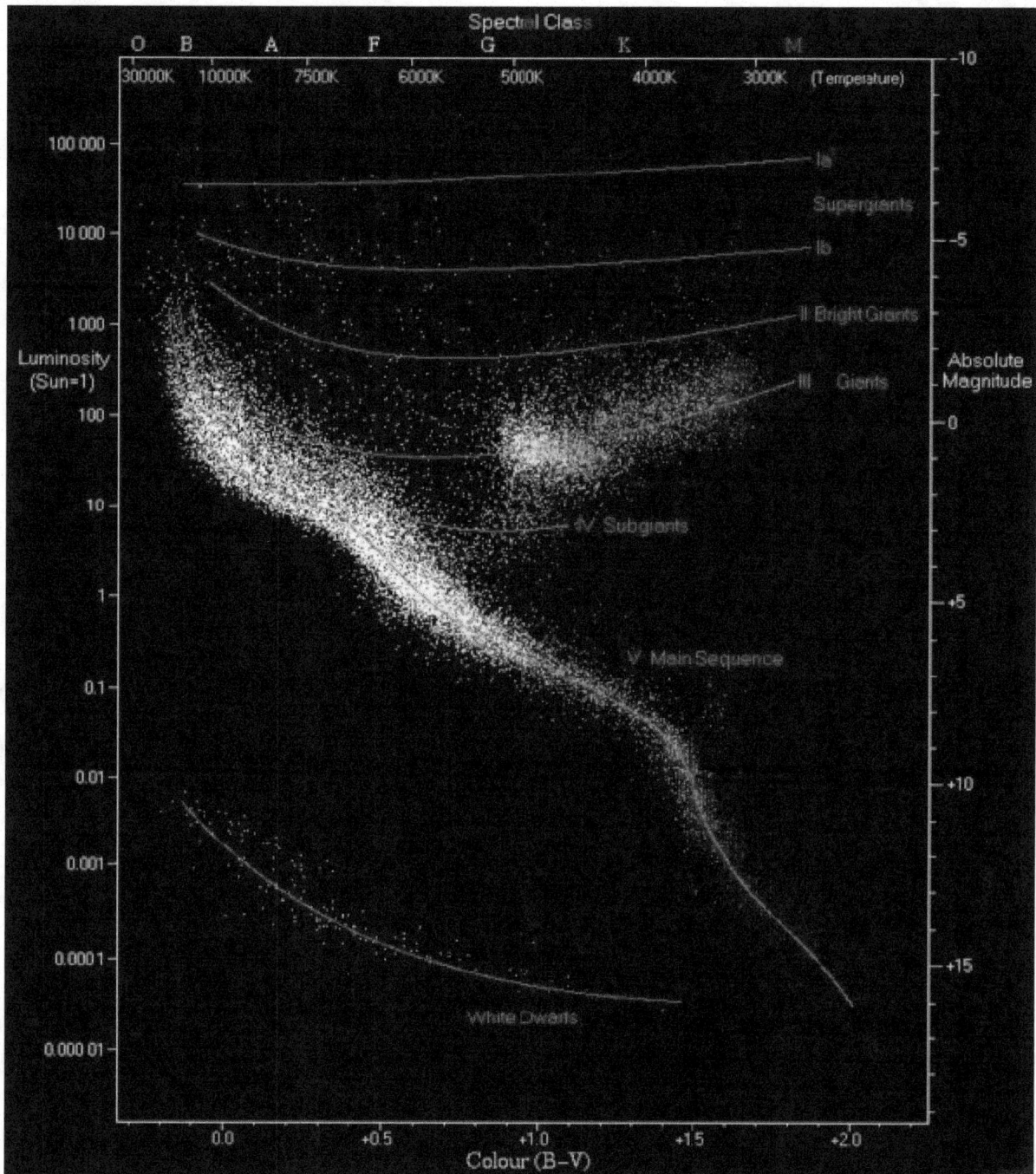

A Hertzsprung–Russell diagram plots the actual brightness (or absolute magnitude) of a star against its color index (represented as B–V). The main sequence is visible as a prominent diagonal band that runs from the upper left to the lower right. This plot shows 22,000 stars from the Hipparcos Catalogue together with 1,000 low-luminosity stars (red and white dwarfs) from the Gliese Catalogue of Nearby Stars.

In 1933, Bengt Strömgren introduced the term Hertzsprung–Russell diagram to denote a luminosity-spectral class diagram.[9] This name reflected the parallel development of this technique by both Hertzsprung and Russell earlier in the century.[7]

As evolutionary models of stars were developed during the 1930s, it was shown that, for stars of a uniform chemical composition, a relationship exists between a star's mass and its luminosity and radius. That is, for a given mass and composition, there is a unique solution for determining the star's radius and luminosity. This became known as the

*Hot and brilliant O-type main-sequence stars in star-forming regions. These are all regions of star formation that contain many hot young stars including several bright stars of spectral type O.[5]*

Vogt-Russell theorem; named after Heinrich Vogt and Henry Norris Russell. By this theorem, once a star's chemical composition and its position on the main sequence is known, so too is the star's mass and radius. (However, it was subsequently discovered that the theorem breaks down somewhat for stars of non-uniform composition.)[10]

A refined scheme for stellar classification was published in 1943 by W. W. Morgan and P. C. Keenan.[11] The MK classification assigned each star a spectral type—based on the Harvard classification—and a luminosity class. The Harvard classification had been developed by assigning a different letter to each star based on the strength of the hydrogen spectral line, before the relationship between spectra and temperature was known. When ordered by temperature and when duplicate classes were removed, the spectral types of stars followed, in order of decreasing temperature with colors ranging from blue to red, the sequence O, B, A, F, G, K and M. (A popular mnemonic for memorizing this sequence of stellar classes is "Oh Be A Fine Girl/Guy, Kiss Me".) The luminosity class ranged from I to V, in order of decreasing luminosity. Stars of luminosity class V belonged to the main sequence.[12]

## 20.2    Formation

Main article: Star formation

When a protostar is formed from the collapse of a giant molecular cloud of gas and dust in the local interstellar medium, the initial composition is homogeneous throughout, consisting of about 70% hydrogen, 28% helium and trace amounts of other elements, by mass.[13] The initial mass of the star depends on the local conditions within the cloud. (The mass distribution of newly formed stars is described empirically by the initial mass function.)[14] During the initial collapse, this pre-main-sequence star generates energy through gravitational contraction. Upon reaching a suitable density, energy generation is begun at the core using an exothermic nuclear fusion process that converts hydrogen into helium.[12]

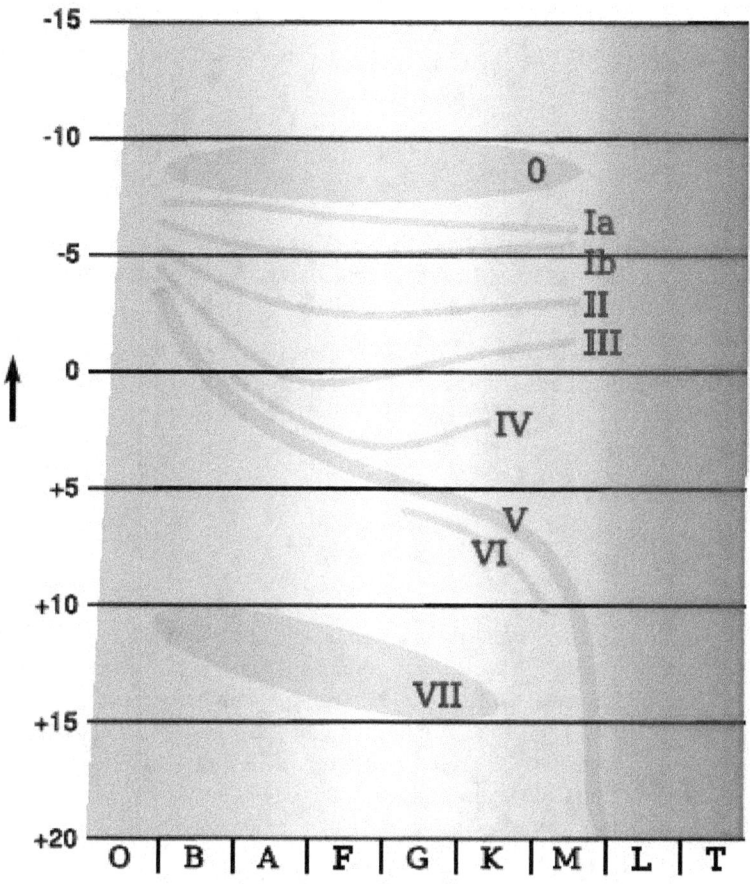

Hertzsprung–Russell diagram
Spectral type
Brown dwarfs
White dwarfs
Red dwarfs
Subdwarfs
Main sequence
("dwarfs")
Subgiants
Giants
Bright giants
Supergiants
Hypergiants
absolute
magni-
tude
(MV)

Once nuclear fusion of hydrogen becomes the dominant energy production process and the excess energy gained from gravitational contraction has been lost,[15] the star lies along a curve on the Hertzsprung–Russell diagram (or HR diagram) called the standard main sequence. Astronomers will sometimes refer to this stage as "zero age main sequence", or ZAMS.[16] The ZAMS curve can be calculated using computer models of stellar properties at the point when stars begin hydrogen fusion. From this point, the brightness and surface temperature of stars typically increase with age.[17]

A star remains near its initial position on the main sequence until a significant amount of hydrogen in the core has been consumed, then begins to evolve into a more luminous star. (On the HR diagram, the evolving star moves up and to the right of the main sequence.) Thus the main sequence represents the primary hydrogen-burning stage of a star's lifetime.[12]

## 20.3    Properties

The majority of stars on a typical HR diagram lie along the main-sequence curve. This line is pronounced because both the spectral type and the luminosity depend only on a star's mass, at least to zeroth-order approximation, as long as it is fusing hydrogen at its core—and that is what almost all stars spend most of their "active" lives doing.[18]

The temperature of a star determines its spectral type via its effect on the physical properties of plasma in its photosphere. A star's energy emission as a function of wavelength is influenced by both its temperature and composition. A key indicator of this energy distribution is given by the color index, $B - V$, which measures the star's magnitude in blue ($B$) and green-yellow ($V$) light by means of filters.[note 1] This difference in magnitude provides a measure of a star's temperature.

## 20.4    Dwarf terminology

Main-sequence stars are called dwarf stars, but this terminology is partly historical and can be somewhat confusing. For the cooler stars, dwarfs such as red dwarfs, orange dwarfs, and yellow dwarfs are indeed much smaller and dimmer than other stars of those colors. However, for hotter blue and white stars, the size and brightness difference between so-called *dwarf* stars that are on the main sequence and the so-called *giant* stars that are not becomes smaller; for the hottest stars it is not directly observable. For those stars the terms *dwarf* and *giant* refer to differences in spectral lines which indicate if a star is on the main sequence or off it. Nevertheless, very hot main-sequence stars are still sometimes called dwarfs, even though they have roughly the same size and brightness as the "giant" stars of that temperature.[19]

The common use of *dwarf* to mean main sequence is confusing in another way, because there are dwarf stars which are not main-sequence stars. For example, a white dwarf is the dead core of a star that is left after the star has shed its outer layers, that is much smaller than a main-sequence star—roughly the size of Earth. These represent the final evolutionary stage of many main-sequence stars.[20]

## 20.5    Parameters

By treating the star as an idealized energy radiator known as a black body, the luminosity $L$ and radius $R$ can be related to the effective temperature $T_{\text{eff}}$ by the Stefan–Boltzmann law:

$$L = 4\pi\sigma R^2 Teff^4$$

where $\sigma$ is the Stefan–Boltzmann constant. As the position of a star on the HR diagram shows its approximate luminosity, this relation can be used to estimate its radius.[21]

The mass, radius and luminosity of a star are closely interlinked, and their respective values can be approximated by three relations. First is the Stefan–Boltzmann law, which relates the luminosity $L$, the radius $R$ and the surface temperature $Teff$. Second is the mass–luminosity relation, which relates the luminosity $L$ and the mass $M$. Finally, the relationship between $M$ and $R$ is close to linear. The ratio of $M$ to $R$ increases by a factor of only three over 2.5 orders of magnitude of $M$. This relation is roughly proportional to the star's inner temperature $TI$, and its extremely slow increase reflects the fact that the rate of energy generation in the core strongly depends on this temperature, whereas it has to fit the mass–luminosity relation. Thus, a too high or too low temperature will result in stellar instability.

A better approximation is to take $\epsilon = L/M$, the energy generation rate per unit mass, as $\epsilon$ is proportional to $TI^{15}$, where $TI$ is the core temperature. This is suitable for stars at least as massive as the Sun, exhibiting the CNO cycle, and gives the better fit $R \propto M^{0.78}$.[22]

### 20.5.1  Sample parameters

The table below shows typical values for stars along the main sequence. The values of luminosity ($L$), radius ($R$) and mass ($M$) are relative to the Sun—a dwarf star with a spectral classification of G2 V. The actual values for a star may vary by as much as 20–30% from the values listed below.[23]

## 20.6   Energy generation

See also: Stellar nucleosynthesis

All main-sequence stars have a core region where energy is generated by nuclear fusion. The temperature and density of this core are at the levels necessary to sustain the energy production that will support the remainder of the star. A reduction of energy production would cause the overlaying mass to compress the core, resulting in an increase in the fusion rate because of higher temperature and pressure. Likewise an increase in energy production would cause the star to expand, lowering the pressure at the core. Thus the star forms a self-regulating system in hydrostatic equilibrium that is stable over the course of its main sequence lifetime.[29]

Main-sequence stars employ two types of hydrogen fusion processes, and the rate of energy generation from each type depends on the temperature in the core region. Astronomers divide the main sequence into upper and lower parts, based on which of the two is the dominant fusion process. In the lower main sequence, energy is primarily generated as the result of the proton-proton chain, which directly fuses hydrogen together in a series of stages to produce helium.[30] Stars in the upper main sequence have sufficiently high core temperatures to efficiently use the CNO cycle. (See the chart.) This process uses atoms of carbon, nitrogen and oxygen as intermediaries in the process of fusing hydrogen into helium.

At a stellar core temperature of 18 Million Kelvin, the PP process and CNO cycle are equally efficient, and each type generates half of the star's net luminosity. As this is the core temperature of a star with about 1.5 $M\odot$, the upper main sequence consists of stars above this mass. Thus, roughly speaking, stars of spectral class F or cooler belong to the lower main sequence, while A-type stars or hotter are upper main-sequence stars.[17] The transition in primary energy production from one form to the other spans a range difference of less than a single solar mass. In the Sun, a one solar-mass star, only 1.5% of the energy is generated by the CNO cycle.[31] By contrast, stars with 1.8 $M\odot$ or above generate almost their entire energy output through the CNO cycle.[32]

The observed upper limit for a main-sequence star is 120–200 $M\odot$.[33] The theoretical explanation for this limit is that stars above this mass can not radiate energy fast enough to remain stable, so any additional mass will be ejected in a series of pulsations until the star reaches a stable limit.[34] The lower limit for sustained proton–proton nuclear fusion is about 0.08 $M\odot$ or 80 times the mass of Jupiter.[30] Below this threshold are sub-stellar objects that can not sustain hydrogen fusion, known as brown dwarfs.[35]

## 20.7   Structure

Main article: Stellar structure
Because there is a temperature difference between the core and the surface, or photosphere, energy is transported outward. The two modes for transporting this energy are radiation and convection. A radiation zone, where energy is transported by radiation, is stable against convection and there is very little mixing of the plasma. By contrast, in a convection zone the energy is transported by bulk movement of plasma, with hotter material rising and cooler material descending. Convection is a more efficient mode for carrying energy than radiation, but it will only occur under conditions that create a steep temperature gradient.[29][36]

In massive stars (above 10 $M\odot$)[37] the rate of energy generation by the CNO cycle is very sensitive to temperature, so the fusion is highly concentrated at the core. Consequently, there is a high temperature gradient in the core region, which results in a convection zone for more efficient energy transport.[30] This mixing of material around the core removes the

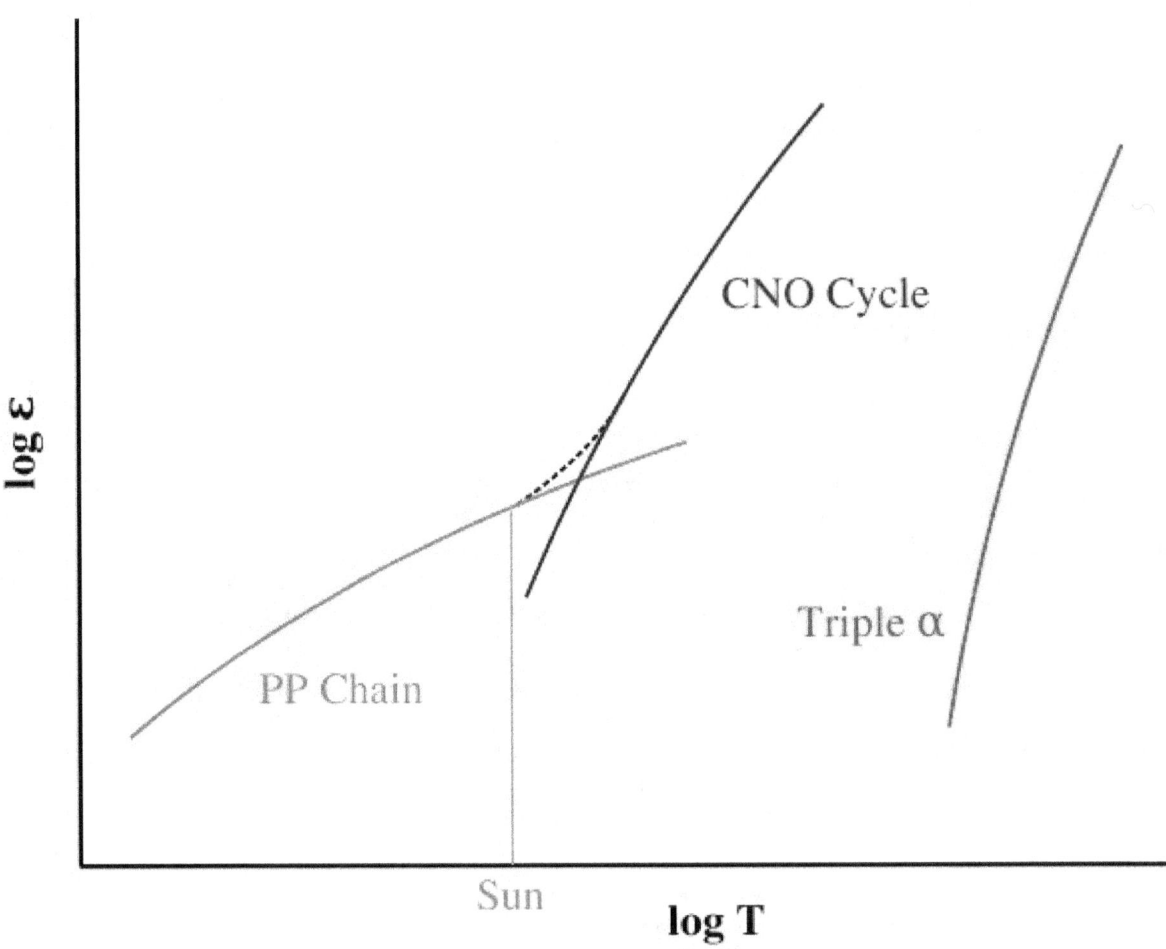

*This graph shows the logarithm of the relative energy output (ε) for the proton-proton (PP), CNO and triple-α fusion processes at different temperatures. The dashed line shows the combined energy generation of the PP and CNO processes within a star. At the Sun's core temperature, the PP process is more efficient.*

helium ash from the hydrogen-burning region, allowing more of the hydrogen in the star to be consumed during the main-sequence lifetime. The outer regions of a massive star transport energy by radiation, with little or no convection.[29]

Intermediate-mass stars such as Sirius may transport energy primarily by radiation, with a small core convection region.[38] Medium-sized, low-mass stars like the Sun have a core region that is stable against convection, with a convection zone near the surface that mixes the outer layers. This results in a steady buildup of a helium-rich core, surrounded by a hydrogen-rich outer region. By contrast, cool, very low-mass stars (below 0.4 $M_\odot$) are convective throughout.[14] Thus the helium produced at the core is distributed across the star, producing a relatively uniform atmosphere and a proportionately longer main sequence lifespan.[29]

## 20.8   Luminosity-color variation

As non-fusing helium ash accumulates in the core of a main-sequence star, the reduction in the abundance of hydrogen per unit mass results in a gradual lowering of the fusion rate within that mass. Since it is the outflow of fusion-supplied energy that supports the higher layers of the star, the core is compressed, producing higher temperatures and pressures. Both factors increase the rate of fusion thus moving the equilibrium towards a smaller, denser, hotter core producing more energy whose increased outflow pushes the higher layers further out. Thus there is a steady increase in the luminosity and

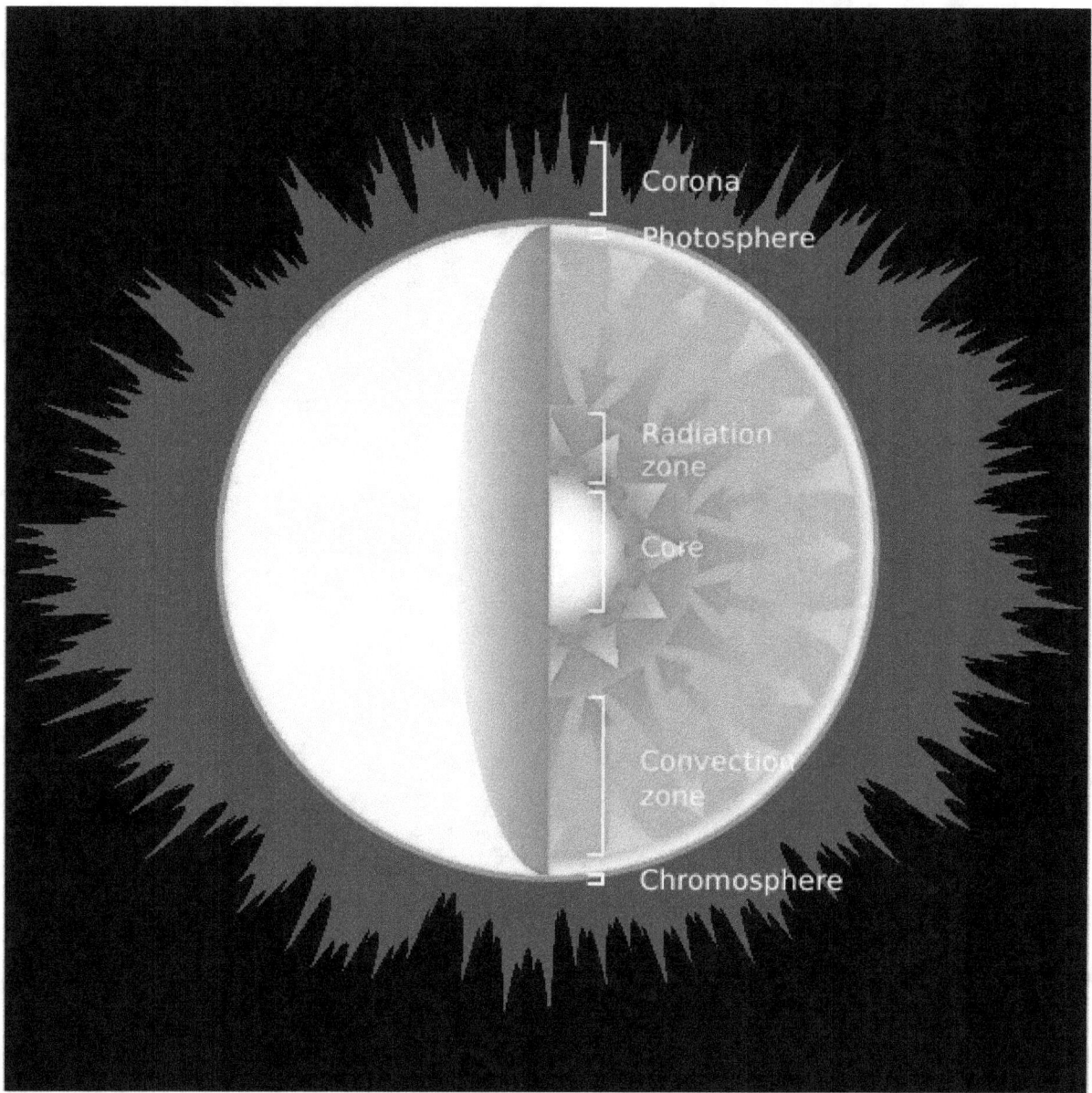

*This diagram shows a cross-section of a Sun-like star, showing the internal structure.*

radius of the star over time.[17] For example, the luminosity of the early Sun was only about 70% of its current value.[39] As a star ages this luminosity increase changes its position on the HR diagram. This effect results in a broadening of the main sequence band because stars are observed at random stages in their lifetime. That is, the main sequence band develops a thickness on the HR diagram; it is not simply a narrow line.[40]

Other factors that broaden the main sequence band on the HR diagram include uncertainty in the distance to stars and the presence of unresolved binary stars that can alter the observed stellar parameters. However, even perfect observation would show a fuzzy main sequence because mass is not the only parameter that affects a star's color and luminosity. Variations in chemical composition caused by the initial abundances, the star's evolutionary status,[41] interaction with a close companion,[42] rapid rotation,[43] or a magnetic field can all slightly change a main-sequence star's HR diagram position, to name just a few factors. As an example, there are metal-poor stars (with a very low abundance of elements with higher atomic numbers than helium) that lie just below the main sequence and are known as subdwarfs. These stars are fusing hydrogen in their cores and so they mark the lower edge of main sequence fuzziness caused by variance in chemical composition.[44]

A nearly vertical region of the HR diagram, known as the instability strip, is occupied by pulsating variable stars known as Cepheid variables. These stars vary in magnitude at regular intervals, giving them a pulsating appearance. The strip intersects the upper part of the main sequence in the region of class A and F stars, which are between one and two solar masses. Pulsating stars in this part of the instability strip that intersects the upper part of the main sequence are called Delta Scuti variables. Main-sequence stars in this region experience only small changes in magnitude and so this variation is difficult to detect.[45] Other classes of unstable main-sequence stars, like Beta Cephei variables, are unrelated to this instability strip.

## 20.9   Lifetime

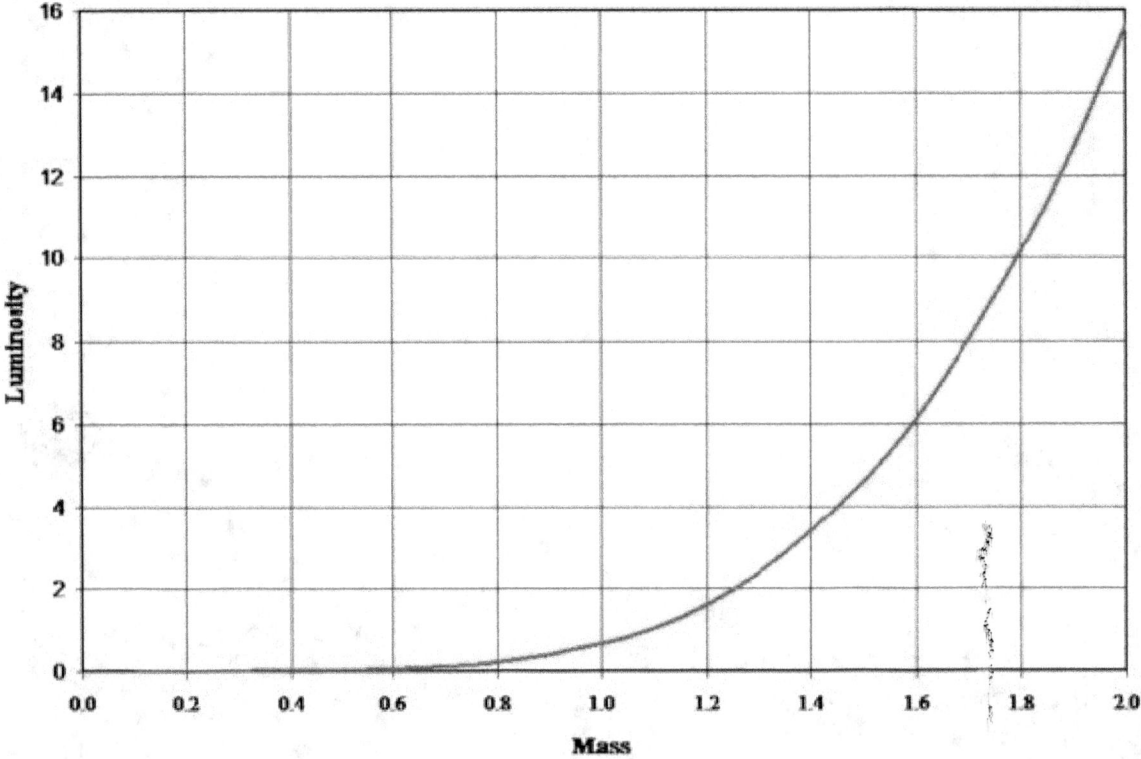

*This plot gives an example of the mass-luminosity relationship for zero-age main-sequence stars. The mass and luminosity are relative to the present-day Sun.*

The total amount of energy that a star can generate through nuclear fusion of hydrogen is limited by the amount of hydrogen fuel that can be consumed at the core. For a star in equilibrium, the energy generated at the core must be at least equal to the energy radiated at the surface. Since the luminosity gives the amount of energy radiated per unit time, the total life span can be estimated, to first approximation, as the total energy produced divided by the star's luminosity.[46]

For a star with at least 0.5 $M\odot$, once the hydrogen supply in its core is exhausted and it expands to become a red giant, it can start to fuse helium atoms to form carbon. The energy output of the helium fusion process per unit mass is only about a tenth the energy output of the hydrogen process, and the luminosity of the star increases.[47] This results in a much shorter length of time in this stage compared to the main sequence lifetime. (For example, the Sun is predicted to spend 130 million years burning helium, compared to about 12 billion years burning hydrogen.)[48] Thus, about 90% of the observed stars above 0.5 $M\odot$ will be on the main sequence.[49] On average, main-sequence stars are known to follow an empirical mass-luminosity relationship.[50] The luminosity (L) of the star is roughly proportional to the total mass (M) as the following power law:

$L \propto M^{3.5}$

This relationship applies to main-sequence stars in the range 0.1–50 $M_\odot$.[51]

The amount of fuel available for nuclear fusion is proportional to the mass of the star. Thus, the lifetime of a star on the main sequence can be estimated by comparing it to solar evolutionary models. The Sun has been a main-sequence star for about 4.5 billion years and it will become a red giant in 6.5 billion years,[52] for a total main sequence lifetime of roughly $10^{10}$ years. Hence:[53]

$$\tau_{MS} \approx 10^{10} \text{years} \cdot \left[ \frac{M}{M_\odot} \right] \cdot \left[ \frac{L_\odot}{L} \right] = 10^{10} \text{years} \cdot \left[ \frac{M}{M_\odot} \right]^{-2.5}$$

where $M$ and $L$ are the mass and luminosity of the star, respectively, $M_\odot$ is a solar mass, $L_\odot$ is the solar luminosity and $\tau_{MS}$ is the star's estimated main sequence lifetime.

Although more massive stars have more fuel to burn and might be expected to last longer, they also must radiate a proportionately greater amount with increased mass. Thus, the most massive stars may remain on the main sequence for only a few million years, while stars with less than a tenth of a solar mass may last for over a trillion years.[54]

The exact mass-luminosity relationship depends on how efficiently energy can be transported from the core to the surface. A higher opacity has an insulating effect that retains more energy at the core, so the star does not need to produce as much energy to remain in hydrostatic equilibrium. By contrast, a lower opacity means energy escapes more rapidly and the star must burn more fuel to remain in equilibrium.[55] Note, however, that a sufficiently high opacity can result in energy transport via convection, which changes the conditions needed to remain in equilibrium.[17]

In high-mass main-sequence stars, the opacity is dominated by electron scattering, which is nearly constant with increasing temperature. Thus the luminosity only increases as the cube of the star's mass.[47] For stars below 10 $M_\odot$, the opacity becomes dependent on temperature, resulting in the luminosity varying approximately as the fourth power of the star's mass.[51] For very low-mass stars, molecules in the atmosphere also contribute to the opacity. Below about 0.5 $M_\odot$, the luminosity of the star varies as the mass to the power of 2.3, producing a flattening of the slope on a graph of mass versus luminosity. Even these refinements are only an approximation, however, and the mass-luminosity relation can vary depending on a star's composition.[14]

## 20.10   Evolutionary tracks

See also: Stellar evolution

Once a main-sequence star consumes the hydrogen at its core, the loss of energy generation causes its gravitational collapse to resume. Stars with less than 0.23 $M_\odot$,[3] are predicted to directly become white dwarfs once energy generation by nuclear fusion of hydrogen at their core comes to a halt. In stars between this threshold and 10 $M_\odot$, the hydrogen surrounding the helium core reaches sufficient temperature and pressure to undergo fusion, forming a hydrogen-burning shell. In consequence of this change, the outer envelope of the star expands and decreases in temperature, turning it into a red giant. At this point the star is evolving off the main sequence and entering the giant branch. The path which the star now follows across the HR diagram, to the upper right of the main sequence, is called an evolutionary track.

The helium core of a red giant continues to collapse until it is entirely supported by electron degeneracy pressure—a quantum mechanical effect that restricts how closely matter can be compacted. For stars of more than about 0.5 $M_\odot$,[56] the core eventually reaches a temperature where it becomes hot enough to burn helium into carbon via the triple alpha process.[57][58] Stars with more than 5–7.5 $M_\odot$ can additionally fuse elements with higher atomic numbers.[59][60] For stars with ten or more solar masses, this process can lead to an increasingly dense core that finally collapses, ejecting the star's overlying layers in a Type II supernova explosion,[4] Type Ib supernova or Type Ic supernova.

When a cluster of stars is formed at about the same time, the life span of these stars will depend on their individual masses. The most massive stars will leave the main sequence first, followed steadily in sequence by stars of ever lower masses. Thus the stars will evolve in order of their position on the main sequence, proceeding from the most massive at the left toward the right of the HR diagram. The current position where stars in this cluster are leaving the main sequence is

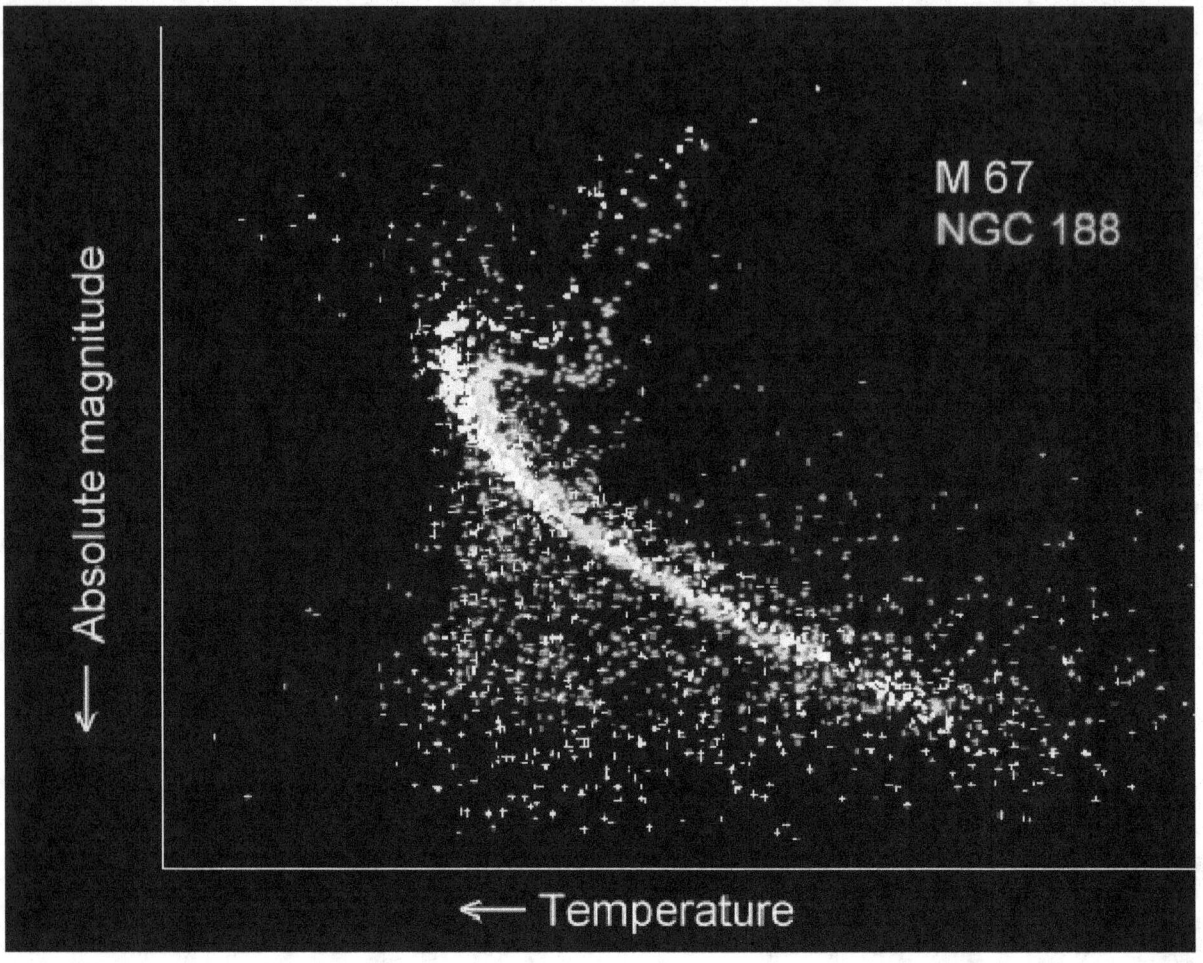

*This shows the Hertzsprung–Russell diagrams for two open clusters. NGC 188 (blue) is older, and shows a lower turn off from the main sequence than that seen in M67 (yellow). The dots outside the two sequences are mostly foreground and background stars with no relation to the clusters.*

known as the turn-off point. By knowing the main sequence lifespan of stars at this point, it becomes possible to estimate the age of the cluster.[611]

## 20.11   See also

- Hydrogen-burning process

- Red dwarf

- Supergiant

## 20.12   Notes

[1]  By measuring the difference between these values, this eliminates the need to correct the magnitudes for distance. However, see extinction.

[2]  The Sun is a typical type G2V star.

# 20.13 References

[1] Harding E. Smith (1999-04-21). "The Hertzsprung-Russell Diagram". *Gene Smith's Astronomy Tutorial*. Center for Astrophysics & Space Sciences, University of California, San Diego. Retrieved 2009-10-29.

[2] Richard Powell (2006). "The Hertzsprung Russell Diagram". *An Atlas of the Universe*. Retrieved 2009-10-29.

[3] Adams, Fred C.; Laughlin, Gregory (April 1997). "A Dying Universe: The Long Term Fate and Evolution of Astrophysical Objects".*Reviews of Modern Physics***69**(2): 337–372. arXiv:astro-ph/9701131. Bibcode:1997RvMP...69..337A.doi:10.1103/Rev

[4] Gilmore, Gerry (2004). "The Short Spectacular Life of a Superstar". *Science* **304** (5697): 1915–1916. doi:10.1126/science.1100370. PMID 15218132. Retrieved 2007-05-01.

[5] "The Brightest Stars Don't Live Alone". *ESO Press Release*. Retrieved 27 July 2012.

[6] Longair, Malcolm S. (2006). *The Cosmic Century: A History of Astrophysics and Cosmology*. Cambridge University Press. pp. 25–26. ISBN 0-521-47436-1.

[7] Brown, Laurie M.; Pais, Abraham; Pippard, A. B., eds. (1995). *Twentieth Century Physics*. Bristol; New York: Institute of Physics, American Institute of Physics. p. 1696. ISBN 0-7503-0310-7. OCLC 33102501.

[8] Russell, H. N. (1913). ""Giant" and "dwarf" stars". *The Observatory* **36**: 324–329. Bibcode:1913Obs....36..324R.

[9] Strömgren, Bengt (1933). "On the Interpretation of the Hertzsprung-Russell-Diagram". *Zeitschrift für Astrophysik* **7**: 222–248. Bibcode:1933ZA......7..222S.

[10] Schatzman, Evry L.; Praderie, Francoise (1993). *The Stars*. Springer. pp. 96–97. ISBN 3-540-54196-9.

[11] Morgan, W. W.; Keenan, P. C.; Kellman, E. (1943). *An atlas of stellar spectra, with an outline of spectral classification*. Chicago, Illinois: The University of Chicago press. Retrieved 2008-08-12.

[12] Unsöld, Albrecht (1969). *The New Cosmos*. Springer-Verlag New York Inc. p. 268. ISBN 0-387-90886-2.

[13] Gloeckler, George; Geiss, Johannes (2004). "Composition of the local interstellar medium as diagnosed with pickup ions". *Advances in Space Research* **34** (1): 53–60. Bibcode:2004AdSpR..34...53G. doi:10.1016/j.asr.2003.02.054.

[14] Kroupa, Pavel (2002). "The Initial Mass Function of Stars: Evidence for Uniformity in Variable Systems". *Science* **295** (5552): 82–91. arXiv:astro-ph/0201098. Bibcode:2002Sci...295...82K. doi:10.1126/science.1067524. PMID 11778039. Retrieved 2007-12-03.

[15] Schilling, Govert (2001). "New Model Shows Sun Was a Hot Young Star".*Science***293**(5538): 2188–2189. doi:10.1126/science PMID 11567116. Retrieved 2007-02-04.

[16] "Zero Age Main Sequence". *The SAO Encyclopedia of Astronomy*. Swinburne University. Retrieved 2007-12-09.

[17] Clayton, Donald D. (1983). *Principles of Stellar Evolution and Nucleosynthesis*. University of Chicago Press. ISBN 0-226-10953-4.

[18] "Main Sequence Stars". Australia Telescope Outreach and Education. Retrieved 2007-12-04.

[19] Moore, Patrick (2006). *The Amateur Astronomer*. Springer. ISBN 1-85233-878-4.

[20] "White Dwarf". *COSMOS—The SAO Encyclopedia of Astronomy*. Swinburne University. Retrieved 2007-12-04.

[21] "Origin of the Hertzsprung-Russell Diagram". University of Nebraska. Retrieved 2007-12-06.

[22] "A course on stars' physical properties, formation and evolution" (PDF). University of St. Andrews. Retrieved 2010-05-18.

[23] Siess, Lionel (2000). "Computation of Isochrones". Institut d'Astronomie et d'Astrophysique, Université libre de Bruxelles. Retrieved 2007-12-06.—Compare, for example, the model isochrones generated for a ZAMS of 1.1 solar masses. This is listed in the table as 1.26 times the solar luminosity. At metallicity Z=0.01 the luminosity is 1.34 times solar luminosity. At metallicity Z=0.04 the luminosity is 0.89 times the solar luminosity.

[24] Zombeck, Martin V. (1990). *Handbook of Space Astronomy and Astrophysics* (2nd ed.). Cambridge University Press. ISBN 0-521-34787-4. Retrieved 2007-12-06.

[25]  "SIMBAD Astronomical Database". Centre de Données astronomiques de Strasbourg. Retrieved 2008-11-21.

[26]  Luck, R. Earle; Heiter, Ulrike (2005). "Stars within 15 Parsecs: Abundances for a Northern Sample". *The Astronomical Journal* **129** (2): 1063–1083. Bibcode:2005AJ....129.1063L. doi:10.1086/427250.

[27]  "LTT 2151 – High proper-motion Star". Centre de Données astronomiques de Strasbourg. Retrieved 2008-08-12.

[28]  Staff (2008-01-01). "List of the Nearest Hundred Nearest Star Systems". Research Consortium on Nearby Stars. Retrieved 2008-08-12.

[29]  Brainerd, Jerome James (2005-02-16). "Main-Sequence Stars". The Astrophysics Spectator. Retrieved 2007-12-04.

[30]  Karttunen, Hannu (2003). *Fundamental Astronomy*. Springer. ISBN 3-540-00179-4.

[31]  Bahcall, John N.; Pinsonneault, M. H.; Basu, Sarbani (2001-07-10). "Solar Models: Current Epoch and Time Dependences, Neutrinos, and Helioseismological Properties". *The Astrophysical Journal* **555** (2): 990–1012. arXiv:astro-ph/0212331. Bibcode:

[32]  Salaris, Maurizio; Cassisi, Santi (2005). *Evolution of Stars and Stellar Populations*. John Wiley and Sons. p. 128. ISBN 0-470-09220-3.

[33]  Oey, M. S.; Clarke, C. J. (2005). "Statistical Confirmation of a Stellar Upper Mass Limit". *The Astrophysical Journal* **620** (1): L43–L46. arXiv:astro-ph/0501135. Bibcode:2005ApJ...620L..43O. doi:10.1086/428396.

[34]  Ziebarth, Kenneth (1970). "On the Upper Mass Limit for Main-Sequence Stars". *Astrophysical Journal* **162**: 947–962. Bibcode:1970ApJ...162..947Z. doi:10.1086/150726.

[35]  Burrows, A.; Hubbard, W. B.; Saumon, D.; Lunine, J. I. (March 1993). "An expanded set of brown dwarf and very low mass star models". *Astrophysical Journal, Part 1* **406** (1): 158–171. Bibcode:1993ApJ...406..158B. doi:10.1086/172427.

[36]  Aller, Lawrence H. (1991). *Atoms, Stars, and Nebulae*. Cambridge University Press. ISBN 0-521-31040-7.

[37]  Bressan, A. G.; Chiosi, C.; Bertelli, G. (1981). "Mass loss and overshooting in massive stars". *Astronomy and Astrophysics* **102** (1): 25–30. Bibcode:1981A&A...102...25B.

[38]  Lochner, Jim; Gibb, Meredith; Newman, Phil (2006-09-06). "Stars". NASA. Retrieved 2007-12-05.

[39]  Gough, D. O. (1981). "Solar interior structure and luminosity variations". *Solar Physics* **74** (1): 21–34. Bibcode:1981SoPh...74.. doi:10.1007/BF00151270.

[40]  Padmanabhan, Thanu (2001). *Theoretical Astrophysics*. Cambridge University Press. ISBN 0-521-56241-4.

[41]  Wright, J. T. (2004). "Do We Know of Any Maunder Minimum Stars?". *The Astronomical Journal* **128** (3): 1273–1278. arXiv:astro-ph/0406338. Bibcode:2004AJ....128.1273W. doi:10.1086/423221. Retrieved 2007-12-06.

[42]  Tayler, Roger John (1994). *The Stars: Their Structure and Evolution*. Cambridge University Press. ISBN 0-521-45885-4.

[43]  Sweet, I. P. A.; Roy, A. E. (1953). "The structure of rotating stars". *Monthly Notices of the Royal Astronomical Society* **113**: 701–715. Bibcode:1953MNRAS.113..701S. doi:10.1093/mnras/113.6.701.

[44]  Burgasser, Adam J.; Kirkpatrick, J. Davy; Lepine, Sebastien (July 5–9, 2004). *Spitzer Studies of Ultracool Subdwarfs: Metal-poor Late-type M, L and T Dwarfs*. Proceedings of the 13th Cambridge Workshop on Cool Stars, Stellar Systems and the Sun (Hamburg, Germany: Dordrecht, D. Reidel Publishing Co): 237. Retrieved 2007-12-06.

[45]  Green, S. F.; Jones, Mark Henry; Burnell, S. Jocelyn (2004). *An Introduction to the Sun and Stars*. Cambridge University Press. ISBN 0-521-54622-2.

[46]  Richmond, Michael W. (2004-11-10). "Stellar evolution on the main sequence". Rochester Institute of Technology. Retrieved 2007-12-03.

[47]  Prialnik, Dina (2000). *An Introduction to the Theory of Stellar Structure and Evolution*. Cambridge University Press. ISBN 0-521-65937-X.

[48]  Schröder, K.-P.; Connon Smith, Robert (May 2008). "Distant future of the Sun and Earth revisited". *Monthly Notices of the Royal Astronomical Society* **386** (1): 155–163. arXiv:0801.4031. Bibcode:2008MNRAS.386..155S. doi:10.1111/j.1365-2966.2008.13022.x.

[49] Arnett, David (1996). *Supernovae and Nucleosynthesis: An Investigation of the History of Matter, from the Big Bang to the Present.* Princeton University Press. ISBN 0-691-01147-8.—Hydrogen fusion produces $8\times10^{18}$ erg/g while helium fusion produces $8\times10^{17}$ erg/g.

[50] For a detailed historical reconstruction of the theoretical derivation of this relationship by Eddington in 1924, see: Lecchini, Stefano (2007). *How Dwarfs Became Giants. The Discovery of the Mass-Luminosity Relation.* Bern Studies in the History and Philosophy of Science. ISBN 3-9522882-6-8.

[51] Rolfs, Claus E.; Rodney, William S. (1988). *Cauldrons in the Cosmos: Nuclear Astrophysics.* University of Chicago Press. ISBN 0-226-72457-3.

[52] Sackmann, I.-Juliana; Boothroyd, Arnold I.; Kraemer, Kathleen E. (November 1993). "Our Sun. III. Present and Future". *Astrophysical Journal* **418**: 457–468. Bibcode:1993ApJ...418..457S. doi:10.1086/173407.

[53] Hansen, Carl J.; Kawaler, Steven D. (1994). *Stellar Interiors: Physical Principles, Structure, and Evolution.* Birkhäuser. p. 28. ISBN 0-387-94138-X.

[54] Laughlin, Gregory; Bodenheimer, Peter; Adams, Fred C. (1997). "The End of the Main Sequence". *The Astrophysical Journal* **482** (1): 420–432. Bibcode:1997ApJ...482..420L. doi:10.1086/304125.

[55] Imamura, James N. (1995-02-07). "Mass-Luminosity Relationship". University of Oregon. Archived from the original on December 14, 2006. Retrieved 2007-01-08.

[56] Fynbo, Hans O. U.; et al. (2004). "Revised rates for the stellar triple-$\alpha$ process from measurement of 12C nuclear resonances". *Nature* **433** (7022): 136–139. doi:10.1038/nature03219. PMID 15650733.

[57] Sitko, Michael L. (2000-03-24). "Stellar Structure and Evolution". University of Cincinnati. Archived from the original on March 26, 2005. Retrieved 2007-12-05.

[58] Staff (2006-10-12). "Post-Main Sequence Stars". Australia Telescope Outreach and Education. Retrieved 2008-01-08.

[59] Girardi, L.; Bressan, A.; Bertelli, G.; Chiosi, C. (2000). "Evolutionary tracks and isochrones for low- and intermediate-mass stars: From 0.15 to 7 $M_{sun}$, and from Z=0.0004 to 0.03". *Astronomy and Astrophysics Supplement* **141** (3): 371–383. arXiv:astro-ph/9910164. Bibcode:2000A&AS..141..371G. doi:10.1051/aas:2000126.

[60] Poelarends, A. J. T.; Herwig, F.; Langer, N.; Heger, A. (March 2008). "The Supernova Channel of Super-AGB Stars". *The Astrophysical Journal* **675** (1): 614–625. arXiv:0705.4643. Bibcode:2008ApJ...675..614P. doi:10.1086/520872.

[61] Krauss, Lawrence M.; Chaboyer, Brian (2003). "Age Estimates of Globular Clusters in the Milky Way: Constraints on Cosmology". *Science* **299** (5603): 65–69. Bibcode:2003Sci...299...65K. doi:10.1126/science.1075631. PMID 12511641.

## 20.14 Further reading

### 20.14.1 General

- Kippenhahn, Rudolf, *100 Billion Suns*, Basic Books, New York, 1983.

### 20.14.2 Technical

- Arnett, David, *Supernovae and Nucleosynthesis*, Princeton University Press, Princeton, 1996.

- Bahcall, John N., *Neutrino Astrophysics*, Cambridge University Press, Cambridge, 1989.

- Bahcall, John N., Pinsonneault, M.H., and Basu, Sarbani, *"Solar Models: Current Epoch and Time Dependences, Neutrinos, and Helioseismological Properties,"* The Astrophysical Journal, 555, 990, 2001.

- Barnes, C. A., Clayton, D. D., and Schramm, D. N.(eds.), *Essays in Nuclear Astrophysics*, Cambridge University Press, Cambridge, 1982.

- Bowers, Richard L., and Deeming, Terry, *Astrophysics I: Stars*, Jones and Bartlett, Publishers, Boston, 1984.

- Bradley W. Carroll and Dale A. Ostlie (2007). *An Introduction to Modern Astrophysics*. Person Education Addison-Wesley San Francisco. ISBN 0-80530402-9.

- Chabrier, Gilles, and Baraffe, Isabelle, *"Theory of Low-Mass Stars and Substellar Objects,"* Annual Review of Astronomy and Astrophysics, 38, 337, 2000.

- Chandrasekhar, S., *An Introduction to the study of stellar Structure*, Dover Publications, Inc., New York, 1967.

- Clayton, Donald D., *Principles of Stellar Evolution and Nucleosynthesis*, University of Chicago Press, Chicago, 1983.

- Cox, J. P., and Giuli, R. T., *Principles of Stellar Structure*, Gordon and Breach, New York, 1968.

- Fowler, William ., Caughlan, Georgeanne R., and Zimmerman, Barbara A., *"Thermonuclear Reaction Rates, I,"* Annual Review of Astronomy and Astrophysics, 5, 525, 1967.

- Fowler, William A., Caughlan, Georgeanne R., and Zimmerman, Barbara A., *"Thermonuclear Reaction Rates, II, "* Annual Review of Astronomy and Astrophysics, 13, 69, 1975.

- Hansen, Carl J., Kawaler, Steven D., and Trimble, *Virginia Stellar Interiors: Physical Principles, Structure, and Evolution, Second Edition*, Springer-Verlag, New York, 2004.

- Harris, Michael J., Fowler, William A., Caughlan, Georgeanne R., and Zimmerman, Barbara A., *"Thermonuclear Reaction Rates, III,"* Annual Review of Astronomy and Astrophysics, 21, 165, 1983.

- Iben, Icko, Jr, *"Stellar Evolution Within and Off the Main Sequence,"* Annual Review of Astronomy and Astrophysics, 5, 571, 1967.

- Iglesias, Carlos A, and Rogers, Forrest J., *"Updated Opal Opacities,"* The Astrophysical Journal, 464, 943, 1996.

- Kippenhahn, Rudolf, and Weigert, Alfred, *Stellar Structure and Evolution*, Springer-Verlag, Berlin, 1990.

- Liebert, James, and Probst, Ronald G., *"Very Low Mass Stars"*, Annual Review of Astronomy and Astrophysics, 25, 437, 1987.

- Padmanabhan, T., *Theoretical Astrophysics*, Cambridge University Press, Cambridge, 2002.

- Prialnik, Dina, *An Introduction to the Theory of Stellar Structure and Evolution*, Cambridge University Press, Cambridge, 2000.

- Novotny, Eva, *Introduction to Stellar Atmospheres and Interior*, Oxford University Press, New York, 1973.

- Shore, Steven N., *The Tapestry of Modern Astrophysics*, John Wiley and Sons, Hoboken, 2003.

# Chapter 21

# Black hole

For other uses, see Black hole (disambiguation).

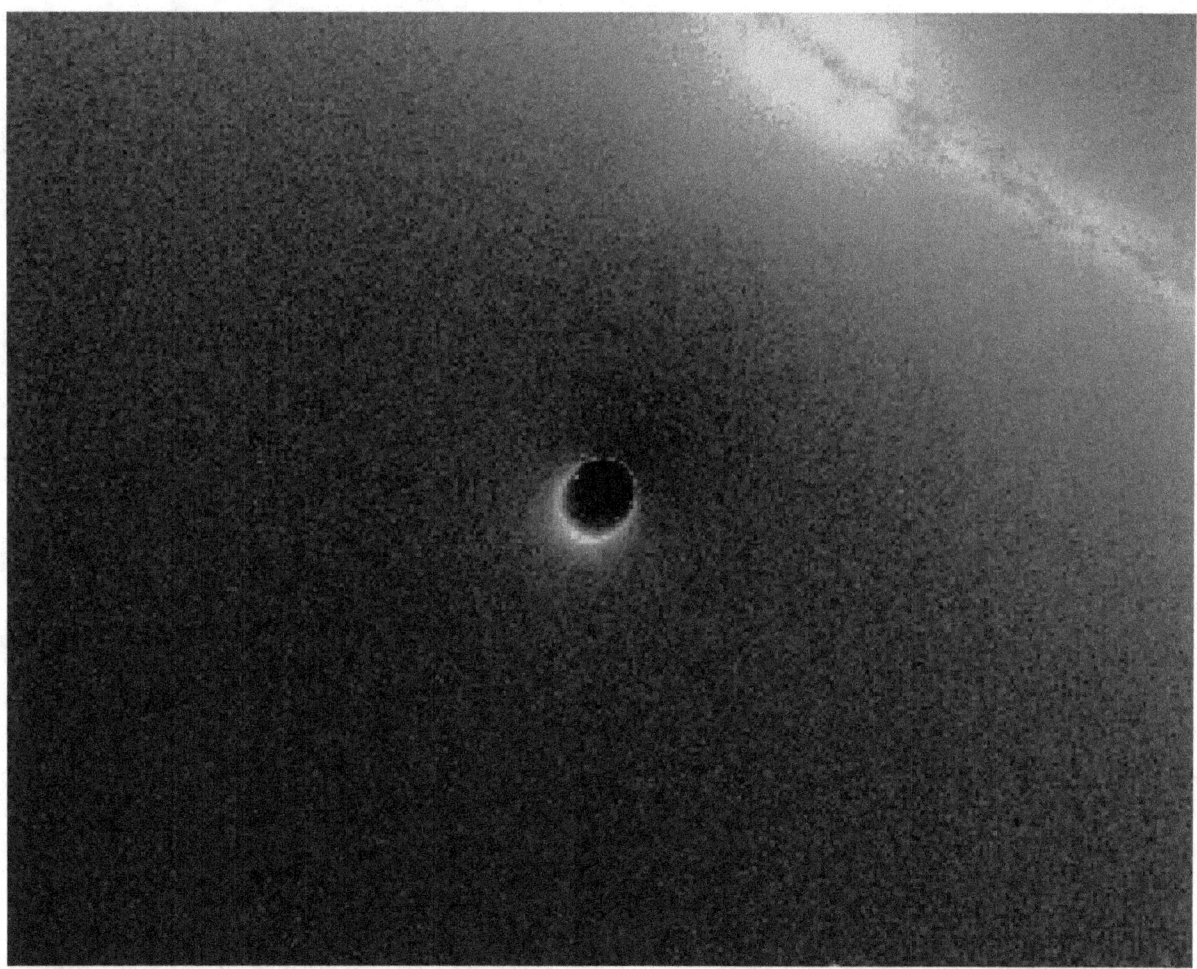

*Simulation of gravitational lensing by a black hole, which distorts the image of a galaxy in the background*

A **black hole** is a geometrically defined region of spacetime exhibiting such strong gravitational effects that nothing—including particles and electromagnetic radiation such as light—can escape from inside it.[1] The theory of general relativity predicts that a sufficiently compact mass can deform spacetime to form a black hole.[2][3] The boundary of the region from which no escape is possible is called the event horizon. Although crossing the event horizon has enormous effect on the fate of the object crossing it, it appears to have no locally detectable features. In many ways a black hole acts like an ideal black body, as it reflects no light.[4][5] Moreover, quantum field theory in curved spacetime predicts that event horizons emit Hawking radiation, with the same spectrum as a black body of a temperature inversely proportional to its mass. This temperature is on the order of billionths of a kelvin for black holes of stellar mass, making it essentially impossible to observe.

Objects whose gravitational fields are too strong for light to escape were first considered in the 18th century by John Michell and Pierre-Simon Laplace. The first modern solution of general relativity that would characterize a black hole was found by Karl Schwarzschild in 1916, although its interpretation as a region of space from which nothing can escape was first published by David Finkelstein in 1958. Long considered a mathematical curiosity, it was during the 1960s that theoretical work showed black holes were a generic prediction of general relativity. The discovery of neutron stars sparked interest in gravitationally collapsed compact objects as a possible astrophysical reality.

Black holes of stellar mass are expected to form when very massive stars collapse at the end of their life cycle. After a black hole has formed, it can continue to grow by absorbing mass from its surroundings. By absorbing other stars and merging with other black holes, supermassive black holes of millions of solar masses ($M\odot$) may form. There is general consensus that supermassive black holes exist in the centers of most galaxies.

Despite its invisible interior, the presence of a black hole can be inferred through its interaction with other matter and with electromagnetic radiation such as visible light. Matter falling onto a black hole can form an accretion disk heated by friction, forming some of the brightest objects in the universe. If there are other stars orbiting a black hole, their orbit can be used to determine its mass and location. Such observations can be used to exclude possible alternatives (such as neutron stars). In this way, astronomers have identified numerous stellar black hole candidates in binary systems, and established that the radio source known as Sagittarius A*, at the core of our own Milky Way galaxy, contains a supermassive black hole of about 4.3 million solar masses.

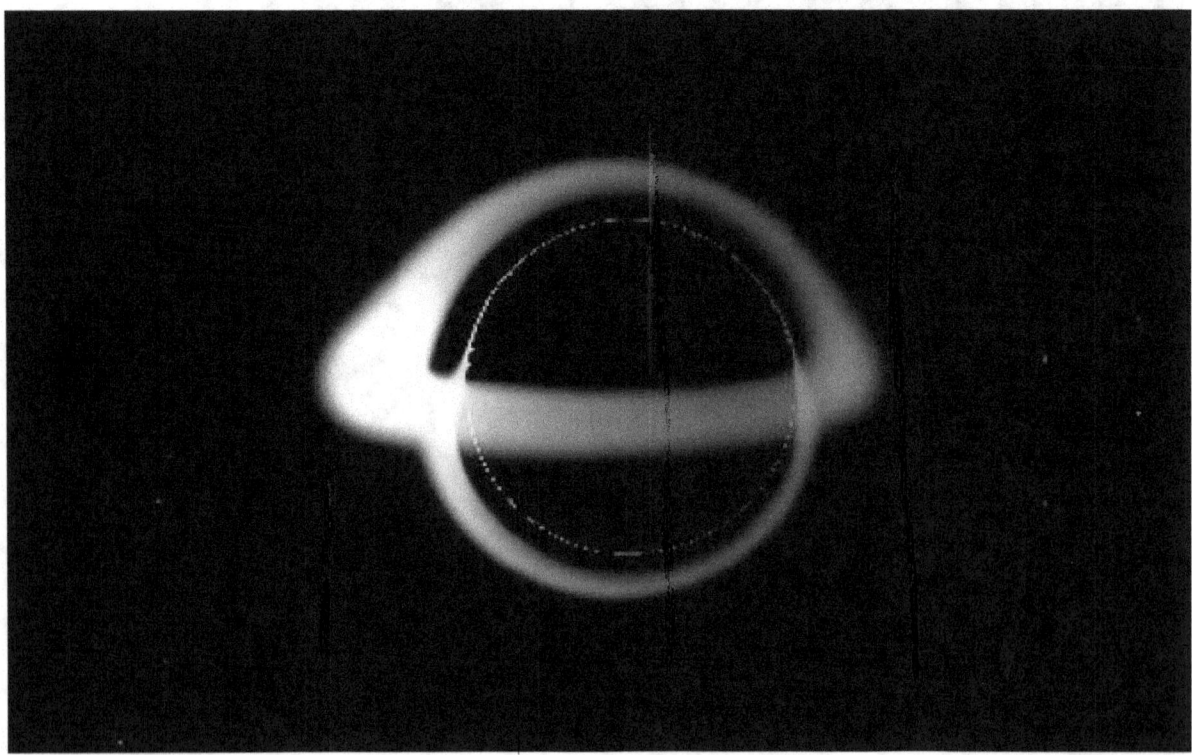

*Predicted appearance of non-rotating black hole with toroidal ring of ionised matter, such as has been proposed*[6] *as a model for Sagittarius A*. The asymmetry is due to the Doppler effect resulting from the enormous orbital speed needed for centrifugal balance of the very strong gravitational attraction of the hole.*

## 21.1 History

*Simulated view of a black hole in front of the Large Magellanic Cloud. Note the gravitational lensing effect, which produces two enlarged but highly distorted views of the Cloud. Across the top, the Milky Way disk appears distorted into an arc.*

The idea of a body so massive that even light could not escape was first put forward by John Michell in a letter written to Henry Cavendish in 1783 of the Royal Society:

> If the semi-diameter of a sphere of the same density as the Sun were to exceed that of the Sun in the proportion of 500 to 1, a body falling from an infinite height towards it would have acquired at its surface greater velocity than that of light, and consequently supposing light to be attracted by the same force in proportion to its vis inertiae, with other bodies, all light emitted from such a body would be made to return towards it by its own proper gravity.
> — John Michell[7]

In 1796, mathematician Pierre-Simon Laplace promoted the same idea in the first and second editions of his book *Exposition du système du Monde* (it was removed from later editions).[8][9] Such "dark stars" were largely ignored in the nineteenth century, since it was not understood how a massless wave such as light could be influenced by gravity.[10]

### 21.1.1 General relativity

In 1915, Albert Einstein developed his theory of general relativity, having earlier shown that gravity does influence light's motion. Only a few months later, Karl Schwarzschild found a solution to the Einstein field equations, which describes the gravitational field of a point mass and a spherical mass.[11] A few months after Schwarzschild, Johannes Droste, a student of Hendrik Lorentz, independently gave the same solution for the point mass and wrote more extensively about its properties.[12][13] This solution had a peculiar behaviour at what is now called the Schwarzschild radius, where it became singular, meaning that some of the terms in the Einstein equations became infinite. The nature of this surface was not quite understood at the time. In 1924, Arthur Eddington showed that the singularity disappeared after a change of coordinates (see Eddington–Finkelstein coordinates), although it took until 1933 for Georges Lemaître to realize that this meant the singularity at the Schwarzschild radius was an unphysical coordinate singularity.[14] Arthur Eddington did however comment on the possibility of a star with mass compressed to the Schwarzschild radius in a 1926 book, noting that Einstein's theory allows us to rule out overly large densities for visible stars like Betelgeuse because "a star of 250 million km. radius could not possibly have so high a density as the sun. Firstly, the force of gravitation would be so great that light would be unable to escape from it, the rays falling back to the star like a stone to the earth. Secondly, the red shift of the spectral lines would be so great that the spectrum would be shifted out of existence. Thirdly, the mass would produce so much curvature of the space-time metric that space would close up around the star, leaving us outside (i.e., nowhere)."[15][16]

In 1931, Subrahmanyan Chandrasekhar calculated, using special relativity, that a non-rotating body of electron-degenerate matter above a certain limiting mass (now called the Chandrasekhar limit at 1.4 $M_\odot$) has no stable solutions.[17] His arguments were opposed by many of his contemporaries like Eddington and Lev Landau, who argued that some yet unknown mechanism would stop the collapse.[18] They were partly correct: a white dwarf slightly more massive than the Chandrasekhar limit will collapse into a neutron star,[19] which is itself stable because of the Pauli exclusion principle. But in 1939, Robert Oppenheimer and others predicted that neutron stars above approximately 3 $M_\odot$ (the Tolman–Oppenheimer–Volkoff limit) would collapse into black holes for the reasons presented by Chandrasekhar, and concluded that no law of physics was likely to intervene and stop at least some stars from collapsing to black holes.[20]

Oppenheimer and his co-authors interpreted the singularity at the boundary of the Schwarzschild radius as indicating that this was the boundary of a bubble in which time stopped. This is a valid point of view for external observers, but not for infalling observers. Because of this property, the collapsed stars were called "frozen stars",[21] because an outside observer would see the surface of the star frozen in time at the instant where its collapse takes it inside the Schwarzschild radius.

### 21.1.2 Golden age

See also: Golden age of general relativity

In 1958, David Finkelstein identified the Schwarzschild surface as an event horizon, "a perfect unidirectional membrane: causal influences can cross it in only one direction".[22] This did not strictly contradict Oppenheimer's results, but extended them to include the point of view of infalling observers. Finkelstein's solution extended the Schwarzschild solution for the future of observers falling into a black hole. A complete extension had already been found by Martin Kruskal, who was urged to publish it.[23]

These results came at the beginning of the golden age of general relativity, which was marked by general relativity and black holes becoming mainstream subjects of research. This process was helped by the discovery of pulsars in 1967,[24][25] which, by 1969, were shown to be rapidly rotating neutron stars.[26] Until that time, neutron stars, like black holes, were regarded as just theoretical curiosities; but the discovery of pulsars showed their physical relevance and spurred a further interest in all types of compact objects that might be formed by gravitational collapse.

In this period more general black hole solutions were found. In 1963, Roy Kerr found the exact solution for a rotating black hole. Two years later, Ezra Newman found the axisymmetric solution for a black hole that is both rotating and electrically charged.[27] Through the work of Werner Israel,[28] Brandon Carter,[29][30] and David Robinson[31] the no-hair theorem emerged, stating that a stationary black hole solution is completely described by the three parameters of the Kerr–Newman metric; mass, angular momentum, and electric charge.[32]

At first, it was suspected that the strange features of the black hole solutions were pathological artifacts from the symmetry conditions imposed, and that the singularities would not appear in generic situations. This view was held in particular by Vladimir Belinsky, Isaak Khalatnikov, and Evgeny Lifshitz, who tried to prove that no singularities appear in generic solutions. However, in the late 1960s Roger Penrose[33] and Stephen Hawking used global techniques to prove that singularities appear generically.[34]

Work by James Bardeen, Jacob Bekenstein, Carter, and Hawking in the early 1970s led to the formulation of black hole thermodynamics.[35] These laws describe the behaviour of a black hole in close analogy to the laws of thermodynamics by relating mass to energy, area to entropy, and surface gravity to temperature. The analogy was completed when Hawking, in 1974, showed that quantum field theory predicts that black holes should radiate like a black body with a temperature proportional to the surface gravity of the black hole.[36]

The first use of the term "black hole" in print was by journalist Ann Ewing in her article *"Black Holes' in Space"*, dated 18 January 1964, which was a report on a meeting of the American Association for the Advancement of Science.[37] John Wheeler used the term "black hole" at a lecture in 1967, leading some to credit him with coining the phrase. After Wheeler's use of the term, it was quickly adopted in general use.

## 21.2 Properties and structure

The no-hair theorem states that, once it achieves a stable condition after formation, a black hole has only three independent physical properties: mass, charge, and angular momentum.[32] Any two black holes that share the same values for these properties, or parameters, are indistinguishable according to classical (i.e. non-quantum) mechanics.

These properties are special because they are visible from outside a black hole. For example, a charged black hole repels other like charges just like any other charged object. Similarly, the total mass inside a sphere containing a black hole can be found by using the gravitational analog of Gauss's law, the ADM mass, far away from the black hole.[38] Likewise, the angular momentum can be measured from far away using frame dragging by the gravitomagnetic field.

When an object falls into a black hole, any information about the shape of the object or distribution of charge on it is evenly distributed along the horizon of the black hole, and is lost to outside observers. The behavior of the horizon in this situation is a dissipative system that is closely analogous to that of a conductive stretchy membrane with friction and electrical resistance—the membrane paradigm.[39] This is different from other field theories like electromagnetism, which do not have any friction or resistivity at the microscopic level, because they are time-reversible. Because a black hole eventually achieves a stable state with only three parameters, there is no way to avoid losing information about the initial conditions: the gravitational and electric fields of a black hole give very little information about what went in. The information that is lost includes every quantity that cannot be measured far away from the black hole horizon, including approximately conserved quantum numbers such as the total baryon number and lepton number. This behavior is so puzzling that it has been called the black hole information loss paradox.[40][41]

### 21.2.1 Physical properties

The simplest static black holes have mass but neither electric charge nor angular momentum. These black holes are often referred to as Schwarzschild black holes after Karl Schwarzschild who discovered this solution in 1916.[11] According to Birkhoff's theorem, it is the only vacuum solution that is spherically symmetric.[42] This means that there is no observable difference between the gravitational field of such a black hole and that of any other spherical object of the same mass. The popular notion of a black hole "sucking in everything" in its surroundings is therefore only correct near a black hole's horizon; far away, the external gravitational field is identical to that of any other body of the same mass.[43]

Solutions describing more general black holes also exist. Charged black holes are described by the Reissner–Nordström metric, while the Kerr metric describes a rotating black hole. The most general stationary black hole solution known is the Kerr–Newman metric, which describes a black hole with both charge and angular momentum.[44]

While the mass of a black hole can take any positive value, the charge and angular momentum are constrained by the mass. In Planck units, the total electric charge $Q$ and the total angular momentum $J$ are expected to satisfy

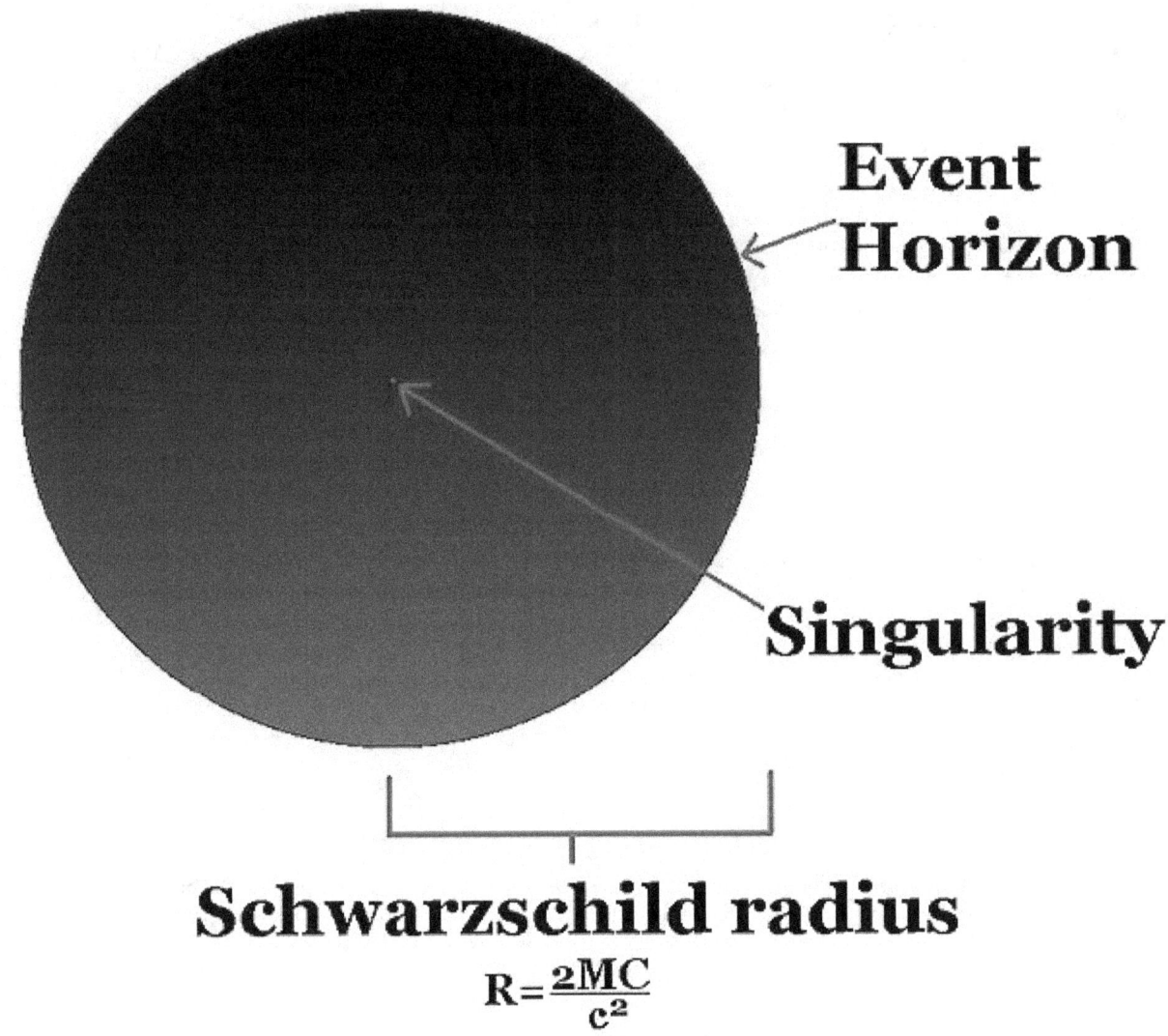

*A simple illustration of a non-spinning black hole*

$$Q^2 + \left(\tfrac{J}{M}\right)^2 \leq M^2$$

for a black hole of mass $M$. Black holes saturating this inequality are called extremal. Solutions of Einstein's equations that violate this inequality exist, but they do not possess an event horizon. These solutions have so-called naked singularities that can be observed from the outside, and hence are deemed *unphysical*. The cosmic censorship hypothesis rules out the formation of such singularities, when they are created through the gravitational collapse of realistic matter.[2] This is supported by numerical simulations.[45]

Due to the relatively large strength of the electromagnetic force, black holes forming from the collapse of stars are expected to retain the nearly neutral charge of the star. Rotation, however, is expected to be a common feature of compact objects. The black-hole candidate binary X-ray source GRS 1915+105[46] appears to have an angular momentum near the maximum allowed value.

Black holes are commonly classified according to their mass, independent of angular momentum $J$ or electric charge $Q$. The size of a black hole, as determined by the radius of the event horizon, or Schwarzschild radius, is roughly proportional to the mass $M$ through

$$r_{sh} = \frac{2GM}{c^2} \approx 2.95 \, \frac{M}{M_{Sun}} \, \text{km},$$

where $r_{sh}$ is the Schwarzschild radius and $MSun$ is the mass of the Sun.[47] This relation is exact only for black holes with zero charge and angular momentum; for more general black holes it can differ up to a factor of 2.

## 21.2.2 Event horizon

Main article: Event horizon

The defining feature of a black hole is the appearance of an event horizon—a boundary in spacetime through which matter and light can only pass inward towards the mass of the black hole. Nothing, not even light, can escape from inside the event horizon. The event horizon is referred to as such because if an event occurs within the boundary, information from that event cannot reach an outside observer, making it impossible to determine if such an event occurred.[49]

As predicted by general relativity, the presence of a mass deforms spacetime in such a way that the paths taken by particles bend towards the mass.[50] At the event horizon of a black hole, this deformation becomes so strong that there are no paths that lead away from the black hole.

To a distant observer, clocks near a black hole appear to tick more slowly than those further away from the black hole.[51] Due to this effect, known as gravitational time dilation, an object falling into a black hole appears to slow down as it approaches the event horizon, taking an infinite time to reach it.[52] At the same time, all processes on this object slow down, for a fixed outside observer, causing emitted light to appear redder and dimmer, an effect known as gravitational redshift.[53] Eventually, the falling object becomes so dim that it can no longer be seen.

On the other hand, an indestructible observer falling into a black hole does not notice any of these effects as he crosses the event horizon. According to his own clock, which appears to him to tick normally, he crosses the event horizon after a finite time without noting any singular behaviour. In particular, he is unable to determine exactly when he crosses it, as it is impossible to determine the location of the event horizon from local observations.[54]

The shape of the event horizon of a black hole is always approximately spherical.[Note 2][57] For non-rotating (static) black holes the geometry is precisely spherical, while for rotating black holes the sphere is somewhat oblate.

## 21.2.3 Singularity

Main article: Gravitational singularity

At the center of a black hole as described by general relativity lies a gravitational singularity, a region where the spacetime curvature becomes infinite.[58] For a non-rotating black hole, this region takes the shape of a single point and for a rotating black hole, it is smeared out to form a ring singularity lying in the plane of rotation.[59] In both cases, the singular region has zero volume. It can also be shown that the singular region contains all the mass of the black hole solution.[60] The singular region can thus be thought of as having infinite density.

Observers falling into a Schwarzschild black hole (i.e., non-rotating and not charged) cannot avoid being carried into the singularity, once they cross the event horizon. They can prolong the experience by accelerating away to slow their descent, but only up to a point; after attaining a certain ideal velocity, it is best to free fall the rest of the way.[61] When they reach the singularity, they are crushed to infinite density and their mass is added to the total of the black hole. Before that happens, they will have been torn apart by the growing tidal forces in a process sometimes referred to as spaghettification or the "noodle effect".[62]

In the case of a charged (Reissner–Nordström) or rotating (Kerr) black hole, it is possible to avoid the singularity. Extending these solutions as far as possible reveals the hypothetical possibility of exiting the black hole into a different spacetime with the black hole acting as a wormhole.[63] The possibility of traveling to another universe is however only theoretical, since any perturbation will destroy this possibility.[64] It also appears to be possible to follow closed timelike curves (going back to one's own past) around the Kerr singularity, which lead to problems with causality like the grandfather paradox.[65] It is expected that none of these peculiar effects would survive in a proper quantum treatment of rotating and charged black holes.[66]

The appearance of singularities in general relativity is commonly perceived as signaling the breakdown of the theory.[67] This breakdown, however, is expected; it occurs in a situation where quantum effects should describe these actions, due to the extremely high density and therefore particle interactions. To date, it has not been possible to combine quantum and gravitational effects into a single theory, although there exist attempts to formulate such a theory of quantum gravity. It is generally expected that such a theory will not feature any singularities.[68][69]

### 21.2.4   Photon sphere

Main article: Photon sphere

The photon sphere is a spherical boundary of zero thickness such that photons moving along tangents to the sphere will be trapped in a circular orbit. For non-rotating black holes, the photon sphere has a radius 1.5 times the Schwarzschild radius. The orbits are dynamically unstable, hence any small perturbation (such as a particle of infalling matter) will grow over time, either setting it on an outward trajectory escaping the black hole or on an inward spiral eventually crossing the event horizon.[70]

While light can still escape from inside the photon sphere, any light that crosses the photon sphere on an inbound trajectory will be captured by the black hole. Hence any light reaching an outside observer from inside the photon sphere must have been emitted by objects inside the photon sphere but still outside of the event horizon.[70]

Other compact objects, such as neutron stars, can also have photon spheres.[71] This follows from the fact that the gravitational field of an object does not depend on its actual size, hence any object that is smaller than 1.5 times the Schwarzschild radius corresponding to its mass will indeed have a photon sphere.

### 21.2.5   Ergosphere

Main article: Ergosphere

Rotating black holes are surrounded by a region of spacetime in which it is impossible to stand still, called the ergosphere. This is the result of a process known as frame-dragging; general relativity predicts that any rotating mass will tend to slightly "drag" along the spacetime immediately surrounding it. Any object near the rotating mass will tend to start moving in the direction of rotation. For a rotating black hole, this effect becomes so strong near the event horizon that an object would have to move faster than the speed of light in the opposite direction to just stand still.[72]

The ergosphere of a black hole is bounded by the (outer) event horizon on the inside and an oblate spheroid, which coincides with the event horizon at the poles and is noticeably wider around the equator. The outer boundary is sometimes called the *ergosurface*.

Objects and radiation can escape normally from the ergosphere. Through the Penrose process, objects can emerge from the ergosphere with more energy than they entered. This energy is taken from the rotational energy of the black hole causing it to slow down.[73]

## 21.3   Formation and evolution

Considering the exotic nature of black holes, it may be natural to question if such bizarre objects could exist in nature or to suggest that they are merely pathological solutions to Einstein's equations. Einstein himself wrongly thought that black holes would not form, because he held that the angular momentum of collapsing particles would stabilize their motion at some radius.[74] This led the general relativity community to dismiss all results to the contrary for many years. However, a minority of relativists continued to contend that black holes were physical objects,[75] and by the end of the 1960s, they had persuaded the majority of researchers in the field that there is no obstacle to forming an event horizon.

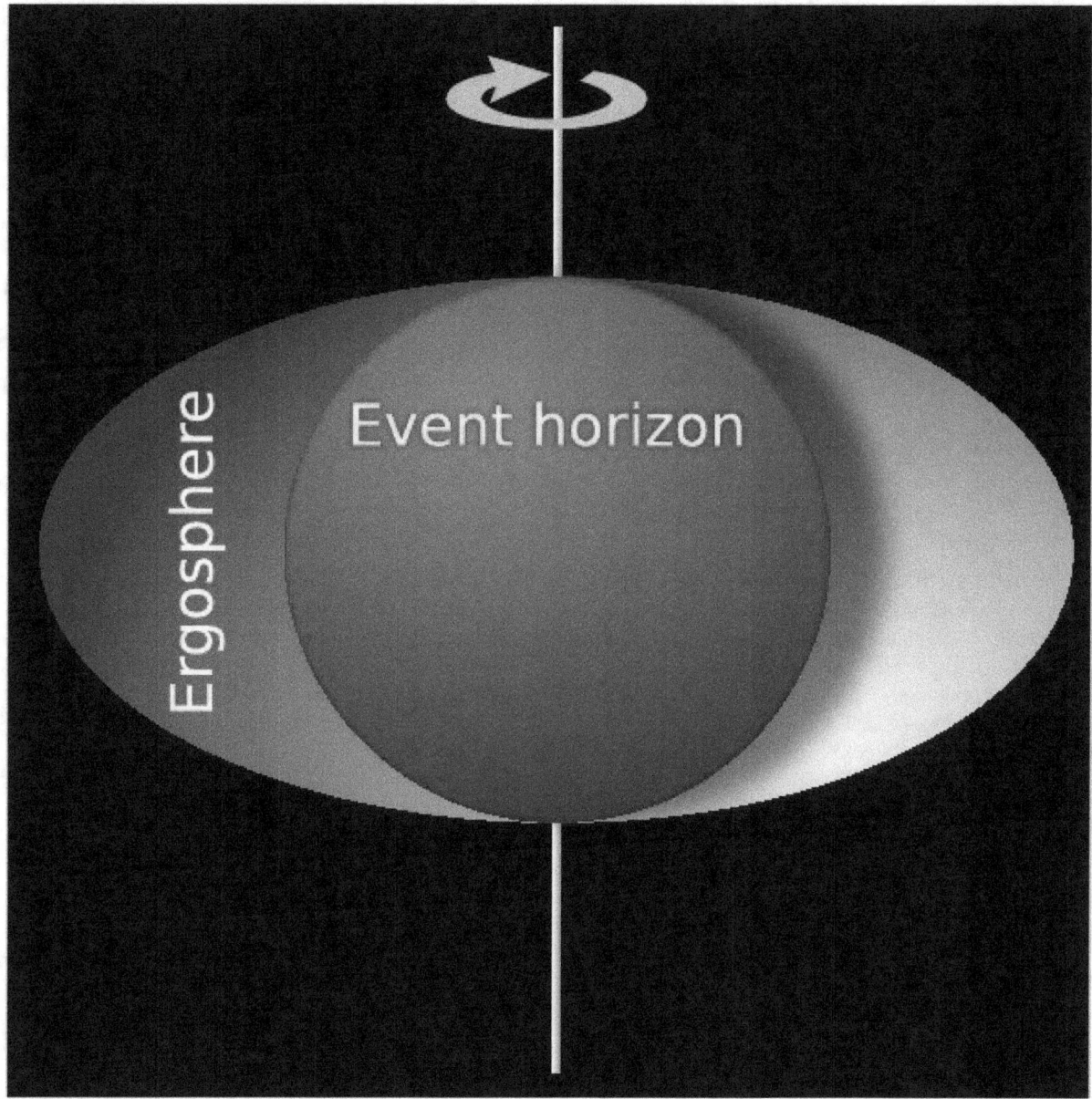

*The ergosphere is an oblate spheroid region outside of the event horizon, where objects cannot remain stationary.*

Once an event horizon forms, Penrose proved that a singularity will form somewhere inside it.[133] Shortly afterwards, Hawking showed that many cosmological solutions describing the Big Bang have singularities without scalar fields or other exotic matter (see Penrose–Hawking singularity theorems). The Kerr solution, the no-hair theorem and the laws of black hole thermodynamics showed that the physical properties of black holes were simple and comprehensible, making them respectable subjects for research.[76] The primary formation process for black holes is expected to be the gravitational collapse of heavy objects such as stars, but there are also more exotic processes that can lead to the production of black holes.

## 21.3.1 Gravitational collapse

Main article: Gravitational collapse

Gravitational collapse occurs when an object's internal pressure is insufficient to resist the object's own gravity. For stars this usually occurs either because a star has too little "fuel" left to maintain its temperature through stellar nucleosynthesis, or because a star that would have been stable receives extra matter in a way that does not raise its core temperature. In either case the star's temperature is no longer high enough to prevent it from collapsing under its own weight.[77] The collapse may be stopped by the degeneracy pressure of the star's constituents, condensing the matter in an exotic denser state. The result is one of the various types of compact star. The type of compact star formed depends on the mass of the remnant—the matter left over after the outer layers have been blown away, such from a supernova explosion or by pulsations leading to a planetary nebula. Note that this mass can be substantially less than the original star—remnants exceeding 5 $M\odot$ are produced by stars that were over 20 $M\odot$ before the collapse.[77]

If the mass of the remnant exceeds about 3–4 $M\odot$ (the Tolman–Oppenheimer–Volkoff limit[20])—either because the original star was very heavy or because the remnant collected additional mass through accretion of matter—even the degeneracy pressure of neutrons is insufficient to stop the collapse. No known mechanism (except possibly quark degeneracy pressure, see quark star) is powerful enough to stop the implosion and the object will inevitably collapse to form a black hole.[77]

The gravitational collapse of heavy stars is assumed to be responsible for the formation of stellar mass black holes. Star formation in the early universe may have resulted in very massive stars, which upon their collapse would have produced black holes of up to $10^3$ $M\odot$. These black holes could be the seeds of the supermassive black holes found in the centers of most galaxies.[78]

While most of the energy released during gravitational collapse is emitted very quickly, an outside observer does not actually see the end of this process. Even though the collapse takes a finite amount of time from the reference frame of infalling matter, a distant observer sees the infalling material slow and halt just above the event horizon, due to gravitational time dilation. Light from the collapsing material takes longer and longer to reach the observer, with the light emitted just before the event horizon forms delayed an infinite amount of time. Thus the external observer never sees the formation of the event horizon; instead, the collapsing material seems to become dimmer and increasingly red-shifted, eventually fading away.[79]

### Primordial black holes in the Big Bang

Gravitational collapse requires great density. In the current epoch of the universe these high densities are only found in stars, but in the early universe shortly after the big bang densities were much greater, possibly allowing for the creation of black holes. The high density alone is not enough to allow the formation of black holes since a uniform mass distribution will not allow the mass to bunch up. In order for primordial black holes to form in such a dense medium, there must be initial density perturbations that can then grow under their own gravity. Different models for the early universe vary widely in their predictions of the size of these perturbations. Various models predict the creation of black holes, ranging from a Planck mass to hundreds of thousands of solar masses.[80] Primordial black holes could thus account for the creation of any type of black hole.

## 21.3.2   High-energy collisions

Gravitational collapse is not the only process that could create black holes. In principle, black holes could be formed in high-energy collisions that achieve sufficient density. As of 2002, no such events have been detected, either directly or indirectly as a deficiency of the mass balance in particle accelerator experiments.[81] This suggests that there must be a lower limit for the mass of black holes. Theoretically, this boundary is expected to lie around the Planck mass ($m_P = \sqrt{\hbar c/G} \approx 1.2\times10^{19}$ GeV/$c^2 \approx 2.2\times10^{-8}$ kg), where quantum effects are expected to invalidate the predictions of general relativity.[82] This would put the creation of black holes firmly out of reach of any high-energy process occurring on or near the Earth. However, certain developments in quantum gravity suggest that the Planck mass could be much lower: some braneworld scenarios for example put the boundary as low as 1 TeV/$c^2$.[83] This would make it conceivable for micro black holes to be created in the high-energy collisions occurring when cosmic rays hit the Earth's atmosphere, or possibly in the Large Hadron Collider at CERN. Yet these theories are very speculative, and the creation of black holes in these processes is deemed unlikely by many specialists.[84] Even if micro black holes should be formed in these collisions, it is expected that they would evaporate in about $10^{-25}$ seconds, posing no threat to the Earth.[85]

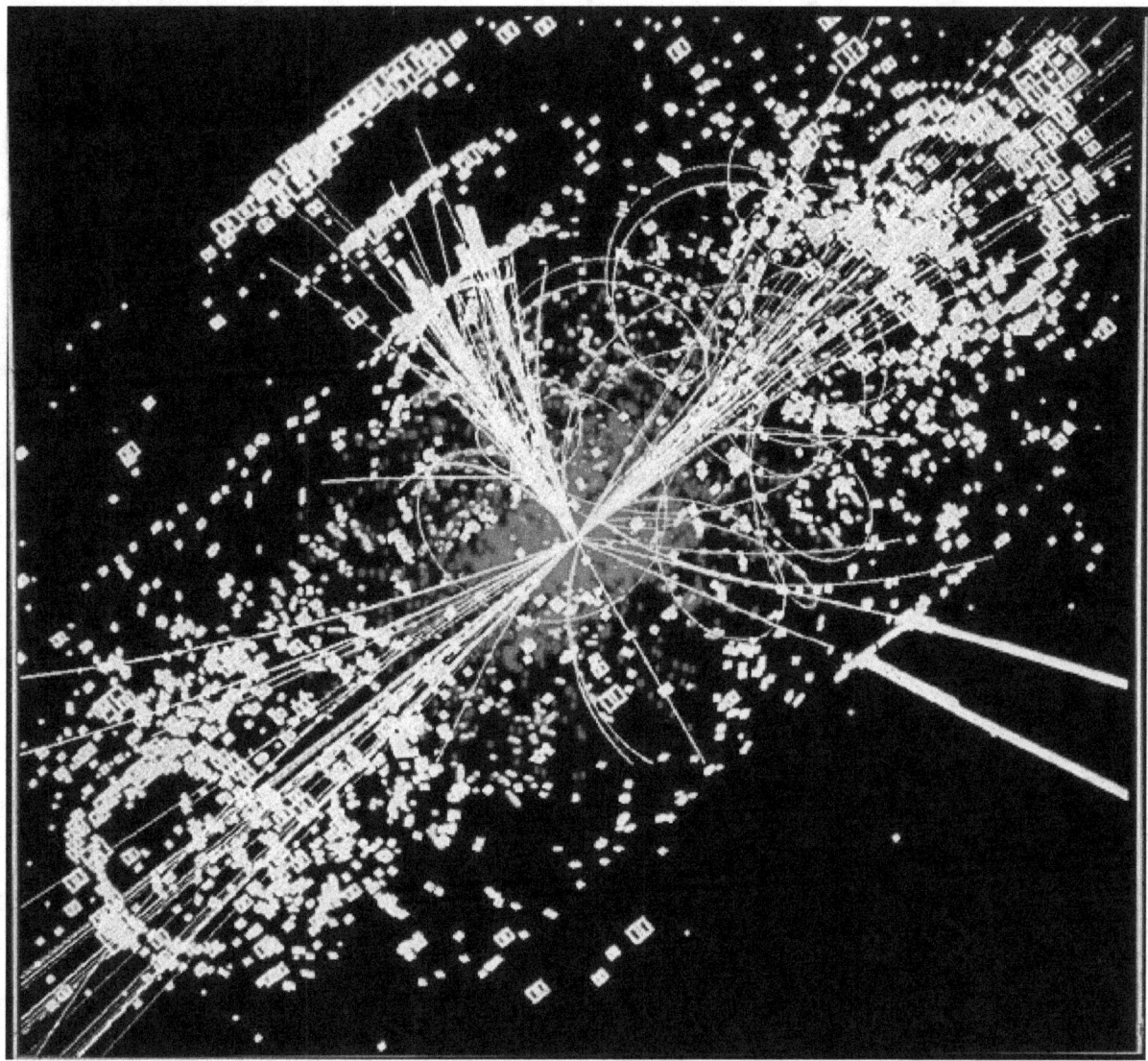

*A simulated event in the CMS detector, a collision in which a micro black hole may be created.*

### 21.3.3  Growth

Once a black hole has formed, it can continue to grow by absorbing additional matter. Any black hole will continually absorb gas and interstellar dust from its direct surroundings and omnipresent cosmic background radiation. This is the primary process through which supermassive black holes seem to have grown.[78] A similar process has been suggested for the formation of intermediate-mass black holes in globular clusters.[86]

Another possibility is for a black hole to merge with other objects such as stars or even other black holes. Although not necessary for growth, this is thought to have been important, especially for the early development of supermassive black holes, which could have formed from the coagulation of many smaller objects.[78] The process has also been proposed as the origin of some intermediate-mass black holes.[87][88]

### 21.3.4  Evaporation

Main article: Hawking radiation

In 1974, Hawking predicted that black holes are not entirely black but emit small amounts of thermal radiation;[36] this effect has become known as Hawking radiation. By applying quantum field theory to a static black hole background, he determined that a black hole should emit particles in a perfect black body spectrum. Since Hawking's publication, many others have verified the result through various approaches.[89] If Hawking's theory of black hole radiation is correct, then black holes are expected to shrink and evaporate over time because they lose mass by the emission of photons and other particles.[36] The temperature of this thermal spectrum (Hawking temperature) is proportional to the surface gravity of the black hole, which, for a Schwarzschild black hole, is inversely proportional to the mass. Hence, large black holes emit less radiation than small black holes.[90]

A stellar black hole of 1 $M\odot$ has a Hawking temperature of about 100 nanokelvins. This is far less than the 2.7 K temperature of the cosmic microwave background radiation. Stellar-mass or larger black holes receive more mass from the cosmic microwave background than they emit through Hawking radiation and thus will grow instead of shrink. To have a Hawking temperature larger than 2.7 K (and be able to evaporate), a black hole needs to have less mass than the Moon. Such a black hole would have a diameter of less than a tenth of a millimeter.[91]

If a black hole is very small the radiation effects are expected to become very strong. Even a black hole that is heavy compared to a human would evaporate in an instant. A black hole with the mass of a car would have a diameter of about $10^{-24}$ m and take a nanosecond to evaporate, during which time it would briefly have a luminosity more than 200 times that of the Sun. Lower-mass black holes are expected to evaporate even faster; for example, a black hole of mass 1 TeV/$c^2$ would take less than $10^{-88}$ seconds to evaporate completely. For such a small black hole, quantum gravitation effects are expected to play an important role and could even—although current developments in quantum gravity do not indicate so[92]—hypothetically make such a small black hole stable.[93]

## 21.4   Observational evidence

By their very nature, black holes do not directly emit any signals other than the hypothetical Hawking radiation; since the Hawking radiation for an astrophysical black hole is predicted to be very weak, this makes it impossible to directly detect astrophysical black holes from the Earth. A possible exception to the Hawking radiation being weak is the last stage of the evaporation of light (primordial) black holes; searches for such flashes in the past have proven unsuccessful and provide stringent limits on the possibility of existence of light primordial black holes.[95] NASA's Fermi Gamma-ray Space Telescope launched in 2008 will continue the search for these flashes.[96]

Astrophysicists searching for black holes thus have to rely on indirect observations. A black hole's existence can sometimes be inferred by observing its gravitational interactions with its surroundings. A project run by MIT's Haystack Observatory is attempting to observe the event horizon of a black hole directly. Initial results are encouraging.[97]

### 21.4.1   Accretion of matter

See also: Accretion disc

Due to conservation of angular momentum, gas falling into the gravitational well created by a massive object will typically form a disc-like structure around the object. Artist's impressions such as the accompanying representation of a black hole with corona commonly depict the black hole as if it were a flat-space material body hiding the part of the disc just behind it, but detailed mathematical modelling[99] shows that the image of the disc would actually be distorted by light bending in such a way that the upper side of the disc is entirely visible, while there is even a partially visible secondary image of the underside.

Within such a disc, friction will cause angular momentum to be transported outward, allowing matter to fall further inward, releasing potential energy and increasing the temperature of the gas.[100]

In the case of compact objects such as white dwarfs, neutron stars, and black holes, the gas in the inner regions becomes so hot that it will emit vast amounts of radiation (mainly X-rays), which may be detected by telescopes. This process of accretion is one of the most efficient energy-producing processes known; up to 40% of the rest mass of the accreted material can be emitted in radiation.[100] (In nuclear fusion only about 0.7% of the rest mass will be emitted as energy.) In many cases, accretion discs are accompanied by relativistic jets emitted along the poles, which carry away much of the energy. The mechanism for the creation of these jets is currently not well understood.

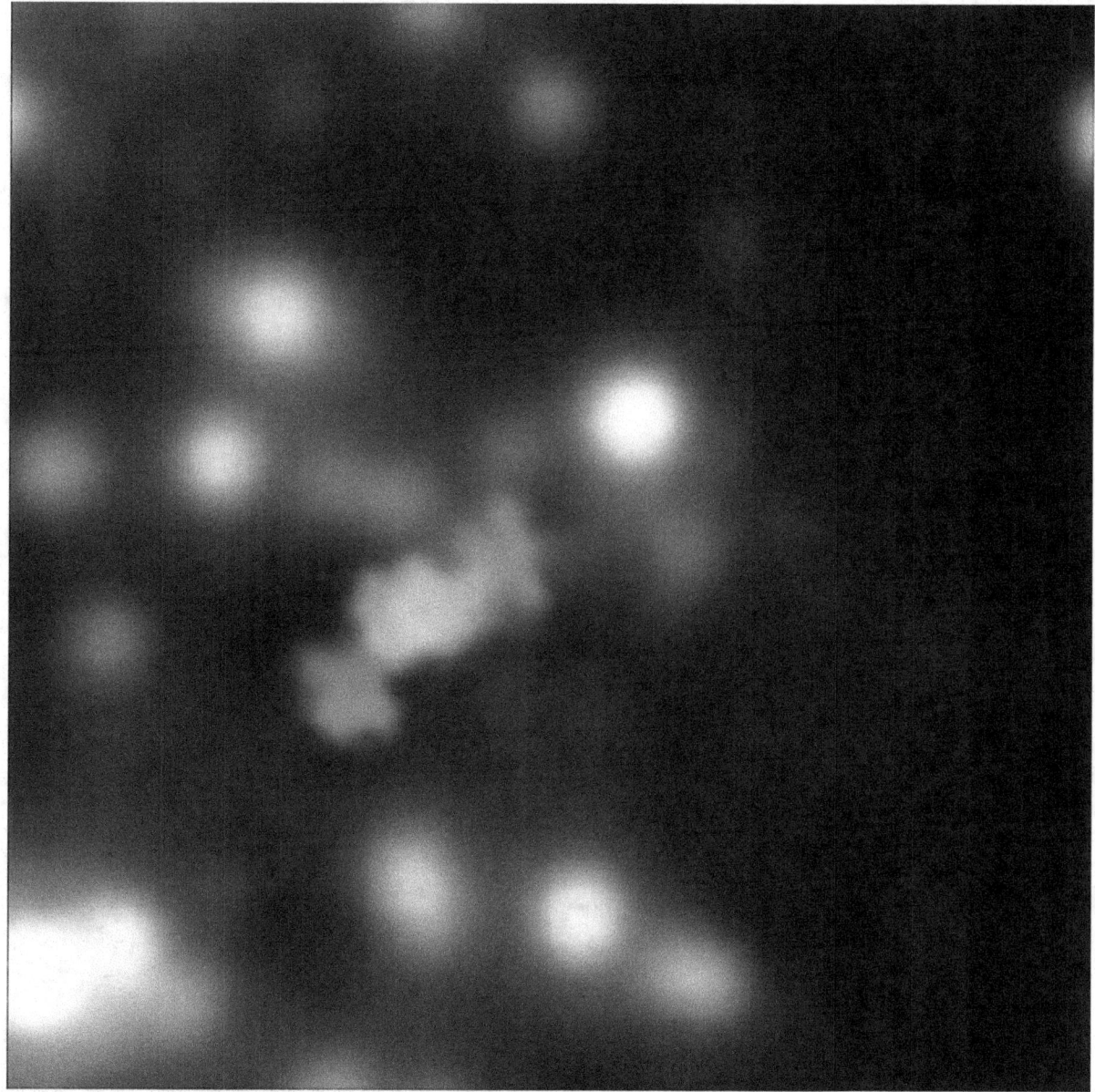

*Gas cloud ripped apart by black hole at the centre of the Milky Way.*[94]

As such many of the universe's more energetic phenomena have been attributed to the accretion of matter on black holes. In particular, active galactic nuclei and quasars are believed to be the accretion discs of supermassive black holes.[101] Similarly, X-ray binaries are generally accepted to be binary star systems in which one of the two stars is a compact object accreting matter from its companion.[101] It has also been suggested that some ultraluminous X-ray sources may be the accretion disks of intermediate-mass black holes.[102]

## 21.4.2   X-ray binaries

See also: X-ray binary

 X-ray binaries are binary star systems that are luminous in the X-ray part of the spectrum. These X-ray emissions are generally thought to be caused by one of the component stars being a compact object accreting matter from the other (regular) star. The presence of an ordinary star in such a system provides a unique opportunity for studying the central

*Black hole with corona, X-ray source (artist's concept).*[1981]

*Predicted view from outside the horizon of a Schwarzschild black hole lit by a thin accretion disc*

object and determining if it might be a black hole.

If such a system emits signals that can be directly traced back to the compact object, it cannot be a black hole. The absence of such a signal does, however, not exclude the possibility that the compact object is a neutron star. By studying the companion star it is often possible to obtain the orbital parameters of the system and obtain an estimate for the mass of the compact object. If this is much larger than the Tolman–Oppenheimer–Volkoff limit (that is, the maximum mass a neutron star can have before collapsing) then the object cannot be a neutron star and is generally expected to be a black

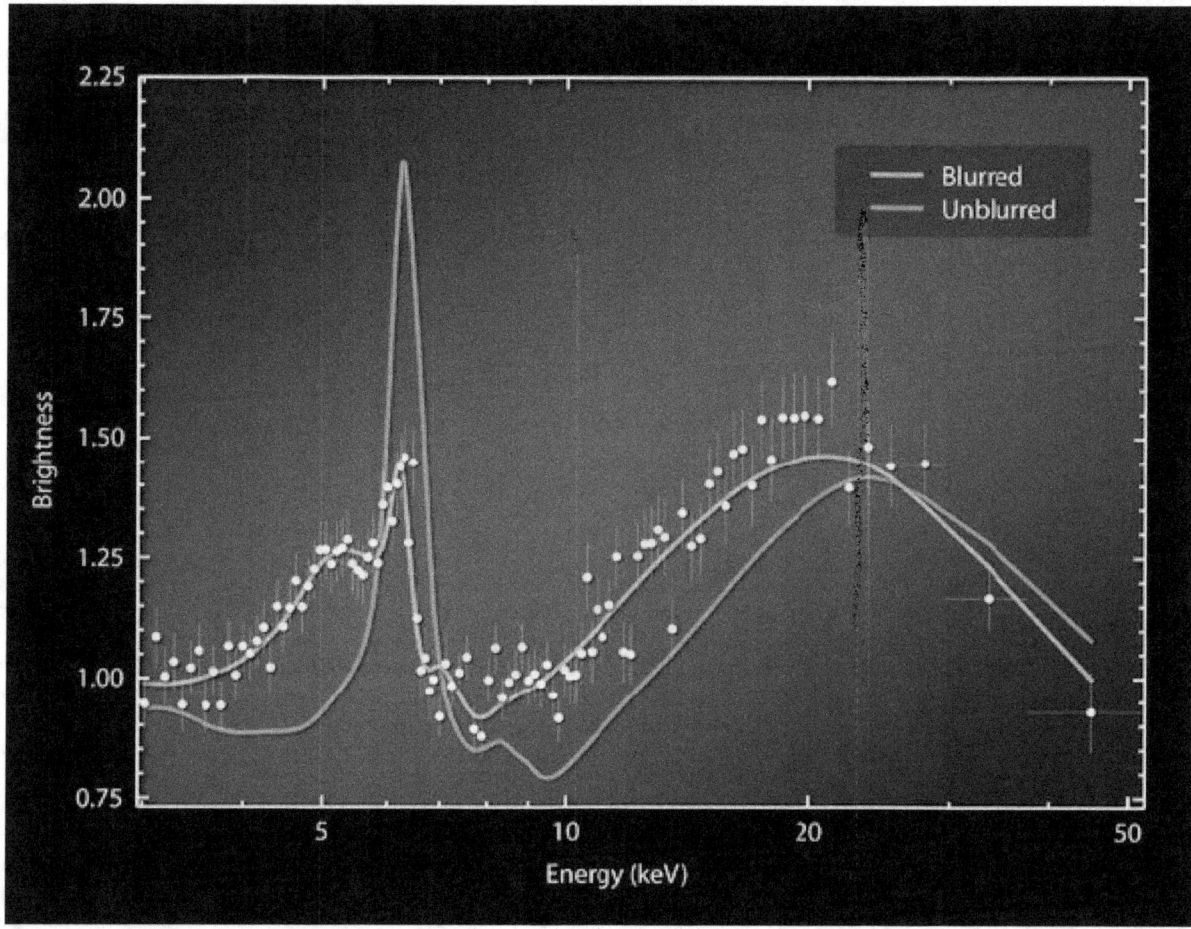

*Blurring of X-rays near Black hole (NuSTAR: 12 August 2014).[98]*

hole.[101]

The first strong candidate for a black hole, Cygnus X-1, was discovered in this way by Charles Thomas Bolton,[103] Louise Webster and Paul Murdin[104] in 1972.[105][106] Some doubt, however, remained due to the uncertainties resultant from the companion star being much heavier than the candidate black hole.[101] Currently, better candidates for black holes are found in a class of X-ray binaries called soft X-ray transients.[101] In this class of system the companion star is relatively low mass allowing for more accurate estimates in the black hole mass. Moreover, these systems are only active in X-ray for several months once every 10–50 years. During the period of low X-ray emission (called quiescence), the accretion disc is extremely faint allowing for detailed observation of the companion star during this period. One of the best such candidates is V404 Cyg.

### Quiescence and advection-dominated accretion flow

The faintness of the accretion disc during quiescence is suspected to be caused by the flow entering a mode called an advection-dominated accretion flow (ADAF). In this mode, almost all the energy generated by friction in the disc is swept along with the flow instead of radiated away. If this model is correct, then it forms strong qualitative evidence for the presence of an event horizon.[107] Because, if the object at the center of the disc had a solid surface, it would emit large amounts of radiation as the highly energetic gas hits the surface, an effect that is observed for neutron stars in a similar state.[100]

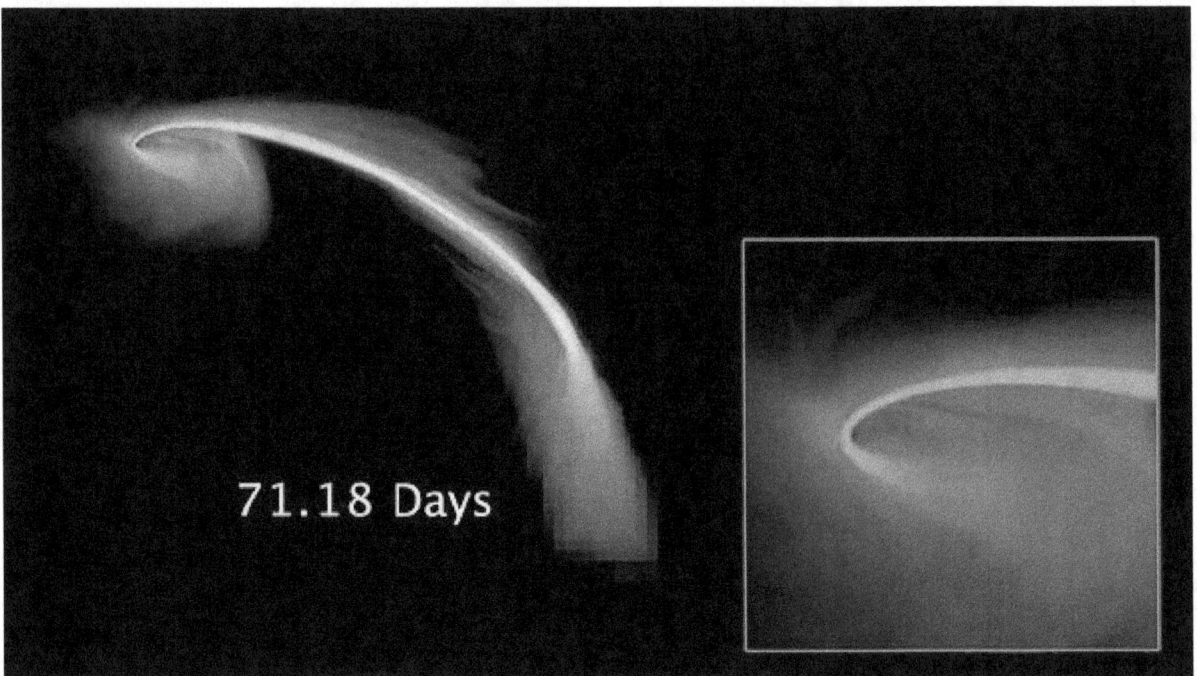

71.18 Days

*A computer simulation of a star being consumed by a black hole. The blue dot indicates the location of the black hole.*

**Quasi-periodic oscillations**

Main article: Quasi-periodic oscillations

The X-ray emission from accretion disks sometimes flickers at certain frequencies. These signals are called quasi-periodic oscillations and are thought to be caused by material moving along the inner edge of the accretion disk (the innermost stable circular orbit). As such their frequency is linked to the mass of the compact object. They can thus be used as an alternative way to determine the mass of potential black holes.[108]

### 21.4.3  Galactic nuclei

See also: Active galactic nucleus

Astronomers use the term "active galaxy" to describe galaxies with unusual characteristics, such as unusual spectral line emission and very strong radio emission. Theoretical and observational studies have shown that the activity in these active galactic nuclei (AGN) may be explained by the presence of supermassive black holes, which can be millions of times more massive than stellar ones. The models of these AGN consist of a central black hole that may be millions or billions of times more massive than the Sun; a disk of gas and dust called an accretion disk; and two jets that are perpendicular to the accretion disk.[109][110]

Although supermassive black holes are expected to be found in most AGN, only some galaxies' nuclei have been more carefully studied in attempts to both identify and measure the actual masses of the central supermassive black hole candidates. Some of the most notable galaxies with supermassive black hole candidates include the Andromeda Galaxy, M32, M87, NGC 3115, NGC 3377, NGC 4258, NGC 4889, NGC 1277, OJ 287, APM 08279+5255 and the Sombrero Galaxy.[112]

It is now widely accepted that the center of nearly every galaxy, not just active ones, contains a supermassive black hole.[113] The close observational correlation between the mass of this hole and the velocity dispersion of the host galaxy's bulge, known as the M-sigma relation, strongly suggests a connection between the formation of the black hole and the galaxy itself.[114]

*A Chandra X-Ray Observatory image of Cygnus X-1, which was the first strong black hole candidate discovered*

Currently, the best evidence for a supermassive black hole comes from studying the proper motion of stars near the center of our own Milky Way.[116] Since 1995 astronomers have tracked the motion of 90 stars in a region called Sagittarius A*. By fitting their motion to Keplerian orbits they were able to infer in 1998 that 2.6 million $M\odot$ must be contained in a volume with a radius of 0.02 light-years.[117] Since then one of the stars—called S2—has completed a full orbit. From the orbital data they were able to place better constraints on the mass and size of the object causing the orbital motion of stars in the Sagittarius A* region, finding that there is a spherical mass of 4.3 million $M\odot$ contained within a radius of less than 0.002 lightyears.[116] While this is more than 3000 times the Schwarzschild radius corresponding to that mass, it is at least consistent with the central object being a supermassive black hole, and no "realistic cluster [of stars] is physically tenable".[117]

*This animation compares the X-ray 'heartbeats' of GRS 1915 and IGR J17091, two black holes that ingest gas from companion stars.*

### 21.4.4   Effects of strong gravity

Another way that the black hole nature of an object may be tested in the future is through observation of effects caused by strong gravity in their vicinity. One such effect is gravitational lensing: The deformation of spacetime around a massive object causes light rays to be deflected much like light passing through an optic lens. Observations have been made of weak gravitational lensing, in which light rays are deflected by only a few arcseconds. However, it has never been directly observed for a black hole.[118] One possibility for observing gravitational lensing by a black hole would be to observe stars in orbit around the black hole. There are several candidates for such an observation in orbit around Sagittarius A*.[118]

Another option would be the direct observation of gravitational waves produced by an object falling into a black hole, for example a compact object falling into a supermassive black hole through an extreme mass ratio inspiral. Matching the observed waveform to the predictions of general relativity would allow precision measurements of the mass and angular momentum of the central object, while at the same time testing general relativity.[119] These types of events are a primary target for the proposed Laser Interferometer Space Antenna.

### 21.4.5   Alternatives

See also: Exotic star

The evidence for stellar black holes strongly relies on the existence of an upper limit for the mass of a neutron star. The size of this limit heavily depends on the assumptions made about the properties of dense matter. New exotic phases of matter could push up this bound.[101] A phase of free quarks at high density might allow the existence of dense quark stars,[120] and some supersymmetric models predict the existence of Q stars.[121] Some extensions of the standard model posit the existence of preons as fundamental building blocks of quarks and leptons, which could hypothetically form preon stars.[122] These hypothetical models could potentially explain a number of observations of stellar black hole candidates. However, it can be shown from general arguments in general relativity that any such object will have a maximum mass.[101]

Since the average density of a black hole inside its Schwarzschild radius is inversely proportional to the square of its mass, supermassive black holes are much less dense than stellar black holes (the average density of a $10^8$ $M\odot$ black hole

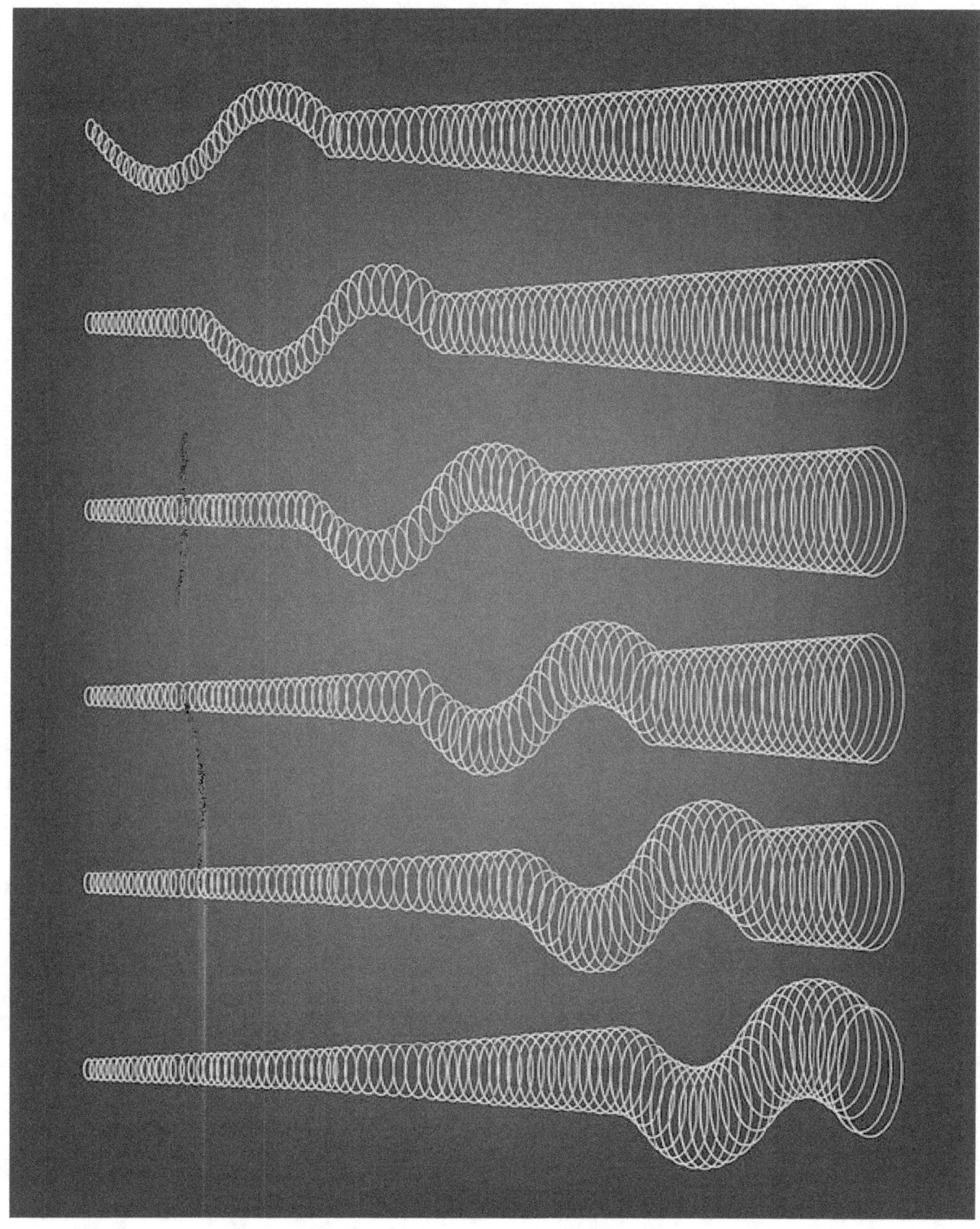

*Magnetic waves, called Alfvén S-waves, flow from the base of black hole jets.*

is comparable to that of water).[101] Consequently, the physics of matter forming a supermassive black hole is much better understood and the possible alternative explanations for supermassive black hole observations are much more mundane. For example, a supermassive black hole could be modelled by a large cluster of very dark objects. However,

*Detection of unusually bright X-Ray flare from Sagittarius A\*, a black hole in the center of the Milky Way galaxy on 5 January 2015.*[111]

such alternatives are typically not stable enough to explain the supermassive black hole candidates.[101]

The evidence for stellar and supermassive black holes implies that in order for black holes not to form, general relativity must fail as a theory of gravity, perhaps due to the onset of quantum mechanical corrections. A much anticipated feature of a theory of quantum gravity is that it will not feature singularities or event horizons (and thus no black holes).[123] In 2002,[124] much attention has been drawn by the fuzzball model in string theory. Based on calculations in specific situations in string theory, the proposal suggests that generically the individual states of a black hole solution do not have an event horizon or singularity, but that for a classical/semi-classical observer the statistical average of such states does appear just like an ordinary black hole in general relativity.[125]

## 21.5   Open questions

*Simulation of gas cloud after close approach to the black hole at the centre of the Milky Way.[115]*

### 21.5.1   Entropy and thermodynamics

Further information: Black hole thermodynamics
In 1971, Hawking showed under general conditions[Note 3] that the total area of the event horizons of any collection of

$$S = \frac{1}{4} \frac{c^3 k}{G\hbar} A$$

*The formula for the Bekenstein–Hawking entropy (S) of a black hole, which depends on the area of the black hole (A). The constants are the speed of light (c), the Boltzmann constant (k), Newton's constant (G), and the reduced Planck constant (ħ).*

classical black holes can never decrease, even if they collide and merge.[126] This result, now known as the second law of black hole mechanics, is remarkably similar to the second law of thermodynamics, which states that the total entropy of a system can never decrease. As with classical objects at absolute zero temperature, it was assumed that black holes had zero entropy. If this were the case, the second law of thermodynamics would be violated by entropy-laden matter entering a black hole, resulting in a decrease of the total entropy of the universe. Therefore, Bekenstein proposed that a black hole should have an entropy, and that it should be proportional to its horizon area.[127]

The link with the laws of thermodynamics was further strengthened by Hawking's discovery that quantum field theory predicts that a black hole radiates blackbody radiation at a constant temperature. This seemingly causes a violation of the second law of black hole mechanics, since the radiation will carry away energy from the black hole causing it to shrink. The radiation, however also carries away entropy, and it can be proven under general assumptions that the sum of the entropy of the matter surrounding a black hole and one quarter of the area of the horizon as measured in Planck units is in fact always increasing. This allows the formulation of the first law of black hole mechanics as an analogue of the first law of thermodynamics, with the mass acting as energy, the surface gravity as temperature and the area as entropy.[127]

One puzzling feature is that the entropy of a black hole scales with its area rather than with its volume, since entropy is normally an extensive quantity that scales linearly with the volume of the system. This odd property led Gerard 't Hooft and Leonard Susskind to propose the holographic principle, which suggests that anything that happens in a volume of spacetime can be described by data on the boundary of that volume.[128]

Although general relativity can be used to perform a semi-classical calculation of black hole entropy, this situation is theoretically unsatisfying. In statistical mechanics, entropy is understood as counting the number of microscopic configurations of a system that have the same macroscopic qualities (such as mass, charge, pressure, etc.). Without a satisfactory theory of quantum gravity, one cannot perform such a computation for black holes. Some progress has been made in various approaches to quantum gravity. In 1995, Andrew Strominger and Cumrun Vafa showed that counting the microstates of a specific supersymmetric black hole in string theory reproduced the Bekenstein–Hawking entropy.[129] Since then, similar results have been reported for different black holes both in string theory and in other approaches to quantum gravity like loop quantum gravity.[130]

### 21.5.2   Information loss paradox

Main article: Black hole information paradox

Because a black hole has only a few internal parameters, most of the information about the matter that went into forming the black hole is lost. Regardless of the type of matter which goes into a black hole, it appears that only information concerning the total mass, charge, and angular momentum are conserved. As long as black holes were thought to persist forever this information loss is not that problematic, as the information can be thought of as existing inside the black hole, inaccessible from the outside. However, black holes slowly evaporate by emitting Hawking radiation. This radiation does not appear to carry any additional information about the matter that formed the black hole, meaning that this information appears to be gone forever.[131]

The question whether information is truly lost in black holes (the black hole information paradox) has divided the theoretical physics community (see Thorne–Hawking–Preskill bet). In quantum mechanics, loss of information corresponds to the violation of vital property called unitarity, which has to do with the conservation of probability. It has been argued that loss of unitarity would also imply violation of conservation of energy.[132] Over recent years evidence has been building that indeed information and unitarity are preserved in a full quantum gravitational treatment of the problem.[133]

## 21.6   See also

- List of nearest black holes

- Black brane

- Black hole complementarity

- Black hole starship

- Black holes in fiction

- Black string

- BTZ black hole

- Dumb hole

- General relativity

- Kugelblitz (astrophysics)

- List of black holes

- Susskind-Hawking battle

- Timeline of black hole physics

- White hole

- Wormhole

## 21.7   Notes

[1] The set of possible paths, or more accurately the future light cone containing all possible world lines (in this diagram the light cone is represented by the V-shaped region bounded by arrows representing light ray world lines), is tilted in this way in Eddington–Finkelstein coordinates (the diagram is a "cartoon" version of an Eddington–Finkelstein coordinate diagram), but in other coordinates the light cones are not tilted in this way, for example in Schwarzschild coordinates they simply narrow without tilting as one approaches the event horizon, and in Kruskal–Szekeres coordinates the light cones don't change shape or orientation at all.[48]

[2] This is true only for 4-dimensional spacetimes. In higher dimensions more complicated horizon topologies like a black ring are possible.[55][56]

[3] In particular, he assumed that all matter satisfies the weak energy condition.

## 21.8   References

[1] Wald 1984, pp. 299–300

[2] Wald, R. M. (1997). "Gravitational Collapse and Cosmic Censorship". arXiv:gr-qc/9710068 [gr-qc].

[3] Overbye, Dennis (8 June 2015). "Black Hole Hunters". *NASA*. Retrieved 8 June 2015.

[4] Schutz, Bernard F. (2003). *Gravity from the ground up*. Cambridge University Press. p. 110. ISBN 0-521-45506-5.

[5] Davies, P. C. W. (1978). "Thermodynamics of Black Holes" (PDF). *Reports on Progress in Physics* **41** (8): 1313–1355. Bibcode:1978RPPh...41.1313D. doi:10.1088/0034-4885/41/8/004.

[6] O. Straub, F.H. Vincent, M.A. Abramowicz, E. Gourgoulhon, T. Paumard, "Modelling the black hole silhouette in Sgr A* with ion tori, *Astron. Astroph 543 (2012) A8*

[7] Michell, J. (1784). "On the Means of Discovering the Distance, Magnitude, &c. of the Fixed Stars, in Consequence of the Diminution of the Velocity of Their Light, in Case Such a Diminution Should be Found to Take Place in any of Them, and Such Other Data Should be Procured from Observations, as Would be Farther Necessary for That Purpose". *Philosophical Transactions of the Royal Society* **74** (0): 35–57. Bibcode:1784RSPT...74...35M. doi:10.1098/rstl.1784.0008. JSTOR 106576.

[8] Gillispie, C. C. (2000). *Pierre-Simon Laplace, 1749–1827: a life in exact science*. Princeton paperbacks. Princeton University Press. p. 175. ISBN 0-691-05027-9.

[9] Israel, W. (1989). "Dark stars: the evolution of an idea". In Hawking, S. W.; Israel, W. *300 Years of Gravitation*. Cambridge University Press. ISBN 978-0-521-37976-2.

[10] Thorne 1994, pp. 123–124

[11] Schwarzschild, K. (1916). "Über das Gravitationsfeld eines Massenpunktes nach der Einsteinschen Theorie". *Sitzungsberichte der Königlich Preussischen Akademie der Wissenschaften* **7**: 189–196. and Schwarzschild, K. (1916). "Über das Gravitations-feld eines Kugel aus inkompressibler Flüssigkeit nach der Einsteinschen Theorie". *Sitzungsberichte der Königlich Preussischen Akademie der Wissenschaften* **18**: 424–434.

[12] Droste, J. (1917). "On the field of a single centre in Einstein's theory of gravitation, and the motion of a particle in that field" (PDF). *Proceedings Royal Academy Amsterdam* **19** (1): 197–215.

[13] Kox, A. J. (1992). "General Relativity in the Netherlands: 1915–1920". In Eisenstaedt, J.; Kox, A. J. *Studies in the history of general relativity*. Birkhäuser. p. 41. ISBN 978-0-8176-3479-7.

[14] 't Hooft, G. (2009). "Introduction to the Theory of Black Holes" (PDF). Institute for Theoretical Physics / Spinoza Institute. pp. 47–48.

[15] Eddington, Arthur (1926). *The Internal Constitution of the Stars*. Cambridge University Press. p. 6.

[16] Kip Thorne comments on this quote on pages 134-135 of his book *Black Holes and Time Warps*, writing that "The first conclusion was the Newtonian version of light not escaping; the second was a semi-accurate, relativistic description; and the third was typical Eddingtonian hyperbole ... when a star is as small as the critical circumference, the curvature is strong but not infinite, and space is definitely not wrapped around the star. Eddington may have known this, but his description made a good story, and it captured in a whimsical way the spirit of Schwarzschild's spacetime curvature."

[17] Venkataraman, G. (1992). *Chandrasekhar and his limit*. Universities Press. p. 89. ISBN 81-7371-035-X.

[18] Detweiler, S. (1981). "Resource letter BH-1: Black holes". *American Journal of Physics* **49**(5): 394–400. Bibcode:1981AmJPh. doi:10.1119/1.12686.

[19] Harpaz, A. (1994). *Stellar evolution*. A K Peters. p. 105. ISBN 1-56881-012-1.

[20] Oppenheimer, J. R.; Volkoff, G. M. (1939). "On Massive Neutron Cores". *Physical Review* **55**(4): 374–381. Bibcode:1939PhRv. doi:10.1103/PhysRev.55.374.

[21] Ruffini, R.; Wheeler, J. A. (1971). "Introducing the black hole" (PDF). *Physics Today* **24** (1): 30–41. Bibcode:1971PhT....24a..30R. doi:10.1063/1.3022513.

[22] Finkelstein, D. (1958). "Past-Future Asymmetry of the Gravitational Field of a Point Particle". *Physical Review* **110** (4): 965–967. Bibcode:1958PhRv..110..965F. doi:10.1103/PhysRev.110.965.

[23] Kruskal, M. (1960). "Maximal Extension of Schwarzschild Metric". *Physical Review* **119**(5): 1743. Bibcode:1960PhRv..119.1. doi:10.1103/PhysRev.119.1743.

[24] Hewish, A.; et al. (1968). "Observation of a Rapidly Pulsating Radio Source". *Nature* **217**(5130): 709–713. Bibcode:1968Na. doi:10.1038/217709a0

[25] Pilkington, J. D. H.; et al. (1968). "Observations of some further Pulsed Radio Sources". *Nature* **218** (5137): 126–129. Bibcode:1968Natur.218..126P. doi:10.1038/218126a0

[26] Hewish, A. (1970). "Pulsars". *Annual Review of Astronomy and Astrophysics* **8** (1): 265–296. Bibcode:1970ARA&A...8..265H. doi:10.1146/annurev.aa.08.090170.001405.

[27] Newman, E. T.; et al. (1965). "Metric of a Rotating, Charged Mass". *Journal of Mathematical Physics* **6**(6): 918. Bibcode:1965 doi:10.1063/1.1704351

[28] Israel, W. (1967). "Event Horizons in Static Vacuum Space-Times". *Physical Review* **164** (5): 1776. Bibcode:1967PhRv..164.1776I. doi:10.1103/PhysRev.164.1776.

[29] Carter, B. (1971). "Axisymmetric Black Hole Has Only Two Degrees of Freedom". *Physical Review Letters* **26** (6): 331. Bibcode:1971PhRvL..26..331C. doi:10.1103/PhysRevLett.26.331.

[30] Carter, B. (1977). "The vacuum black hole uniqueness theorem and its conceivable generalisations". *Proceedings of the 1st Marcel Grossmann meeting on general relativity*. pp. 243–254.

[31] Robinson, D. (1975). "Uniqueness of the Kerr Black Hole". *Physical Review Letters* **34** (14): 905. Bibcode:1975PhRvL...34..905R. doi:10.1103/PhysRevLett.34.905.

[32] Heusler, M. (1998). "Stationary Black Holes: Uniqueness and Beyond". *Living Reviews in Relativity* **1** (6). doi:10.12942/lrr-1998-6. Archived from the original on 1999-02-03. Retrieved 2011-02-08.

[33] Penrose, R.(1965). "Gravitational Collapse and Space-Time Singularities".*Physical Review Letters***14**(3): 57. Bibcode:1965P doi:10.1103/PhysRevLett.14.57.

[34] Ford, L. H. (2003). "The Classical Singularity Theorems and Their Quantum Loopholes". *International Journal of Theoretical Physics* **42** (6): 1219. doi:10.1023/A:1025754515197.

[35] Bardeen, J. M.; Carter, B.; Hawking, S. W. (1973). "The four laws of black hole mechanics". *Communications in Mathematical Physics* **31** (2): 161–170. Bibcode:1973CMaPh..31..161B. doi:10.1007/BF01645742. MR MR0334798. Zbl 1125.83309.

[36] Hawking, S. W. (1974). "Black hole explosions?". *Nature* **248** (5443): 30–31. Bibcode:1974Natur.248...30H. doi:10.1038/248030a0.

[37] Quinion, M. (26 April 2008). "Black Hole". *World Wide Words*. Retrieved 2008-06-17.

[38] Carroll 2004, p. 253

[39] Thorne, K. S.; Price, R. H. (1986). *Black holes: the membrane paradigm*. Yale University Press. ISBN 978-0-300-03770-8.

[40] Anderson, Warren G. (1996). "The Black Hole Information Loss Problem". *Usenet Physics FAQ*. Retrieved 2009-03-24.

[41] Preskill, J. (1994-10-21). *Black holes and information: A crisis in quantum physics* (PDF). Caltech Theory Seminar.

[42] Hawking & Ellis 1973, Appendix B

[43] Seeds, Michael A.; Backman, Dana E. (2007). *Perspectives on Astronomy*. Cengage Learning. p. 167. ISBN 0-495-11352-2

[44] Shapiro, S. L.; Teukolsky, S. A. (1983). *Black holes, white dwarfs, and neutron stars: the physics of compact objects*. John Wiley and Sons. p. 357. ISBN 0-471-87316-0.

[45] Berger, B. K. (2002). "Numerical Approaches to Spacetime Singularities". *Living Reviews in Relativity* **5**: 1. arXiv:gr-qc/0201056. Bibcode:2002LRR.....5....1B. doi:10.12942/lrr-2002-1. Retrieved 2007-08-04.

[46] McClintock, J. E.; Shafee, R.; Narayan, R.; Remillard, R. A.; Davis, S. W.; Li, L.-X. (2006). "The Spin of the Near-Extreme Kerr Black Hole GRS 1915+105". *Astrophysical Journal* **652** (1): 518–539. arXiv:astro-ph/0606076. Bibcode:2006ApJ...652..518M. doi:10.1086/508457.

[47] Wald 1984, pp. 124–125

[48] Thorne, Misner & Wheeler 1973, p. 848

[49] Wheeler 2007, p. 179

[50] Carroll 2004, Ch. 5.4 and 7.3

[51] Carroll 2004, p. 217

[52] Carroll 2004, p. 218

[53] "Inside a black hole". *Knowing the universe and its secrets*. Retrieved 2009-03-26.

[54] Carroll 2004, p. 222

[55] Emparan, R.; Reall, H. S. (2008). "Black Holes in Higher Dimensions". *Living Reviews in Relativity* **11** (6). arXiv:0801.3471. Bibcode:2008LRR....11....6E. doi:10.12942/lrr-2008-6. Retrieved 2011-02-10.

[56] Obers, N. A. (2009). Papantonopoulos, Eleftherios, ed. "Black Holes in Higher-Dimensional Gravity". *Lecture Notes in Physics*. Lecture Notes in Physics **769**: 211–258. arXiv:0802.0519. doi:10.1007/978-3-540-88460-6. ISBN 978-3-540-88459-0.

[57] hawking & ellis 1973, Ch. 9.3

[58] Carroll 2004, p. 205

[59] Carroll 2004, pp. 264–265

[60] Carroll 2004, p. 252

[61] Lewis, G. F.; Kwan, J. (2007). "No Way Back: Maximizing Survival Time Below the Schwarzschild Event Horizon". *Publications of the Astronomical Society of Australia* **24** (2): 46–52. arXiv:0705.1029. Bibcode:2007PASA...24...46L. doi:10.1071/AS07012.

[62] Wheeler 2007, p. 182

[63] Carroll 2004, pp. 257–259 and 265–266

[64] Droz, S.; Israel, W.; Morsink, S. M. (1996). "Black holes: the inside story". *Physics World* **9** (1): 34–37. Bibcode:1996PhyW....9...34D.

[65] Carroll 2004, p. 266

[66] Poisson, E.; Israel, W. (1990). "Internal structure of black holes". *Physical Review D* **41** (6): 1796. Bibcode:1990PhRvD..41.1796P. doi:10.1103/PhysRevD.41.1796.

[67] Wald 1984, p. 212

[68] Hamade, R. (1996). "Black Holes and Quantum Gravity". *Cambridge Relativity and Cosmology*. University of Cambridge. Retrieved 2009-03-26.

[69] Palmer, D. "Ask an Astrophysicist: Quantum Gravity and Black Holes". NASA. Retrieved 2009-03-26.

[70] Nitta, Daisuke; Chiba, Takeshi; Sugiyama, Naoshi (September 2011). "Shadows of colliding black holes". *Physical Review D* **84** (6). arXiv:1106.2425. Bibcode:2011PhRvD..84f3008N. doi:10.1103/PhysRevD.84.063008

[71] Nemiroff, R. J. (1993). "Visual distortions near a neutron star and black hole". *American Journal of Physics* **61** (7): 619. arXiv:astro-ph/9312003. Bibcode:1993AmJPh..61..619N. doi:10.1119/1.17224.

[72] Carroll 2004, Ch. 6.6

[73] Carroll 2004, Ch. 6.7

[74] Einstein, A. (1939). "On A Stationary System With Spherical Symmetry Consisting of Many Gravitating Masses". *Annals of Mathematics* **40** (4): 922–936. doi:10.2307/1968902. JSTOR 1968902.

[75] Kerr, R. P. (2009). "The Kerr and Kerr-Schild metrics". In Wiltshire, D. L.; Visser, M.; Scott, S. M. *The Kerr Spacetime*. Cambridge University Press. arXiv:0706.1109. ISBN 978-0-521-88512-6.

[76] Hawking, S. W.; Penrose, R. (January 1970). "The Singularities of Gravitational Collapse and Cosmology". *Proceedings of the Royal Society A* **314** (1519): 529–548. Bibcode:1970RSPSA.314..529H. doi:10.1098/rspa.1970.0021. JSTOR 2416467.

[77] Carroll 2004, Section 5.8

[78] Rees, M. J.; Volonteri, M. (2007). "Massive black holes: formation and evolution". In Karas, V.; Matt, G. *Black Holes from Stars to Galaxies—Across the Range of Masses*. Cambridge University Press. pp. 51–58. arXiv:astro-ph/0701512. ISBN 978-0-521-86347-6.

[79] Penrose, R. (2002). "Gravitational Collapse: The Role of General Relativity" (PDF). *General Relativity and Gravitation* **34** (7): 1141. Bibcode:2002GReGr..34.1141P. doi:10.1023/A:1016578408204.

[80] Carr, B. J. (2005). "Primordial Black Holes: Do They Exist and Are They Useful?". In Suzuki, H.; Yokoyama, J.; Suto, Y.; Sato, K. *Inflating Horizon of Particle Astrophysics and Cosmology*. Universal Academy Press. arXiv:astro-ph/0511743. ISBN 4-946443-94-0.

[81] Giddings, S. B.; Thomas, S. (2002). "High energy colliders as black hole factories: The end of short distance physics". *Physical Review D* **65** (5): 056010. arXiv:hep-ph/0106219. Bibcode:2002PhRvD..65e6010G. doi:10.1103/PhysRevD.65.056010.

[82] Harada, T. (2006). "Is there a black hole minimum mass?". *Physical Review D* **74** (8): 084004. arXiv:gr-qc/0609055. Bibcode:2006PhRvD..74h4004H. doi:10.1103/PhysRevD.74.084004.

[83] Arkani-Hamed, N.; Dimopoulos, S.; Dvali, G. (1998). "The hierarchy problem and new dimensions at a millimeter". *Physics Letters B* **429** (3–4): 263. arXiv:hep-ph/9803315. Bibcode:1998PhLB..429..263A. doi:10.1016/S0370-2693(98)00466-3.

[84] LHC Safety Assessment Group. "Review of the Safety of LHC Collisions" (PDF). CERN.

[85] Cavaglià, M. (2010). "Particle accelerators as black hole factories?". *Einstein-Online* (Max Planck Institute for Gravitational Physics (Albert Einstein Institute)) **4**: 1010.

[86] Vesperini, E.; McMillan, S. L. W.; d'Ercole, A.; et al. (2010). "Intermediate-Mass Black Holes in Early Globular Clusters". *The Astrophysical Journal Letters* **713** (1): L41–L44. arXiv:1003.3470. Bibcode:2010ApJ...713L..41V. doi:10.1088/2041-8205/713/1/L41.

[87] Zwart, S. F. P.; Baumgardt, H.; Hut, P.; et al. (2004). "Formation of massive black holes through runaway collisions in dense young star clusters". *Nature* **428** (6984): 724–6. arXiv:astro-ph/0402622. Bibcode:2004Natur.428..724P. doi:10.1038/nature02448. PMID 15085124.

[88] O'Leary, R. M.; Rasio, F. A.; Fregeau, J. M.; et al. (2006). "Binary Mergers and Growth of Black Holes in Dense Star Clusters". *The Astrophysical Journal* **637** (2): 937. arXiv:astro-ph/0508224. Bibcode:2006ApJ...637..937O. doi:10.1086/498446.

[89] Page, D. N. (2005). "Hawking radiation and black hole thermodynamics". *New Journal of Physics* **7**: 203. arXiv:hep-th/0409024. Bibcode:2005NJPh....7..203P. doi:10.1088/1367-2630/7/1/203.

[90] Carroll 2004. Ch. 9.6

[91] "Evaporating black holes?". *Einstein online*. Max Planck Institute for Gravitational Physics. 2010. Retrieved 2010-12-12.

[92] Giddings, S. B.; Mangano, M. L. (2008). "Astrophysical implications of hypothetical stable TeV-scale black holes". *Physical Review D* **78** (3): 035009. arXiv:0806.3381. Bibcode:2008PhRvD..78c5009G. doi:10.1103/PhysRevD.78.035009.

[93] Peskin, M. E. (2008). "The end of the world at the Large Hadron Collider?". *Physics* **1**: 14. Bibcode:2008PhyOJ...1...14P. doi:10.1103/Physics.1.14.

[94] "Ripped Apart by a Black Hole". *ESO Press Release*. Retrieved 19 July 2013.

[95] Fichtel, C. E.; Bertsch, D. L.; Dingus, B. L.; et al. (1994). "Search of the energetic gamma-ray experiment telescope (EGRET) data for high-energy gamma-ray microsecond bursts". *Astrophysical Journal* **434** (2): 557–559. Bibcode:1994ApJ...434..557F. doi:10.1086/174758.

[96] Naeye, R. "Testing Fundamental Physics". NASA. Retrieved 2008-09-16.

[97] "Event Horizon Telescope". MIT Haystack Observatory. Retrieved 6 April 2012.

[98] "NASA's NuSTAR Sees Rare Blurring of Black Hole Light". *NASA*. 12 August 2014. Retrieved 12 August 2014.

[99] "Short-cut method of solution of geodesic equations for Schwarzchild black hole", J.A. Marck. Class.Quant. Grav. 13 (1996) 393-402.

[100] McClintock, J. E.; Remillard, R. A. (2006). "Black Hole Binaries". In Lewin, W.; van der Klis, M. *Compact Stellar X-ray Sources*. Cambridge University Press. arXiv:astro-ph/0306213. ISBN 0-521-82659-4. section 4.1.5.

[101] Celotti, A.; Miller, J. C.; Sciama, D. W. (1999). "Astrophysical evidence for the existence of black holes". *Classical and Quantum Gravity* **16** (12A): A3–A21. arXiv:astro-ph/9912186. doi:10.1088/0264-9381/16/12A/301.

[102] Winter, L. M.; Mushotzky, R. F.; Reynolds, C. S. (2006). "XMM-Newton Archival Study of the Ultraluminous X-Ray Population in Nearby Galaxies". *The Astrophysical Journal* **649** (2): 730. arXiv:astro-ph/0512480. Bibcode:2006ApJ...649..730W. doi:10.1086/506579.

[103] Bolton, C. T. (1972). "Identification of Cygnus X-1 with HDE 226868". *Nature* **235** (5336): 271–273. Bibcode:1972Natur.235..271B. doi:10.1038/235271b0.

[104] Webster, B. L.; Murdin, P. (1972). "Cygnus X-1—a Spectroscopic Binary with a Heavy Companion ?". *Nature* **235** (5332): 37–38. Bibcode:1972Natur.235...37W. doi:10.1038/235037a0.

[105] Rolston, B. (10 November 1997). "The First Black Hole". *The bulletin*. University of Toronto. Archived from the original on 2008-05-02. Retrieved 2008-03-11.

[106] Shipman, H. L.; Yu, Z; Du, Y.W (1 January 1975). "The implausible history of triple star models for Cygnus X-1 Evidence for a black hole". *Astrophysical Letters* **16** (1): 9–12. Bibcode:1975ApL.....16....9S. doi:10.1016/S0304-8853(99)00384-4.

[107] Narayan, R.; McClintock, J. (2008). "Advection-dominated accretion and the black hole event horizon". *New Astronomy Reviews* **51** (10–12): 733. arXiv:0803.0322. Bibcode:2008NewAR..51..733N. doi:10.1016/j.newar.2008.03.002.

[108] "NASA scientists identify smallest known black hole" (Press release). Goddard Space Flight Center. 2008-04-01. Retrieved 2009-03-14.

[109] Krolik, J. H. (1999). *Active Galactic Nuclei*. Princeton University Press. Ch. 1.2. ISBN 0-691-01151-6.

[110] Sparke, L. S.; Gallagher, J. S. (2000). *Galaxies in the Universe: An Introduction*. Cambridge University Press. Ch. 9.1. ISBN 0-521-59740-4.

[111] Chou, Felicia; Anderson, Janet; Watzke, Megan (5 January 2015). "RELEASE 15-001 - NASA's Chandra Detects Record-Breaking Outburst from Milky Way's Black Hole". *NASA*. Retrieved 6 January 2015.

[112] Kormendy, J.; Richstone, D. (1995). "Inward Bound—The Search For Supermassive Black Holes In Galactic Nuclei". *Annual Reviews of Astronomy and Astrophysics* **33**(1): 581–624. Bibcode:1995ARA&A..33..581K.doi:10.1146/annurev.aa.33.09019

[113] King, A. (2003). "Black Holes, Galaxy Formation, and the MBH-σ Relation". *The Astrophysical Journal Letters* **596** (1): 27–29. arXiv:astro-ph/0308342. Bibcode:2003ApJ...596L..27K. doi:10.1086/379143.

[114] Ferrarese, L.; Merritt, D. (2000). "A Fundamental Relation Between Supermassive Black Holes and their Host Galaxies". *The Astrophysical Journal Letters* **539** (1): 9–12. arXiv:astro-ph/0006053. Bibcode:2000ApJ...539L...9F. doi:10.1086/312838.

[115] "A Black Hole's Dinner is Fast Approaching". *ESO Press Release*. Retrieved 6 February 2012.

[116] Gillessen, S.; Eisenhauer, F.; Trippe, S.; et al. (2009). "Monitoring Stellar Orbits around the Massive Black Hole in the Galactic Center". *The Astrophysical Journal* **692** (2): 1075. arXiv:0810.4674. Bibcode:2009ApJ...692.1075G. doi:10.1088/0004-637X/692/2/1075.

[117] Ghez, A. M.; Klein, B. L.; Morris, M.; et al. (1998). "High Proper-Motion Stars in the Vicinity of Sagittarius A*: Evidence for a Supermassive Black Hole at the Center of Our Galaxy". *The Astrophysical Journal* **509** (2): 678. arXiv:astro-ph/9807210. Bibcode:1998ApJ...509..678G. doi:10.1086/306528.

[118] Bozza, V. (2010). "Gravitational Lensing by Black Holes". *General Relativity and Gravitation* **42** (42): 2269–2300. arXiv:0911.2187. Bibcode:2010GReGr..42.2269B. doi:10.1007/s10714-010-0988-2.

[119] Barack, L.; Cutler, C. (2004). "LISA capture sources: Approximate waveforms, signal-to-noise ratios, and parameter estimation accuracy".*Physical Review D***69**(69): 082005. arXiv:gr-qc/0310125. Bibcode:2004PhRvD..69h2005B.doi:10.1103/PhysRev

[120] Kovacs, Z.; Cheng, K. S.; Harko, T. (2009). "Can stellar mass black holes be quark stars?". *Monthly Notices of the Royal Astronomical Society* **400** (3): 1632–1642. arXiv:0908.2672. Bibcode:2009MNRAS.400.1632K. doi:10.1111/j.1365-2966.2009.15571.x.

[121] Kusenko, A. (2006). "Properties and signatures of supersymmetric Q-balls". arXiv:hep-ph/0612159.

[122] Hansson, J.; Sandin, F. (2005). "Preon stars: a new class of cosmic compact objects". *Physics Letters B* **616** (1–2): 1. arXiv:astro-ph/0410417. Bibcode:2005PhLB..616....1H. doi:10.1016/j.physletb.2005.04.034.

[123] Kiefer, C. (2006). "Quantum gravity: general introduction and recent developments". *Annalen der Physik* **15** (1–2): 129. arXiv:gr-qc/0508120. Bibcode:2006AnP...518..129K. doi:10.1002/andp.200510175.

[124] "[NKS, Mathur states, 't Hooft-Polyakov monopoles, and Ward-Takahashi identities] - A New Kind of Science: The NKS Forum". *wolframscience.com*. Retrieved 12 April 2015.

[125] Skenderis, K.; Taylor, M. (2008). "The fuzzball proposal for black holes". *Physics Reports* **467** (4–5): 117. arXiv:0804.0552. Bibcode:2008PhR...467..117S. doi:10.1016/j.physrep.2008.08.001.

[126] Hawking, S. W. (1971). "Gravitational Radiation from Colliding Black Holes". *Physical Review Letters* **26** (21): 1344–1346. Bibcode:1971PhRvL..26.1344H. doi:10.1103/PhysRevLett.26.1344.

[127] Wald, R. M. (2001). "The Thermodynamics of Black Holes". *Living Reviews in Relativity* **4** (6): 12119. arXiv:gr-qc/9912119. Bibcode:1999gr.qc....12119W. doi:10.12942/lrr-2001-6. Retrieved 2011-02-10.

[128] 't Hooft, G. (2001). "The Holographic Principle". In Zichichi, A. *Basics and highlights in fundamental physics*. Subnuclear series **37**. World Scientific. arXiv:hep-th/0003004. ISBN 978-981-02-4536-8.

[129] Strominger, A.; Vafa, C. (1996). "Microscopic origin of the Bekenstein-Hawking entropy". *Physics Letters B* **379** (1–4): 99. arXiv:hep-th/9601029. Bibcode:1996PhLB..379...99S. doi:10.1016/0370-2693(96)00345-0.

[130] Carlip, S. (2009). "Black Hole Thermodynamics and Statistical Mechanics". *Lecture Notes in Physics*. Lecture Notes in Physics **769**: 89. arXiv:0807.4520. doi:10.1007/978-3-540-88460-6_3. ISBN 978-3-540-88459-0.

[131] Hawking, S. W. "Does God Play Dice?". *www.hawking.org.uk*. Retrieved 2009-03-14.

[132] Giddings, S. B. (1995). "The black hole information paradox". *Particles, Strings and Cosmology*. Johns Hopkins Workshop on Current Problems in Particle Theory 19 and the PASCOS Interdisciplinary Symposium 5. arXiv:hep-th/9508151.

[133] Mathur, S. D. (2011). *The information paradox: conflicts and resolutions*. XXV International Symposium on Lepton Photon Interactions at High Energies. arXiv:1201.2079.

## 21.9  Further reading

**Popular reading**

- Ferguson, Kitty (1991). *Black Holes in Space-Time*. Watts Franklin. ISBN 0-531-12524-6.

- Hawking, Stephen (1988). *A Brief History of Time*. Bantam Books, Inc. ISBN 0-553-38016-8.

- Hawking, Stephen; Penrose, Roger (1996). *The Nature of Space and Time*. Princeton University Press. ISBN 0-691-03791-4.

- Melia, Fulvio (2003). *The Black Hole at the Center of Our Galaxy*. Princeton U Press. ISBN 978-0-691-09505-9.

- Melia, Fulvio (2003). *The Edge of Infinity. Supermassive Black Holes in the Universe*. Cambridge U Press. ISBN 978-0-521-81405-8.

- Pickover, Clifford (1998). *Black Holes: A Traveler's Guide*. Wiley, John & Sons, Inc. ISBN 0-471-19704-1.

- Thorne, Kip S. (1994). *Black Holes and Time Warps*. Norton, W. W. & Company, Inc. ISBN 0-393-31276-3.

- Wheeler, J. Craig (2007). *Cosmic Catastrophes* (2nd ed.). Cambridge University Press. ISBN 0-521-85714-7.

**University textbooks and monographs**

- Carroll, Sean M. (2004). *Spacetime and Geometry*. Addison Wesley. ISBN 0-8053-8732-3., the lecture notes on which the book was based are available for free from Sean Carroll's website.

- Carter, B. (1973). "Black hole equilibrium states". In DeWitt, B. S.; DeWitt, C. *Black Holes*.

- Chandrasekhar, Subrahmanyan (1999). *Mathematical Theory of Black Holes*. Oxford University Press. ISBN 0-19-850370-9.

- Frolov, V. P.; Novikov, I. D. (1998). "Black hole physics".

- Frolov, Valeri P.; Zelnikov, Andrei (2011). *Introduction to Black Hole Physics*. Oxford: Oxford University Press. ISBN 978-0-19-969229-3. Zbl 1234.83001.

- Hawking, S. W.; Ellis, G. F. R. (1973). *Large Scale Structure of space time*. Cambridge University Press. ISBN 0-521-09906-4.

- Melia, Fulvio (2007). *The Galactic Supermassive Black Hole*. Princeton U Press. ISBN 978-0-691-13129-0.

- Misner, Charles; Thorne, Kip S.; Wheeler, John (1973). *Gravitation*. W. H. Freeman and Company. ISBN 0-7167-0344-0.

- Taylor, Edwin F.; Wheeler, John Archibald (2000). *Exploring Black Holes*. Addison Wesley Longman. ISBN 0-201-38423-X.

- Wald, Robert M. (1984). *General Relativity*. University of Chicago Press. ISBN 978-0-226-87033-5.

- Wald, Robert M. (1992). *Space, Time, and Gravity: The Theory of the Big Bang and Black Holes*. University of Chicago Press. ISBN 0-226-87029-4.

- Black holes Teviet Creighton, Richard H. Price Scholarpedia 3(1):4277. doi:10.4249/scholarpedia.4277

**Review papers**

- Gallo, Elena; Marolf, Donald (2009). "Resource Letter BH-2: Black Holes". *American Journal of Physics* **77** (4): 294. arXiv:0806.2316. Bibcode:2009AmJPh..77..294G. doi:10.1119/1.3056569.

- Hughes, Scott A. (2005). "Trust but verify: The case for astrophysical black holes". arXiv:hep-ph/0511217. Lecture notes from 2005 SLAC Summer Institute.

## 21.10  External links

- 

- Black Holes on *In Our Time* at the BBC. (listen now)

- Stanford Encyclopedia of Philosophy: "Singularities and Black Holes" by Erik Curiel and Peter Bokulich.

- Black Holes: Gravity's Relentless Pull—Interactive multimedia Web site about the physics and astronomy of black holes from the Space Telescope Science Institute

- Frequently Asked Questions (FAQs) on Black Holes

- "Schwarzschild Geometry"

- Advanced Mathematics of Black Hole Evaporation

- Hubble site

**Videos**

- 16-year-long study tracks stars orbiting Milky Way black hole

- Movie of Black Hole Candidate from Max Planck Institute

- Nature.com 2015-04-20 3D simulations of colliding black holes

# Chapter 22

# Neutron star

For the story by Larry Niven, see Neutron Star (short story).

A **neutron star** is a type of compact star that can result from the gravitational collapse of a massive star after a supernova. Neutron stars are the densest and smallest stars known to exist in the universe; with a radius of only about 12–13 km (7 mi), they can have a mass of about twice that of the Sun.

Neutron stars are composed almost entirely of neutrons, which are subatomic particles without net electrical charge and with slightly larger mass than protons. Neutron stars are very hot and are supported against further collapse by quantum degeneracy pressure due to the phenomenon described by the Pauli exclusion principle, which states that no two neutrons (or any other fermionic particles) can occupy the same place and quantum state simultaneously.

A neutron star has a mass of at least 1.1 and perhaps up to 3 solar masses ($M\odot$),[1][2] though the highest observed mass is 2.01 $M\odot$ Neutron stars typically have a surface temperature around ~$6\times10^5$ K.[3][4][5][6][lower-alpha 1] Neutron stars have overall densities of $3.7\times10^{17}$ to $5.9\times10^{17}$ kg/m$^3$ ($2.6\times10^{14}$ to $4.1\times10^{14}$ times the density of the Sun),[lower-alpha 2] which is comparable to the approximate density of an atomic nucleus of $3\times10^{17}$ kg/m$^3$.[7] The neutron star's density varies from below $1\times10^9$ kg/m$^3$ in the crust—increasing with depth—to above $6\times10^{17}$ or $8\times10^{17}$ kg/m$^3$ deeper inside (denser than an atomic nucleus).[8] A normal-sized matchbox containing neutron-star material would have a mass of approximately 5 trillion tons or ~1000 km$^3$ of Earth rock.

In general, compact stars of less than 1.39 $M\odot$ (the Chandrasekhar limit) are white dwarfs, whereas compact stars with a mass between 1.4 $M\odot$ and 3 $M\odot$ (the Tolman–Oppenheimer–Volkoff limit) should be neutron stars. The maximum observed mass of neutron stars is about 2 $M\odot$. The smallest observed mass of a stellar black hole is about 5 $M\odot$, though compact stars with more than 10 $M\odot$ will overcome the neutron degeneracy pressure and gravitational collapse will usually occur to produce a black hole.[9] Between 3 $M\odot$ and 5 $M\odot$, hypothetical intermediate-mass stars such as quark stars and electroweak stars have been proposed, but none have been shown to exist. The equations of state of matter at such high densities are not precisely known because of the theoretical and empirical difficulties.

Some neutron stars rotate very rapidly (up to 716 times a second,[10][11] or approximately 43,000 revolutions per minute) and emit beams of electromagnetic radiation as pulsars. Indeed, the discovery of pulsars in 1967 first suggested that neutron stars exist. Gamma-ray bursts may be produced from rapidly rotating, high-mass stars that collapse to form a neutron star, or from the merger of binary neutron stars. There are thought to be around 100 million neutron stars in the galaxy, but they can only be easily detected in certain instances, such as if they are a pulsar or part of a binary system. Non-rotating and non-accreting neutron stars are virtually undetectable; however, the Hubble Space Telescope has observed one thermally radiating neutron star, called RX J185635-3754.

## 22.1 Formation

Any main-sequence star with an initial mass of above 8 $M\odot$ has the potential to become a neutron star. As the star evolves away from the main sequence, subsequent nuclear burning produces an iron-rich core. When all nuclear fuel in the core has been exhausted, the core must be supported by degeneracy pressure alone. Further deposits of material from

235

*Radiation from the pulsar PSR B1509-58, a rapidly spinning neutron star, makes nearby gas glow in X-rays (gold, from Chandra) and illuminates the rest of the nebula, here seen in infrared (blue and red, from WISE).*

shell burning cause the core to exceed the Chandrasekhar limit. Electron-degeneracy pressure is overcome and the core collapses further, sending temperatures soaring to over $5\times10^9$ K. At these temperatures, photodisintegration (the breaking up of iron nuclei into alpha particles by high-energy gamma rays) occurs. As the temperature climbs even higher, electrons and protons combine to form neutrons, releasing a flood of neutrinos. When densities reach nuclear density of $4\times10^{17}$ kg/m$^3$, neutron degeneracy pressure halts the contraction. The infalling outer atmosphere of the star is flung outwards, becoming a Type II or Type Ib supernova. The remnant left is a neutron star. If it has a mass greater than about 5 $M\odot$, it collapses further to become a black hole. Other neutron stars are formed within close binaries.

As the core of a massive star is compressed during a Type II, Type Ib or Type Ic supernova, and collapses into a neutron star, it retains most of its angular momentum. Because it has only a tiny fraction of its parent's radius (and therefore its moment of inertia is sharply reduced), a neutron star is formed with very high rotation speed, and then gradually slows down. Neutron stars are known that have rotation periods from about 1.4 ms to 30 s. The neutron star's density also gives it very high surface gravity, with typical values ranging from $10^{12}$ to $10^{13}$ m/s$^2$ (more than $10^{11}$ times of that of Earth).[6]

One measure of such immense gravity is the fact that neutron stars have an escape velocity ranging from 100,000 km/s to 150,000 km/s, that is, from a third to half the speed of light. Matter falling onto the surface of a neutron star would be accelerated to tremendous speed by the star's gravity. The force of impact would likely destroy the object's component atoms, rendering all its matter identical, in most respects, to the rest of the star.

## 22.2 Properties

*Gravitational light deflection at a neutron star. Due to relativistic light deflection more than half of the surface is visible (each chequered patch here represents 30 degrees by 30 degrees).[12] In natural units, the mass of the depicted star is 1 and its radius 4, or twice its Schwarzschild radius.[12]*

The gravitational field at the star's surface is about $2 \times 10^{11}$ times stronger than on Earth. Such a strong gravitational field acts as a gravitational lens and bends the radiation emitted by the star such that parts of the normally invisible rear surface become visible.[12] If the radius of the neutron star is $3GM/c^2$ or less, then the photons may be trapped in an orbit, thus making the whole surface of that neutron star visible, along with destabilizing orbits at that and less than that of the radius. A fraction of the mass of a star that collapses to form a neutron star is released in the supernova explosion from which it forms (from the law of mass-energy equivalence, $E = mc^2$). The energy comes from the gravitational binding energy of a neutron star.

Neutron star relativistic equations of state provided by Jim Lattimer include a graph of radius vs. mass for various models.[13] The most likely radii for a given neutron star mass are bracketed by models AP4 (smallest radius) and MS2 (largest radius). BE is the ratio of gravitational binding energy mass equivalent to observed neutron star gravitational mass of "M" kilograms with radius "R" meters.[14]

$$BE = \frac{0.60\,\beta}{1 - \frac{\beta}{2}} \quad \beta = GM/Rc^2$$

Given current values

$$G = 6.6742 \times 10^{-11} \text{ m}^3\text{kg}^{-1}\text{s}^{-2} \text{ [15]}$$

$$c^2 = 8.98755 \times 10^{16} \text{ m}^2\text{s}^{-2}$$

$$M_{solar} = 1.98844 \times 10^{30} \text{ kg}$$

and star masses "M" commonly reported as multiples of one solar mass,

$$M_x = \frac{M}{M_\odot}$$

then the relativistic fractional binding energy of a neutron star is

$$BE = \frac{885.975\,M_x}{R - 738.313\,M_x}$$

A 2 $M\odot$ neutron star would not be more compact than 10,970 meters radius (AP4 model). Its mass fraction gravitational binding energy would then be 0.187, −18.7% (exothermic). This is not near 0.6/2 = 0.3, −30%.

A neutron star is so dense that one teaspoon (5 milliliters) of its material would have a mass over $5.5 \times 10^{12}$ kg (that is 1100 tonnes per 1 nanolitre), about 900 times the mass of the Great Pyramid of Giza.[lower-alpha 3] Hence, the gravitational force of a typical neutron star is such that if an object were to fall from a height of one meter, it would only take one microsecond to hit the surface of the neutron star, and would do so at around 2000 kilometers per second, or 7.2 million kilometers per hour.[16]

The temperature inside a newly formed neutron star is from around $10^{11}$ to $10^{12}$ kelvin.[8] However, the huge number of neutrinos it emits carry away so much energy that the temperature falls within a few years to around $10^6$ kelvin.[8] Even at 1 million kelvin, most of the light generated by a neutron star is in X-rays.

The pressure increases from $3 \times 10^{33}$ to $1.6 \times 10^{35}$ Pa from the inner crust to the center.[17]

The equation of state for a neutron star is still not known. It is assumed that it differs significantly from that of a white dwarf, whose equation of state is that of a degenerate gas that can be described in close agreement with special relativity. However, with a neutron star the increased effects of general relativity can no longer be ignored. Several equations of state have been proposed (FPS, UU, APR, L, SLy, and others) and current research is still attempting to constrain the theories to make predictions of neutron star matter.[6][18] This means that the relation between density and mass is not fully known, and this causes uncertainties in radius estimates. For example, a 1.5 $M\odot$ neutron star could have a radius of 10.7, 11.1, 12.1 or 15.1 kilometres (for EOS FPS, UU, APR or L respectively).[18]

## 22.3  Structure

Current understanding of the structure of neutron stars is defined by existing mathematical models, but it might be possible to infer through studies of neutron-star oscillations. Similar to asteroseismology for ordinary stars, the inner structure might be derived by analyzing observed frequency spectra of stellar oscillations.[6]

Current models indicate that matter at the surface of a neutron star is composed of ordinary atomic nuclei crushed into a solid lattice with a sea of electrons flowing through the gaps between them. It is possible that the nuclei at the surface are iron, due to iron's high binding energy per nucleon.[19] It is also possible that heavy elements, such as iron, simply sink beneath the surface, leaving only light nuclei like helium and hydrogen.[19] If the surface temperature exceeds $10^6$ kelvin (as in the case of a young pulsar), the surface should be fluid instead of the solid phase observed in cooler neutron stars (temperature $<10^6$ kelvin).[19]

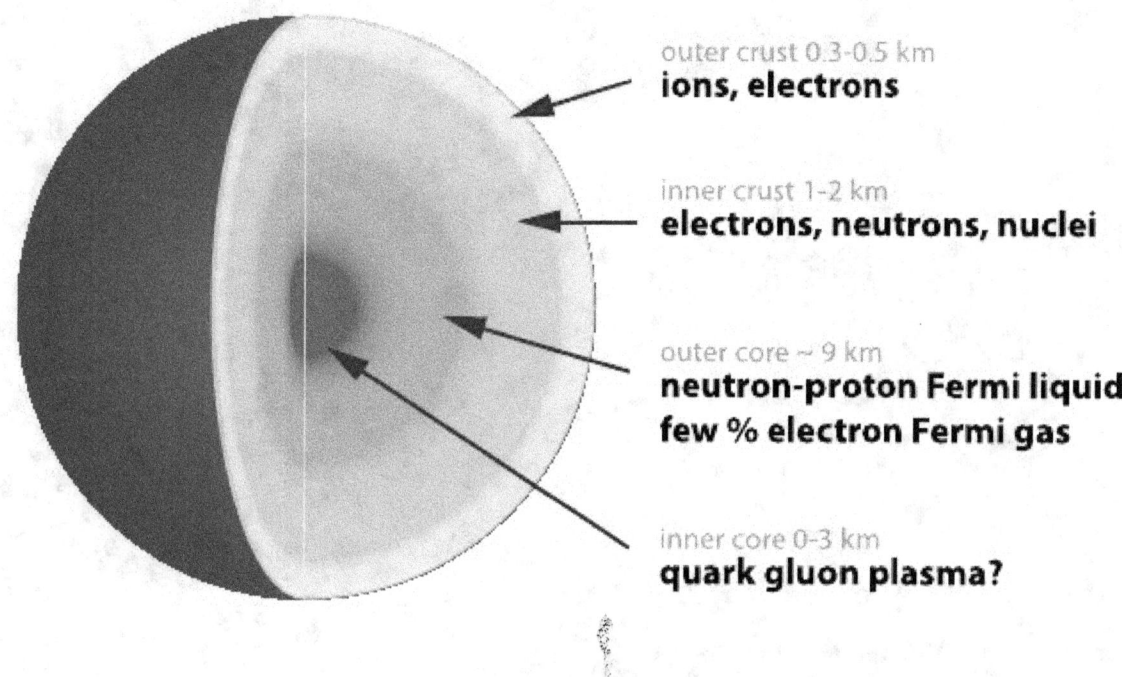

outer crust 0.3-0.5 km
**ions, electrons**

inner crust 1-2 km
**electrons, neutrons, nuclei**

outer core ~ 9 km
**neutron-proton Fermi liquid
few % electron Fermi gas**

inner core 0-3 km
**quark gluon plasma?**

*Cross-section of neutron star. Densities are in terms of $\rho_0$ the saturation nuclear matter density, where nucleons begin to touch.*

The "atmosphere" of the star is hypothesized to be at most several micrometers thick, and its dynamic is fully controlled by the star's magnetic field. Below the atmosphere one encounters a solid "crust". This crust is extremely hard and very smooth (with maximum surface irregularities of ~5 mm), because of the extreme gravitational field.[20] The expected hierarchy of phases of nuclear matter in the inner crust has been characterized as nuclear pasta.[21]

Proceeding inward, one encounters nuclei with ever increasing numbers of neutrons; such nuclei would decay quickly on Earth, but are kept stable by tremendous pressures. As this process continues at increasing depths, neutron drip becomes overwhelming, and the concentration of free neutrons increases rapidly. In this region, there are nuclei, free electrons, and free neutrons. The nuclei become increasingly small (gravity and pressure overwhelming the strong force) until the core is reached, by definition the point where they disappear altogether.

The composition of the superdense matter in the core remains uncertain. One model describes the core as superfluid neutron-degenerate matter (mostly neutrons, with some protons and electrons). More exotic forms of matter are possible, including degenerate strange matter (containing strange quarks in addition to up and down quarks), matter containing high-energy pions and kaons in addition to neutrons,[6] or ultra-dense quark-degenerate matter.

## 22.4 History of discoveries

In 1934, Walter Baade and Fritz Zwicky proposed the existence of the neutron star,[22][lower-alpha 4] only a year after the discovery of the neutron by Sir James Chadwick.[25] In seeking an explanation for the origin of a supernova, they tentatively proposed that in supernova explosions ordinary stars are turned into stars that consist of extremely closely packed neutrons that they called neutron stars. Baade and Zwicky correctly proposed at that time that the release of the gravitational binding energy of the neutron stars powers the supernova: "In the supernova process, mass in bulk is annihilated". Neutron stars were thought to be too faint to be detectable and little work was done on them until November 1967, when Franco Pacini (1939–2012) pointed out that if the neutron stars were spinning and had large magnetic fields, then electromagnetic waves would be emitted. Unbeknown to him, radio astronomer Antony Hewish and his research assistant Jocelyn Bell at Cambridge were shortly to detect radio pulses from stars that are now believed to be highly magnetized, rapidly spinning neutron stars, known as pulsars.

*The first direct observation of a neutron star in visible light. The neutron star is RX J185635-3754.*

In 1965, Antony Hewish and Samuel Okoye discovered "an unusual source of high radio brightness temperature in the Crab Nebula".[26] This source turned out to be the Crab Pulsar that resulted from the great supernova of 1054.

In 1967, Iosif Shklovsky examined the X-ray and optical observations of Scorpius X-1 and correctly concluded that the radiation comes from a neutron star at the stage of accretion.[27]

In 1967, Jocelyn Bell and Antony Hewish discovered regular radio pulses from CP 1919. This pulsar was later interpreted as an isolated, rotating neutron star. The energy source of the pulsar is the rotational energy of the neutron star. The majority of known neutron stars (about 2000, as of 2010) have been discovered as pulsars, emitting regular radio pulses.

In 1971, Riccardo Giacconi, Herbert Gursky, Ed Kellogg, R. Levinson, E. Schreier, and H. Tananbaum discovered 4.8

second pulsations in an X-ray source in the constellation Centaurus, Cen X-3. They interpreted this as resulting from a rotating hot neutron star. The energy source is gravitational and results from a rain of gas falling onto the surface of the neutron star from a companion star or the interstellar medium.

In 1974, Antony Hewish was awarded the Nobel Prize in Physics "for his decisive role in the discovery of pulsars" without Jocelyn Bell who shared in the discovery.

In 1974, Joseph Taylor and Russell Hulse discovered the first binary pulsar, PSR B1913+16, which consists of two neutron stars (one seen as a pulsar) orbiting around their center of mass. Einstein's general theory of relativity predicts that massive objects in short binary orbits should emit gravitational waves, and thus that their orbit should decay with time. This was indeed observed, precisely as general relativity predicts, and in 1993, Taylor and Hulse were awarded the Nobel Prize in Physics for this discovery.

In 1982, Don Backer and colleagues discovered the first millisecond pulsar, PSR B1937+21. This objects spins 642 times per second, a value that placed fundamental constraints on the mass and radius of neutron stars. Many millisecond pulsars were later discovered, but PSR B1937+21 remained the fastest-spinning known pulsar for 24 years, until PSR J1748-2446ad (which spins more than 700 times a second) was discovered.

In 2003, Marta Burgay and colleagues discovered the first double neutron star system where both components are detectable as pulsars, PSR J0737-3039. The discovery of this system allows a total of 5 different tests of general relativity, some of these with unprecedented precision.

In 2010, Paul Demorest and colleagues measured the mass of the millisecond pulsar PSR J1614–2230 to be $1.97 \pm 0.04$ $M_\odot$, using Shapiro delay.[28] This was substantially higher than any previously measured neutron star mass (1.67 $M_\odot$, see PSR J1903+0327), and places strong constraints on the interior composition of neutron stars.

In 2013, John Antoniadis and colleagues measured the mass of PSR J0348+0432 to be $2.01 \pm 0.04$ $M_\odot$, using white dwarf spectroscopy.[29] This confirmed the existence of such massive stars using a different method. Furthermore, this allowed, for the first time, a test of general relativity using such a massive neutron star.

## 22.5  Rotation

Neutron stars rotate extremely rapidly after their formation due to the conservation of angular momentum; like spinning ice skaters pulling in their arms, the slow rotation of the original star's core speeds up as it shrinks. A newborn neutron star can rotate several times a second; sometimes, the neutron star absorbs orbiting matter from a companion star, increasing the rotation to several hundred times per second, reshaping the neutron star into an oblate spheroid.

Over time, neutron stars slow down (spin down) because their rotating magnetic fields radiate energy; older neutron stars may take several seconds for each revolution.

The rate at which a neutron star slows its rotation is usually constant and very small: the observed rates of decline are between $10^{-10}$ and $10^{-21}$ seconds for each rotation. Therefore, for a typical slow down rate of $10^{-15}$ seconds per rotation, a neutron star now rotating in 1 second will rotate in 1.000003 seconds after a century, or 1.03 seconds after 1 million years.

Sometimes a neutron star will *spin up* or undergo a *glitch*, a sudden small increase of its rotation speed. Glitches are thought to be the effect of a starquake — as the rotation of the star slows down, the shape becomes more spherical. Due to the stiffness of the "neutron" crust, this happens as discrete events when the crust ruptures, similar to tectonic earthquakes. After the starquake, the star will have a smaller equatorial radius, and because angular momentum is conserved, rotational speed increases. Recent work, however, suggests that a starquake would not release sufficient energy for a neutron star glitch; it has been suggested that glitches may instead be caused by transitions of vortices in the superfluid core of the star from one metastable energy state to a lower one.[30]

Neutron stars have been observed to "pulse" radio and x-ray emissions believed to be caused by particle acceleration near the magnetic poles, which need not be aligned with the rotation axis of the star. Through mechanisms not yet entirely understood, these particles produce coherent beams of radio emission. External viewers see these beams as pulses of radiation whenever the magnetic pole sweeps past the line of sight. The pulses come at the same rate as the rotation of the neutron star, and thus, appear periodic. Neutron stars that emit such pulses are called pulsars.

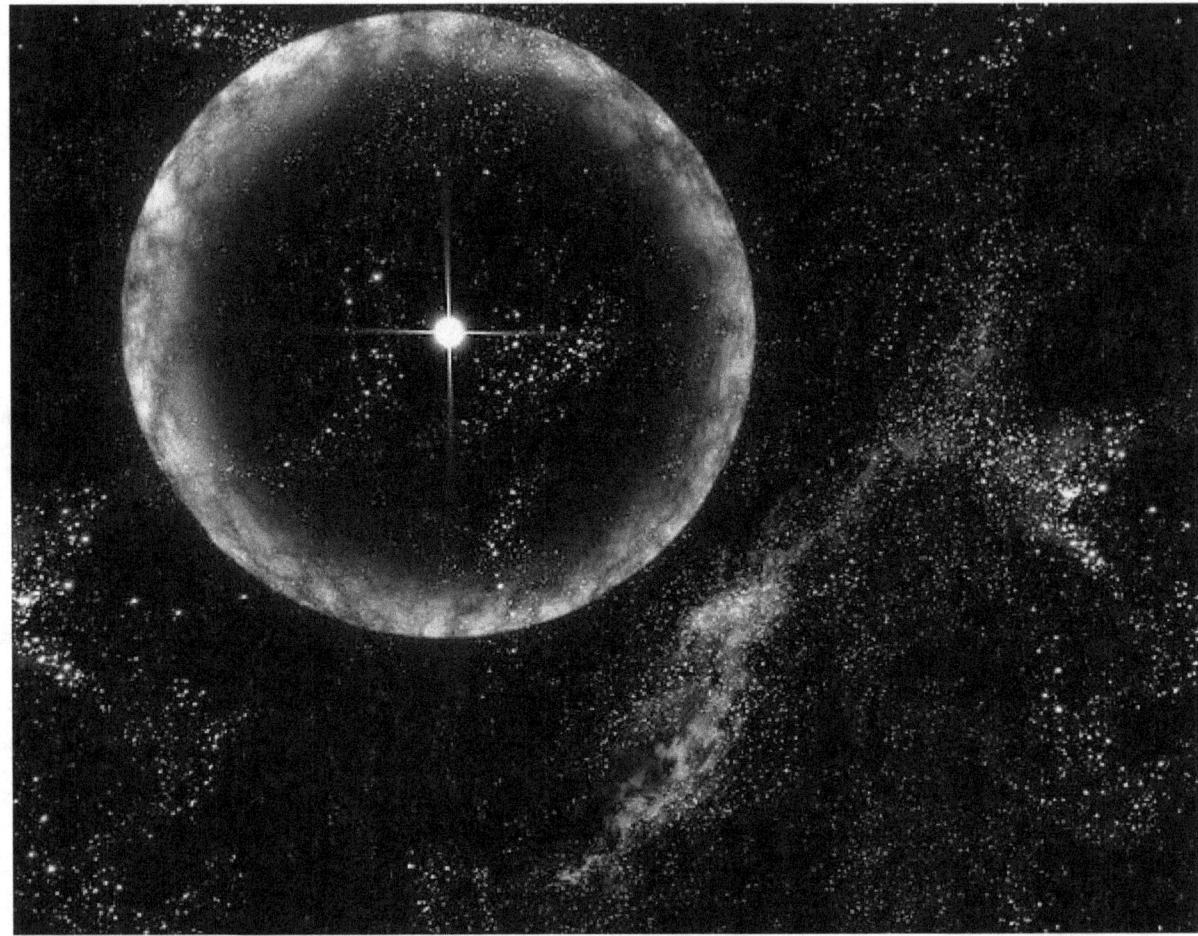

*NASA artist's conception of a "starquake", or "stellar quake".*

The most rapidly rotating neutron star currently known, PSR J1748-2446ad, rotates at 716 rotations per second.[31] A recent paper reported the detection of an X-ray burst oscillation (an indirect measure of spin) at 1122 Hz from the neutron star XTE J1739-285.[32] However, at present, this signal has only been seen once, and should be regarded as tentative until confirmed in another burst from this star.

## 22.6   Population and distances

At present, there are about 2000 known neutron stars in the Milky Way and the Magellanic Clouds, the majority of which have been detected as radio pulsars. Neutron stars are mostly concentrated along the disk of the Milky Way although the spread perpendicular to the disk is large because the supernova explosion process can impart high speeds (400 km/s) to the newly formed neutron star.

Some of the closest neutron stars are RX J1856.5-3754 about 400 light years away and PSR J0108-1431 at about 424 light years.[33] RX J1856.5-3754 is a member of a close group of neutron stars called The Magnificent Seven. Another nearby neutron star that was detected transiting the backdrop of the constellation Ursa Minor has been nicknamed Calvera by its Canadian and American discoverers, after the villain in the 1960 film *The Magnificent Seven*. This rapidly moving object was discovered using the ROSAT/Bright Source Catalog.

## 22.7 Binary neutron stars

About 5% of all known neutron stars are members of a binary system. The formation and evolution scenario of binary neutron stars is a rather exotic and complicated process.[34] The companion stars may be either ordinary stars, white dwarfs or other neutron stars. According to modern theories of binary evolution it is expected that neutron stars also exist in binary systems with black hole companions. Such binaries are expected to be prime sources for emitting gravitational waves. Neutron stars in binary systems often emit X-rays, which is caused by the heating of material (gas) accreted from the companion star. Material from the outer layers of a (bloated) companion star is sucked towards the neutron star as a result of its very strong gravitational field. As a result of this process binary neutron stars may also coalesce into black holes if the accretion of mass takes place under extreme conditions.[35] It has been proposed that coalescence of binaries consisting of two neutron stars may be responsible for producing short gamma-ray bursts. Such events may also be responsible for producing all chemical elements beyond iron,[36] as opposed to the supernova nucleosynthesis theory.

*Circinus X-1: X-ray light rings from a binary neutron star (24 June 2015: Chandra X-ray Observatory).*

## 22.8   Subtypes

- Neutron star

    - Protoneutron star (PNS), theorized.[37]

    - Radio-quiet neutron stars

    - Radio loud neutron star

        - Single pulsars–general term for neutron stars that emit directed pulses of radiation towards us at regular intervals (due to their strong magnetic fields).

            - Rotation-powered pulsar *("radio pulsar")*

                - Magnetar–a neutron star with an extremely strong magnetic field (1000 times more than a regular neutron star), and long rotation periods (5 to 12 seconds).

                    - Soft gamma repeater (SGR)

                    - Anomalous X-ray pulsar (AXP)

        - Binary pulsars

            - Low-mass X-ray binaries (LMXB)

            - Intermediate-mass X-ray binaries (IMXB)

            - High-mass X-ray binaries (HMXB)

            - Accretion-powered pulsar *("X-ray pulsar")*

                - X-ray burster–a neutron star with a low mass binary companion from which matter is accreted resulting in irregular bursts of energy from the surface of the neutron star.

                - Millisecond pulsar (MSP) *("recycled pulsar")*

                    - Sub-millisecond pulsar[38]

    - Exotic star

        - Quark star–currently a hypothetical type of neutron star composed of quark matter, or strange matter. As of 2008, there are three candidates.

        - Electroweak star–currently a hypothetical type of extremely heavy neutron star, in which the quarks are converted to leptons through the electroweak force, but the gravitational collapse of the star is prevented by radiation pressure. As of 2010, there is no evidence for their existence.

        - Preon star–currently a hypothetical type of neutron star composed of preon matter. As of 2008, there is no evidence for the existence of preons.

## 22.9   Giant nucleus

A neutron star has some of the properties of an atomic nucleus, including density (within an order of magnitude) and being composed of nucleons. In popular scientific writing, neutron stars are therefore sometimes described as giant nuclei. However, in other respects, neutron stars and atomic nuclei are quite different. In particular, a nucleus is held together by the strong interaction, whereas a neutron star is held together by gravity, and thus the density and structure of neutron stars is more variable. It is generally more useful to consider such objects as stars.

## 22.10   Examples of neutron stars

- PSR J0108-1431 – closest neutron star

- LGM-1 – the first recognized radio-pulsar

- PSR B1257+12 – the first neutron star discovered with planets (a millisecond pulsar)

- SWIFT J1756.9-2508 – a millisecond pulsar with a stellar-type companion with planetary range mass (below brown dwarf)

- PSR B1509-58 source of the "Hand of God" photo shot by the Chandra X-ray Observatory.

- PSR J0348+0432 – the most massive neutron star with a well-constrained mass, $2.01 \pm 0.04 M_\odot$ .

## 22.11 Gallery

- Play media

  Video - Neutron stars contain 500,000 Earth-masses in 25 km (16 mi) dia. sphere.

- Play media

  Video - Neutron stars colliding (animation).

- Play media

  Video - Neutron star collision.

## 22.12 See also

- Dragon's Egg

- Neutronium

- Preon-degenerate matter

- Rotating radio transients

## 22.13 Notes

[1] A neutron star's density increases as its mass increases, and its radius decreases non-linearly. (NASA mass radius graph)

[2] $3.7 \times 10^{17}$ kg/m$^3$ derives from mass $2.68 \times 10^{30}$ kg / volume of star of radius 12 km; $5.9 \times 10^{17}$ kg m$^{-3}$ derives from mass $4.2 \times 10^{30}$ kg per volume of star radius 11.9 km

[3] The average density of material in a neutron star of radius 10 km is $1.1 \times 10^{12}$ kg cm$^{-3}$. Therefore, 5 ml of such material is $5.5 \times 10^{12}$ kg, or 5 500 000 000 metric tons. This is about 15 times the total mass of the human world population. Alternatively, 5 ml from a neutron star of radius 20 km radius (average density $8.35 \times 10^{10}$ kg cm$^{-3}$) has a mass of about 400 million metric tons, or about the mass of all humans.

[4] Even before the discovery of neutron, in 1931, neutron stars were *anticipated* by Lev Landau, who wrote about stars where "atomic nuclei come in close contact, forming one gigantic nucleus"[23]). However, the widespread opinion that Landau *predicted* neutron stars proves to be wrong.[24]

## 22.14   References

[1] Özel, Feryal; Psaltis, Dimitrios; Narayan, Ramesh; Santos Villarreal, Antonio (September 2012). "On the Mass Distribution and Birth Masses of Neutron Stars". *The Astrophysical Journal* **757** (1): 13. arXiv:1201.1006. Bibcode:2012ApJ...757...55O. doi:10.1088/0004-637X/757/1/55. Retrieved 14 May 2015.

[2] Chamel, N.; Haensel, P.; Zdunik, J.L.; Fantina, A.F. (19 November 2013). "On the Maximum Mass of Neutron Stars" (PDF). *International Journal of Modern Physics* **1** (28): 1330018. arXiv:1307.3995. Bibcode:2013IJMPE..2230018C. doi:10.1142/S021830131330018X.Retrieved14May2015.

[3] Bulent Kiziltan (2011). *Reassessing the Fundamentals: On the Evolution, Ages and Masses of Neutron Stars*. Universal-Publishers. ISBN 1-61233-765-1.

[4] Neutron star mass measurements

[5] "Nasa Ask an Astrophysist: Maximum Mass of a Neutron Star".

[6] Paweł Haensel; A Y Potekhin; D G Yakovlev (2007). *Neutron Stars*. Springer. ISBN 0-387-33543-9.

[7] "Calculating a Neutron Star's Density". Retrieved 2006-03-11. NB $3 \times 10^{17}$ kg/m$^3$ is $3 \times 10^{14}$ g/cm$^3$

[8] "Introduction to neutron stars". Retrieved 2007-11-11.

[9] . a 10 $M\odot$ star will collapse into a black hole.

[10] Hessels, Jason; Ransom, Scott M.; Stairs, Ingrid H.; Freire, Paulo C. C.; et al. (2006). "A Radio Pulsar Spinning at 716 Hz". *Science* **311** (5769): 1901–1904. arXiv:astro-ph/0601337. Bibcode:2006Sci...311.1901H. doi:10.1126/science.1123430. PMID 16410486.

[11] Naeye, Robert (2006-01-13). "Spinning Pulsar Smashes Record". *Sky & Telescope*. Retrieved 2008-01-18.

[12] Zahn, Corvin (1990-10-09). "Tempolimit Lichtgeschwindigkeit" (in German). Retrieved 2009-10-09. Durch die gravitative Lichtablenkung ist mehr als die Hälfte der Oberfläche sichtbar. Masse des Neutronensterns: 1, Radius des Neutronensterns: 4, ... dimensionslosen Einheiten (c, G = 1)

[13] Neutron Star Masses and Radii, p. 9/20, bottom

[14] J. M. Lattimer and M. Prakash, "Neutron Star Structure and the Equation of State" Astrophysical J. 550(1) 426 (2001); http://arxiv.org/abs/astro-ph/0002232

[15] Measurement of Newton's Constant Using a Torsion Balance with Angular Acceleration Feedback , Phys. Rev. Lett. 85(14) 2869 (2000)

[16] Miscellaneous Facts

[17] Neutron degeneracy pressure (Archive). Physics Forums. Retrieved on 2011-10-09.

[18] NASA. Neutron Star Equation of State Science Retrieved 2011-09-26 Archived February 20, 2013 at the Wayback Machine

[19] V. S. Beskin (1999). "*Radiopulsars*". УФН. Т.169, №11, p.1173-1174

[20] neutron star

[21] Pons, José A.; Viganò, Daniele; Rea, Nanda (2013). "Too much "pasta" for pulsars to spin down". *Nature Physics* **9** (7): 431–434. arXiv:1304.6546. Bibcode:2013NatPh...9..431P. doi:10.1038/nphys2640.

[22] Baade, Walter & Zwicky, Fritz (1934). "Remarks on Super-Novae and Cosmic Rays".*Phys. Rev.***46**(1): 76–77. Bibcode:1934 doi:10.1103/PhysRev.46.76.2.

[23] Landau L.D. (1932). "On the theory of stars". *Phys. Z. Sowjetunion* **1**: 285–288.

[24] P. Haensel, A. Y. Potekhin, & D. G. Yakovlev (2007). *Neutron Stars 1: Equation of State and Structure* (New York: Springer), page 2 http://adsabs.harvard.edu/abs/2007ASSL..326.....H

[25] Chadwick, James (1932). "On the possible existence of a neutron". *Nature* **129** (3252): 312. Bibcode:1932Natur.129Q.312C. doi:10.1038/129312a0.

[26] Hewish, A. & Okoye, S. E. (1965). "Evidence of an unusual source of high radio brightness temperature in the Crab Nebula". *Nature* **207** (4992): 59–60. Bibcode:1965Natur.207...59H. doi:10.1038/207059a0.

[27] Shklovsky, I.S. (April 1967). "On the Nature of the Source of X-Ray Emission of SCO XR-1". *Astrophys. J.* **148** (1): L1–L4. Bibcode:1967ApJ...148L...1S. doi:10.1086/180001.

[28] Demorest, PB; Pennucci, T; Ransom, SM; Roberts, MS; et al. (2010). "A two-solar-mass neutron star measured using Shapiro delay". *Nature* **467** (7319): 1081–1083. arXiv:1010.5788. Bibcode:2010Natur.467.1081D. doi:10.1038/nature09466. PMID 20981094.

[29] Antoniadis, J (2012). "A Massive Pulsar in a Compact Relativistic Binary". *Science* **340** (6131): 1233232. arXiv:1304.6875. Bibcode:2013Sci...340..448A2010. doi:10.1126/science.1233232.

[30] Alpar, M Ali (January 1, 1998). "Pulsars, glitches and superfluids". Physicsworld.com.

[31] [astro-ph/0601337] A Radio Pulsar Spinning at 716 Hz

[32] University of Chicago Press – Millisecond Variability from XTE J1739285 – 10.1086/513270

[33] Posselt, B.; Neuhäuser, R.; Haberl, F. (March 2009). "Searching for substellar companions of young isolated neutron stars". *Astronomy and Astrophysics* **496**(2): 533–545. arXiv:0811.0398. Bibcode:2009A&A...496..533P.doi:10.1051/0004-6361/2008

[34] Tauris & van den Heuvel (2006), in Compact Stellar X-ray Sources. Eds. Lewin and van der Klis, Cambridge University Press http://adsabs.harvard.edu/abs/2006csxs.book..623T

[35] Compact Stellar X-ray Sources (2006). Eds. Lewin and van der Klis, Cambridge University

[36] Urry, Meg (July 20, 2013). "Gold comes from stars". CNN.

[37] Neutrino-Driven Protoneutron Star Winds, Todd A. Thompson.

[38] Nakamura, T. (1989). "Binary Sub-Millisecond Pulsar and Rotating Core Collapse Model for SN1987A". *Progress of Theoretical Physics* **81** (5): 1006–1020. Bibcode:1989PThPh..81.1006N. doi:10.1143/PTP.81.1006.

- "ASTROPHYSICS: ON OBSERVED PULSARS". *scienceweek.com*. Retrieved 6 August 2004.

- Norman K. Glendenning; R. Kippenhahn; I. Appenzeller; G. Borner; et al. (2000). *Compact Stars* (2nd ed.).

- Kaaret; Prieskorn; in 't Zand; Brandt; et al. (2006). "Evidence for 1122 Hz X-Ray Burst Oscillations from the Neutron-Star X-Ray Transient XTE J1739-285". *The Astrophysical Journal* **657** (2): L97. arXiv:astro-ph/0611716. Bibcode:2007ApJ...657L..97K. doi:10.1086/513270.

## 22.15 External links

- Introduction to neutron stars

- Neutron Stars for Undergraduates and its Errata

- NASA on pulsars

- "NASA Sees Hidden Structure Of Neutron Star In Starquake". SpaceDaily.com. April 26, 2006

- "Mysterious X-ray sources may be lone neutron stars". *New Scientist*.

- "Massive neutron star rules out exotic matter". *New Scientist*. According to a new analysis, exotic states of matter such as free quarks or BECs do not arise inside neutron stars.

- "Neutron star clocked at mind-boggling velocity". *New Scientist*. A neutron star has been clocked traveling at more than 1500 kilometers per second.

# Chapter 23

# Electron capture

This article is about the radioactive decay mode. For the fragmentation method used in mass spectrometry, see Electron capture ionization. For the detector used in gas chromatography, see Electron-capture dissociation.

**Electron capture** (**K-electron capture**, also **K-capture**, or **L-electron capture**, **L-capture**) is a process in which the proton-rich nucleus of an electrically neutral atom absorbs an inner atomic electron, usually from the K or L electron shell. This process thereby changes a nuclear proton to a neutron and simultaneously causes the emission of an electron neutrino.

The daughter nuclide, if it is in an excited state, then transitions to its ground state. Usually, a gamma ray is emitted during this transition, but nuclear de-excitation may also take place by internal conversion.

Following capture of an inner electron from the atom, an outer electron replaces the electron that was captured and one or more characteristic X-ray photons is emitted in this process. Electron capture sometimes also results in the Auger effect, where an electron is ejected from the atom's electron shell due to interactions between the atom's electrons in the process of seeking a lower energy electron state.

Following electron capture, the atomic number is reduced by one, the neutron number is increased by one, and there is no change in atomic mass. Simple electron capture results in a neutral atom, since the loss of the electron in the electron shell is balanced by a loss of positive nuclear positive charge. However, a positive atomic ion may result from further Auger electron emission.

Electron capture is an example of weak interaction, one of the four fundamental forces.

Electron capture is the primary decay mode for isotopes with a relative superabundance of protons in the nucleus, but with insufficient energy difference between the isotope and its prospective daughter (the isobar with one less positive charge) for the nuclide to decay by emitting a positron. Electron capture is always an alternate decay mode for radioactive isotopes that do have sufficient energy to decay by positron emission. It is sometimes called **inverse beta decay**, though this term can also refer to the interaction of an electron antineutrino with a proton.[1]

If the energy difference between the parent atom and the daughter atom is less than 1.022 MeV, positron emission is forbidden as not enough decay energy is available to allow it, and thus electron capture is the sole decay mode. For example, rubidium-83 (37 protons, 46 neutrons) will decay to krypton-83 (36 protons, 47 neutrons) solely by electron capture (the energy difference, or decay energy, is about 0.9 MeV).

A free proton cannot normally be changed to a free neutron by this process; the proton and neutron must be part of a larger nucleus.

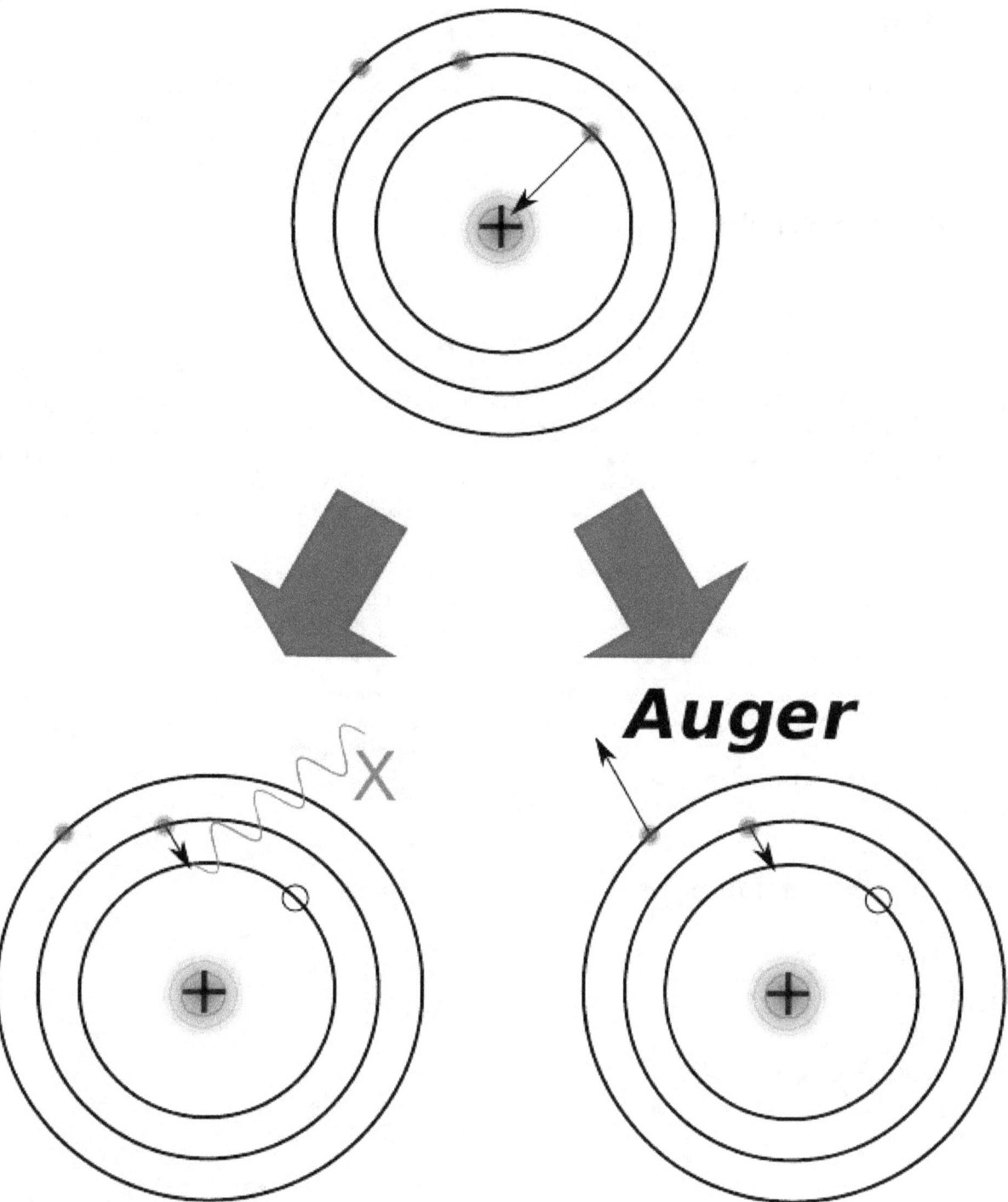

*Scheme of two types of electron capture. Top: The nucleus absorbs an electron. Lower left: An outer electron replaces the "missing" electron. An x-ray, equal in energy to the difference between the two electron shells, is emitted. Lower right: In the Auger effect, the energy released when the outer electron replaces the inner electron is transferred to an outer electron. The outer electron is ejected from the atom, leaving a positive ion.*

## 23.1   History

The theory of electron capture was first discussed by Gian-Carlo Wick in a 1934 paper, and then developed by Hideki Yukawa and others. K-electron capture was first observed by Luis Alvarez, in vanadium-48. He reported it in a 1937 paper in *Physical Review*.[2][3][4] Alvarez went on to study electron capture in gallium-67 and other nuclides.[2][5][6]

## 23.2   Reaction details

The electron that is captured is one of the atom's own electrons, and not a new, incoming electron, as might be suggested by the way the above reactions are written. Radioactive isotopes that decay by pure electron capture can be inhibited from radioactive decay if they are fully ionized ("stripped" is sometimes used to describe such ions). It is hypothesized that such elements, if formed by the r-process in exploding supernovae, are ejected fully ionized and so do not undergo radioactive decay as long as they do not encounter electrons in outer space. Anomalies in elemental distributions are thought to be partly a result of this effect on electron capture. Inverse decays can also be induced by full ionisation; for instance, $^{163}$Ho decays into $^{163}$Dy by electron capture; however, a fully ionised $^{163}$Dy decays into a bound state of $^{163}$Ho by the process of bound-state $\beta^-$ decay.[7]

Chemical bonds can also affect the rate of electron capture to a small degree (in general, less than 1%) depending on the proximity of electrons to the nucleus. For example in $^7$Be, a difference of 0.9% has been observed between half-lives in metallic and insulating environments.[8] This relatively large effect is due to the fact that beryllium is a small atom whose valence electrons are close to the nucleus.

Around the elements in the middle of the periodic table, isotopes that are lighter than stable isotopes of the same element tend to decay through electron capture, while isotopes heavier than the stable ones decay by electron emission. Electron capture happens most often in the heavier neutron-deficient elements where the mass change is smallest and positron emission isn't always possible. When the loss of mass in a nuclear reaction is greater than zero but less than 2m[0-1e-], the process cannot occur by positron emission but is spontaneous for electron capture.

## 23.3   Common examples

Some common radioisotopes that decay by electron capture include:

For a full list, see the table of nuclides.

## 23.4   References

[1] "The Reines-Cowan Experiments: Detecting the Poltergeist" (PDF). *Los Alamos National Laboratory* **25**: 3. 1997.

[2] Luis W. Alvarez, W. Peter Trower (1987). "Chapter 3: K-Electron Capture by Nuclei (with the commentary of Emilio Segré)" In *Discovering Alvarez: selected works of Luis W. Alvarez, with commentary by his students and colleagues.* University of Chicago Press, pp. 11–12, ISBN 978-0-226-81304-2.

[3] "Luis Alvarez, The Nobel Prize in Physics 1968", biography, nobelprize.org. Accessed October 7, 2009.

[4] Alvarez, Luis W. (1937). "Nuclear K Electron Capture". *Physical Review* **52**: 134–135. Bibcode:1937PhRv...52..134A. doi:10.1103/PhysRev.52.134.

[5] Alvarez, Luis W. (1937). "Electron Capture and Internal Conversion in Gallium 67".*Physical Review***53**: 606. Bibcode:1938 doi:10.1103/PhysRev.53.606.

[6] Alvarez, Luis W. (1938). "The Capture of Orbital Electrons by Nuclei".*Physical Review***54**: 486–497. Bibcode:1938PhRv...54 doi:10.1103/PhysRev.54.486.

[7] Fritz Bosch (1995). "Manipulation of Nuclear Lifetimes in Storage Rings" (PDF). *Physica Scripta* **T59**: 221–229.

[8] B. Wang; et al. (2006). "Change of the $^7$Be electron capture half-life in metallic environments". *The European Physical Journal A* **28**: 375–377. Bibcode:2006EPJA...28..375W. doi:10.1140/epja/i2006-10068-x. (subscription required)

## 23.5 External links

-  **The LIVEChart of Nuclides - IAEA** with filter on electron capture

# Chapter 24

# Pair-instability supernova

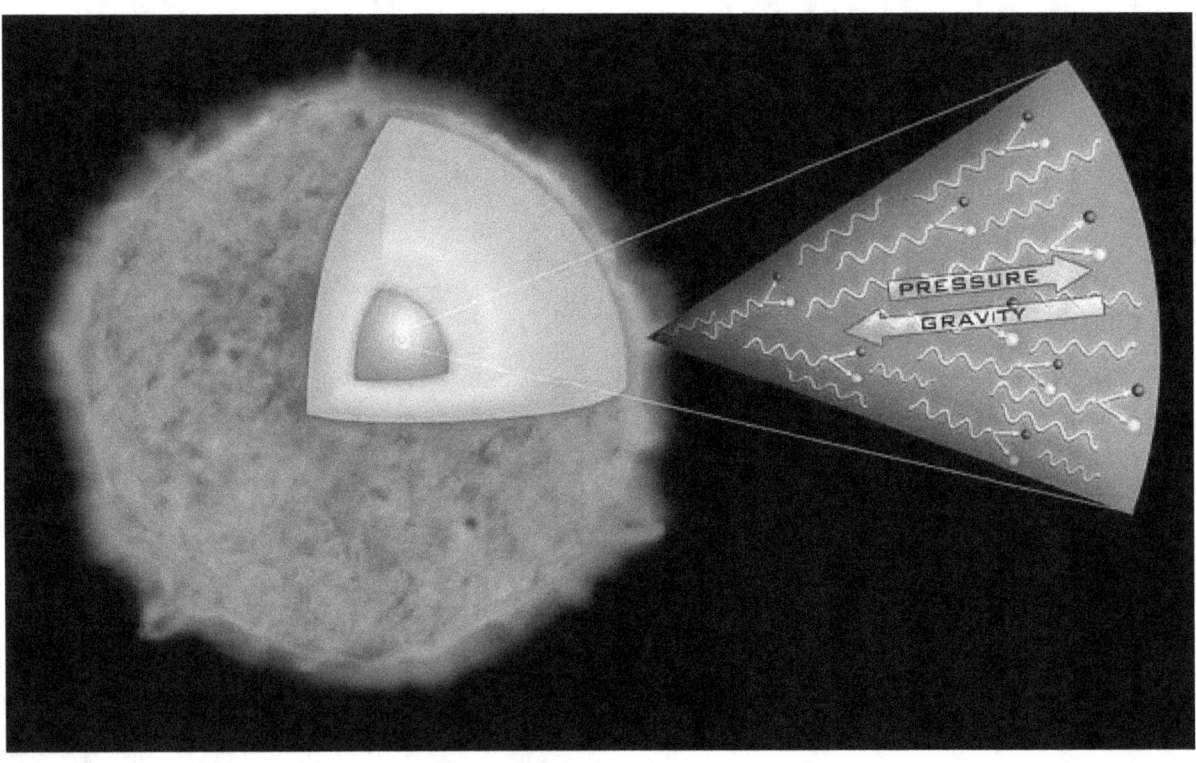

*This illustration explains the pair-instability supernova process that astronomers think triggered the explosion in SN 2006gy. When a star is very massive, the gamma-rays produced in its core can become so energetic that some of their energy is drained away into production of particle and anti-particle pairs. The resulting drop in pressure causes the star to partially collapse under its own huge gravity. After this violent collapse, runaway thermonuclear reactions (not shown here) ensue and the star explodes, spewing the remains into space.*

A **pair-instability supernova** occurs when pair production, the production of free electrons and positrons in the collision between atomic nuclei and energetic gamma rays, reduces thermal pressure inside a supermassive star's core. This pressure drop leads to a partial collapse, then greatly accelerated burning in a runaway thermonuclear explosion which blows the star completely apart without leaving a black hole remnant behind.[1] Pair-instability supernovae can only happen in stars with a mass range from around 130 to 250 solar masses and low to moderate metallicity (low abundance of elements other than hydrogen and helium, a situation common in Population III stars). The recently observed objects SN 2006gy, SN 2007bi,[2] SN 2213-1745 and SN 1000+0216[3] are hypothesized to have been pair-instability supernovae.

## 24.1 Physics

### 24.1.1 Photon pressure

Light in thermal equilibrium has a black body spectrum with an energy density proportional to the fourth power of the temperature (hence the Stefan-Boltzmann law). The wavelength of maximum emission from a blackbody is inversely proportional to its temperature. That is, the frequency, and the energy, of the greatest population of photons of black body radiation is directly proportional to the temperature, and reaches the gamma ray energy range at temperatures above $3 \times 10^8$ K.

In very large hot stars, pressure from gamma rays in the stellar core keeps the upper layers of the star supported against gravitational pull from the core. If the energy density of gamma rays is suddenly reduced, then the outer layers of the star will collapse inwards. The sudden heating and compression of the core generates gamma rays energetic enough to be converted into an avalanche of electron-positron pairs, further reducing the pressure. When the collapse stops, the positrons find electrons and the pressure from gamma rays is driven up, again. The population of positrons provides a brief reservoir of new gamma rays as the expanding supernova's core pressure drops.

### 24.1.2 Pair creation and annihilation

Sufficiently energetic gamma rays can interact with nuclei, electrons, or one another to produce electron-positron pairs, and electron-positron pairs can annihilate, producing gamma rays. From Einstein's equation $E = mc^2$, gamma rays must have more energy than the mass of the electron–positron pairs to produce these pairs.

At the high densities of a stellar core, pair production and annihilation occur rapidly, thereby keeping gamma rays, electrons, and positrons in thermal equilibrium. The higher the temperature, the higher the gamma ray energies, and the larger the amount of energy transferred.

### 24.1.3 Pair-instability

As temperatures and gamma ray energies increase, more and more gamma ray energy is absorbed in creating electron-positron pairs. This reduction in gamma ray energy density reduces the radiation pressure that supports the outer layers of the star. The star contracts, compressing and heating the core, thereby increasing the proportion of energy absorbed by pair creation. Pressure nonetheless increases, but in a pair-instability collapse, the increase in pressure is not enough to resist the increase in gravitational forces as the star becomes denser.

## 24.2 Stellar susceptibility

For a star to undergo pair-instability supernova, the loss in total outward pressure resulting from the increased creation of positron/electron pairs by gamma ray collisions must be sufficiently great to allow the inward gravitational pressure to overwhelm the remaining outward pressure. Among stellar mechanisms not responsive to the reduction in outward pressure effected by pair creation, rotational speed and metallicity are the most important.

Stars exhibiting these characteristics still contract as gravity's inward pressure increases relative to the star's total outward pressure. Unlike their slower or less metal-rich cousins, however, these stars continue to exert outward pressure sufficient to prevent contractions so great that gravity entirely overwhelms its opposition and collapses the star.

Stars formed by collision mergers having a metallicity Z between 0.02 and 0.001 may end their lives as pair-instability supernovae if their mass is in the appropriate range.[4]

Very large high metallicity stars are probably unstable due to the Eddington limit, and would tend to shed mass during the formation process.

## 24.3    Stellar behavior

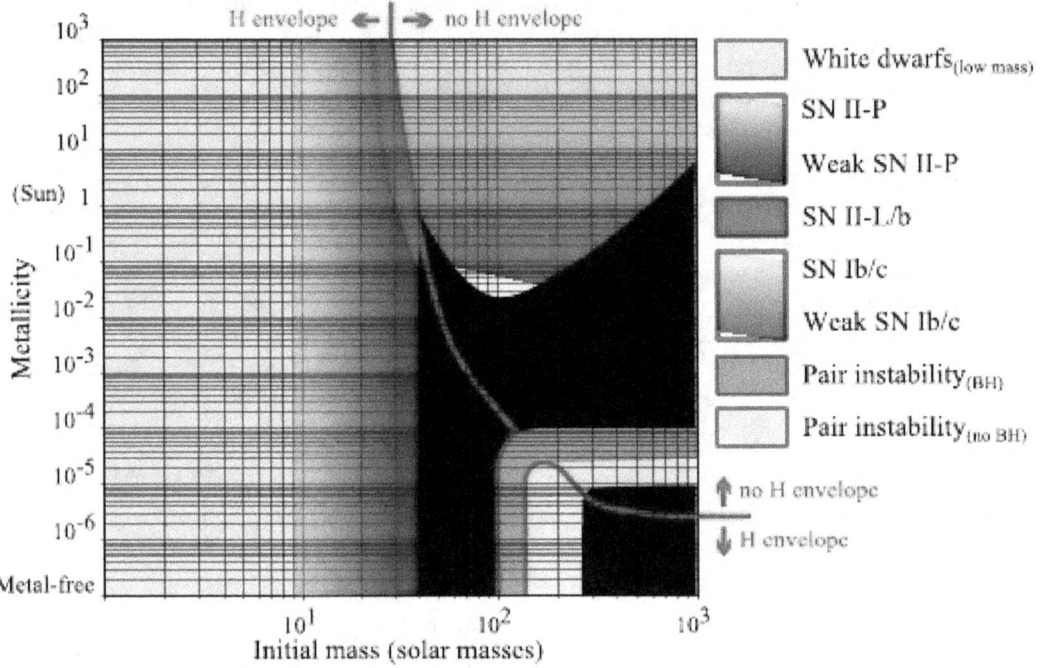

*Supernovae as initial mass-metallicity*

Several sources describe the stellar behavior for large stars in pair-instability conditions.[5][6]

### 24.3.1    Below 100 solar masses

Gamma rays produced by stars of fewer than 100 or so solar masses are not energetic enough to produce electron-positron pairs. Some of these stars will undergo supernovae at the end of their lives, but the causative mechanisms are unrelated to pair-instability.

### 24.3.2    100 to 130 solar masses

These stars are large enough to produce gamma rays with enough energy to create electron-positron pairs, but the resulting net reduction in counter-gravitational pressure is insufficient to cause the core-overpressure required for supernova. Instead, the contraction caused by pair-creation provokes increased thermonuclear activity within the star that repulses the inward pressure and returns the star to equilibrium. It is thought that stars of this size undergo a series of these pulses until they shed sufficient mass to drop below 100 solar masses, at which point they are no longer hot enough to support pair-creation. Pulsing of this nature may have been responsible for the variations in brightness experienced by Eta Carinae in 1843, though this explanation is not universally accepted.

### 24.3.3    130 to 250 solar masses

For very high mass stars, with mass at least 130 and up to perhaps roughly 250 solar masses, a true pair-instability supernova can occur. In these stars, the first time that conditions support pair creation instability, the situation runs out of control. The collapse proceeds to efficiently compress the star's core; the overpressure is sufficient to allow runaway nuclear fusion to burn it in a few seconds, creating a thermonuclear explosion.[6] With more thermal energy released than the star's gravitational binding energy, it is completely disrupted; no black hole or other remnant is left behind.

In addition to the immediate energy release, a large fraction of the star's core is transformed to nickel-56, a radioactive isotope which decays with a half-life of 6.1 days into cobalt-56. Cobalt-56 has a half-life of 77 days and then further decays to the stable isotope iron-56 (see Supernova nucleosynthesis). For the hypernova SN 2006gy, studies indicate that perhaps 40 solar masses of the original star were released as Ni-56, almost the entire mass of the star's core regions.[5] Collision between the exploding star core and gas it ejected earlier, and radioactive decay, release most of the visible light.

### 24.3.4   250 solar masses or more

A different reaction mechanism, photodisintegration, results after collapse starts in stars of at least 250 solar masses. This endothermic (energy-absorbing) reaction causes the star to continue collapse into a black hole rather than exploding due to thermonuclear reactions.

## 24.4   Appearance

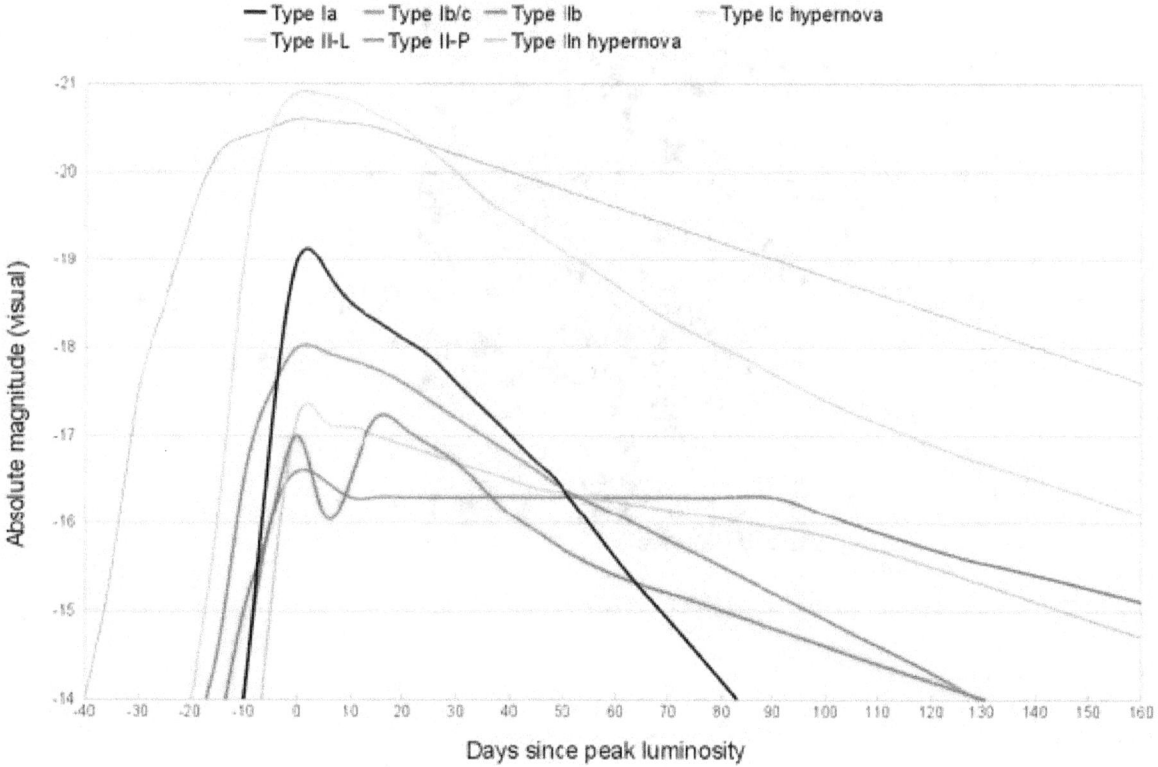

*Light curves compared to normal supernovae*

### 24.4.1   Luminosity

Pair instability supernovae are popularly thought to be highly luminous. This is only the case for the most massive progenitors since the luminosity depends strongly on the ejected mass of radioactive $Ni_{56}$. They can have peak luminosities of over $10^{37}$ W, brighter than type Ia supernovae, but at lower masses peak luminosities are less than $10^{35}$ W, comparable to or less than typical type II supernovae.[7]

### 24.4.2   Spectrum

The spectra of pair instability supernovae depend on the nature of the progenitor star. Thus they can appear as type II or type Ib/c supernova spectra. Progenitors with a significant remaining hydrogen envelope will produce a type II supernova, those with no hydrogen but significant helium will produce a type Ib, and those with no hydrogen and virtually no helium will produce a type Ic.[7]

### 24.4.3   Light curves

In contrast to the spectra, the light curves are quite different from the common types of supernova. The light curves are highly extended, with peak luminosity occurring months after onset.[7] This is due to the extreme amounts of $^{56}$Ni expelled, and the optically dense ejecta, as the star is entirely disrupted.

### 24.4.4   Remnant

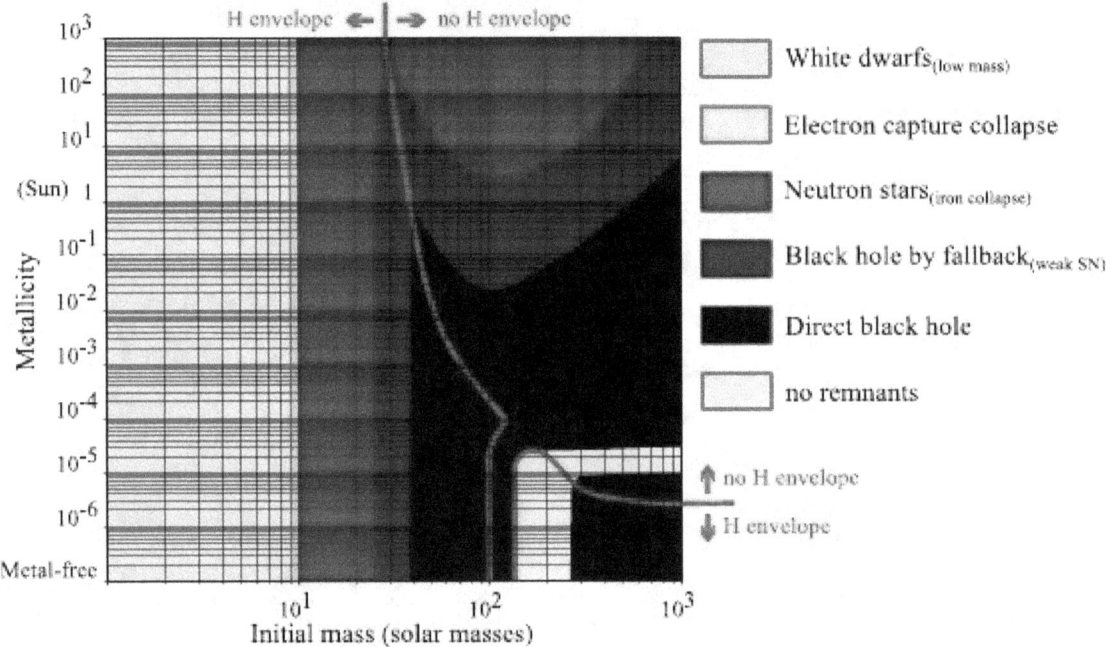

*Remnants of single massive stars*

Pair instability supernovae completely destroy the progenitor star and do not leave behind a neutron star or black hole. The entire mass of the star is ejected, so a nebular remnant is produced and many solar masses of heavy elements are returned to interstellar space.

## 24.5   See also

- Pulsational pair-instability supernova
- Pair creation
- Thermonuclear supernova

- Thermal runaway

# 24.6 References

[1] Fraley, Gary S. (1968). "Supernovae Explosions Induced by Pair-Production Instability". *Astrophysics and Space Science* **2** (1): 96–114. Bibcode:1968Ap&SS...2...96F. doi:10.1007/BF00651498.

[2] Gal-Yam, A.; Mazzali, P.; Ofek, E. O.; et al. (3 December 2009), "Supernova 2007bi as a pair-instability explosion". *Nature* **462**: 624–627, arXiv:1001.1156, Bibcode:2009Natur.462..624G, doi:10.1038/nature08579, PMID 19956255

[3] Cooke, J.; Sullivan, M.; Gal-Yam, A.; Barton, E. J.; Carlberg, R. G.; Ryan-Weber, E. V.; Horst, C.; Omori, Y.; Díaz, C. G. (2012). "Superluminous supernovae at redshifts of 2.05 and 3.90". *Nature* **491** (7423): 228–231. doi:10.1038/nature11521. PMID 23123848.

[4] Belkus, H.; Van Bever, J.; Vanbeveren, D. (2007). "The Evolution of Very Massive Stars". *The Astrophysical Journal* **659** (2): 1576–1581. arXiv:astro-ph/0701334. Bibcode:2007ApJ...659.1576B. doi:10.1086/512181.

[5] Smith, Nathan; Li, Weidong; Foley, Ryan J.; Wheeler, J. Craig; et al. (2007). "SN 2006gy: Discovery of the Most Luminous Supernova Ever Recorded, Powered by the Death of an Extremely Massive Star like η Carinae". *The Astrophysical Journal* **666** (2): 1116–1128. arXiv:astro-ph/0612617. Bibcode:2007ApJ...666.1116S. doi:10.1086/519949.

[6] Fryer, C.L.; Woosley, S. E.; Heger, A. (2001). "Pair-Instability Supernovae, Gravity Waves, and Gamma-Ray Transients". *The Astrophysical Journal* **550** (1). arXiv:astro-ph/0007176. Bibcode:2001ApJ...550..372F. doi:10.1086/319719.

[7] Kasen, D.; Woosley, S. E.; Heger, A. (2011). "Pair Instability Supernovae: Light Curves, Spectra, and Shock Breakout" (pdf). *The Astrophysical Journal* **734** (2): 102. arXiv:1101.3336. Bibcode:2011ApJ...734..102K. doi:10.1088/0004-637X/734/2/102.

# Chapter 25

# Photodisintegration

**Photodisintegration** (also called **phototransmutation**) is a physical process in which an extremely high energy gamma ray is absorbed by an atomic nucleus and causes it to enter an excited state, which immediately decays by emitting a subatomic particle. A single proton, neutron or alpha particle[1] is effectively knocked out of the nucleus by the incoming gamma ray. Photodisintegration is endothermic (energy absorbing) for atomic nuclei lighter than iron and sometimes exothermic (energy releasing) for atomic nuclei heavier than iron. Photodisintegration is responsible for the nucleosynthesis of at least some heavy, proton rich elements via p-process which takes place in supernovae.

## 25.1 Photodisintegration of deuterium

A photodisintegration reaction

was used by James Chadwick and Maurice Goldhaber to measure the proton-neutron mass difference.[2] This experiment proves that a neutron is not a bound state of a proton and an electron,[3] as had been proposed by Ernest Rutherford.

## 25.2 Photodisintegration of beryllium

The photodisintegration of beryllium by gamma rays emitted by antimony-124 is used as a source for thermal neutrons.[4][5]

## 25.3 Hypernovae

In explosions of very large stars (250 or more times the mass of Earth's Sun), photodisintegration is a major factor in the supernova event. As the star reaches the end of its life, it reaches temperatures and pressures where photodisintegration's energy-absorbing effects temporarily reduce pressure and temperature within the star's core. This causes the core to start to collapse as energy is taken away by photodisintegration, and the collapsing core leads to the formation of a black hole. A portion of mass escapes in the form of relativistic jets, which could have "sprayed" the first metals into the universe.[6][7]

## 25.4 Photofission

Photofission is a similar but distinct process, in which a nucleus, after absorbing a gamma ray, undergoes nuclear fission (splits into two fragments of nearly equal mass).

# 25.5 References

[1] Clayton, D. D. (1984). *Principles of Stellar Evolution and Nucleosynthesis*. University of Chicago Press. p. 519. ISBN 978-0-22-610953-4.

[2] Chadwick, J.; Goldhaber, M. (1934). "A nuclear 'photo-effect': disintegration of the diplon by γ rays". *Nature* **134** (3381): 237–238. Bibcode:1934Natur.134..237C. doi:10.1038/134237a0.

[3] Livesy, D. L. (1966). *Atomic and Nuclear Physics*. Waltham, MA: Blaisdell. p. 347. LCCN 65017961.

[4] Lalovic, M.; Werle, H. (1970). "The energy distribution of antimonyberyllium photoneutrons". *Journal of Nuclear Energy* **24** (3): 123–132. Bibcode:1970JNuE...24..123L. doi:10.1016/0022-3107(70)90058-4.

[5] Ahmed, S. N. (2007). *Physics and Engineering of Radiation Detection*. p. 51. ISBN 978-0-12-045581-2.

[6] Fryer, C. L.; Woosley, S. E.; Heger, A. (2001). "Pair-Instability Supernovae, Gravity Waves, and Gamma-Ray Transients". *The Astrophysical Journal* **550** (1): 372–382. arXiv:astro-ph/0007176. Bibcode:2001ApJ...550..372F. doi:10.1086/319719.

[7] Heger, A.; Fryer, C. L.; Woosley, S. E.; Langer, N.; Hartmann, D. H. (2003). "How Massive Single Stars End Their Life". *The Astrophysical Journal* **591** (1): 288–300. arXiv:astro-ph/0212469. Bibcode:2003ApJ...591..288H. doi:10.1086/375341.

# Chapter 26

# Metallicity

In astronomy and physical cosmology, the **metallicity** or **Z**, is the fraction of mass of a star or other kind of astronomical object, beyond hydrogen (**X**) and helium (**Y**).[1][2] Most of the physical matter in the universe is in the form of hydrogen and helium, so astronomers conveniently use the blanket term "metals" to refer to all other elements.[3] For example, stars or nebulae that are relatively rich in carbon, nitrogen, oxygen, and neon would be "metal-rich" in astrophysical terms, even though those elements are non-metals in chemistry. This term should not be confused with the usual physical definition of solid metals.

Metallicity within stars and other astronomical objects is an approximate estimation of their chemical abundances that change over time by the mechanisms of stellar evolution,[4] and therefore provide an indication of their age.[5] In cosmological terms, the universe is also chemically evolving. According to the Big Bang Theory, the early universe first consisted of hydrogen and helium, with trace amounts of lithium and beryllium, but no heavier elements. Through the process of stellar evolution, where stars at the end of their lives discard most of their mass by stellar winds or explode as supernovae, the metal content of the Galaxy and the universe increases.[6] It is postulated that older generations of stars generally have lower metallicities than those of younger generations.[7]

Observed changes in the chemical abundances of different types of stars, based on the spectral peculiarities that were later attributed to metallicity, led astronomer Walter Baade in 1944 to propose the existence of two different populations of stars.[8] These became commonly known as **Population I** (metal-rich) and **Population II** (metal-poor) stars. A third stellar population was introduced in 1978, known as **Population III** stars.[9][10][11] These extremely metal-poor stars were theorised to have been the 'first-born' stars created in the universe.

## 26.1   Definition

Stellar composition, as determined by spectroscopy, is usually simply defined by the parameters **X**, **Y** and **Z**. Here **X** is the fractional percentage of hydrogen, **Y** is the fractional percentage of helium, and all the remaining chemical elements as the fractional percentage, **Z**. It is simply defined as;

$$X + Y + Z = 1.00$$

In most stars, nebulae and other astronomical sources, hydrogen and helium are the two dominant elements. The hydrogen mass fraction is generally expressed as $X \equiv \frac{m_H}{M}$ where $M$ is the total mass of the system and $m_H$ the fractional mass of the hydrogen it contains. Similarly, the helium mass fraction is denoted as $Y \equiv \frac{m_{He}}{M}$ . The remainder of the elements are collectively referred to as 'metals', and the metallicity—the mass fraction of elements heavier than helium—can be calculated as

$$Z = \sum_{i > He} \frac{m_i}{M} = 1 - X - Y.$$

*The globular cluster M80. Stars in globular clusters are mainly older metal-poor members of Population II.*

For the Sun, these parameters are often assumed to have the following approximate values,[12] although recent research shows that lower values for $Z_{sun}$ might be more appropriate:[13][14]

The metallicity of many astronomical objects cannot be measured directly. Instead, proxies are used to obtain an indirect estimate. For example, an observer might measure the iron content of a galaxy (for example using the brightness of an iron emission line) directly, then compare that value with models to estimate the total metallicity.

### 26.1.1 Calculation

The overall stellar metallicity is often defined using the total iron-content of the star "[Fe/H]", since iron is not only the most abundant heavy element, but it is among the easiest to measure with spectral data in the visible spectrum. The abundance ratio is defined as the logarithm of the ratio of a star's iron abundance compared to that of the Sun and is expressed thus:

$$[\text{Fe/H}] = \log_{10}\left(\frac{N_{Fe}}{N_H}\right)_{star} - \log_{10}\left(\frac{N_{Fe}}{N_H}\right)_{sun}$$

where $N_{Fe}$ and $N_H$ are the number of iron and hydrogen atoms per unit of volume respectively. The unit often used for metallicity is the "dex" which is a (now-deprecated) contraction of 'decimal exponent'.[15] By this formulation, stars with a higher metallicity than the Sun have a positive logarithmic value, whereas those with a lower metallicity than the Sun have a negative value. The logarithm is based on powers of 10; stars with a value of +1 have ten times the metallicity of the Sun ($10^1$). Conversely, those with a value of −1 have one-tenth ($10^{-1}$), while those with a value of −2 have a hundredth ($10^{-2}$), and so on.[3] Young Population I stars have significantly higher iron-to-hydrogen ratios than older Population II stars. Primordial Population III stars are estimated to have a metallicity of less than −6.0, that is, less than a millionth of the abundance of iron in the Sun.

The same sort of notation is used to express differences in the individual elements from the solar proportion. For example, the notation "[O/Fe]" represents the difference in the logarithm of the star's oxygen abundance compared to that of the Sun and the logarithm of the star's iron abundance compared to the Sun:

$$[\text{O/Fe}] = \log_{10}\left(\frac{N_O}{N_{Fe}}\right)_{star} - \log_{10}\left(\frac{N_O}{N_{Fe}}\right)_{sun}$$

$$= \left[\log_{10}\left(\frac{N_O}{N_H}\right)_{star} - \log_{10}\left(\frac{N_O}{N_H}\right)_{sun}\right] - \left[\log_{10}\left(\frac{N_{Fe}}{N_H}\right)_{star} - \log_{10}\left(\frac{N_{Fe}}{N_H}\right)_{sun}\right].$$

The point of this notation is that if a mass of gas is diluted with pure hydrogen, then its [Fe/H] value will decrease (because there are fewer iron atoms per hydrogen atom after the dilution), but for all other elements X, the [X/Fe] ratios will remain unchanged. By contrast, if a mass of gas is polluted with some amount of pure oxygen, then its [Fe/H] will remain unchanged but its [O/Fe] ratio will increase. In general, a given stellar nucleosynthetic process alters the proportions of only a few elements or isotopes, so a star or gas sample with nonzero [X/Fe] values may be showing the signature of particular nuclear processes.

### 26.1.2 Relation between Z and [Fe/H]

These two ways of expressing the *metallic* content of a star are related through the equation:

$$\log_{10}\left(\frac{Z/X}{Z_{sun}/X_{sun}}\right) = [\text{M/H}]$$

where [M/H] is the star's total metal abundance (i.e. all elements heavier than helium) defined as a more general expression than the one for [Fe/H]:

$$[\text{M/H}] = \log_{10}\left(\frac{N_M}{N_H}\right)_{star} - \log_{10}\left(\frac{N_M}{N_H}\right)_{sun}.$$

The iron abundance and the total metal abundance are often assumed to be related through a constant A as:

$$[\text{M/H}] = A * [\text{Fe/H}]$$

where A assumes values between 0.9 and 1. Using the formulas presented above, the relation between Z and [Fe/H] can finally be written as:

$$\log_{10}\left(\frac{Z/X}{Z_{sun}/X_{sun}}\right) = A * [\text{Fe/H}].$$

## 26.2 See also

- Abundance of the chemical elements

- Galaxy formation and evolution

- GRB 090423, the most distant seen, presumably from a low-metallicity progenitor

- Cosmos Redshift 7, a galaxy that reportedly contains Population III stars

- Metallicity distribution function

- Stellar evolution

## 26.3 References

[1] D. Kunth & G. Östlin (2000). "The Most Metal-poor Galaxies" **10** (1). The Astronomy and Astrophysics Review. Retrieved 3 February 2015.

[2] W. Sutherland (26 March 2013). "The Galaxy. Chapter 4. Galactic Chemical Evolution" (PDF). Retrieved 13 January 2015.

[3] John C. Martin. "What we learn from a star's metal content". *New Analysis RR Lyrae Kinematics in the Solar Neighborhood*. Retrieved September 7, 2005.

[4] McWilliam, Andrew (26 March 2013). "Abundance Ratios and galactic Chemical Evolution". Retrieved 13 January 2015.

[5] McWilliam, Andrew (1997-01-01). "Abundance Ratios and galactic Chemical Evolution : Age-Metallicity Relation". Retrieved 2015-01-13.

[6] F. Hoyle (1954). "On Nuclear Reactions Occurring in Very Hot Stars. I. the Synthesis of Elements from Carbon to Nickel.". *Astrophysical Journal Supplement* **1**: 121–146. Bibcode:1954ApJS....1..121H. doi:10.1086/190005.

[7] McWilliam, Andrew (1997-01-01). "Abundance Ratios and galactic Chemical Evolution : Introduction". Retrieved 2015-01-13.

[8] W. Baade (1944). "The Resolution of Messier 32, NGC 205, and the Central Region of the Andromeda Nebula.". *Astrophysical Journal* **100**: 121–146. Bibcode:1944ApJ...100..137B. doi:10.1086/144650.

[9] M.J. Rees (1978). "Origin of pregalactic microwave background". *Nature* **275**: 35–37. Bibcode:1978Natur.275...35R. doi:10.1038/275035a0.

[10] S.D.M. White; M.J. Rees (1978). "Core condensation in heavy halos - A two-stage theory for galaxy formation and clustering". *Monthly Notices Royal Astronomical Society* **183**: 341–358. Bibcode:1978MNRAS.183..341W. doi:10.1093/mnras/183.3.341.

[11] J.L. Puget; J. Heyvaerts (1980). "Population III stars and the shape of the cosmological black body radiation". *Astronomy and Astrophysics* **83**: L10–L12. Bibcode:1980A&A....83L..10P.

[12] A. Unsöld; B. Baschek; R.C. Smith; C.A. Hein (1983). *The New Cosmos*. Springer New York. doi:10.1007/978-1-4757-1791-4. ISBN 978-0-387-90886-1.

[13] "The new solar abundances - Part I: the observations". Communications in Asteroseismology. January 2006. Retrieved 2013-06-25.

[14] "Solar Heavy-Element Abundance: Constraints from Frequency Separation Ratios of Low-Degree p-Modes". The Astrophysical Journal. November 2007. Retrieved 2013-06-30.

[15] R. Rowlett; et al. (July 2005). "How Many? A Dictionary of Units of Measurement". University of North Carolina. Retrieved 3 February 2015.

- Salvaterra, R.; Ferrara, A.; Schneider, R. (2004). "Induced formation of primordial low-mass stars". *New Astronomy* **10** (2): 113. arXiv:astro-ph/0304074. Bibcode:2004NewA...10..113S. doi:10.1016/j.newast.2004.06.003.

- A. Heger; S. E. Woosley (2002). "The Nucleosynthetic Signature of Population III". *Astrophysical Journal* **567** (1): 532–543. arXiv:astro-ph/0107037. Bibcode:2002ApJ...567..532H. doi:10.1086/338487.

## 26.4   Sources

- Page 593-In Quest of the Universe Fourth Edition Karl F. Kuhn Theo Koupelis. Jones and Bartlett Publishers Canada. 2004. ISBN 0-7637-0810-0

- Bromm, Volker; Larson, Richard B. (2004). "THE FIRST STARS". *Annual Review of Astronomy and Astrophysics* **42**: 79–118. arXiv:astro-ph/0311019. Bibcode:2004ARA&A..42...79B.doi:10.1146/annurev.astro.42.053102.1

÷

# Chapter 27

# Supergiant

This article is about the type of star. For supergiant planets, see giant planet. For supergiant amphipods, see Alicella. For the development company, see Supergiant Games.

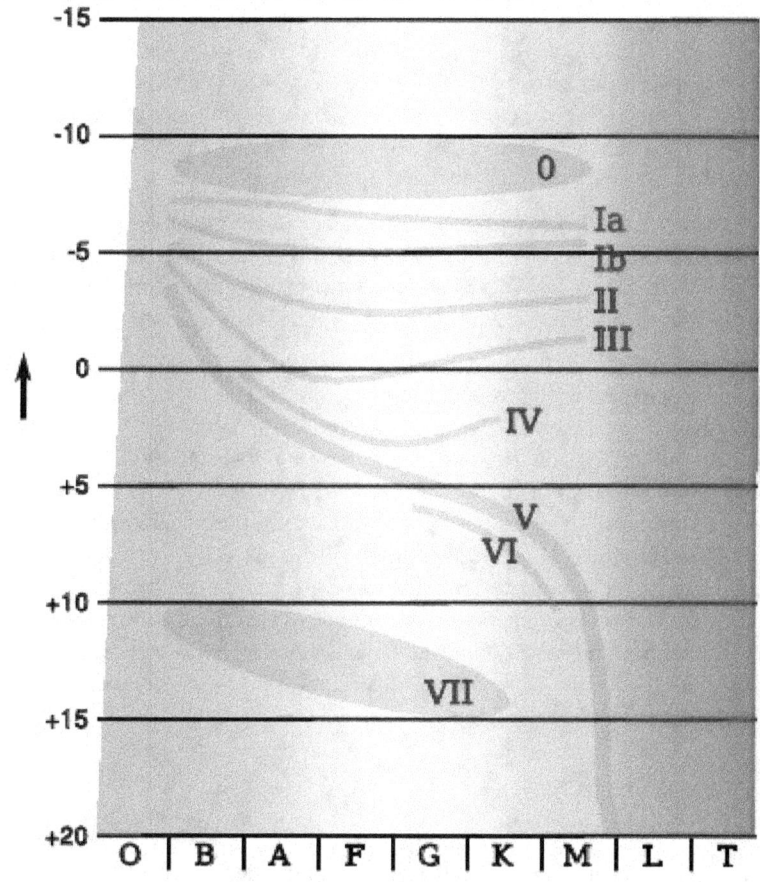

Hertzsprung–Russell diagram
Spectral type
Brown dwarfs
White dwarfs
Red dwarfs
Subdwarfs
Main sequence

("dwarfs")
Subgiants
Giants
Bright giants
Supergiants
Hypergiants
absolute
magni-
tude
(MV)

**Supergiants** are among the most massive and most luminous stars. They occupy the top region of the Hertzsprung–Russell diagram with bolometric absolute magnitudes between −5 and −12 and temperatures from about 3,500K to over 20,000K.

# 27.1    Properties

Supergiants have masses from 8 to 12 times the Sun ($M\odot$) upwards, and luminosities from about 10,000 to over a million times the Sun ($L\odot$). They vary greatly in radius, usually from 30 to 500, or even in excess of 1,000 solar radii ($R\odot$). They are massive enough to begin core helium burning gently before the core becomes degenerate, without a flash, and without the strong dredge-ups that lower-mass stars experience. They go on to successively ignite heavier elements, usually all the way to iron. Also because of their high masses they are destined to explode as supernovae.

The Stefan-Boltzmann law dictates that the relatively cool surfaces of red supergiants radiate much less energy per unit area than those of blue supergiants; thus, for a given luminosity red supergiants are larger than their blue counterparts. Radiation pressure limits the largest cool supergiants to around 1,500 $R\odot$ and the most massive hot supergiants to around a million $L\odot$ (MV around −9). Stars near and occasionally beyond these limits become unstable, pulsate, and experience rapid mass loss.

Supergiants are categorized on the basis of their spectra. Supergiants occur in every spectral class from young blue class O supergiants to highly evolved red class M supergiants. Because they are enlarged compared to main-sequence and giant stars of the same spectral type, they have lower surface gravities and changes can be observed in their line profiles. Supergiants are also evolved stars with higher levels of heavy elements than main-sequence stars. This is the basis of the MK luminosity system which assigns stars to luminosity classes purely from observing their spectra. In addition to the line changes due to low surface gravity and fusion products, the most luminous stars have high mass-loss rates and resulting clouds of expelled circumstellar materials which can produce emission lines, P Cygni profiles, or forbidden lines. The MK system assigns stars to luminosity classes: **Ib** for supergiants; **Ia** for luminous supergiants; and **0** (zero) or **Ia⁺** for hypergiants. In reality there is very much of a continuum rather than well defined bands for these classifications, and classifications such as **Iab** are used for intermediate luminosity supergiants. Supergiant spectra are frequently annotated to indicate spectral peculiarities, for example B2Iae or F8Iabpec.

## 27.1.1    Categorisation of stars

Although the term supergiant does not have a single concrete definition, there are several other categories of evolved star which are not, not always, or no longer, generally treated as supergiants. Definitions based solely on spectral type, solely on luminosity, or on criteria such as size, mass, composition, internal structure, or life stage, will all include somewhat different stars under the heading of supergiant.

Asymptotic-giant-branch (AGB) stars are highly evolved lower-mass red giants with luminosities almost as high as red supergiants, but because of their low mass, being in a different stage of development (helium shell burning), and their lives ending in a different way (planetary nebula and white dwarf rather than supernova), astrophysicists prefer to keep them separate. The dividing line becomes blurred at around 7–10 $M\odot$ (or as high as 12 $M\odot$ in some models[11]) where stars start to undergo limited fusion of elements heavier than helium. Specialists studying these stars often refer to them as super AGB stars, since they have many properties in common with AGB such as thermal pulsing. Others describe them as low-mass supergiants since they start to burn elements heavier than helium and can explode as supernovae.[2] These intermediate stars develop oxygen–magnesium–neon cores that either lead to the rare oxygen–neon white dwarf or an electron-capture supernova.

Wolf–Rayet stars are also high-mass luminous evolved stars, hotter than most supergiants and smaller, visually less bright but often more luminous because of their high temperatures. They have spectra dominated by helium and other heavier elements, usually showing little or no hydrogen, which is a clue to their nature as stars even more evolved than supergiants. Just as the AGB stars occur in almost the same region of the HR diagram as red supergiants, Wolf–Rayet stars can occur in the same region of the HR diagram as the hottest blue supergiants and main-sequence stars.

The most massive and luminous main-sequence stars are almost indistinguishable from the supergiants they quickly evolve into. They have almost identical temperatures and very similar luminosities, and only the most detailed analyses can distinguish the spectral features that show they have evolved away from the narrow early O-type main-sequence to the nearby area of early O-type supergiants. Such early O-type supergiants share many features with WNLh Wolf–Rayet stars and are sometimes designated as slash stars, intermediates between the two types.

Luminous blue variables (LBVs) are a type of star that occur in the same region of the HR diagram as blue supergiants, but are generally classified separately. They are evolved, expanded, massive, and luminous stars, often hypergiants, but they have very specific spectral variability which defies the assignment of a standard spectral type. LBVs only observed at a particular time, or over a period of time when they are stable, may simply be designated as hot supergiants, or as candidate LBVs due to their luminosity.

Hypergiants are frequently treated as a different category of star from supergiants, although in all important respects they are just a more luminous category of supergiant. They are evolved, expanded, massive and luminous stars like supergiants, but at the most massive and luminous extreme, and with particular additional properties of undergoing high mass-loss due to their extreme luminosities and instability. Generally only the more evolved supergiants show hypergiant properties since their instability increases after high mass-loss and some increase in luminosity.

Some B(e)-type stars have temperatures and luminosities equivalent to blue supergiants, although other B(e)-type stars are clearly different. Some researchers distinguish the B(e) objects as separate from supergiants, while others prefer to define the particularly massive and luminous B(e)-type stars a subgroup of supergiants. The latter has become more common with the understanding that the B(e) phenomenon arises separately in a number of distinct types of stars, including some that are clearly just a phase in the life of supergiants.

## 27.1.2 Variability

While most supergiants show some degree of photometric variability, such as Alpha Cygni variables, semiregular variables, and irregular variables, there are certain well defined types of variables amongst the supergiants. The instability strip crosses the region of supergiants, and specifically many Classical Cepheid variables are supergiants. The same region of instability extends to include the even more luminous yellow hypergiants, an extremely rare and short-lived class of luminous supergiant. Many R Coronae Borealis variables are yellow supergiants although not all, but this variability is due to their unusual chemical composition rather than a physical instability.

Further types of variable stars, such as RV Tauri variables and PV Telescopii variables, are often described as supergiants. Although the rare RV Tau stars are frequently assigned spectral types with a supergiant luminosity class on account of their low surface gravity, they are lower-mass lower-luminosity post-AGB stars. Likewise the even rarer PV Tel variables are often described as supergiants, but have lower luminosities than supergiants and peculiar B(e) class spectra extremely deficient in hydrogen. Possibly they are also post-AGB objects, or perhaps "born-again" AGB stars.

The LBVs already mentioned are variable with multiple semi-regular periods and less predictable eruptions and giant outbursts. Although they are essentially supergiants in nature, extremely luminous, massive, evolved stars with expanded outer layers, they are so distinctive and unusual that they are often treated as a separate category without being referred to as supergiants or given a supergiant spectral type. Often their spectral type will be given just as "LBV" because they have peculiar and highly variable spectral features, with temperatures varying from about 8,000 K in outburst up to 20,000 K or more when "quiescent".

## 27.2   Evolution

Main article: Stellar evolution

O type main-sequence stars and the most massive of the B type blue-white stars become supergiants. Because of their extreme masses they have short lifespans of 30 million years down to a few hundred thousand years.[3] They are mainly observed in young galactic structures such as open clusters, the arms of spiral galaxies, and in irregular galaxies. They are less abundant in spiral galaxy bulges, and are rarely observed in elliptical galaxies, or globular clusters, which are composed mainly of old stars.

Supergiants develop when massive main-sequence stars run out of hydrogen in their cores. They then start to expand, just like lower-mass stars, but unlike lower-mass stars, they begin to fuse helium in the core almost immediately. This means that they do not increase their luminosity as dramatically as lower-mass stars and they progress nearly horizontally across the HR diagram to become red supergiants. Also, unlike lower-mass stars, red supergiants are massive enough to fuse elements heavier than helium, so they do not puff off their atmospheres as planetary nebulae when their helium becomes depleted. Furthermore, they cannot lose enough mass to form a white dwarf, so will leave behind a neutron star or black hole remnant, usually after a core collapse supernova explosion.

Stars more massive than about 40 $M\odot$ cannot expand into a red supergiant. They burn too quickly and lose their outer layers too quickly, so they reach the blue supergiant stage, or perhaps yellow hypergiant, and then return to become hotter stars. The most massive stars, above about 100 $M\odot$, hardly move at all from their position as O main-sequence stars. These stars convect so efficiently that they mix hydrogen from the surface right down to the core. They continue to fuse hydrogen until it is almost entirely depleted throughout the star, then very rapidly evolve through a series of stages of very similar hot and luminous stars, if supergiants, slash stars, WNh-, WN-, and possibly WC- or WO-type stars. They are expected to explode as supernovae, but it is not clear how far they evolve before this happens. The existence of these supergiants still burning hydrogen in their cores may necessitate a slightly more complex definition of supergiant: a massive star with increased size and luminosity due to fusion products building up, but still with some hydrogen remaining.[4]

The first stars in the universe are thought to have been considerably brighter and more massive than the stars in the modern universe. These stars were part of the theorized population III of stars. Their existence is necessary to explain observations of elements other than hydrogen and helium in quasars. Although they may have been larger and more luminous than any supergiant known today, their structure was quite different, with reduced convection and less mass loss. Their very short lives are likely to have ended in violent photodisintegration or pair instability supernovae.

## 27.3   Supernova progenitors

Main article: Supernova

Most type II supernova progenitors are thought to be red supergiants, while the less common type Ib/c supernovae are produced by hotter Wolf–Rayet stars that have completely lost more of their hydrogen atmosphere.[5] Almost by definition, supergiants are destined to end their lives violently. Stars that are large enough to start fusing elements heavier than helium just do not seem to have any way to lose enough mass to avoid catastrophic core collapse, although some of them may collapse almost without trace into their own central black holes.

However, the simple "onion" models showing red supergiants inevitably developing to an iron core and then exploding have been shown to be much too simplistic. The progenitor for the unusual type II Supernova 1987A was a blue supergiant,[6] thought to have already passed through the red supergiant phase of its life, and this is now known to be far from an exceptional situation. Much research is now focused on how blue supergiants can explode as a supernova and when red supergiants can survive to become hotter supergiants again.[7]

*Direct image of the star UY Scuti, a red supergiant which is one of the largest known stars.*

## 27.4 Well known examples

Supergiants are rare and short-lived stars, but their high luminosity means that there are many naked eye examples, including some of the brightest stars in the sky. Rigel is the brightest star in the constellation Orion and a typical blue-white supergiant, Deneb is the brightest star in Cygnus and a white supergiant, Delta Cephei is the famous prototype Cepheid variable and a yellow supergiant, while Betelgeuse and Antares are red supergiants. μ Cephei is one of the reddest stars visible to the naked eye and one of the largest in the galaxy. Rho Cassiopeiae is a naked eye variable, a yellow hypergiant, and one of the most luminous naked eye stars.

## 27.5 See also

- Giant star

- Red giant

- Red supergiant

- Blue supergiant

- Hypergiant

- List of stellar angular diameters

- Yellow supergiant

## 27.6 References

[1] Siess, L. (2006). "Evolution of massive AGB stars". *Astronomy and Astrophysics* **448**(2): 717–729. Bibcode:2006A&A...448. doi:10.1051/0004-6361:20053043.

[2] Poelarends, A. J. T.; Herwig, F.; Langer, N.; Heger, A. (2008). "The Supernova Channel of Super-AGB Stars". *The Astrophysical Journal* **675**: 614. arXiv:0705.4643. Bibcode:2008ApJ...675..614P. doi:10.1086/520872.

[3] Richmond, Michael. "Stellar evolution on the main sequence". Retrieved 2006-08-24.

[4] Sylvia Ekström; Cyril Georgy; Georges Meynet; Jose Groh; Anahí Granada (2013). "Red supergiants and stellar evolution". arXiv:1303.1629v1 [astro-ph.SR].

[5] Groh, Jose H.; Georges Meynet; Cyril Georgy; Sylvia Ekstrom (2013). "Fundamental properties of core-collapse Supernova and GRB progenitors: Predicting the look of massive stars before death". arXiv:1308.4681v1 [astro-ph.SR].

[6] Lyman, J. D.; Bersier, D.; James, P. A. (2013). "Bolometric corrections for optical light curves of core-collapse supernovae". *Monthly Notices of the Royal Astronomical Society* **437** (4): 3848. doi:10.1093/mnras/stt2187.

[7] Van Dyk, S. D.; Li, W.; Filippenko, A. V. (2003). "A Search for Core-Collapse Supernova Progenitors in Hubble Space Telescope Images". *Publications of the Astronomical Society of the Pacific* **115** (803): 1. arXiv:astro-ph/0210347. Bibcode:2003PAS doi:10.1086/345748.

- Tempesti, Piero, ed. (1979). *Enciclopedia dell'Astronomia*. Curcio.

- http://alobel.freeshell.org/rcas.html

- http://www.solstation.com/x-objects/rho-cas.htm

# Chapter 28

# Hypergiant

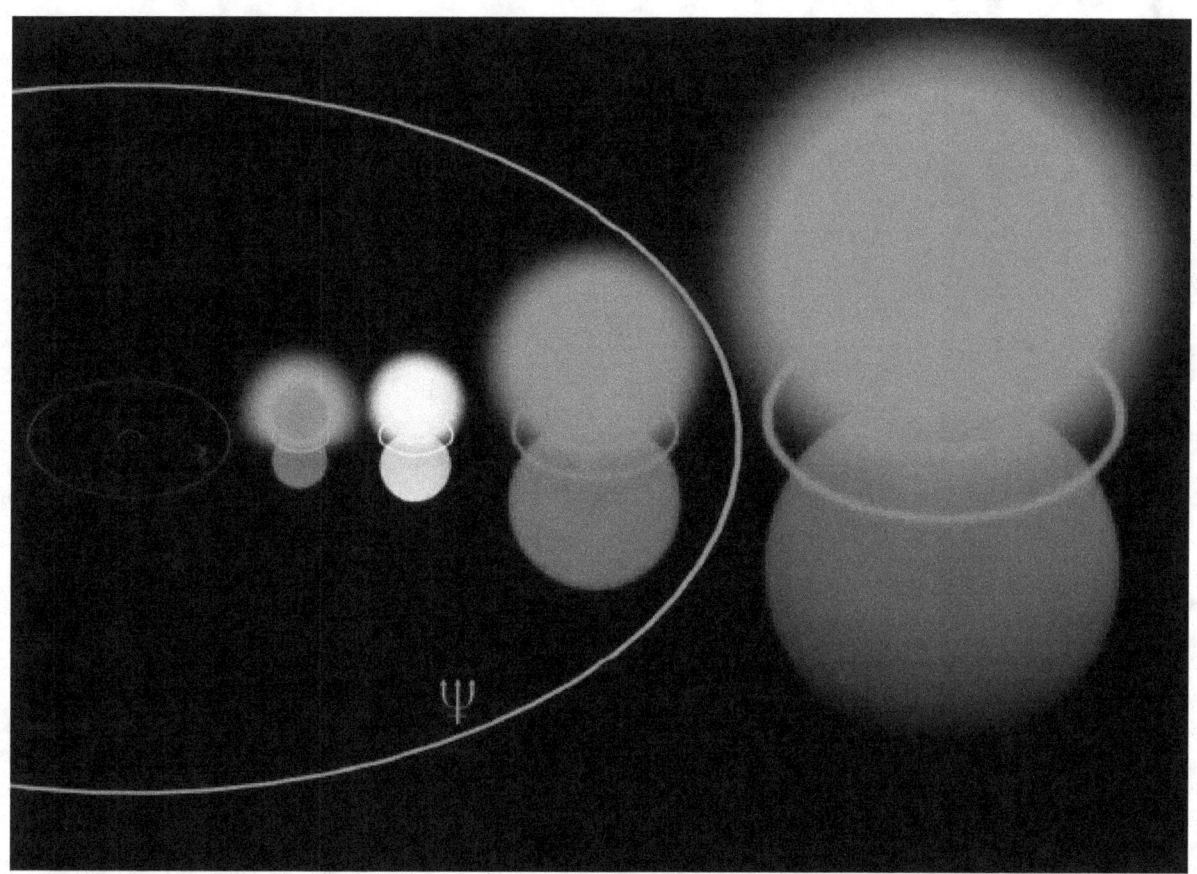

*Comparison of a blue hypergiant, yellow hypergiant, red supergiant, and red hypergiant superimposed on an outline of the Solar System. The blue half-ring centered near the left edge represents the orbit of Neptune.*

A **hypergiant** (luminosity class **0** or **Ia⁺**) is a star with an enormous luminosity showing signs of a very high rate of mass loss.

The word "hypergiant" is commonly used as a loose term for the most luminous stars found, even though there are more precise definitions. In 1956, the astronomers Feast and Thackeray used the term super-supergiant (later changed into hypergiant) for stars with an absolute magnitude brighter than $MV = -7$ ($MB_{ol}$ will be larger for very cool and very hot stars, for example at least −9.7 for a B0 hypergiant). In 1971, Keenan suggested that the term would be used only for supergiants showing at least one broad emission component in Hα, indicating an extended stellar atmosphere or a relatively large mass loss rate. The Keenan criterion is the one most commonly used by scientists today.[1] Additionally, hypergiants are expected to have characteristic broadening and red-shifting of their spectral lines producing a distinctive shape known as a P Cygni profile. The use of hydrogen emission is not helpful for defining the coolest hypergiants, and these are largely classified on luminosity since mass loss is almost inevitable for the class.

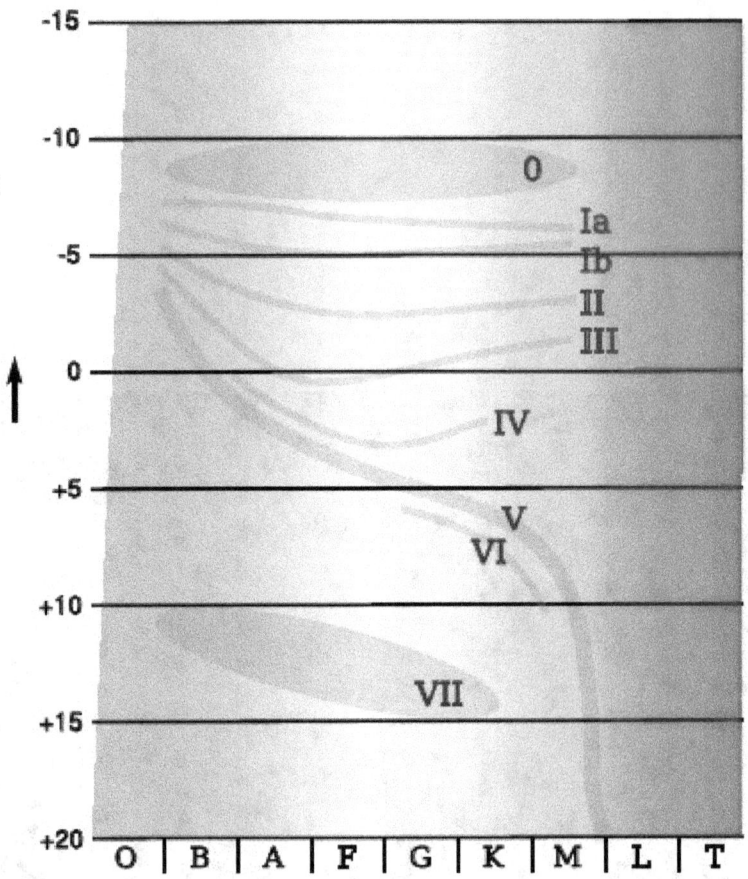

Hertzsprung–Russell diagram
Spectral type
Brown dwarfs
White dwarfs
Red dwarfs
Subdwarfs
Main sequence
("dwarfs")
Subgiants
Giants
Bright giants
Supergiants
Hypergiants
absolute
magni-
tude
(MV)

Many astronomers do not use the term hypergiant, except occasionally for specific well-defined groups such as the yellow hypergiants, so it is common to see the term RSG (red supergiant) or B(e) supergiant (blue supergiant with emission spectra) being used to refer to stars that this article defines as hypergiants. There is an MKK luminosity class 0 (zero) for hypergiants, but this is rarely seen in published spectral classifications. More commonly, hypergiants will be classed as Ia-0, Ia⁺, or even just Iae based solely on the observed spectra. As noted, red supergiants rarely receive these extra spectral classifications. Initial observation of a highly luminous star is insufficient for it to be defined as a hypergiant. That requires the detection of the spectral signatures of atmospheric instability and high mass loss. So it is quite possible for non-hypergiant supergiant stars to have the same or higher luminosity as a hypergiant of the same spectral class.

## 28.1  Formation

Stars with an initial mass above about 25 $M\odot$ quickly move away from the main sequence and increase somewhat in luminosity to become blue supergiants. They cool and enlarge at approximately constant luminosity to become a red supergiant, then contract and increase in temperature as the outer layers are blown away. They may "bounce" backwards and forwards executing one or more "blue loops", still at a fairly steady luminosity, until they explode as a supernova or completely shed their outer layers to become a Wolf–Rayet star. Stars with an initial mass above about 40 $M\odot$ are simply too luminous to develop a stable extended atmosphere and so they never cool sufficiently to become red supergiants. The most massive stars, especially rapidly rotating stars with enhanced convection and mixing, may skip these steps and move directly to the Wolf–Rayet stage.

This means that stars at the top of the Hertzsprung–Russell diagram where hypergiants are found may be newly evolved from the main sequence and still with high mass, or much more evolved post-red supergiant stars that have lost a significant fraction of their initial mass, and these objects cannot be distinguished simply on the basis of their luminosity and temperature. High-mass stars with a high proportion of remaining hydrogen are more stable, while older stars with lower masses and a higher proportion of heavy elements have less stable atmospheres due to increased radiation pressure and decreased gravitational attraction. These are thought to be the hypergiants, near the Eddington limit and rapidly losing mass.

The yellow hypergiants are thought to be generally post-red supergiant stars that have already lost most of their atmospheres and hydrogen. A few more stable high mass yellow supergiants with approximately the same luminosity are known and thought to be evolving towards the red supergiant phase, but these are rare as this is expected to be a rapid transition. Because yellow hypergiants are post-red supergiant stars, there is a fairly hard upper limit to their luminosity at around 500,000 - 750,000 $L\odot$, but blue hypergiants can be much more luminous, sometimes several million $L\odot$.

Almost all hypergiants exhibit variations in luminosity over time due to instabilities within their interiors, but these are small except for two distinct instability regions where luminous blue variables (LBVs) and yellow hypergiants are found. Because of their high masses, the lifetime of a hypergiant is very short in astronomical timescales: only a few million years compared to around 10 billion years for stars like the Sun. Hypergiants are only created in the largest and densest areas of star formation and because of their short lives, only a small number are known despite their extreme luminosity that allows them to be identified even in neighbouring galaxies. The time spent in some phases such as LBVs can be as short as a few thousand years.[2][3]

## 28.2  Stability

As the luminosity of stars increases greatly with mass, the luminosity of hypergiants often lies very close to the Eddington limit, which is the luminosity at which the radiation pressure expanding the star outward equals the force of the star's gravity collapsing the star inward. This means that the radiative flux passing through the photosphere of a hypergiant may be nearly strong enough to lift off the photosphere. Above the Eddington limit, the star would generate so much radiation that parts of its outer layers would be thrown off in massive outbursts; this would effectively restrict the star from shining at higher luminosities for longer periods.

A good candidate for hosting a continuum-driven wind is Eta Carinae, one of the most massive stars ever observed. With an estimated mass of around 130 solar masses and a luminosity four million times that of the Sun, astrophysicists speculate that Eta Carinae may occasionally exceed the Eddington limit.[4] The last time might have been a series of outbursts observed in 1840–1860, reaching mass loss rates much higher than our current understanding of what stellar

*Great nebula in Carina, surrounding Eta Carinae*

winds would allow.[5]

As opposed to line-driven stellar winds (that is, ones driven by absorbing light from the star in huge numbers of narrow spectral lines), continuum driving does not require the presence of "metallic" atoms — atoms other than hydrogen and helium, which have few such lines — in the photosphere. This is important, since most massive stars also are very metal-poor, which means that the effect must work independently of the metallicity. In the same line of reasoning, the continuum driving may also contribute to an upper mass limit even for the first generation of stars right after the Big Bang, which did not contain any metals at all.

Another theory to explain the massive outbursts of, for example, Eta Carinae is the idea of a deeply situated hydrodynamic explosion, blasting off parts of the star's outer layers. The idea is that the star, even at luminosities below the Eddington limit, would have insufficient heat convection in the inner layers, resulting in a density inversion potentially leading to a massive explosion. The theory has, however, not been explored very much, and it is uncertain whether this really can happen.[6]

Another theory associated with hypergiant stars is the potential to form a pseudo-photosphere, that is a spherical optically dense surface that is actually formed by the stellar wind rather than being the true surface of the star. Such a pseudo-photosphere would be significantly cooler than the deeper surface below the outward-moving dense wind. This has been hypothesized to account for the "missing" intermediate-luminosity LBVs and the presence of yellow hypergiants at approximately the same luminosity and cooler temperatures. The yellow hypergiants are actually the LBVs having formed a pseudo-photosphere and so apparently having a lower temperature.[7]

## 28.3 Relationships with Ofpe, WNL, LBV, and other supergiant stars

Hypergiants are evolved, high luminosity, high-mass stars that occur in the same or similar regions of the HR diagram to stars with different classifications. It is not always clear whether the different classifications represent stars with different initial conditions, stars at different stages of an evolutionary track, or is just an artifact of our observations. Model details vary[8][9] but there are many areas of agreement. Some of these distinctions are not necessarily helpful in establishing relationships between different types of stars or the differences between them since they have been developed based on differing criteria and for different purposes.

Although most supergiant stars are less luminous than hypergiants of the same temperature, a few fall in the same luminosity range. Ordinary supergiants lack the strong H emission and broadened spectral lines that indicate rapid mass loss in the hypergiants. Lower mass supergiants do not return from the red supergiant phase, either exploding as supernovae or leaving behind a white dwarf.

Luminous blue variables are a class of highly luminous hot stars that display characteristic spectral variation. They often lie in a "quiescent" zone with hotter stars generally being more luminous, but periodically undergo large surface eruptions and move to a narrow zone where stars of all luminosities have approximately the same temperature, around 8,000K. This "active" zone is near the hot edge of the unstable "void" where yellow hypergiants are found, with some overlap. It is not clear whether yellow hypergiants ever manage to get past the instability void to become LBVs or explode as a supernova.

Blue hypergiants are found in the same parts of the HR diagram as LBVs but do not necessarily show the LBV variations. Some but not all LBVs show the characteristics of hypergiant spectra at least some of the time, but many authors would exclude all LBVs from the hypergiant class and treat them separately. Blue hypergiants that do not show LBV characteristics may be progenitors of LBVs, or vice versa, or both. Lower mass LBVs may be a transitional stage to or from cool hypergiants or are different type of object.

Wolf–Rayet stars are extremely hot stars that have lost much or all of their outer layers. WNL is a term used for late stage (i.e. cooler) Wolf–Rayet stars with spectra dominated by nitrogen. Although these are generally thought to be the stage reached by hypergiant stars after sufficient mass loss, it is possible that a small group of hydrogen-rich WNL stars are actually progenitors of blue hypergiants or LBVs. These are the closely related Ofpe (O-type spectra plus H, He, and N emission lines, and other peculiarities) and WN9 (the coolest nitrogen Wolf–Rayet stars) which may be a brief intermediate stage between high mass main-sequence stars and hypergiants or LBVs. Quiescent LBVs have been observed with WNL spectra and apparent Ofpe/WNL stars have changed to show blue hypergiant spectra. High rotation rates cause massive stars to shed their atmospheres quickly and prevent the passage from main sequence to supergiant, so these directly become Wolf–Rayet stars. Wolf Rayet stars, slash stars, cool slash stars (aka WN10/11), Ofpe, Of⁺, and Of° stars are not considered hypergiants. Although they are luminous and often have strong emission lines, they have characteristic spectra of their own.

## 28.4 Known hypergiants

Hypergiants are difficult to study due to their rarity. Many hypergiants have highly variable spectra, but they are grouped here into broad spectral classes.

### 28.4.1 Luminous blue variables

Some luminous blue variables are classified as hypergiants, during at least part of their cycle of variation:

- Eta Carinae, inside the Keyhole Nebula (NGC 3372) in the southern constellation of Carina. Eta Carinae is extremely massive, possibly as much as 120 to 150 times the mass of the Sun, and is four to five million times as luminous. Possibly a different type of object from the LBVs, or extreme for a LBV.

- P Cygni, in the northern constellation of Cygnus. Prototype for the characteristic LBV spectral lines.

- S Doradus, in a nearby galaxy called the Large Magellanic Cloud, in the southern constellation of Dorado. Prototype variable, LBVs are still often called S Doradus variables.

- The Pistol Star (V4627 Sgr), near the center of the Milky Way, in the constellation of Sagittarius. The Pistol Star is possibly as much as 150 times more massive than the Sun, and is about 1.7 million times more luminous. Considered a candidate LBV, but variability has not been confirmed.

- LBV 1806-20 in the cluster Cl* 1806-20 on the other side of the Milky Way.

- V4029 Sagittarii

- V905 Scorpii

- HD 269700,[7][10] R116 in the LMC

- HD 6884,[11] (R40 in SMC)

## 28.4.2   Blue hypergiants

Usually B-class, occasionally late O or early A:

- Zeta[1] Scorpii, the brightest star of the OB association Scorpius OB1 and a LBV candidate.

- V1429 Aquilae, (= MWC 314) in the constellation of Aquila, LBV candidate with a supergiant companion.

- V430 Scuti[12]

- V452 Scuti, poorly studied LBV candidate[13]

- HD 80077, LBV candidate[12]

- Cygnus OB2-12, which some authors consider an LBV because of its extreme luminosity, although it has not shown the characteristic variability.

- HDE 269128 (R81 in LMC), LBV candidate, eclipsing binary system.[14]

- HD 268835 (R66 in LMC)

- V4030 Sagittarii

- V1768 Cygni[12]

- BP Crucis (Wray 977 or GX 301-2), binary with a pulsar companion[12]

- HT Sagittae[12]

- V2140 Cygni[12]

- HD 37974[15] (R126 in LMC)

- HD 32034[16] (R62 in LMC)

- HD 269781[16] (in LMC)

- HD 269661[16] (R111 in LMC)

*A hypergiant star and its proplyd compared to the Solar System*

- HD 269604[16] (in LMC)

In Westerlund 1:[17]

- W5 (possible Wolf–Rayet)

- W7

- W13 (binary?)

- W33

- W42a

In Galactic Center Region:[18]

- Star 13, type O, LBV candidate

- Star 18, type O, LBV candidate

*Field around yellow hypergiant star HR 5171*

### 28.4.3   Yellow hypergiants

Yellow hypergiants with late A -K spectra.

- Rho Cassiopeiae, in the northern constellation of Cassiopeia, is about 500,000 times as luminous as the Sun.
- V509 Cassiopeiae
- HD 33579 (in LMC)
- IRC+10420 (V1302 Aql)
- IRAS 18357-0604[19]
- HD 7583 (R45 in SMC)
- V766 Centauri (=HR5171A)[20]
- V1427 Aquilae, may just be a closer post-AGB star[21]
- IRAS 17163-3907[22]
- V382 Carinae
- Variable A (in M33)
- HD 268757[15] (R59 in LMC)

In Westerlund 1:[17]

- W4

- W8a

- W12a

- W16a

- W32

- W265

Plus at least two probable cool hypergiants in the recently discovered Scutum Red Supergiant Clusters: F15 and possibly F13 in RSGC1 and Star 49 in RSGC2.

### 28.4.4 Red hypergiants

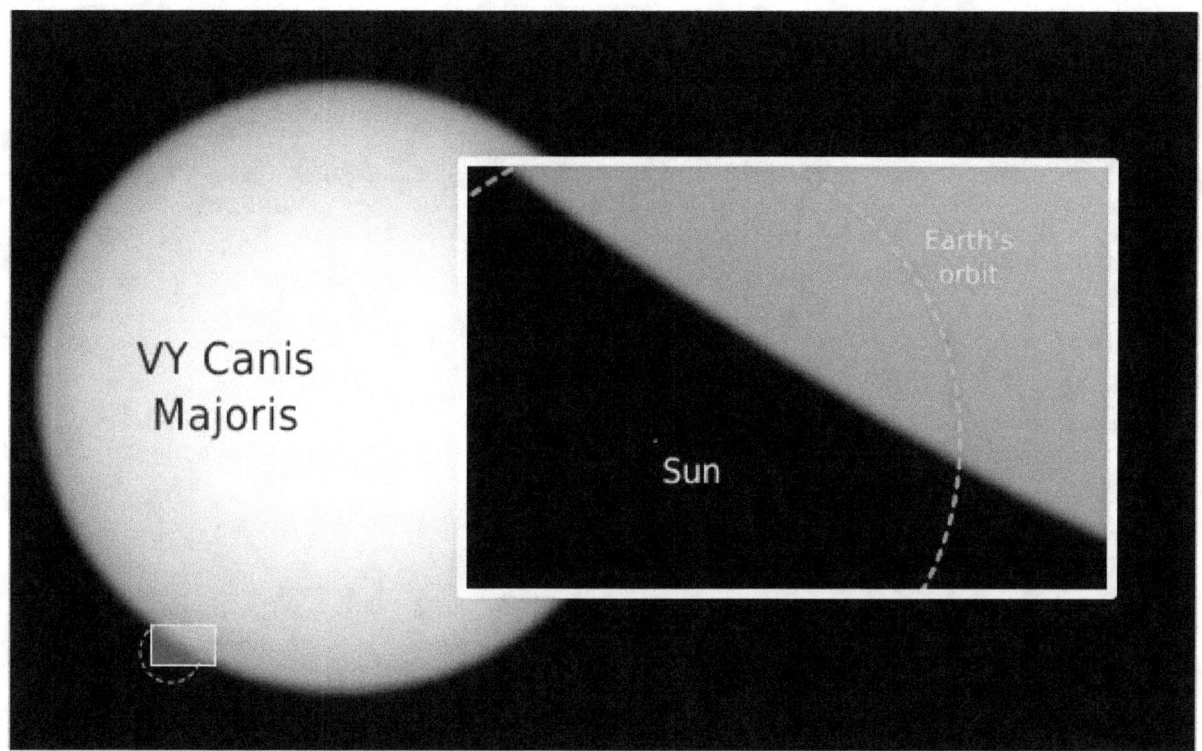

*Size comparison between the Sun and VY Canis Majoris, a hypergiant which is one of the largest known stars*

M type spectra, the largest known stars.

- NML Cygni

- WOH G64

- VX Sagittarii

- VV Cephei

- S Persei

- RW Cephei

- VY Canis Majoris

A survey expected to capture virtually all Magellanic Cloud red hypergiants[23] detected around a dozen M class stars $M_v$−7 and brighter, around a quarter of a million times more luminous than the Sun, and from about 1,000 times the radius of the Sun upwards.

## 28.5   See also

- List of most massive known stars

- Yellow hypergiant

- Hypernova

## 28.6   References

[1] de Jager, C. (1998). "The Yellow Hypergiants". *The Astronomy and Astrophysics Review* 8(3): 145–180. Bibcode:1998A&AR doi:10.1007/s001590050009.

[2] Cyril Georgy; Sylvia Ekström; Georges Meynet; Philip Massey; Levesque; Raphael Hirschi; Patrick Eggenberger; André Maeder (2012). "Grids of stellar models with rotation II. WR populations and supernovae/GRB progenitors at Z = 0.014". *Astronomy & Astrophysics* **542**: A29. arXiv:1203.5243v2 [astro-ph.SR]. Bibcode:2012A&A...542A..29G. doi:10.1051/0004-6361/201118340.

[3] Brott, I.; Evans, C. J.; Hunter, I.; De Koter, A.; Langer, N.; Dufton, P. L.; Cantiello, M.; Trundle, C.; Lennon, D. J.; De Mink, S. E.; Yoon, S. -C.; Anders, P. (2011). "Rotating massive main-sequence stars". *Astronomy & Astrophysics* **530**: A116. arXiv:1102.0766. Bibcode:2011A&A...530A.116B. doi:10.1051/0004-6361/201016114.

[4] Owocki, S. P.; Van Marle, Allard Jan (2007). "Luminous Blue Variables & Mass Loss near the Eddington Limit". *Proceedings of the International Astronomical Union* 3: 71–83. arXiv:0801.2519. Bibcode:2008IAUS..250...71O.doi:10.1017/S1743921308

[5] Owocki, S. P.; Gayley, K. G.; Shaviv, N. J. (2004). "A porosity-length formalism for photon-tiring limited mass loss from stars above the Eddington limit". *The Astrophysical Journal* **616** (1): 525–541. arXiv:astro-ph/0409573. Bibcode:2004ApJ...616..5 doi:10.1086/424910.

[6] Smith, N.; Owocki, S. P. (2006). "On the role of continuum driven eruptions in the evolution of very massive stars and population III stars". *The Astrophysical Journal* **645** (1): L45–L48. arXiv:astro-ph/0606174. Bibcode:2006ApJ...645L..45S. doi:10.1086/506523.

[7] Vink, J. S. (2012). "Eta Carinae and the Luminous Blue Variables". *Eta Carinae and the Supernova Impostors*. Astrophysics and Space Science Library **384**. pp. 221–247. doi:10.1007/978-1-4614-2275-4_10. ISBN 978-1-4614-2274-7.

[8] Langer, R. B.; Garcia-Segura, G. (1996). *Massive Stars: the Pre-Supernova Evolution of internal and Circumstellar Structure*. *Reviews in Modern Astronomy*. 11 Stars and Galaxies. p. 57. ISBN 978-3-9805176-1-4.

[9] Stothers, N.; Chin, C.-W. (1996). "Evolution of Massive Stars into Luminous Blue Variables and Wolf-Rayet stars for a range of metallicities". *The Astrophysical Journal* **468**: 842–850. Bibcode:1996ApJ...468..842S. doi:10.1086/177740.

[10] Van Genderen, A. M.; Sterken, C. (1999). "Light variations of massive stars (alpha Cygni variables). XVII. The LMC supergiants R 74 (LBV), R 78, HD 34664 = S 22 (B[e]/LBV), R 84 and R 116 (LBV?)". *Astronomy and Astrophysics* **349**: 537. Bibcode:1999A&A...349..537V.

[11] Sterken, C.; de Groot, M.; van Genderen, A. M. (1998). "Cyclicities in the Light Variations of Luminous Blue Variables II. R40 developing an S Doradus phase". *Astronomy and Astrophysics* (Astronomy & Astrophysics) **333**: 565. Bibcode:1998A&A...333

[12] Clark, J. S.; Najarro, F.; Negueruela, I.; Ritchie, B. W.; Urbaneja, M. A.; Howarth, I. D. (2012). "On the nature of the galactic early-B hypergiants". *Astronomy & Astrophysics* (pdf) **541**: A145. arXiv:1202.3991v1. Bibcode:2012A&A...541A.145C. doi:10.1051/0004-6361/201117472.

[13] Miroshnichenko, A. S.; Chentsov, E. L.; Klochkova, V. G. (2000). "AS314: A dusty A-type hypergiant". *Astronomy and Astrophysics Supplement Series* **144** (3): 379. Bibcode:2000A&AS..144..379M. doi:10.1051/aas:2000216.

[14] Wolf, B.; Kaufer, A.; Rivinius, T.; Stahl, O.; Szeifert, T.; Tubbesing, S.; Schmid, H. M. (2000). "Spectroscopic Monitoring of Luminous Hot Stars of the Magellanic Clouds". *Thermal and Ionization Aspects of Flows from Hot Stars* **204**: 43. Bibcode:2000ASPC..204...43W.

[15] Van Genderen, A. M.; Jones, A.; Sterken, C. (2006). "Light variations of alpha Cygni variables in the Magellanic Clouds". *The Journal of Astronomical Data* **12**: 4. Bibcode:2006JAD....12....4V.

[16] Kathryn F. Neugent; Philip Massey; Brian Skiff; Georges Meynet (April 2012). "Yellow and Red Supergiants in the Magellanic Clouds". *Astrophysical Journal* **749** (2): 177. arXiv:1202.4225. Bibcode:2012ApJ...749..177N. doi:10.1088/0004-637X/749/2/177.

[17] Clark, J. S.; Negueruela, I.; Crowther, P. A.; Goodwin, S. P. (2005). "On the massive stellar population of the super star cluster Westerlund 1". *Astronomy and Astrophysics* **434** (3): 949. arXiv:astro-ph/0504342. Bibcode:2005A&A...434..949C. doi:10.1051/0004-6361:20042413.

[18] Stolovy, S. R.; Cotera, A.; Dong, H.; Morris, M. R.; Wang, Q. D.; Stolovy, S. R.; Lang, C. (2010). "Isolated Wolf-Rayet Stars and O Supergiants in the GalacticCenter Region Identified via Paschen-a Excess". *Astronomy & Astrophysics* **725**: 188. arXiv:1009.2769v3. Bibcode:2010ApJ...725..188M. doi:10.1088/0004-637X/725/1/188.

[19] Clark, J. S.; Negueruela, I.; Gonzalez-Fernandez, C. (2013). "IRAS 18357-0604 - an analogue of the galactic yellow hypergiant IRC +10420?". *Astronomy & Astrophysics* **561**: A15. arXiv:1311.3956v1 [astro-ph.SR]. Bibcode:2014A&A...561A..15C. doi:10.1051/0004-6361/201322772.

[20] Schuster, M. T.; Humphreys, R. M.; Marengo, M. (2006). "The Circumstellar Environments of NML Cygni and the Cool Hypergiants". *The Astronomical Journal* **131**: 603. arXiv:astro-ph/0510010. Bibcode:2006AJ....131..603S. doi:10.1086/498395.

[21] Jura, M.; Velusamy, T.; Werner, M. W. (2001). "What Next for the Likely Presupernova HD 179821?". *The Astrophysical Journal* **556**: 408. arXiv:astro-ph/0103282. Bibcode:2001ApJ...556..408J. doi:10.1086/321553.

[22] Lagadec, E.; Zijlstra, A. A.; Oudmaijer, R. D.; Verhoelst, T.; Cox, N. L. J.; Szczerba, R.; Mékarnia, D.; Van Winckel, H. (2011). "A double detached shell around a post-red supergiant: IRAS 17163-3907, the Fried Egg nebula". *Astronomy & Astrophysics* **534**: L10. Bibcode:2011A&A...534L..10L. doi:10.1051/0004-6361/201117521.

[23] Levesque, E. M.; Massey, P.; Olsen, K. A. G.; Plez, B.; Meynet, G.; Maeder, A. (2006). "The Effective Temperatures and Physical Properties of Magellanic Cloud Red Supergiants: The Effects of Metallicity". *The Astrophysical Journal* **645** (2): 1102. arXiv:astro-ph/0603596. Bibcode:2006ApJ...645.1102L. doi:10.1086/504417.

# Chapter 29

# Supernova impostor

NGC 3184 showing SN impostor SN 2010dn.[1]

**Supernova impostors** are stellar explosions that appear at first to be a type of supernova but do not destroy their progenitor stars. As such, they are a class of extra-powerful novae. They are also known as Type V supernovae, Eta Carinae analogs, and giant eruptions of luminous blue variables LBV.[2]

## 29.1 Appearance, origin and mass loss

Supernova impostors appear as remarkably faint supernovae of spectral type IIn—which have hydrogen in their spectrum and narrow spectral lines that indicate relatively low gas speeds. These impostors exceed their pre-outburst states by several magnitudes, with typical peak absolute visual magnitudes of $-11$ to $-14$, making these outbursts as bright as the most luminous stars. The trigger mechanism of these outbursts remains unexplained, though it is thought to be caused by violating the classical Eddington luminosity limit, initiating severe mass loss. If the ratio of radiated energy to kinetic energy is near unity, as in Eta Carinae, then we might expect an ejected mass of about 0.16 solar masses.

## 29.2 Examples

Possible examples of supernova impostors include the 1843 eruption of Eta Carinae, P Cygni, SN 1961V,[3] SN 1954J, SN 1997bs, SN 2008S in NGC 6946, and SN 2010dn[1] where detections of the surviving progenitor stars are claimed.

One supernova impostor that made news after the fact was the one observed on October 20, 2004, in the galaxy UGC 4904 by Japanese amateur astronomer Koichi Itagaki. This LBV star exploded just two years later, on October 11, 2006, as supernova SN 2006jc.[4]

## 29.3 References

[1] Smith, Nathan; Weidong, Li; Silverman, Jeffrey; Ganeshalingam, Mo; Filippenko, Alexei (2010). "Luminous Blue Variable eruptions and related transients: Diversity of progenitors and outburst properties". *Solar and Stellar Astrophysics* **1010**: 3718. arXiv:1010.3718. Bibcode:2011MNRAS.415..773S. doi:10.1111/j.1365-2966.2011.18763.x.

[2] Smith, Nathan; Ganeshalingam, Mohan; Chornock, Ryan; Filippenko, Alexei; Weidong, Li; et al. (2009). "SN 2008S: A Cool Super-Eddington Wind in a Supernova Impostor". *Astrophysical Journal Letters* **697** (1): L49–L53. arXiv:0811.3929. Bibcode:2009ApJ...697L..49S. doi:10.1088/0004-637X/697/1/L49.

[3] Kochanek, C.S.; Szczygiel, D.M.; Stanek, K.Z. (2010). "The Supernova Impostor Impostor SN 1961V: Spitzer Shows That Zwicky Was Right (Again)". *Solar and Stellar Astrophysics*. arXiv:1010.3704. Bibcode:2011ApJ...737...76K. doi:10.1088/0004-637X/737/2/76.

[4] "NASA – Supernova Imposter Goes Supernova". Nasa.gov. Retrieved 2010-01-13.

# Chapter 30

# Wolf–Rayet star

**Wolf–Rayet stars** (often referred to as WR stars) are a heterogeneous set of stars with unusual spectra showing prominent broad emission lines of highly ionised helium and nitrogen or carbon. The spectra indicate very high surface temperatures of 30,000 K to around 200,000 K, surface enhancement of heavy elements, and strong stellar winds.

Classic (or Population I) **Wolf–Rayet stars** are evolved, massive stars, O-type stars over 20 solar masses when they were on the main sequence, that have now completely lost their outer hydrogen and are fusing helium or heavier elements in the core. A subset of WR stars are the central stars of planetary nebulae (CSPNe), post Asymptotic Giant Branch stars that were similar to the Sun while on the main sequence, but have now ceased fusion and shed their atmospheres to show a bare carbon-oxygen core. Another group (type WNh) show hydrogen lines in their spectra and are young extremely massive stars still fusing hydrogen at the core, with nitrogen mixed to the surface and strong radiation-driven mass loss. They are all highly luminous due to their high temperatures, thousands of times the bolometric luminosity of the Sun ($L\odot$) for the CSPNe, hundreds of thousands $L\odot$ for the Population I WR stars, to over a million $L\odot$ for the WNh stars, although not exceptionally bright visually since most of their radiation output is in the ultraviolet.[1][2][3]

The naked-eye stars Gamma Velorum and Theta Muscae, as well as the most massive known star, R136a1 in 30 Doradus, are all Wolf–Rayet stars.

## 30.1   Observation history

In 1867, using the 40 cm Foucault telescope at the Paris Observatory, astronomers Charles Wolf and Georges Rayet[4] discovered three stars in the constellation Cygnus (HD 191765, HD 192103 and HD 192641, now designated as WR 134, WR135, and WR137 respectively) that displayed broad emission bands on an otherwise continuous spectrum.[5] Most stars only display absorption lines or bands in their spectra, as a result of overlying elements absorbing light energy at specific frequencies, so these were clearly unusual objects.

The nature of the emission bands in the spectra of a Wolf–Rayet star remained a mystery for several decades. Edward C. Pickering theorized that the lines were caused by an unusual state of hydrogen, and it was found that this "Pickering series" of lines followed a pattern similar to the Balmer series, when half-integral quantum numbers were substituted. It was later shown that the lines resulted from the presence of helium; a gas that was discovered in 1868.[6] Pickering noted similarities between Wolf–Rayet spectra and nebular spectra, and this similarity led to the conclusion that some or all Wolf Rayet stars were the central stars of planetary nebulae.[7]

By 1929, the width of the emission bands was being attributed to Doppler broadening, and hence that the gas surrounding these stars must be moving with velocities of 300–2400 km/s along the line of sight. The conclusion was that a Wolf–Rayet star is continually ejecting gas into space, producing an expanding envelope of nebulous gas. The force ejecting the gas at the high velocities observed is radiation pressure.[8] It was well known that many stars with Wolf Rayet type spectra were the central stars of planetary nebulae, but also that many were not associated with an obvious planetary nebula or any visible nebulousity at all.[9]

*Hubble Space Telescope image of nebula M1-67 around Wolf–Rayet star WR 124*

In addition to helium, emission lines of carbon, oxygen and nitrogen were identified in the spectra of Wolf–Rayet stars.[10] In 1938, the International Astronomical Union classified the spectra of Wolf–Rayet stars into types WN and WC, depending on whether the spectrum was dominated by lines of nitrogen or carbon-oxygen respectively.[11]

In 1969, several CSPNe with strong OVI emissions lines were grouped under a new "OVI sequence", or just OVI type.[12] These were subsequently referred to as [WO] stars.[13] Similar stars not associated with planetary nebulae were described shortly after and the WO classification was eventually also adopted for population I WR stars.[13][14]

The understanding that certain late, and sometimes not-so-late, WN stars with hydrogen lines in their spectra are at a different stage of evolution from hydrogen-free WR stars has led to the introduction of the term *WNh* to distinguish these stars generally from other WN stars. They were previously referred to as WNL stars, although there are late-type WN stars without hydrogen as well as WR stars with hydrogen as early as WN5.[15]

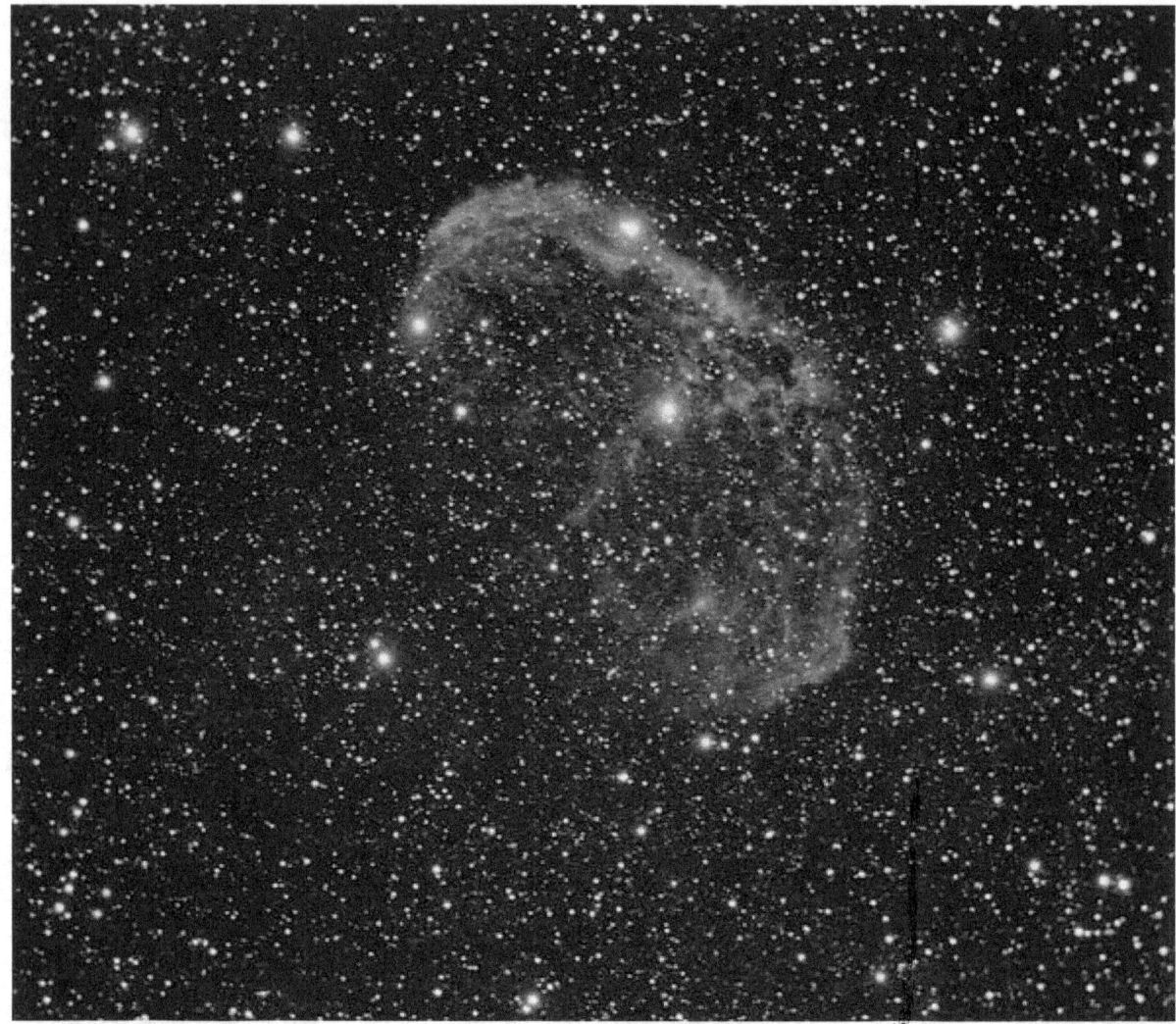

*WR 136 is a WN6 star where the atmosphere shed during the red supergiant phase has been shocked by the hot fast WR winds to form a visible bubble nebula*

## 30.2   Classification

Wolf–Rayet stars were named on the basis of the strong broad emission lines in their spectra, identified with helium, nitrogen, carbon, silicon, and oxygen, but with hydrogen lines usually weak or absent. The first system of classification split these into stars with dominant lines of ionised nitrogen (NIII, NIV, and NV) and those with dominant lines of ionised carbon (CIII and CIV) and sometimes oxygen (OIII - OVI), referred to as WN and WC respectively.[16] The two classes WN and WC were further split into temperature sequences WN5-WN8 and WC6-WC8 based on the relative strengths of the 541.1nm HeII and 587.5 nm HeI lines. Wolf–Rayet emission lines frequently have a broadened absorption wing (P Cygni profile) suggesting circumstellar material. A WO sequence has also been separated from the WC sequence for even hotter stars where emission of ionised oxygen dominates that of ionised carbon, although the actual proportions of those elements in the stars are likely to be comparable.[9] WC and WO spectra are formally distinguished based on the presence or absence of CIII emission.[17] WC spectra also generally lack the OVI lines that are strong in WO spectra.[18]

The WN spectral sequence was expanded to include WN2 - WN9, and the definitions refined based on the relative strengths of the NIII lines at 463.4-464.1 nm and 531.4 nm, the NIV lines at 347.9-348.4 nm and 405.8 nm, and the NV lines at 460.3 nm, 461.9 nm, and 493.3-494.4 nm.[19] These lines are well separated from areas of strong and variable He emission and the line strengths are well correlated with temperature. Stars with spectra intermediate between WN and

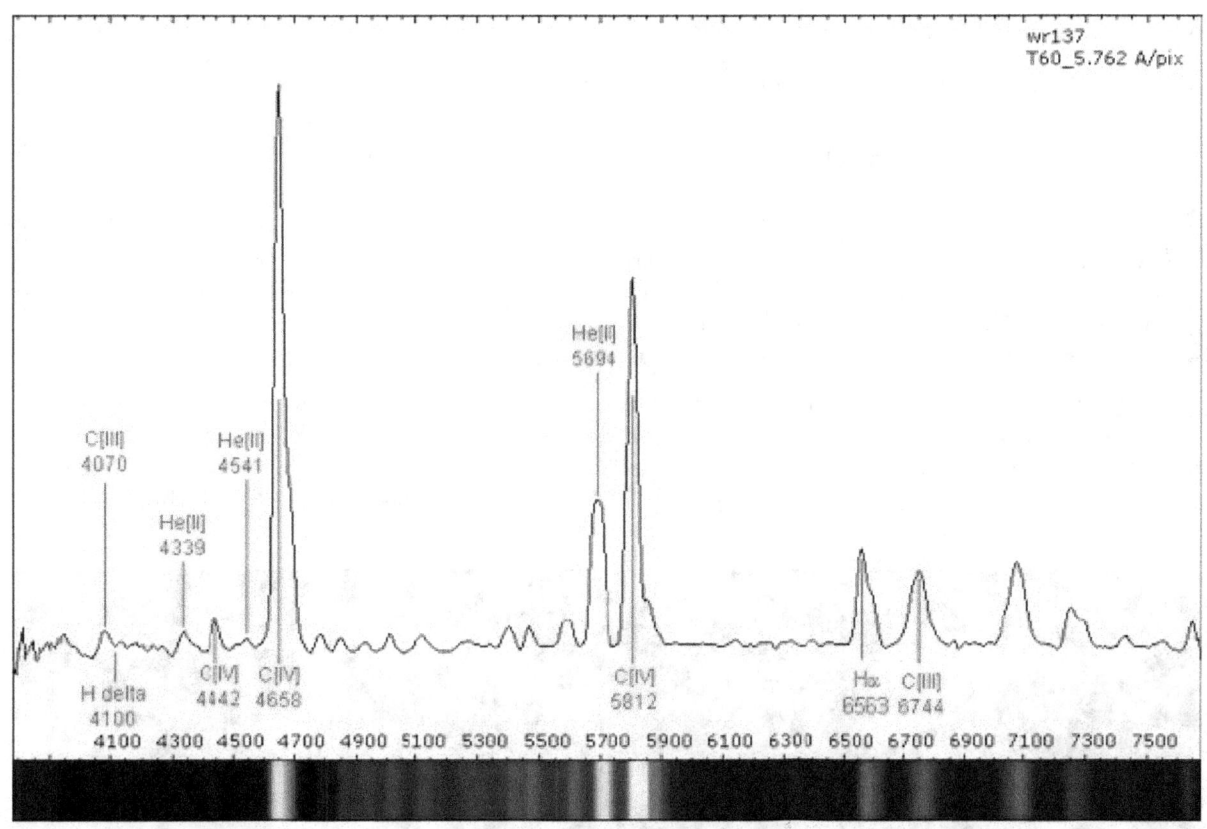

*Spectrum of WR137, a WC7 star[3] and one of the three original WR stars*

Ofpe have been classified as WN10 and WN11 although this nomenclature is not universally accepted.[20]

The type WN1 was proposed for stars with neither NIV nor NV lines, to accommodate Brey 1 and Brey 66 which appeared to be intermediate between WN2 and WN2.5.[21] The relative line strengths and widths for each WN sub-class were later quantified, and the ratio between the 541.1 nm HeII and 587.5m, HeI lines was introduced as the primary indicator of the ionisation level and hence of the spectral sub-class. The need for WN1 disappeared and both Brey 1 and Brey 66 are now classified as WN3b. The somewhat obscure WN2.5 and WN4.5 classes were dropped.[22]

The WC spectral sequence was expanded to include WC4 - WC11, although some older papers have also used WC1 - WC3. The primary emission lines used to distinguish the WC sub-types are CII 426.7 nm, CIII at 569.6 nm, CIII/IV465.0 nm, CIV at 580.1-581.2 nm, and the OV (and OIII) blend at 557.2-559.8 nm.[17] The sequence was extended to include WC10 and WC11, and the subclass criteria were quantified based primarily on the relative strengths of carbon lines to rely on ionisation factors even if there were abundance variations between carbon and oxygen.[18]

For WO-type stars the main lines used are CIV at 580.1 nm, OIV at 340.0 nm, OV (and OIII) blend at 557.2-559.8 nm, OVI at 381.1-383.4 nm, OVII at 567.0 nm, and OVIII at 606.8 nm. The sequence was expanded to include WO5 and quantified based the relative strengths of the OVI/CIV and OVI/OV lines.[23] A later scheme, designed for consistency across classical WR stars and CSPNe, returned to the WO1 to WO4 sequence and adjusted the divisions.[18]

Detailed modern studies of Wolf Rayet stars can identify additional spectral features, indicated by suffixes to the main spectral classification:[22]

- h for hydrogen emission;

- ha for hydrogen emission and absorption;

- w for weak lines;

- s for strong lines;

- b for broad strong lines;

- d for dust (occasionally vd, pd, or ed for variable, periodic, or episodic dust).[24]

The classification of Wolf Rayet spectra is complicated by the frequent association of the stars with dense nebulosity, dust clouds, or binary companions. A suffix of "+OB" is used to indicate the presence of absorption lines in the spectrum likely to be associated with a more normal companion star, or "+abs" for absorption lines with an unknown origin.[22]

The hotter WR spectral sub-classes are described as early and the cooler ones as late, consistent with other spectral types. WNE and WCE refer to early type spectra while WNL and WCL refer to late type spectra, with the dividing line approximately at sub-class six or seven. There is no such thing as a late WO-type star. There is a strong tendency for WNE stars to be hydrogen-poor while the spectra of WNL stars frequently include hydrogen lines.[17][25]

Spectral types for the central stars of planetary nebulae are often qualified by surrounding them with square brackets (e.g. [WC2]). They are almost all of the WC sequence, with a very small number of [WO] and [WN] or [WC/WN] types. Temperatures tend to the extremes when compared to population I WR stars, so [WC2] and [WC3] are common and the sequence has been extended to [WC12]. The [WC11] and [WC12] types have distinctive spectra with narrow emission lines and no HeII and CIV lines.[26][27]

*GK Persei (Nova Persei 1901), which showed Wolf Rayet features in its spectrum*[8]

Certain supernovae observed prior to their peak brightness show WR spectra.[28] This is not surprising given the nature of the supernova at this point: a rapidly expanding helium-rich ejecta similar to an extreme Wolf Rayet wind. The WR spectral features only last a matter of hours, the high ionisation features fading by maximum to leave only weak neutral hydrogen and helium emission, before being replaced with a traditional supernova spectrum. It has been proposed to label these spectral types with an "X", for example XWN5(h).[29] Similarly, classical novae develop spectra consisting of broad emission bands similar to a Wolf Rayet star. This is caused by the same physical mechanism: rapid expansion of dense gases around an extremely hot central source.[9]

## 30.3 Nomenclature

*WR 22 in the Carina Nebula*

The first three Wolf Rayet stars to be identified, coincidentally all with hot O companions, had already been numbered in the HD catalogue. These stars and others were referred to as Wolf–Rayet stars from their initial discovery but specific naming conventions for them would not be created until 1962 in the "fourth" catalogue of galactic Wolf Rayet stars.[30] The first three catalogues were not specifically lists of Wolf Rayet stars and they used only existing nomenclature.[31][32][33] The fourth catalogue numbered the Wolf Rayet stars sequentially in order of right ascension. The fifth catalogue used the same numbers prefixed with MR after the author of the fourth catalogue, plus an additional sequence of numbers prefixed with LS for new discoveries.[19] Neither of these numbering schemes is in common use.

The sixth Catalogue of Galactic Wolf Rayet stars was the first to actually bear that name, as well as to describe the previous five catalogues by that name. It also introduced the WR numbers widely used ever since for galactic WR stars. These are again a numerical sequence from WR 1 to WR 158 in order of right ascension.[34] The seventh catalogue and its annex use the same numbering scheme and insert new stars into the sequence using lower case letter suffixes, for example WR 102ka for one of the numerous WR stars discovered in the galactic centre.[17][35] Modern high volume identification surveys use their own numbering schemes for the large numbers of new discoveries.[2] An IAU working group has accepted recommendations to expand the numbering system from the Catalogue of Galactic Wolf Rayet stars so that additional discoveries are given the closest existing WR number plus a numeric suffix in order of discovery. This applies to all discoveries since the 2006 annex, although some of these have already been named under the previous nomenclature; thus WR 42e is now numbered WR 42-1.[36]

Wolf Rayet stars in external galaxies are numbered using different schemes. In the Large Magellanic Cloud, the most widespread and complete nomenclature for WR stars is from the fourth Catalogue of Population I Wolf Rayet stars in the Large Magellanic Cloud, prefixed by BAT-99, for example BAT-99 105.[37] Many of these stars are also referred to by their third catalogue number, for example Brey 77.[38] As of 2015, 152 WR stars are catalogued in the LMC, mostly WN but including three of the extremely rare WO class.[39][40] Many of these stars are often referred to by their RMC (Radcliffe observatory Magellanic Cloud) numbers, frequently abbreviated to just R, for example R136a1.

In the Small Magellanic Cloud SMC WR numbers are used, usually referred to as AB numbers, for example AB7.[41] There are only twelve known WR stars in the SMC, a very low number thought to be due to the low metallicity of that galaxy.[42][42][43]

## 30.4   Nebulae

Main article: Wolf–Rayet nebula

A significant proportion of WR stars are surrounded by nebulosity associated directly with the star, not just the normal background nebulosity associated with any massive star forming region, and not a planetary nebula formed by a post-AGB star. The nebulosity presents a variety of forms and classification has been difficult. Many were originally catalogued as planetary nebulae and only careful study can distinguish a planetary nebula around a low mass post-AGB star from a similar nebula around a more massive core helium burning star.[44]

## 30.5   Properties

Wolf–Rayet stars are a normal stage in the evolution of very massive stars, in which strong, broad emission lines of helium and nitrogen ("WN" sequence), carbon ("WC" sequence), and oxygen ("WO" sequence) are visible. Due to their strong emission lines they can be identified in nearby galaxies. About 500 Wolf–Rayets are catalogued in our own Milky Way Galaxy.[2][17][35] This number has changed dramatically during the last few years as the result of photometric and spectroscopic surveys in the near-infrared dedicated to discovering this kind of object in the Galactic plane.[45] It is expected that there are fewer than 1,000 WR stars in the rest of the Local Group galaxies, with around 150 known in the Magellanic Clouds, 206 in M33,[46] and 154 in M31.[47] Outside the local group, whole galaxy surveys have found thousands more WR stars and candidates. For example, over a thousand WR stars have been detected in M101, from magnitude 21 to 25.[48] WR stars are expected to be particularly common in starburst galaxies and especially Wolf–Rayet galaxies.[49]

The characteristic emission lines are formed in the extended and dense high-velocity wind region enveloping the very hot stellar photosphere, which produces a flood of UV radiation that causes fluorescence in the line-forming wind region. This ejection process uncovers in succession, first the nitrogen-rich products of CNO cycle burning of hydrogen (WN stars), and later the carbon-rich layer due to He burning (WC and WO-type stars).[14]

It can be seen that the WNh stars are completely different objects from the WN stars without hydrogen. Despite the similar spectra, they are much more massive, much larger, and some of the most luminous stars known. They have been detected as early as WN5h in the Magellanic clouds. The nitrogen seen in the spectrum of WNh stars is still the product

*AB7 is one of the highest excitation nebulae in the Magellanic Clouds, two satellite galaxies of our own Milky Way*

of CNO cycle fusion in the core, but it appears at the surface of the most massive stars due to rotational and convectional mixing while still in the core hydrogen burning phase, rather than after the outer envelope is lost during core helium fusion.[15]

Some Wolf–Rayet stars of the carbon sequence ("WC"), especially those belonging to the latest types, are noticeable due to their production of dust. Usually this takes places on those belonging to binary systems as a product of the collision of the stellar winds forming the pair,[17] as is the case of the famous binary WR 104; however this process occurs on single ones too.[3]

A few (roughly 10%) of the central stars of planetary nebulae are, despite their much lower (typically ~0.6 solar) masses, also observationally of the WR-type; i.e. they show emission line spectra with broad lines from helium, carbon and oxygen. Denoted [WR], they are much older objects descended from evolved low-mass stars and are closely related to white dwarfs, rather than to the very young, very massive population I stars that comprise the bulk of the WR class. These are now generally excluded from the class denoted as Wolf–Rayet stars, or referred to as Wolf–Rayet-type stars.[25]

### 30.5.1  Metallicity

The numbers and properties of Wolf Rayet stars vary with the chemical composition of their progenitor stars. A primary driver of this difference is the rate of mass loss at different levels of metallicity. Higher metallicity leads to high mass loss, which affects the evolution of massive stars and also the properties of Wolf Rayet stars. Higher levels of mass loss cause stars to lose their outer layers before an iron core develops and collapses, so that the more massive red supergiants evolve back to hotter temperatures before exploding as a supernova, and the most massive stars never become red supergiants. In the Wolf Rayet stage, higher mass loss leads to stronger depletion of the layers outside the convective core, lower hydrogen surface abundances and more rapid stripping of helium to produce a WC spectrum.

These trends can be observed in the various galaxies of the local group, where metallicity varies from near-solar levels in the milky way, somewhat lower in M31, lower still in the Large Magellanic Cloud, and much lower in the Small Magellanic Cloud. Strong metallicity variations are seen across individual galaxies, with M33 and the milky way showing higher metallicities closer to the centre, and M31 showing higher metallicity in the disk than in the halo. Thus the SMC is seen to have few WR stars compared to its stellar formation rate and no WC stars at all (one star has a WO spectral type), the milky way has roughly equal numbers of WN and WC stars and a large total number of WR stars, and the other main galaxies have somewhat fewer WR stars and more WN than WC types. LMC, and especially SMC, Wolf Rayets have weaker emission and a tendency to higher atmospheric hydrogen fractions. SMC WR stars almost universally show some hydrogen and even absorption lines even at the earliest spectral types, due to weaker winds not entirely masking the photosphere.[52]

The maximum mass of a main-sequence star that can evolve through a red supergiant phase and back to a WNL star is calculated to be around 20 $M\odot$ in the milky way, 32 $M\odot$ in the LMC, and over 50 $M\odot$ in the SMC. The more evolved WNE and WC stages are only reached by stars with an initial mass over 25 $M\odot$ at near-solar metallicity, over 60 $M\odot$ in the LMC. Normal single star evolution is not expected to produce any WNE or WC stars at SMC metallicity.[44]

### 30.5.2  Rotation

Mass loss is influenced by a star's rotation rate, especially strongly at low metallicity. Fast rotation contributes to mixing of core fusion products through the rest of the star, enhancing surface abundances of heavy elements, and driving mass loss. Rotation causes stars to remain on the main sequence longer than non-rotating stars, evolve more quickly away from the red supergiant phase, or even evolve directly from the main sequence to hotter temperatures for very high masses, high metallicity or very rapid rotation.

Stellar mass loss produces a loss of angular momentum and this quickly brakes the rotation of massive stars. Very massive stars at near-solar metallicity should be braked almost to a standstill while still on the main sequence, while at SMC metallicity they can continue to rotate rapidly even at the highest observed masses. Rapid rotation of massive stars may account for the unexpected properties and numbers of SMC WR stars, for example their relatively high temperatures and luminosities.[52]

### 30.5.3  Binaries

Massive stars in binary systems can develop into Wolf Rayet stars due to stripping by a companion rather than inherent mass loss due to a stellar wind. This process is relatively insensitive to the metallicity or rotation of the individual stars and is expected to produce a consistent set of WR stars across all the local group galaxies. As a result, the fraction of WR stars produced through the binary channel, and therefore the number of WR stars observed to be in binaries, should be higher in low metallicity environments. Calculations suggest that the binary fraction of WR stars observed in the SMC should be as high as 98%, although less than half are actually observed to have a massive companion. The binary fraction in the milky way is around, in line with theoretical calculations.[53]

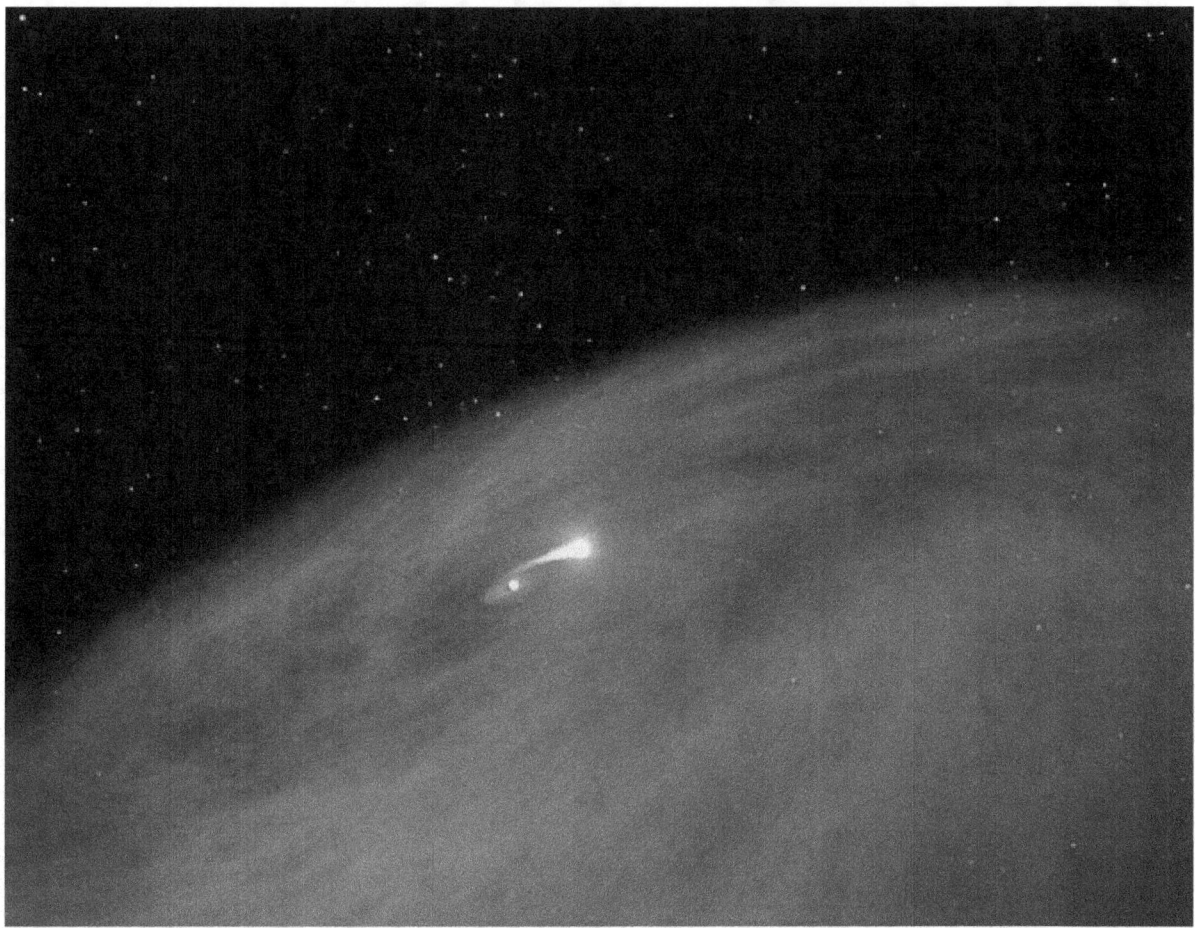

*Artist's illustration of gas disk around massive hydrogen-rich WR 122*

# 30.6   Evolution

Theories about how WR stars form, develop, and die have been slow to form compared to the explanation of less extreme stellar evolution. They are rare, distant, and often obscured, and even into the 21st century many aspects of their lives are unclear.

### 30.6.1   Early ideas

Several astronomers, among them Rublev (1965)[55] and Conti (1976)[56] originally proposed that the WR stars as a class are descended from massive O-stars in which the strong stellar winds characteristic of extremely luminous stars have ejected the unprocessed outer H-rich layers. This has proved to be essentially correct, but with much complexity between a main sequence O-type star and the final WR star.

Early modelling of the evolution of massive stars showed that they evolve away from the main sequence, not towards hotter temperature and a WR state, but by expanding and cooling to become blue and then red supergiants. These supergiants are only modestly more luminous than the main-sequence stars they originate from, but are progressively more unstable as their cores become hotter and their atmospheres more extended. Simple models of nuclear fusion showed that these red supergiants burned heavier elements in their cores until exploding as a supernova, but not becoming WR stars.

Further models showed that there was an upper limit to the stability of luminous stars. Sufficiently massive stars do not become red supergiants, instead shedding their atmospheres so quickly that they remain as blue supergiants, eventually shedding their atmospheres completely and entering the "Wolf Rayet funnel", an area of the Hertzsprung–Russell diagram

*HD 184738, also known as Campbell's Star. This is actually a planetary nebula and the central star is an old low-mass star unlike the main class of massive Wolf–Rayet stars.[54]*

where WR stars become progressively smaller and hotter as they shed more and more of their outer layers.[57][58] The suggestion was that earlier and hotter stars were the later stages of evolution from the later and cooler WR stars, but the results of this evolutionary sequence didn't match observations very well.

## 30.6.2   Current models

Most WR stars, the classical population I type, are now understood as being a natural stage in the evolution of the most massive stars (not counting the less common planetary nebula central stars), either after a period as a red supergiant, after a period as a blue supergiant, or directly from the most massive main-sequence stars. Only the lower mass red supergiants are expected to explode as a supernova at that stage, while more massive red supergiants progress back to hotter temperatures as they expel their atmospheres. Some explode while at the yellow hypergiant or LBV stage, but

*M1-67 is the youngest wind-nebula around a Wolf–Rayet star, called WR124, in the Milky Way.*
*Credit: ESO*

many become Wolf Rayet stars. They have lost or burnt almost all of their hydrogen and are now fusing helium in their cores, or heavier elements for a very brief period at the end of their lives.[59]

Massive main-sequence stars create a very hot core which fuses hydrogen very rapidly via the CNO process and results in strong convection throughout the whole star. This causes mixing of helium to the surface, a process that is enhanced by rotation, possibly by differential rotation where the core is spun up to a faster rotation than the surface. Such stars also show nitrogen enhancement at the surface at a very young age, caused by changes in the proportions of carbon and nitrogen due to the CNO cycle. The enhancement of heavy elements in the atmosphere, as well as increases in luminosity, create strong stellar winds which are the source of the emission line spectra. These stars develop an Of spectrum, Of* if they are sufficiently hot, which develops into a WNh spectrum as the stellar winds increase further. This explains the high mass and luminosity of the WNh stars, which are still burning hydrogen at the core and have lost little of their initial mass. These will eventually expand into blue supergiants (LBVs?) as hydrogen at the core becomes depleted, or if mixing is efficient enough (e.g. through rapid rotation) they may progress directly to WN stars without hydrogen.

Observations of supernovae reveal that around a quarter of core collapse supernovae are of Type Ib, originating from a progenitor with almost no hydrogen, and Type Ic, originating from a progenitor with almost no hydrogen and very little helium. This corresponds rather well to WC and WO-type stars and as this was investigated it appeared that WR stars were likely to end their lives violently rather than fade away to a white dwarf. Thus every star with an initial mass more than about 9 times the Sun would inevitably result in a supernova explosion, many of them from the WR stage.[25][59][60]

A simple progression of WR stars from low to hot temperatures, resulting finally in WO-type stars, is not supported by observation. WO-type stars are extremely rare and all the known examples are more luminous and more massive than the relatively common WC stars. Alternative theories suggest either that the WO-type stars are only formed from the most massive main-sequence stars,[3] and/or that they form an extremely short-lived end stage of just a few thousand years before exploding, with the WC phase corresponding to the core helium burning phase and the WO phase to nuclear burning stages beyond. It is still unclear whether the WO spectrum is purely the result of ionisation effects at very high temperature, reflects an actual chemical abundance difference, or if both effects occur to varying degrees.[59][61][62][63]

Key:

- O: O-type main-sequence star

- Of: evolved O-type showing N and He emission

- BSG: blue supergiant

- RSG: red supergiant

- YHG: yellow hypergiant

- LBV: luminous blue variable

- WNh: WN plus hydrogen lines

- WNL: "late" WN-class Wolf–Rayet star (about WN6 to WN9)

- WNE: "early" WN-class Wolf–Rayet star (about WN2 to WN6)

- WC: WC-class Wolf–Rayet star

- WO: WO-class Wolf–Rayet star

Wolf–Rayet stars form from massive stars, although the evolved population I stars have lost half or more of their initial masses by the time they show an WR appearance. For example, $\gamma^2$ Velorum A currently has a mass around 9 times the Sun, but began with a mass at least 40 times the Sun.[64] High-mass stars are very rare, both because they form less often and because they have short lives. This means that Wolf–Rayet stars themselves are extremely rare because they only form from the most massive main-sequence stars and because they are a relatively short-lived phase in the lives of those stars. This also explains why type Ibc supernovae are less common than type II, since they result from higher-mass stars.

WNh stars, spectroscopically similar but actually a much less evolved star which has only just started to expel its atmosphere, are an exception and still retain much of their initial mass. The most massive stars currently known are all WNh stars rather than O-type main-sequence stars, an expected situation because such stars show helium and nitrogen at the surface only a few thousand years after they form, possibly before they become visible through the surrounding gas cloud. An alternative explanation is that these stars are so massive that they could not form as normal main-sequence stars, instead being the result of mergers of less extreme stars.[65]

The difficulties of modelling the observed numbers and types of Wolf Rayet stars through single star evolution have led to theories that they form through binary interactions which could accelerate loss of the outer layers of a star through mass exchange. WR 122 is a potential example that has a flat disk of gas encircling the star, almost 2 trillion miles wide, and may have a companion star that stripped its outer envelope.[66]

## 30.6.3  Supernovae

Although it is widely accepted that most or all type Ibc supernovae progenitors were WR stars, no conclusive identification has been made of such a progenitor.[62] WR stars are very luminous due to their high temperatures but not visually bright, especially the hottest examples that are expected to be supernova progenitors. Theory suggests that the progenitors of type Ibc supernovae observed to date would not be bright enough to be detected, although they place constraints on the properties of those progenitors. One candidate is under observation as pre-outburst observations show a likely WR star at the site of iPTF13bvn.[67]

Particularly massive WR stars may produce a super-luminous type Ib/c supernova (hypernova) due to the high mass of radioactive nickel ejected. It is possible for a Wolf–Rayet star to progress to a "collapsar" stage in its death throes if it doesn't lose sufficient mass. This is when the core of the star collapses to form a black hole, either directly or by pulling in the surrounding ejected material. This is thought to be the precursor of a long gamma-ray burst. The compact object of Cygnus X-1 is one possible example.

## 30.7  Examples

The most visible example of a Wolf–Rayet star is Gamma 2 Velorum ($\gamma^2$ Vel), which is a naked eye star for those located south of 40 degrees northern latitude. Due to the exotic nature of its spectrum (bright emission lines in lieu of dark absorption lines) it is dubbed the "Spectral Gem of the Southern Skies". The second brightest is Theta Muscae. Both are multiple stars where the primary component is a Wolf Rayet type.

The most massive star and probably most luminous star currently known, R136a1, is also a Wolf–Rayet star of the WNh type that is still fusing hydrogen in its core. This type of star, which includes many of the most luminous and most massive stars, is very young and usually found only in the centre of the densest star clusters. Occasionally a runaway Wolf–Rayet star such as VFTS 682 is found outside such clusters, probably having been ejected from a multiple system or by interaction with other stars.

Only a minority of planetary nebulae have WR type central stars, but a considerable number of well-known planetary nebulae do have them.

## 30.8  References

[1] Tylenda, R.; Acker, A.; Stenholm, B. (1993). "Wolf-Rayet Nuclei of Planetary Nebulae - Observations and Classification". *Astronomy and Astrophysics Supplement* **102**: 595. Bibcode:1993A&AS..102..595T.

[2] Shara, Michael M.; Faherty, Jacqueline K.; Zurek, David; Moffat, Anthony F. J.; Gerke, Jill; Doyon, René; Artigau, Etienne; Drissen, Laurent (2012). "A Near-Infrared Survey of the Inner Galactic Plane for Wolf-Rayet Stars. Ii. Going Fainter: 71 More New W-R Stars". *The Astronomical Journal* **143** (6): 149. Bibcode:2012AJ....143..149S. doi:10.1088/0004-6256/143/6/149.

[3] Sander, A.; Hamann, W.-R.; Todt, H. (2012). "The Galactic WC stars".*Astronomy & Astrophysics***540**: A144. Bibcode:2012A doi:10.1051/0004-6361/201117830.

[4] Murdin, P. (2001). "Wolf, Charles J E (1827?1918)".*The Encyclopedia of Astronomy and Astrophysics*. p. 4101. Bibcode:200 doi:10.1888/0333750888/4101. ISBN 0333750888.

[5] Huggins, W.; Huggins, Mrs. (1890). "On Wolf and Rayet's Bright-Line Stars in Cygnus". *Proceedings of the Royal Society of London* **49** (296–301): 33–46. doi:10.1098/rspl.1890.0063.

[6] Fowler, A. (1912). "Hydrogen, Spectrum of, Observations of the principal and other series of lines in the". *Monthly Notices of the Royal Astronomical Society* **73** (2): 62–105. Bibcode:1912MNRAS..73...62F. doi:10.1093/mnras/73.2.62.

[7] Wright, W. H. (1914). "The relation between the Wolf-Rayet stars and the planetary nebulae". *The Astrophysical Journal* **40**: 466. Bibcode:1914ApJ....40..466W. doi:10.1086/142138.

[8] Beals, C. S. (1929). "On the nature of Wolf-Rayet emission". *Monthly Notices of the Royal Astronomical Society* **90**: 202–212. Bibcode:1929MNRAS..90..202B. doi:10.1093/mnras/90.2.202.

[9] Beals, C. S. (1940). "On the Physical Characteristics of the Wolf Rayet Stars and their Relation to Other Objects of Early Type (with Plates VIII, IX)". *Journal of the Royal Astronomical Society of Canada* **34**: 169. Bibcode:1940JRASC..34..169B.

[10] Beals, C. S. (1933). "Classification and temperatures of Wolf-Rayet stars".*The Observatory***56**: 196–197. Bibcode:1933Obs.

[11] Swings, P. (1942). "The Spectra of Wolf-Rayet Stars and Related Objects".*The Astrophysical Journal***95**: 112. Bibcode:1942A doi:10.1086/144379.

[12]  Starrfield, S.; Cox, A. N.; Kidman, R. B.; Pensnell, W. D. (1985). "An analysis of nonradial pulsations of the central star of the planetary nebula K1-16". *Astrophysical Journal* **293**: L23. Bibcode:1985ApJ...293L..23S. doi:10.1086/184484.

[13]  Sanduleak, N. (1971). "On Stars Having Strong O VI Emission". *The Astrophysical Journal* **164**: L71. Bibcode:1971ApJ...164L. doi:10.1086/180694.

[14]  Barlow, M. J.; Hummer, D. G. (1982). "The WO Wolf–Rayet stars". *Wolf-Rayet stars: Observations, physics, evolution; Proceedings of the Symposium, Cozumel, Mexico* **99**. pp. 387–392. Bibcode:1982IAUS...99..387B. doi:10.1007/978-94-009-7910-9_51. ISBN 978-90-277-1470-1.

[15]  Smith, Nathan; Conti, Peter S. (2008). "On the Role of the WNH Phase in the Evolution of Very Massive Stars: Enabling the LBV Instability with Feedback". *The Astrophysical Journal* **679** (2): 1467–1477. Bibcode:2008ApJ...679.1467S. doi:10.1086/586885.

[16]  Beals, C. S. (1933). "Classification and temperatures of Wolf-Rayet stars". *The Observatory* **56**: 196. Bibcode:1933Obs....56..196B.

[17]  Van Der Hucht, Karel A. (2001). "The VIIth catalogue of galactic Wolf–Rayet stars". *New Astronomy Reviews* **45** (3): 135–232. Bibcode:2001NewAR..45..135V. doi:10.1016/S1387-6473(00)00112-3.

[18]  Crowther, P. A.; De Marco, O.; Barlow, M. J. (1998). "Quantitative classification of WC and WO stars". *Monthly Notices of the Royal Astronomical Society* **296** (2): 367–378. doi:10.1046/j.1365-8711.1998.01360.x. ISSN 0035-8711.

[19]  Smith, Lindsey F. (1968). "A revised spectral classification system and a new catalogue for galactic Wolf-Rayet stars". *Monthly Notices of the Royal Astronomical Society* **138**: 109–121. Bibcode:1968MNRAS.138..109S. doi:10.1093/mnras/138.1.109.

[20]  Crowther, P. A.; Smith, L. J. (1997). "Fundamental parameters of Wolf-Rayet stars. VI. Large Magellanic Cloud WNL stars". *Astronomy and Astrophysics* **320**: 500. Bibcode:1997A&A...320..500C.

[21]  "Spectroscopic studies of Wolf-Rayet stars. IV - Optical spectrophotometry of the emission lines in galactic and large Magellanic Cloud stars". *The Astrophysical Journal* **337**: 251. Bibcode:1989ApJ...337..251C. doi:10.1086/167101.

[22]  "A three-dimensional classification for WN stars". *Monthly Notices of the Royal Astronomical Society* **281**: 163–191. doi:10.1093

[23]  Kingsburgh, R. L.; Barlow, M. J.; Storey, P. J. (1995). "Properties of the WO Wolf-Rayet stars". *Astronomy and Astrophysics (ISSN 0004-6361)* **295**: 75. Bibcode:1995A&A...295...75K.

[24]  "A Mid-Infrared Spectral Survey of Galactic Wolf-Rayet Stars". *The Astronomical Journal* **121**: 2115–2123. Bibcode:2001AJ. doi:10.1086/319968.

[25]  Crowther, Paul A. (2007). "Physical Properties of Wolf-Rayet Stars". *Annual Review of Astronomy and Astrophysics* **45**: 177–219. Bibcode:2007ARA&A..45..177C. doi:10.1146/annurev.astro.45.051806.110615.

[26]  Hamann, W.-R. (1997). "Spectra of Wolf–Rayet type central stars and their analysis (Invited Review)". *Proceedings of the 180th Symposium of the International Astronomical Union.* Kluwer Academic Publishers. p. 91. Bibcode:1997IAUS..180...91H.

[27]  Hamann, Wolf-Rainer (1996). "Spectral analysis and model atmospheres of WR central stars (Invited paper)". *Astrophysics and Space Science* **238**: 31. Bibcode:1996Ap&SS.238...31H. doi:10.1007/BF00645489.

[28]  "The supernova 1998S in NGC 3877: Another supernova with Wolf-Rayet star features in pre-maximum spectrum". *Astronomy and Astrophysics Supplement Series* **144**: 219–225. doi:10.1051/aas:2000208.

[29]  "Early-time spectra of supernovae and their precursor winds". *Astronomy* **572**: L11. Bibcode:2014A&A...572L..11G. doi:10.1 6361/201424852.

[30]  Roberts, M. S. (1962). "The galactic distribution of the Wolf-rayet stars". *The Astronomical Journal* **67**: 79. Bibcode:1962AJ.. doi:10.1086/108603.

[31]  Campbell, W. W. (1895). "Stars whose spectra contain both bright and dark hydrogen lines". *The Astrophysical Journal* **2**: 177. Bibcode:1895ApJ.....2..177C. doi:10.1086/140127.

[32]  Gaposchkin, Cecilia Payne (1930). *The stars of high luminosity.* Harvard Observatory Monographs **3**. p. 1. Bibcode:1930HarM

[33]  Fleming, Williamina Paton Stevens; Pickering, Edward Charles (1912). "Stars having peculiar spectra". *Annals of the Astronomical Observatory of Harvard College* **56**: 165. Bibcode:1912AnHar..56..165F.

[34] Van Der Hucht, Karel A.; Conti, Peter S.; Lundström, Ingemar; Stenholm, Björn (1981). "The Sixth Catalogue of galactic Wolf-Rayet stars, their past and present".*Space Science Reviews***28**(3): 227–306. Bibcode:1981SSRv...28..227V.doi:10.1007/BF00

[35] Van Der Hucht, K. A. (2006). "New Galactic Wolf-Rayet stars, and candidates". *Astronomy and Astrophysics* **458** (2): 453–459. Bibcode:2006A&A...458..453V. doi:10.1051/0004-6361:20065819.

[36] "Spatial distribution of Galactic Wolf-Rayet stars and implications for the global population". *Monthly Notices of the Royal Astronomical Society* **447**: 2322–2347. doi:10.1093/mnras/stu2525.

[37] Breysacher, J.; Azzopardi, M.; Testor, G. (1999). "The fourth catalogue of Population I Wolf-Rayet stars in the Large Magellanic Cloud". *Astronomy and Astrophysics Supplement Series* **137**: 117–145. Bibcode:1999A&AS..137..117B. doi:10.1051/aas:1999240.

[38] Breysacher, J. (1981). "Spectral Classification of Wolf-Rayet Stars in the Large Magellanic Cloud". *Astronomy and Astrophysics Supplement* **43**: 203. Bibcode:1981A&AS...43..203B.

[39] Hainich, R.; Rühling, U.; Todt, H.; Oskinova, L. M.; Liermann, A.; Gräfener, G.; Foellmi, C.; Schnurr, O.; Hamann, W.-R. (2014). "The Wolf-Rayet stars in the Large Magellanic Cloud. A comprehensive analysis of the WN class". *Astronomy & Astrophysics* **565**: A27. Bibcode:2014A&A...565A..27H. doi:10.1051/0004-6361/201322696.

[40] Massey, Philip; Neugent, Kathryn F.; Morrell, Nidia (2015). "A Modern Search for Wolf-Rayet Stars in the Magellanic Clouds. II. A Second Year of Discoveries". *The Astrophysical Journal* **807**: 81. arXiv:1505.06265v1. Bibcode:2015ApJ...807...81M. doi:10.1088/0004-637X/807/1/81.

[41] Azzopardi, M.; Breysacher, J. (1979). "A search for new Wolf-Rayet stars in the Small Magellanic Cloud". *Astronomy and Astrophysics* **75**: 120. Bibcode:1979A&A....75..120A.

[42] Massey, Philip; Olsen, K. A. G.; Parker, J. Wm. (2003). "The Discovery of a 12th Wolf-Rayet Star in the Small Magellanic Cloud". *Publications of the Astronomical Society of the Pacific* **115** (813): 1265–1268. Bibcode:2003PASP..115.1265M. doi:10.1086/379024.

[43] Bonanos, A. Z.; Lennon, D. J.; Köhlinger, F.; Van Loon, J. Th.; Massa, D. L.; Sewilo, M.; Evans, C. J.; Panagia, N.; Babler, B. L.; Block, M.; Bracker, S.; Engelbracht, C. W.; Gordon, K. D.; Hora, J. L.; Indebetouw, R.; Meade, M. R.; Meixner, M.; Misselt, K. A.; Robitaille, T. P.; Shiao, B.; Whitney, B. A. (2010). "Spitzersage-Smc Infrared Photometry of Massive Stars in the Small Magellanic Cloud". *The Astronomical Journal* **140** (2): 416–429. Bibcode:2010AJ....140..416B. doi:10.1088/0004-6256/140/2/416.

[44] Toalá, J. A.; Guerrero, M. A.; Ramos-Larios, G.; Guzmán, V. (2015). "WISE morphological study of Wolf-Rayet nebulae". *Astronomy & Astrophysics* **578**: A66. arXiv:1503.06878v1. Bibcode:2015A&A...578A..66T. doi:10.1051/0004-6361/201525706.

[45] Shara, Michael M.; Moffat, Anthony F. J.; Gerke, Jill; Zurek, David; Stanonik, Kathryn; Doyon, René; Artigau, Etienne; Drissen, Laurent; Villar-Sbaffi, Alfredo (2009). "A Near-Infrared Survey of the Inner Galactic Plane for Wolf-Rayet Stars. I. Methods and First Results: 41 New Wr Stars". *The Astronomical Journal* **138** (2): 402–420. Bibcode:2009AJ....138..402S. doi:10.1088/0004-6256/138/2/402.

[46] Neugent, Kathryn F.; Massey, Philip (2011). "The Wolf-Rayet Content of M33". *The Astrophysical Journal* **733** (2): 123. Bibcode:2011ApJ...733..123N. doi:10.1088/0004-637X/733/2/123.

[47] Neugent, Kathryn F.; Massey, Philip; Georgy, Cyril (2012). "The Wolf-Rayet Content of M31". *The Astrophysical Journal* **759**: 11. Bibcode:2012ApJ...759...11N. doi:10.1088/0004-637X/759/1/11.

[48] Bibby, Joanne; Shara, M. (2012). "A Study of the Wolf-Rayet Population of M101 using the Hubble Space Telescope". *American Astronomical Society* **219**. Bibcode:2012AAS...21924213B.

[49] Schaerer, Daniel; Vacca, William D. (1998). "New Models for Wolf-Rayet and O Star Populations in Young Starbursts". *The Astrophysical Journal* **497** (2): 618–644. Bibcode:1998ApJ...497..618S. doi:10.1086/305487.

[50] Hamann, W.-R.; Gräfener, G.; Liermann, A. (2006). "The Galactic WN stars". *Astronomy and Astrophysics* **457** (3): 1015–1031. Bibcode:2006A&A...457.1015H. doi:10.1051/0004-6361:20065052.

[51] Barniske, A.; Hamann, W.-R.; Gräfener, G. (2006). "Wolf-Rayet stars of the carbon sequence". *ASP Conference Series* **353**: 243. Bibcode:2006ASPC..353..243B.

[52] Hainich, R.; Pasemann, D.; Todt, H.; Shenar, T.; Sander, A.; Hamann, W.-R. (2015). "Wolf-Rayet stars in the Small Mag-
     ellanic Cloud. I. Analysis of the single WN stars". *Astronomy & Astrophysics* **581**: A21. Bibcode:2015A&A...581A..21H.
     doi:10.1051/0004-6361/201526241. ISSN 0004-6361.

[53] Foellmi, C.; Moffat, A. F. J.; Guerrero, M. A. (2003). "Wolf--Rayet binaries in the Magellanic Clouds and implications for
     massive-star evolution -- I. Small Magellanic Cloud". *Monthly Notices of the Royal Astronomical Society* **338** (2): 360–388.
     doi:10.1046/j.1365-8711.2003.06052.x.

[54] "Quantitative classification of WR nuclei of planetary nebulae".*Astronomy and Astrophysics***403**: 659–673. Bibcode:2003A&A.
     doi:10.1051/0004-6361:20030391.

[55] Rublev, S. V. (1965). "Dynamic State of the Atmospheres of Wolf-Rayet Stars".*Soviet Astronomy***8**: 848. Bibcode:1965SvA...

[56] Conti, P. S. (1976). In: *Mémoires de la Société royale des sciences de Liège (Proc. 20th Colloq. Int. Astrophys. Liège).* 6-Sér.
     tome 9. Liège: Soc. r. sci. Liège. pp. 193–212.

[57] Moffat, A. F. J.; Drissen, L.; Robert, C. (1989). "Observational Connections Between Lbv's and Other Stars, with Em-
     phasis on Wolf–Rayet Stars". *Physics of Luminous Blue Variables.* Astrophysics and Space Science Library **157**. p. 229.
     doi:10.1007/978-94-009-1031-7_27. ISBN 978-94-010-6955-7.

[58] Humphreys, R. M. (1991). "The Wolf–Rayet Connection - Luminous Blue Variables and Evolved Supergiants (review)". *Pro-
     ceedings of the 143rd Symposium of the International Astronomical Union* **143**. p. 485. Bibcode:1991IAUS..143..485H.

[59] Groh, Jose H.; Meynet, Georges; Georgy, Cyril; Ekström, Sylvia (2013). "Fundamental properties of core-collapse supernova
     and GRB progenitors: Predicting the look of massive stars before death". *Astronomy & Astrophysics***558**: A131. Bibcode:2013A

[60] Georges Meynet; Cyril Georgy; Raphael Hirschi; Andre Maeder; Phil Massey; Norbert Przybilla; M-Fernanda Nieva (2011).
     "Red Supergiants, Luminous Blue Variables and Wolf-Rayet stars: The single massive star perspective". *Bulletin de la Société
     Royale des Sciences de Liège.* v1 **80** (39): 266–278. arXiv:1101.5873. Bibcode:2011BSRSL..80..266M.

[61] Tramper, Frank (2013). "The nature of WO stars: VLT/X-Shooter spectroscopy of DR1". *Massive Stars: from α to Ω*: 187.
     Bibcode:2013msao.confE.187T.

[62] Eldridge, John J.; Fraser, Morgan; Smartt, Stephen J.; Maund, Justyn R.; Crockett, R. Mark (2013). "The death of massive stars
     - II. Observational constraints on the progenitors of Type Ibc supernovae". *Monthly Notices of the Royal Astronomical Society*
     **436**: 774–795. Bibcode:2013MNRAS.436..774E. doi:10.1093/mnras/stt1612.

[63] Groh, Jose; Meynet, Georges; Ekstrom, Sylvia; Georgy, Cyril (2014). "The evolution of massive stars and their spectra I. A
     non-rotating 60 Msun star from the zero-age main sequence to the pre-supernova stage". *Astronomy & Astrophysics* **564**: A30.
     arXiv:1401.7322v1. Bibcode:2014A&A...564A..30G. doi:10.1051/0004-6361/201322573.

[64] Oberlack, U.; Wessolowski, U.; Diehl, R.; Bennett, K.; Bloemen, H.; Hermsen, W.; Knödlseder, J.; Morris, D.; Schönfelder,
     V.; von Ballmoos, P. (2000). "COMPTEL limits on 26Al 1.809 MeV line emission from gamma2 Velorum". *Astronomy and
     Astrophysics* **353**: 715. Bibcode:2000A&A...353..715O.

[65] Banerjee, Sambaran; Kroupa, Pavel; Oh, Seungkyung (2012). "The emergence of super-canonical stars in R136-type star-
     burst clusters". *Monthly Notices of the Royal Astronomical Society* **426** (2): 1416–1426. Bibcode:2012MNRAS.426.1416B.
     doi:10.1111/j.1365-2966.2012.21672.x.

[66] Mauerhan, Jon C.; Smith, Nathan; Van Dyk, Schuyler D.; Morzinski, Katie M.; Close, Laird M.; Hinz, Philip M.; Males,
     Jared R.; Rodigas, Timothy J. (2015). "Multiwavelength Observations of NaSt1 (WR 122): Equatorial Mass Loss and X-rays
     from an Interacting Wolf-Rayet Binary". *Monthly Notices of the Royal Astronomical Society* **1502**: 1794. arXiv:1502.01794
     [astro-ph.SR]. Bibcode:2015MNRAS.450.2551M. doi:10.1093/mnras/stv257.

[67] Groh, Jose H.; Georgy, Cyril; Ekström, Sylvia (2013). "Progenitors of supernova Ibc: A single Wolf-Rayet star as the possible
     progenitor of the SN Ib iPTF13bvn". *Astronomy & Astrophysics* **558**: L1. Bibcode:2013A&A...558L...1G. doi:10.1051/0004-
     6361/201322369.

[68] Peña, M.; Rechy-Garcia, J. S.; García-Rojas, J. (2013). "Galactic kinematics of Planetary Nebulae with [WC] central star".
     *Revista Mexicana de Astronomía y Astrofísica Vol. 49* **49**: 87. Bibcode:2013RMxAA..49...87P.

## 30.9 Further reading

- "The Wolf-Rayet Stars". *Annual Review of Astronomy and Astrophysics***6**: 39–78. doi:10.1146/annurev.aa.06.090

- Tuthill, Peter G.; Monnier, John D.; Danchi, William C.; Turner, Nils H. (2003). "High-resolution near-IR imaging of the WCd(+OB) environments: Pinwheels". *Proceedings of the 212th International Union of Astronomy Symposium* **212**. p. 121. Bibcode:2003IAUS..212..121T.

- Monnier, J. D.; Tuthill, P. G.; Danchi, W. C. (1999). "Pinwheel Nebula around WR 98[CLC]a[/CLC]". *The Astrophysical Journal* **525** (2): L97. Bibcode:1999ApJ...525L..97M. doi:10.1086/312352. PMID 10525463.

- Dougherty, S. M.; Beasley, A. J.; Claussen, M. J.; Zauderer, B. A.; Bolingbroke, N. J. (2005). "High-Resolution Radio Observations of the Colliding-Wind Binary WR 140". *The Astrophysical Journal* **623**: 447–459. Bibcode:200

## 30.10 External links

- The Twisted Tale of Wolf-Rayet 104

- Perry Berlind's page on Wolf–Rayet Spectral Classifications

- Online catalog of galactic Wolf–Rayet Stars

- Big Old Stars Don't Die Alone (NASA)

- Hubble observes Nasty 1

# Chapter 31

# Gamma-ray burst

For bursts of gamma rays of terrestrial origin, see Terrestrial gamma-ray flash.

**Gamma-ray bursts** (**GRBs**) are flashes of gamma rays associated with extremely energetic explosions that have been

*Artist's illustration showing the life of a massive star as nuclear fusion converts lighter elements into heavier ones. When fusion no longer generates enough pressure to counteract gravity, the star rapidly collapses to form a black hole. Theoretically, energy may be released during the collapse along the axis of rotation to form a gamma-ray burst.*

observed in distant galaxies. They are the brightest electromagnetic events known to occur in the universe.[1] Bursts can last from ten milliseconds to several hours.[2][3][4] The initial burst is usually followed by a longer-lived "afterglow" emitted at longer wavelengths (X-ray, ultraviolet, optical, infrared, microwave and radio).[5]

Most observed GRBs are believed to consist of a narrow beam of intense radiation released during a supernova or hypernova as a rapidly rotating, high-mass star collapses to form a neutron star, quark star, or black hole. A subclass of GRBs (the "short" bursts) appear to originate from a different process – this may be due to the merger of binary neutron stars. The cause of the precursor burst observed in some of these short events may be due to the development of a resonance between the crust and core of such stars as a result of the massive tidal forces experienced in the seconds leading up to their collision, causing the entire crust of the star to shatter.[6]

The sources of most GRBs are billions of light years away from Earth, implying that the explosions are both extremely energetic (a typical burst releases as much energy in a few seconds as the Sun will in its entire 10-billion-year lifetime) and extremely rare (a few per galaxy per million years[7]). All observed GRBs have originated from outside the Milky Way galaxy, although a related class of phenomena, soft gamma repeater flares, are associated with magnetars within the Milky Way. It has been hypothesized that a gamma-ray burst in the Milky Way, pointing directly towards the Earth, could cause a mass extinction event.[8]

GRBs were first detected in 1967 by the Vela satellites, a series of satellites designed to detect covert nuclear weapons tests. Hundreds of theoretical models were proposed to explain these bursts in the years following their discovery, such as collisions between comets and neutron stars.[9] Little information was available to verify these models until the 1997 detection of the first X-ray and optical afterglows and direct measurement of their redshifts using optical spectroscopy, and thus their distances and energy outputs. These discoveries, and subsequent studies of the galaxies and supernovae associated with the bursts, clarified the distance and luminosity of GRBs. These facts definitively placed them in distant galaxies and also connected long GRBs with the explosion of massive stars, the only possible source for the energy outputs observed.

# 31.1 History

Main article: History of gamma-ray burst research

Gamma-ray bursts were first observed in the late 1960s by the U.S. Vela satellites, which were built to detect gamma radiation pulses emitted by nuclear weapons tested in space. The United States suspected that the USSR might attempt to conduct secret nuclear tests after signing the Nuclear Test Ban Treaty in 1963. On July 2, 1967, at 14:19 UTC, the Vela 4 and Vela 3 satellites detected a flash of gamma radiation unlike any known nuclear weapons signature.[10] Uncertain what had happened but not considering the matter particularly urgent, the team at the Los Alamos Scientific Laboratory, led by Ray Klebesadel, filed the data away for investigation. As additional Vela satellites were launched with better instruments, the Los Alamos team continued to find inexplicable gamma-ray bursts in their data. By analyzing the different arrival times of the bursts as detected by different satellites, the team was able to determine rough estimates for the sky positions of sixteen bursts[10] and definitively rule out a terrestrial or solar origin. The discovery was declassified and published in 1973 as an *Astrophysical Journal* article entitled "Observations of Gamma-Ray Bursts of Cosmic Origin".[11]

Many theories were advanced to explain these bursts, most of which posited nearby sources within the Milky Way Galaxy. Little progress was made until the 1991 launch of the Compton Gamma Ray Observatory and its Burst and Transient Source Explorer (BATSE) instrument, an extremely sensitive gamma-ray detector. This instrument provided crucial data that showed the distribution of GRBs is isotropic—not biased towards any particular direction in space, such as toward the galactic plane or the galactic center.[12] Because of the flattened shape of the Milky Way Galaxy, if the sources were from within our own galaxy they would be strongly concentrated in or near the galactic plane. The absence of any such pattern in the case of GRBs provided strong evidence that gamma-ray bursts must come from beyond the Milky Way.[13][14][15][16] However, some Milky Way models are still consistent with an isotropic distribution.[13][17]

## 31.1.1 Counterpart objects as candidate sources

For decades after the discovery of GRBs, astronomers searched for a counterpart at other wavelengths: i.e., any astronomical object in positional coincidence with a recently observed burst. Astronomers considered many distinct classes of objects, including white dwarfs, pulsars, supernovae, globular clusters, quasars, Seyfert galaxies, and BL Lac objects.[18] All such searches were unsuccessful,[nb 1] and in a few cases particularly well-localized bursts (those whose positions were determined with what was then a high degree of accuracy) could be clearly shown to have no bright objects of any nature

# 2704 BATSE Gamma-Ray Bursts

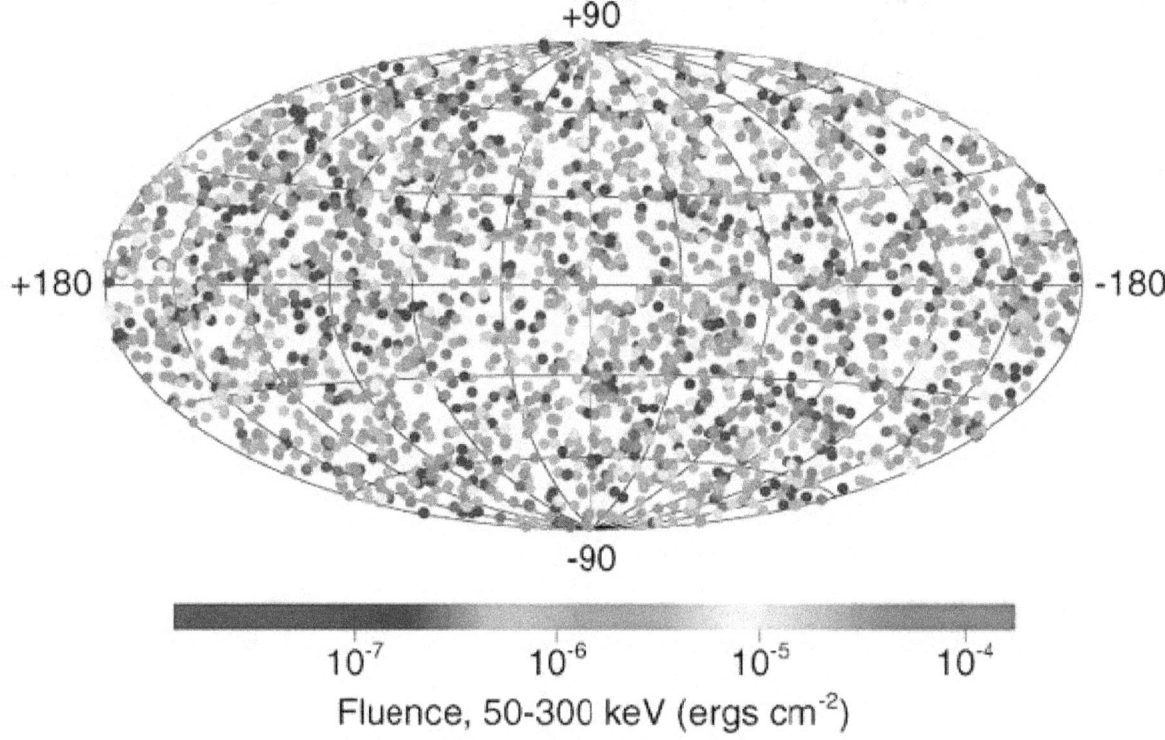

*Positions on the sky of all gamma-ray bursts detected during the BATSE mission. The distribution is isotropic, with no concentration towards the plane of the Milky Way, which runs horizontally through the center of the image.*

consistent with the position derived from the detecting satellites. This suggested an origin of either very faint stars or extremely distant galaxies.[19][20] Even the most accurate positions contained numerous faint stars and galaxies, and it was widely agreed that final resolution of the origins of cosmic gamma-ray bursts would require both new satellites and faster communication.[21]

### 31.1.2   Afterglow

Several models for the origin of gamma-ray bursts postulated that the initial burst of gamma rays should be followed by slowly fading emission at longer wavelengths created by collisions between the burst ejecta and interstellar gas.[22] This fading emission would be called the "afterglow." Early searches for this afterglow were unsuccessful, largely due to the difficulties in observing a burst's position at longer wavelengths immediately after the initial burst. The breakthrough came in February 1997 when the satellite BeppoSAX detected a gamma-ray burst (GRB 970228[nb 2]) and when the X-ray camera was pointed towards the direction from which the burst had originated, it detected fading X-ray emission. The William Herschel Telescope identified a fading optical counterpart 20 hours after the burst.[23] Once the GRB faded, deep imaging was able to identify a faint, distant host galaxy at the location of the GRB as pinpointed by the optical afterglow.[24][25]

Because of the very faint luminosity of this galaxy, its exact distance was not measured for several years. Well before then, another major breakthrough occurred with the next event registered by BeppoSAX, GRB 970508. This event was localized within four hours of its discovery, allowing research teams to begin making observations much sooner than any previous burst. The spectrum of the object revealed a redshift of $z = 0.835$, placing the burst at a distance of roughly 6 billion light years from Earth.[26] This was the first accurate determination of the distance to a GRB, and together with

*The Italian–Dutch satellite BeppoSAX, launched in April 1996, provided the first accurate positions of gamma-ray bursts, allowing follow-up observations and identification of the sources.*

the discovery of the host galaxy of 970228 proved that GRBs occur in extremely distant galaxies.[24][27] Within a few months, the controversy about the distance scale ended: GRBs were extragalactic events originating within faint galaxies at enormous distances. The following year, GRB 980425 was followed within a day by a coincident bright supernova (SN 1998bw), indicating a clear connection between GRBs and the deaths of very massive stars. This burst provided the first strong clue about the nature of the systems that produce GRBs.[28]

BeppoSAX functioned until 2002 and CGRO (with BATSE) was deorbited in 2000. However, the revolution in the study of gamma-ray bursts motivated the development of a number of additional instruments designed specifically to explore the nature of GRBs, especially in the earliest moments following the explosion. The first such mission, HETE-2,[29] launched in 2000 and functioned until 2006, providing most of the major discoveries during this period. One of the most successful space missions to date, Swift, was launched in 2004 and as of 2014 is still operational.[30][31] Swift is equipped with a very sensitive gamma ray detector as well as on-board X-ray and optical telescopes, which can be rapidly and automatically slewed to observe afterglow emission following a burst. More recently, the Fermi mission was launched carrying the Gamma-Ray Burst Monitor, which detects bursts at a rate of several hundred per year, some of which are bright enough to be observed at extremely high energies with Fermi's Large Area Telescope. Meanwhile, on the ground, numerous optical telescopes have been built or modified to incorporate robotic control software that responds immediately to signals sent through the Gamma-ray Burst Coordinates Network. This allows the telescopes to rapidly repoint towards a GRB,

*NASA's Swift Spacecraft launched in November 2004*

often within seconds of receiving the signal and while the gamma-ray emission itself is still ongoing.[32][33]

New developments over the past few years include the recognition of short gamma-ray bursts as a separate class (likely due to merging neutron stars and not associated with supernovae), the discovery of extended, erratic flaring activity at X-ray wavelengths lasting for many minutes after most GRBs, and the discovery of the most luminous (GRB 080319B) and the former most distant (GRB 090423) objects in the universe.[34][35] The most distant known GRB, GRB 090429B, is now the most distant known object in the universe.

## 31.2  Classification

The light curves of gamma-ray bursts are extremely diverse and complex.[36] No two gamma-ray burst light curves are identical,[37] with large variation observed in almost every property: the duration of observable emission can vary from milliseconds to tens of minutes, there can be a single peak or several individual subpulses, and individual peaks can be symmetric or with fast brightening and very slow fading. Some bursts are preceded by a "precursor" event, a weak burst that is then followed (after seconds to minutes of no emission at all) by the much more intense "true" bursting episode.[38] The light curves of some events have extremely chaotic and complicated profiles with almost no discernible patterns.[21]

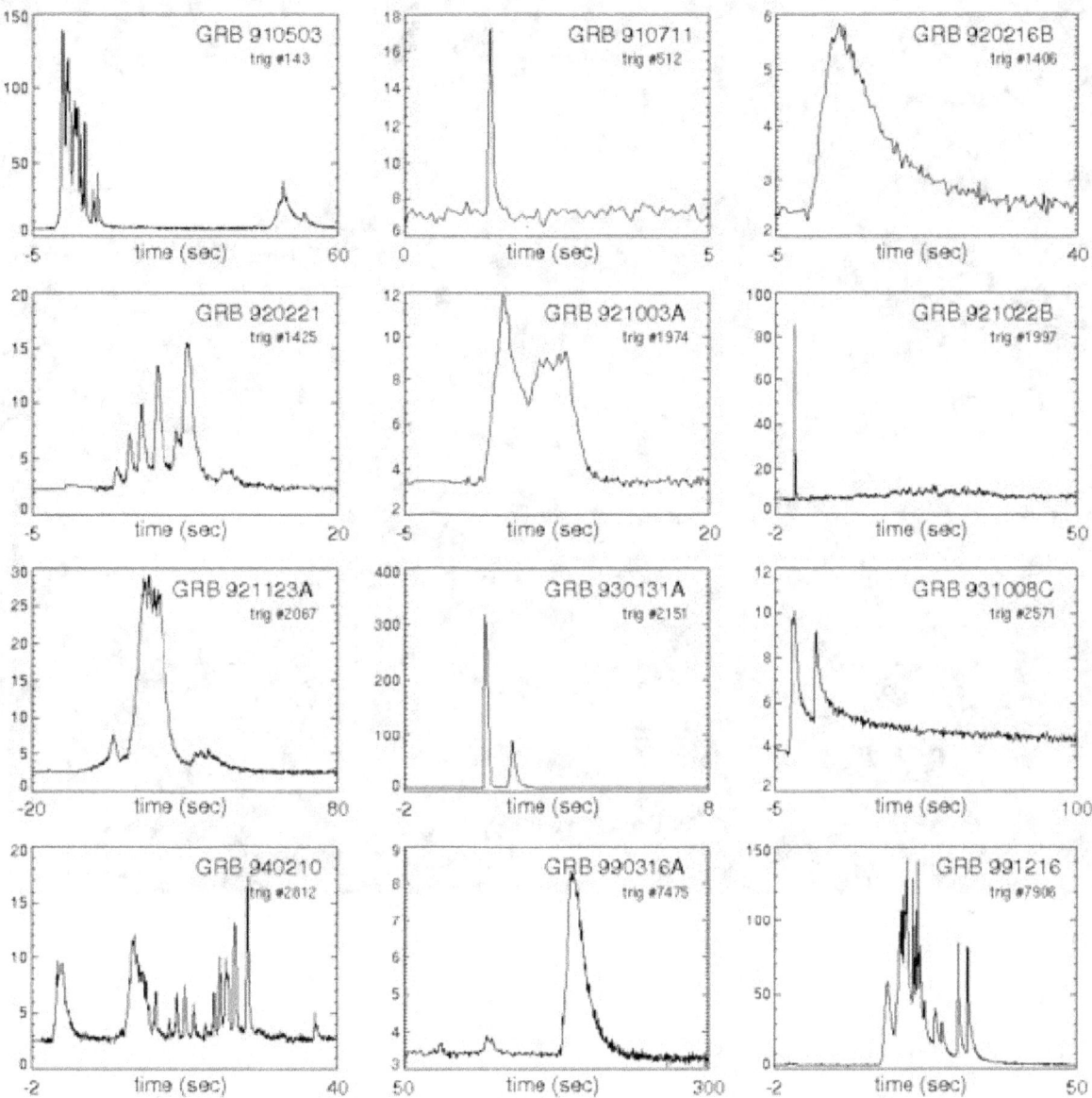

*Gamma-ray burst light curves*

Although some light curves can be roughly reproduced using certain simplified models,[39] little progress has been made in understanding the full diversity observed. Many classification schemes have been proposed, but these are often based solely on differences in the appearance of light curves and may not always reflect a true physical difference in the progenitors of the explosions. However, plots of the distribution of the observed duration[nb 3] for a large number of gamma-ray bursts show a clear bimodality, suggesting the existence of two separate populations: a "short" population with an average duration of about 0.3 seconds and a "long" population with an average duration of about 30 seconds.[40] Both distributions are very broad with a significant overlap region in which the identity of a given event is not clear from duration alone. Additional classes beyond this two-tiered system have been proposed on both observational and theoretical grounds.[41][42][43][44]

*Hubble captures infrared glow of a kilonova blast.*[45]

### 31.2.1   Short gamma-ray bursts

Events with a duration of less than about two seconds are classified as short gamma-ray bursts. These account for about 30% of gamma-ray bursts, but until 2005, no afterglow had been successfully detected from any short event and little was known about their origins.[46] Since then, several dozen short gamma-ray burst afterglows have been detected and localized, several of which are associated with regions of little or no star formation, such as large elliptical galaxies and the central regions of large galaxy clusters.[47][48][49][50] This rules out a link to massive stars, confirming that short events are physically distinct from long events. In addition, there has been no association with supernovae.[51]

The true nature of these objects (or even whether the current classification scheme is accurate) remains unknown, although the leading hypothesis is that they originate from the mergers of binary neutron stars[52] or a neutron star with a black hole. Such mergers might also be expected to produce kilonovae.[53] and evidence for a kilonova associated with GRB 130603B has been seen.[54][55][56] The mean duration of these events of 0.2 seconds suggests a source of very small physical diameter in stellar terms; less than 0.2 light-seconds (about 37,000 miles—four times the Earth's diameter). This further suggests a very compact object as the source. The observation of minutes to hours of X-ray flashes after a short gamma-ray burst is consistent with small particles of a primary object like a neutron star initially swallowed by a black hole in less than two seconds, followed by some hours of lesser energy events, as remaining fragments of tidally disrupted neutron star material (no longer neutronium) remain in orbit to spiral into the black hole, over a longer period of time.[46] A small fraction of short gamma-ray bursts are probably produced by giant flares from soft gamma repeaters in nearby galaxies.[57][58]

### 31.2.2 Long gamma-ray bursts

Most observed events (70%) have a duration of greater than two seconds and are classified as long gamma-ray bursts. Because these events constitute the majority of the population and because they tend to have the brightest afterglows, they have been studied in much greater detail than their short counterparts. Almost every well-studied long gamma-ray burst has been linked to a galaxy with rapid star formation, and in many cases to a core-collapse supernova as well, unambiguously associating long GRBs with the deaths of massive stars.[59] Long GRB afterglow observations, at high redshift, are also consistent with the GRB having originated in star-forming regions.[60]

### 31.2.3 Ultra-long gamma-ray bursts

These events are at the tail end of the long GRB duration distribution, lasting more than 10,000 seconds. They have been proposed to form a separate class, possibly the result of the collapse of a blue supergiant star.[61] Only a small number have been identified to date, their primary characteristic being their gamma ray emission duration. So far, the known and well established ultra long GRBs are GRB 091024A, GRB 101225A, and GRB 111209A.[62][63] A recent study,[64] on the other hand, shows that the existing evidence for a separate ultra-long GRB population with a new type of progenitor is inconclusive, and further multi-wavelength observations are needed to draw a firmer conclusion.

## 31.3 Energetics and beaming

Gamma-ray bursts are very bright as observed from Earth despite their typically immense distances. An average long GRB has a bolometric flux comparable to a bright star of our galaxy despite a distance of billions of light years (compared to a few tens of light years for most visible stars). Most of this energy is released in gamma rays, although some GRBs have extremely luminous optical counterparts as well. GRB 080319B, for example, was accompanied by an optical counterpart that peaked at a visible magnitude of 5.8,[65] comparable to that of the dimmest naked-eye stars despite the burst's distance of 7.5 billion light years. This combination of brightness and distance implies an extremely energetic source. Assuming the gamma-ray explosion to be spherical, the energy output of GRB 080319B would be within a factor of two of the rest-mass energy of the Sun (the energy which would be released were the Sun to be converted entirely into radiation).[34]

No known process in the Universe can produce this much energy in such a short time. Rather, gamma-ray bursts are thought to be highly focused explosions, with most of the explosion energy collimated into a narrow jet traveling at speeds exceeding 99.995% of the speed of light.[66][67] The approximate angular width of the jet (that is, the degree of spread of the beam) can be estimated directly by observing the achromatic "jet breaks" in afterglow light curves: a time after which the slowly decaying afterglow begins to fade rapidly as the jet slows and can no longer beam its radiation as effectively.[68][69] Observations suggest significant variation in the jet angle from between 2 and 20 degrees.[70]

Because their energy is strongly focused, the gamma rays emitted by most bursts are expected to miss the Earth and never be detected. When a gamma-ray burst is pointed towards Earth, the focusing of its energy along a relatively narrow beam causes the burst to appear much brighter than it would have been were its energy emitted spherically. When this effect is taken into account, typical gamma-ray bursts are observed to have a true energy release of about $10^{44}$ J, or about 1/2000 of a Solar mass ($M\odot$) energy equivalent[70]—which is still many times the mass-energy equivalent of the Earth (about 5.5 $\times 10^{41}$ J). This is comparable to the energy released in a bright type Ib/c supernova and within the range of theoretical models. Very bright supernovae have been observed to accompany several of the nearest GRBs.[28] Additional support for focusing of the output of GRBs has come from observations of strong asymmetries in the spectra of nearby type Ic supernova[71] and from radio observations taken long after bursts when their jets are no longer relativistic.[72]

Short (time duration) GRBs appear to come from a lower-redshift (i.e. less distant) population and are less luminous than long GRBs.[73] The degree of beaming in short bursts has not been accurately measured, but as a population they are likely less collimated than long GRBs[74] or possibly not collimated at all in some cases.[75]

*Artist's illustration of a bright gamma-ray burst occurring in a star-forming region. Energy from the explosion is beamed into two narrow, oppositely directed jets.*

## 31.4  Progenitors

Main article: Gamma-ray burst progenitors

 Because of the immense distances of most gamma-ray burst sources from Earth, identification of the progenitors, the systems that produce these explosions, is particularly challenging. The association of some long GRBs with supernovae and the fact that their host galaxies are rapidly star-forming offer very strong evidence that long gamma-ray bursts are associated with massive stars. The most widely accepted mechanism for the origin of long-duration GRBs is the collapsar model,[76] in which the core of an extremely massive, low-metallicity, rapidly rotating star collapses into a black hole in the final stages of its evolution. Matter near the star's core rains down towards the center and swirls into a high-density accretion disk. The infall of this material into a black hole drives a pair of relativistic jets out along the rotational axis, which pummel through the stellar envelope and eventually break through the stellar surface and radiate as gamma rays. Some alternative models replace the black hole with a newly formed magnetar,[77] although most other aspects of the model (the collapse of the core of a massive star and the formation of relativistic jets) are the same.

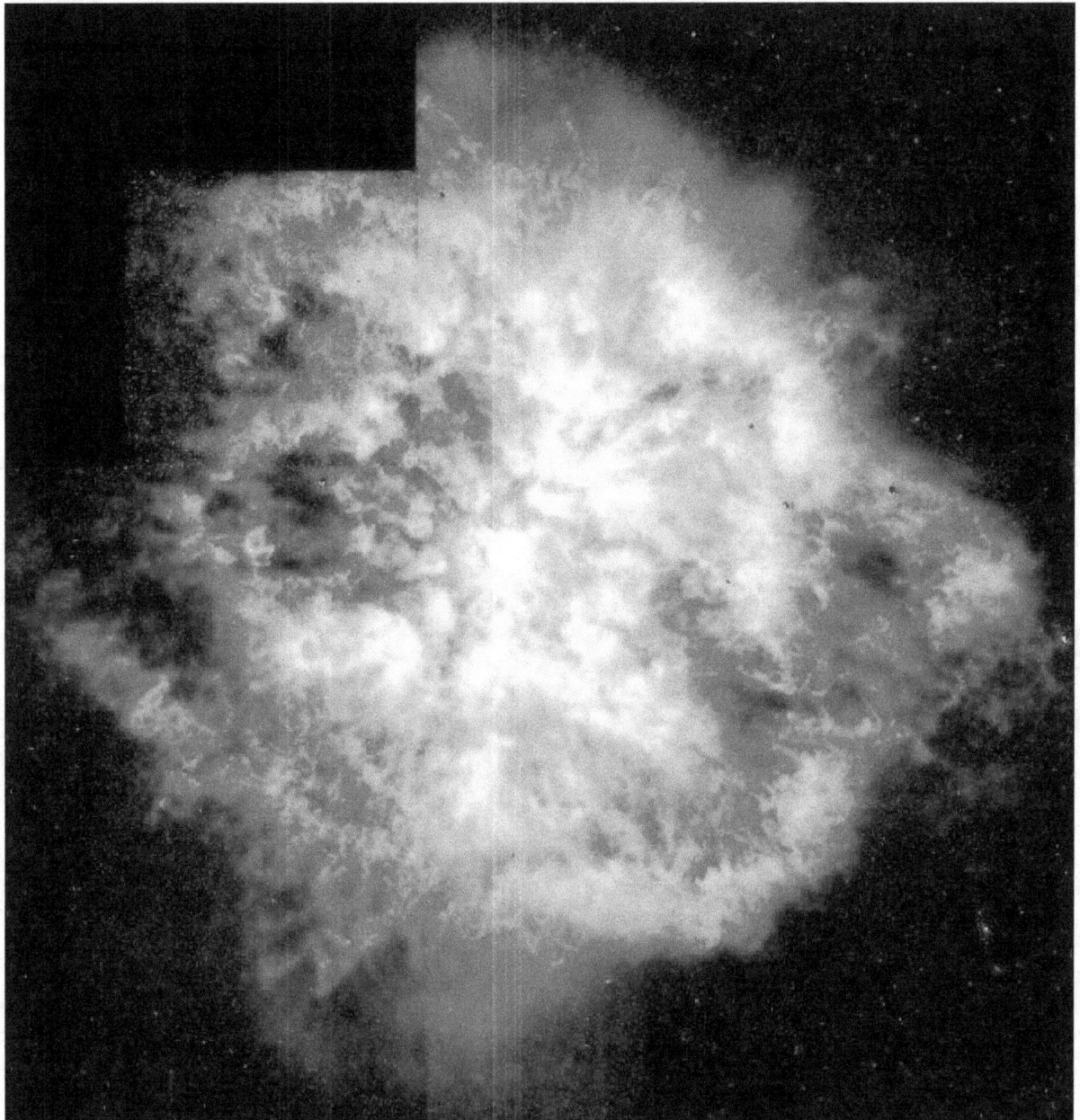

*Hubble Space Telescope image of Wolf–Rayet star WR 124 and its surrounding nebula. Wolf–Rayet stars are candidates for being progenitors of long-duration GRBs.*

The closest analogs within the Milky Way galaxy of the stars producing long gamma-ray bursts are likely the Wolf–Rayet stars, extremely hot and massive stars which have shed most or all of their hydrogen due to radiation pressure. Eta Carinae and WR 104 have been cited as possible future gamma-ray burst progenitors.[78] It is unclear if any star in the Milky Way has the appropriate characteristics to produce a gamma-ray burst.[79]

The massive-star model probably does not explain all types of gamma-ray burst. There is strong evidence that some short-duration gamma-ray bursts occur in systems with no star formation and where no massive stars are present, such as elliptical galaxies and galaxy halos.[73] The favored theory for the origin of most short gamma-ray bursts is the merger of a binary system consisting of two neutron stars. According to this model, the two stars in a binary slowly spiral towards each other due to the release of energy via gravitational radiation[80][81] until the neutron stars suddenly rip each other apart due to tidal forces and collapse into a single black hole. The infall of matter into the new black hole produces an accretion disk and releases a burst of energy, analogous to the collapsar model. Numerous other models have also been proposed to explain short gamma-ray bursts, including the merger of a neutron star and a black hole, the accretion-induced collapse of a neutron star, or the evaporation of primordial black holes.[82][83][84][85]

An alternative explanation proposed by Friedwardt Winterberg is that in the course of a gravitational collapse and in reaching the event horizon of a black hole, all matter disintegrates into a burst of gamma radiation.[86]

### 31.4.1   Tidal disruption events

Main article: Tidal disruption event

This new class of GRB-like events was first discovered through the detection of GRB 110328A by the Swift Gamma-Ray Burst Mission on 28 March 2011. This event had a gamma-ray duration of about 2 days, much longer than even ultra-long GRBs, and was detected in X-rays for many months. It occurred at the center of a small elliptical galaxy at redshift z = 0.3534. There is an ongoing debate as to whether the explosion was the result of stellar collapse or a tidal disruption event accompanied by a relativistic jet, although the latter explanation has become widely favoured.

A tidal disruption event of this sort is when a star interacts with a supermassive black hole shredding the star, and in some cases creating a relativistic jet which produces bright emission of gamma ray radiation. The event GRB 110328A (also denoted Swift J1644+57) was initially argued to be produced by the disruption of main sequence star by a black hole of several million times the mass of the Sun,[87][88][89] although it has subsequently been argued that the disruption of a white dwarf by a black hole of mass about 10 thousand times the Sun may be more likely.[90]

## 31.5   Emission mechanisms

Main article: Gamma-ray burst emission mechanisms

The means by which gamma-ray bursts convert energy into radiation remains poorly understood, and as of 2010 there was still no generally accepted model for how this process occurs.[91] Any successful model of GRB emission must explain the physical process for generating gamma-ray emission that matches the observed diversity of light curves, spectra, and other characteristics.[92] Particularly challenging is the need to explain the very high efficiencies that are inferred from some explosions: some gamma-ray bursts may convert as much as half (or more) of the explosion energy into gamma-rays.[93] Early observations of the bright optical counterparts to GRB 990123 and to GRB 080319B, whose optical light curves were extrapolations of the gamma-ray light spectra,[94][65] have suggested that inverse Compton may be the dominant process in some events. In this model, pre-existing low-energy photons are scattered by relativistic electrons within the explosion, augmenting their energy by a large factor and transforming them into gamma-rays.[95]

The nature of the longer-wavelength afterglow emission (ranging from X-ray through radio) that follows gamma-ray bursts is better understood. Any energy released by the explosion not radiated away in the burst itself takes the form of matter or energy moving outward at nearly the speed of light. As this matter collides with the surrounding interstellar gas, it creates a relativistic shock wave that then propagates forward into interstellar space. A second shock wave, the reverse shock, may propagate back into the ejected matter. Extremely energetic electrons within the shock wave are accelerated by strong local magnetic fields and radiate as synchrotron emission across most of the electromagnetic spectrum.[96][97] This model has generally been successful in modeling the behavior of many observed afterglows at late times (generally, hours to days after the explosion), although there are difficulties explaining all features of the afterglow very shortly after the gamma-ray burst has occurred.[98]

## 31.6   Rate of occurrence and potential effects on life on Earth

All GRBs observed to date have occurred well outside the Milky Way galaxy and have been harmless to Earth. However, if a GRB were to occur within the Milky Way, and its emission were beamed straight towards Earth, the effects could be devastating for the planet. Currently, orbiting satellites detect on average approximately one GRB per day. The closest observed GRB as of March 2014 was GRB 980425, located 40Mpc[99] (130 million light years) away in a (z=0.0085) SBc-type dwarf galaxy.[100] GRB 980425 was far less energetic than the average GRB and was associated with the Type Ib supernova SN 1998bw.[101]

Estimating the exact rate at which GRBs occur is difficult, but for a galaxy of approximately the same size as the Milky Way, the expected rate (for long-duration GRBs) is about one burst every 100,000 to 1,000,000 years.[102] Only a small percentage of these would be beamed towards Earth. Estimates of rate of occurrence of short-duration GRBs are even more uncertain because of the unknown degree of collimation, but are probably comparable.[103]

Since GRBs are thought to involve beamed emission along two jets in opposing directions, only planets in the path of these jets would be subjected to the high energy gamma radiation.[104]

Depending on its distance from Earth, a GRB and its ultraviolet radiation could damage even the most radiation resistant organism known, the bacterium *Deinococcus radiodurans*. These bacteria can endure 2,000 times more radiation than humans. Life surviving an initial onslaught, including those located on the side of the earth facing away from the burst, would have to contend with the potentially lethal after-effect of the depletion of the atmosphere's protective ozone layer by the burst.[105]

### 31.6.1   Hypothetical effects of gamma-ray bursts in the past

GRBs close enough to affect life in some way might occur once every five million years or so – around a thousand times since life on Earth began.[106]

The major Ordovician–Silurian extinction events of 450 million years ago may have been caused by a GRB. The late Ordovician species of trilobite that spent some of its life in the plankton layer near the ocean surface was much harder hit than deep-water dwellers, which tended to stay put within quite restricted areas. Usually it is the more widely spread species that fare better in extinction, and hence this unusual pattern could be explained by a GRB, which would probably devastate creatures living on land and near the ocean surface, but leave deep-sea creatures relatively unharmed.[8]

A case has been made that the cause of the 774–775 carbon-14 spike was the result of a short GRB.

### 31.6.2   Hypothetical effects of gamma-ray bursts in the future

The greatest danger is believed to come from Wolf–Rayet stars, regarded by astronomers as likely GRB candidates. When such stars transition to supernovae, they may emit intense beams of gamma rays, and if Earth were to lie in the beam zone, devastating effects may occur. Gamma rays would not penetrate Earth's atmosphere to impact the surface directly, but they would chemically damage the stratosphere.[8]

For example, if WR 104, at a distance of 8,000 light-years, were to hit Earth with a burst of 10 seconds duration, its gamma rays could deplete about 25 percent of the world's ozone layer. This would result in mass extinction, food chain depletion, and starvation. The side of Earth facing the GRB would receive potentially lethal radiation exposure, which can cause radiation sickness in the short term, and, in the long term, results in serious impacts to life due to ozone layer depletion.[8]

### 31.6.3   Effects after exposure to the gamma-ray burst on Earth's atmosphere

Longer-term, gamma ray energy may cause chemical reactions involving oxygen and nitrogen molecules which may create nitrogen oxide then nitrogen dioxide gas, causing photochemical smog. The GRB may produce enough of the gas to cover the sky and darken it. Gas would prevent sunlight from reaching Earth's surface, producing a "cosmic winter" effect – a similar situation to an impact winter, but not caused by an impact. GRB-produced gas could also even further deplete the ozone layer.

## 31.7   See also

- Gamma-ray astronomy

- List of gamma-ray bursts

    - GRB 020813

    - GRB 130427A

    - GRB 080916C

- Soft gamma repeater

- Gamma-ray Search for Extraterrestrial Intelligence

- Stellar evolution

- Terrestrial gamma-ray flashes

- Fast radio burst

## 31.8   Footnotes

[1]  A notable exception is the 5 March event of 1979, an extremely bright burst that was successfully localized to supernova remnant N49 in the Large Magellanic Cloud. This event is now interpreted as a magnetar giant flare, more related to SGR flares than "true" gamma-ray bursts.

[2]  GRBs are named after the date on which they are discovered: the first two digits being the year, followed by the two-digit month and two-digit day and a letter with the order they were detected during that day. The letter 'A' is appended to the name for the first burst identified, 'B' for the second, and so on. For bursts before the year 2010 this letter was only appended if more than one burst occurred that day.

[3]  The duration of a burst is typically measured by T90, the duration of the period which 90 percent of the burst's energy is emitted. Recently some otherwise "short" GRBs have been shown to be followed by a second, much longer emission episode that when included in the burst light curve results in T90 durations of up to several minutes: these events are only short in the literal sense when this component is excluded.

## 31.9   Notes

[1]  "Gamma Rays". *NASA*.

[2]  Atkinson, Nancy. "New Kind of Gamma Ray Burst is Ultra Long-Lasting". Universetoday.com. Retrieved 2015-05-15.

[3]  Gendre, B.; Stratta, G.; Atteia, J. L.; Basa, S.; Boër, M.; Coward, D. M.; Cutini, S.; d'Elia, V.; Howell, E. J; Klotz, A.; Piro, L. (2013). "The Ultra-Long Gamma-Ray Burst 111209A: The Collapse of a Blue Supergiant?". *The Astrophysical Journal* **766**: 30. arXiv:1212.2392. Bibcode:2013ApJ...766...30G. doi:10.1088/0004-637X/766/1/30.

[4]  Graham, J. F.; Fruchter, A. S. (2013). "The Metal Aversion of LGRBs". *The Astrophysical Journal* **774** (2): 119. arXiv:1211.7068. Bibcode:2013ApJ...774..119G. doi:10.1088/0004-637X/774/2/119.

[5]  Vedrenne & Atteia 2009

[6]  Tsang, David and Read, Jocelyn S. and Hinderer, Tanja and Piro, Anthony L. and Bondarescu, Ruxandra (2012). "Resonant Shattering of Neutron Star Crust". *Physical Review Letters* **108**. p. 5. doi:10.1103/PhysRevLett.108.011102.

[7]  Podsiadlowski 2004

[8]  Melott 2004

[9] Hurley 2003

[10] Schilling 2002, p.12–16

[11] Klebesadel R.W., Strong I.B., and Olson R.A. (1973). "Observations of Gamma-Ray Bursts of Cosmic Origin". *Astrophysical Journal Letters* **182**: L85. Bibcode:1973ApJ...182L..85K. doi:10.1086/181225.

[12] Meegan 1992

[13] Vedrenne & Atteia 2009, p. 16–40

[14] Schilling 2002, p.36–37

[15] Paczyński 1999, p. 6

[16] Piran 1992

[17] Lamb 1995

[18] Hurley 1986, p. 33

[19] Pedersen 1987

[20] Hurley 1992

[21] Fishman & Meegan 1995

[22] Paczynski 1993

[23] van Paradijs 1997

[24] Vedrenne & Atteia 2009, p. 90 – 93

[25] Schilling 2002, p. 102

[26] Reichart 1995

[27] Schilling 2002, p. 118–123

[28] Galama 1998

[29] Ricker 2003

[30] McCray 2008

[31] Gehrels 2004

[32] Akerlof 2003

[33] Akerlof 1999

[34] Bloom 2009

[35] Reddy 2009

[36] Katz 2002, p. 37

[37] Marani 1997

[38] Lazatti 2005

[39] Simić 2005

[40] Kouveliotou 1994

[41] Horvath 1998

[42] Hakkila 2003

[43] Chattopadhyay 2007

[44] Virgili 2009

[45] "Hubble captures infrared glow of a kilonova blast". *Image Gallery*. ESA/Hubble. Retrieved 14 August 2013.

[46] In a Flash NASA Helps Solve 35-year-old Cosmic Mystery. NASA (2005-10-05) The 30% figure is given here, as well as afterglow discussion.

[47] Bloom 2006

[48] Hjorth 2005

[49] Berger 2007

[50] Gehrels 2005

[51] Zhang 2009

[52] Nakar 2007

[53] Metzger, B. D.; Martinez-Pinedo, G.; Darbha, S.; Quataert, E.; et al. (August 2010). "Electromagnetic counterparts of compact object mergers powered by the radioactive decay of r-process nuclei". *Monthly Notices of the Royal Astronomical Society* **406** (4): 2650. arXiv:1001.5029. Bibcode:2010MNRAS.406.2650M. doi:10.1111/j.1365-2966.2010.16864.x.

[54] Tanvir, N. R.; Levan, A. J.; Fruchter, A. S.; Hjorth, J.; Hounsell, R. A.; Wiersema, K.; Tunnicliffe, R. L. (2013). "A 'kilonova' associated with the short-duration γ-ray burst GRB 130603B". *Nature* **500** (7464): 547–9. arXiv:1306.4971. Bibcode:2013Natur.500..547T. doi:10.1038/nature12505. PMID 23912055.

[55] Berger, E.; Fong, W.; Chornock, R. (2013). "ANr-PROCESS KILONOVA ASSOCIATED WITH THE SHORT-HARD GRB 130603B". *The Astrophysical Journal* **774** (2): L23. arXiv:1306.3960. Bibcode:2013ApJ...774L..23B. doi:10.1088/2041-8205/774/2/L23.

[56] Nicole Gugliucci (7 August 2013). "Kilonova Alert! Hubble Solves Gamma Ray Burst Mystery". *news.discovery.com*. Discovery Communications. Retrieved 22 January 2015.

[57] Frederiks 2008

[58] Hurley 2005

[59] Woosley & Bloom 2006

[60] Pontzen et al. 2010

[61] Gendre, B.; Stratta, G.; Atteia, J. L.; Basa, S.; Boër, M.; Coward, D. M.; Cutini, S.; d'Elia, V.; Howell, E. J; Klotz, A.; Piro, L. (2013). "The Ultra-Long Gamma-Ray Burst 111209A: The Collapse of a Blue Supergiant?". *The Astrophysical Journal* **766**: 30. arXiv:1212.2392. Bibcode:2013ApJ...766...30G. doi:10.1088/0004-637X/766/1/30.

[62] Boer, Michel; Gendre, Bruce; Stratta, Giulia (2013). "Are Ultra-long Gamma-Ray Bursts different?". *The Astrophysical Journal* **800**: 16. arXiv:1310.4944. Bibcode:2015ApJ...800...16B. doi:10.1088/0004-637X/800/1/16.

[63] Virgili, F. J.; Mundell, C. G.; Pal'Shin, V.; Guidorzi, C.; Margutti, R.; Melandri, A.; Harrison, R.; Kobayashi, S.; Chornock, R.; Henden, A.; Updike, A. C.; Cenko, S. B.; Tanvir, N. R.; Steele, I. A.; Cucchiara, A.; Gomboc, A.; Levan, A.; Cano, Z.; Mottram, C. J.; Clay, N. R.; Bersier, D.; Kopač, D.; Japelj, J.; Filippenko, A. V.; Li, W.; Svinkin, D.; Golenetskii, S.; Hartmann, D. H.; Milne, P. A.; et al. (2013). "Grb 091024A and the Nature of Ultra-Long Gamma-Ray Bursts". *The Astrophysical Journal* **778**: 54. arXiv:1310.0313. Bibcode:2013ApJ...778...54V. doi:10.1088/0004-637X/778/1/54.

[64] Zhang, Bin-Bin; Zhang, Bing; Murase, Kohta; Connaughton, Valerie; Briggs, Michael S. (2013). "How Long does a Burst Burst?". *The Astrophysical Journal* **787**: 66. arXiv:1310.2540v2. Bibcode:2014ApJ...787...66Z. doi:10.1088/0004-637X/787/1

[65] Racusin 2008

[66] Rykoff 2009

[67] Abdo 2009

[68] Sari 1999

[69] Burrows 2006

[70] Frail 2001

[71] Mazzali 2005

[72] Frail 2000

[73] Prochaska 2006

[74] Watson 2006

[75] Grupe 2006

[76] MacFadyen 1999

[77] Metzger 2007

[78] Plait 2008

[79] Stanek 2006

[80] Abbott 2007

[81] Kochanek 1993

[82] Vietri 1998

[83] MacFadyen 2006

[84] Blinnikov 1984

[85] Cline 1996

[86] Winterberg, Friedwardt (2001 Aug 29). "Gamma-Ray Bursters and Lorentzian Relativity". Z. Naturforsch 56a: 889–892.

[87] Science Daily 2011

[88] Levan 2011

[89] Bloom 2011

[90] Krolick & Piran 11

[91] Stern 2007

[92] Fishman, G. 1995

[93] Fan & Piran 2006

[94] Liang et al. 1999, GRB 990123: The Case for Saturated Comptonization, The Astrophysical Journal, 519:L21-L24, 1999 July 1. http://iopscience.iop.org/1538-4357/519/1/L21/fulltext/995164.text.html

[95] Wozniak 2009

[96] Meszaros 1997

[97] Sari 1998

[98] Nousek 2006

[99] Soderberg, A. M.; Kulkarni, S. R.; Berger, E.; Fox, D. W.; Sako, M.; Frail, D. A.; Gal-Yam, A.; Moon, D. S.; Cenko, S. B.; Yost, S. A.; Phillips, M. M.; Persson, S. E.; Freedman, W. L.; Wyatt, P.; Jayawardhana, R.; Paulson, D. (2004). "The sub-energetic γ-ray burst GRB 031203 as a cosmic analogue to the nearby GRB 980425". Nature 430 (7000): 648–650. arXiv:astro-ph/0408096. Bibcode:2004Natur.430..648S. doi:10.1038/nature02757. PMID 15295592.

[100] Le Floc'h, E.; Charmandaris, V.; Gordon, K.; Forrest, W. J.; Brandl, B.; Schaerer, D.; Dessauges-Zavadsky, M.; Armus, L. (2011). "The first Infrared study of the close environment of a long Gamma-Ray Burst". *The Astrophysical Journal* **746**: 7. arXiv:1111.1234. Bibcode:2012ApJ...746....7L. doi:10.1088/0004-637X/746/1/7.

[101] Kippen, R.M.; Briggs, M. S.; Kommers, J. M.; Kouveliotou, C.; Hurley, K.; Robinson, C. R.; Van Paradijs, J.; Hartmann, D. H.; Galama, T. J.; Vreeswijk, P. M. (October 1998). "On the Association of Gamma-Ray Bursts with Supernovae". *The Astrophysical Journal* **506** (1): L27–L30. arXiv:astro-ph/9806364. Bibcode:1998ApJ...506L..27K. doi:10.1086/311634.

[102] "Gamma-ray burst 'hit Earth in 8th Century'". *Rebecca Morelle*. BBC. 2013-01-21. Retrieved January 21, 2013.

[103] Guetta and Piran 2006

[104] Welsh, Jennifer (2011-07-10). "Can gamma-ray bursts destroy life on Earth?". MSN. Retrieved October 27, 2011.

[105] "Death from across the galaxy". World-science.net. Retrieved 2012-12-30.

[106] New Scientist print edition, 15 December 2001, p 10). John Scalo and Craig Wheeler of the University of Texas at Austin

## 31.10   Books

- Vedrenne, G and Atteia, J.-L. (2009). *Gamma-Ray Bursts: The brightest explosions in the Universe*. Springer. ISBN 978-3-540-39085-5.

- Chryssa Kouveliotou, Stanford E. Woosley, Ralph A. M. J., ed. (2012). *Gamma-ray bursts*. Cambridge: Cambridge University Press. ISBN 0-521-66209-5.

## 31.11   References

- Abbott, B.; et al. (2007). "Search for Gravitational Waves Associated with 39 Gamma-Ray Bursts Using Data from the Second, Third, and Fourth LIGO Runs". *Physical Review D* **77** (6): 062004. arXiv:0709.0766. Bibcode:2008 PhRvD..77f2004A.doi:10.1103/PhysRevD.77.062004.

- Abdo, A.A.; et al. (2009). "Fermi Observations of High-Energy Gamma-Ray Emission from GRB 080916C". *Science* **323** (5922): 1688–93. Bibcode:2009Sci...323.1688A. doi:10.1126/science.1169101. PMID 19228997.

- Akerlof, C.; et al. (1999). "Observation of contemporaneous optical radiation from a gamma-ray burst". *Nature* **398** (3): 400–402. arXiv:astro-ph/9903271. Bibcode:1999Natur.398..400A. doi:10.1038/18837.

- Akerlof, C.; et al. (2003). "The ROTSE-III Robotic Telescope System". *Publications of the Astronomical Society of the Pacific* **115** (803): 132–140. arXiv:astro-ph/0210238. Bibcode:2003PASP..115..132A. doi:10.1086/345490.

- Atwood, W.B.; Fermi/LAT Collaboration (2009). "The Large Area Telescope on the Fermi Gamma-ray Space Telescope Mission". *The Astrophysical Journal* **697** (2): 1071. arXiv:0902.1089. Bibcode:2009ApJ...697.1071A. doi:10.1088/0004-637X/697/2/1071.

- Ball, J.A. (1995). "Gamma-Ray Bursts: The ETI Hypothesis". *The Astrophysical Journal*.

- Barthelmy, S.D.; et al. (2005). "The Burst Alert Telescope (BAT) on the SWIFT Midex Mission". *Space Science Reviews* **120** (3–4): 143–164. arXiv:astro-ph/0507410. Bibcode:2005SSRv..120..143B. doi:10.1007/s11214-005-5096-3.

- Berger, E.; et al. (2007). "Galaxy Clusters Associated with Short GRBs. I. The Fields of GRBs 050709, 050724, 050911, and 051221a". *Astrophysical Journal* **660**: 496–503. arXiv:astro-ph/0608498. Bibcode:2007ApJ...660..4 .doi:10.1086/512664.

- Blinnikov, S.; et al. (1984). "Exploding Neutron Stars in Close Binaries". *Soviet Astronomy Letters* **10**: 177. Bibcode:1984SvAL...10..177B.

- Bloom, J.S.; et al. (2006). "Closing in on a Short-Hard Burst Progenitor: Constraints from Early-Time Optical Imaging and Spectroscopy of a Possible Host Galaxy of GRB 050509b". *Astrophysical Journal* **638**: 354–368. arXiv:astro-ph/0505480. Bibcode:2006ApJ...638..354B. doi:10.1086/498107.

- Bloom, J.S.; et al. (2009). "Observations of the Naked-Eye GRB 080319B: Implications of Nature's Brightest Explosion". *Astrophysical Journal* **691**: 723–737. arXiv:0803.3215. Bibcode:2009ApJ...691..723B. doi:10.1088/000

- Bloom, J. S.; et al. (2011). "A Possible Relativistic Jetted Outburst from a Massive Black Hole Fed by a Tidally Disrupted Star". *Science* **332** (6039): 203. arXiv:1104.3257. Bibcode:2011Sci...333..203B. doi:10.1126/science.1207150.

- Burrows, D.N.; et al. (2006). "Jet Breaks in Short Gamma-Ray Bursts. II. The Collimated Afterglow of GRB 051221A". *Astrophysical Journal* **653**: 468–473. arXiv:astro-ph/0604320. Bibcode:2006ApJ...653..468B.doi:10.

- Cline, D.B. (1996). "Primordial black-hole evaporation and the quark–gluon phase transition". *Nuclear Physics A* **610**: 500. Bibcode:1996NuPhA.610..500C. doi:10.1016/S0375-9474(96)00383-1.

- Chattopadhyay, T.; et al. (2007). "Statistical Evidence for Three Classes of Gamma-Ray Bursts". *Astrophysical Journal* **667** (2): 1017. arXiv:0705.4020. Bibcode:2007ApJ...667.1017C. doi:10.1086/520317.

- Ejzak, L.M.; et al. (2007). "Terrestrial Consequences of Spectral and Temporal Variability in Ionizing Photon Events". *Astrophysical Journal* **654**: 373–384. arXiv:astro-ph/0604556. Bibcode:2007ApJ...654..373E.doi:10.

- Fan, Y. and Piran, T. (2006). "Gamma-ray burst efficiency and possible physical processes shaping the early afterglow". *Monthly Notices of the Royal Astronomical Society* **369**: 197–206. arXiv:astro-ph/0601054. Bibcode:2006 doi:10.1111/j.1365-2966.2006.10280.x.

- Fishman, C.J. and Meegan, C.A. (1995). "Gamma-Ray Bursts". *Annual Review of Astronomy and Astrophysics* **33**: 415–458. Bibcode:1995ARA&A..33..415F. doi:10.1146/annurev.aa.33.090195.002215.

- Fishman, G.J. (1995). "Gamma-Ray Bursts: An Overview". NASA. Retrieved 2007-10-12.

- Frail, D.A.; et al. (2001). "Beaming in Gamma-Ray Bursts: Evidence for a Standard Energy Reservoir". *Astrophysical Journal Letters* **562**: L557–L558. arXiv:astro-ph/0102282. Bibcode:2001ApJ...562L..55F. doi:10.1086/338119.

- Frail, D.A.; et al. (2000). "A 450 Day Light Curve of the Radio Afterglow of GRB 970508: Fireball Calorimetry". *Astrophysical Journal* **537**(7): 191–204. arXiv:astro-ph/9910319. Bibcode:2000ApJ...537..191F.doi:10.1086/309

- Frederiks, D.; et al. (2008). "GRB 051103 and GRB 070201 as Giant Flares from SGRs in Nearby Galaxies". In Galassi, Palmer, and Fenimore. *American Institute of Physics Conference Series* **1000**. pp. 271–275. Bibcode:2008AIPC.1000..271F. doi:10.1063/1.2943461.

- Frontera, F. and Piro, L. (1998). *Proceedings of Gamma-Ray Bursts in the Afterglow Era*. Astronomy and Astrophysics Supplement Series.

- Galama, T.J.; et al. (1998). "An unusual supernova in the error box of the gamma-ray burst of 25 April 1998". *Nature* **395** (6703): 670–672. arXiv:astro-ph/9806175. Bibcode:1998Natur.395..670G. doi:10.1038/27150.

- Garner, R. (2008). "NASA's Swift Catches Farthest Ever Gamma-Ray Burst". NASA. Retrieved 2008-11-03.

- Gehrels, N.; et al. (2004). "The Swift Gamma-Ray Burst Mission". *Astrophysical Journal* **611** (2): 1005–1020. Bibcode:2004ApJ...611.1005G. doi:10.1086/422091.

- Gehrels, N.; et al. (2005). "A short gamma-ray burst apparently associated with an elliptical galaxy at redshift z=0.225". *Nature* **437** (7060): 851–854. arXiv:astro-ph/0505630. Bibcode:2005Natur.437..851G. doi:10.1038 PMID16208363.

- Grupe, D.; et al. (2006). "Jet Breaks in Short Gamma-Ray Bursts. I: The Uncollimated Afterglow of GRB 050724". *Astrophysical Journal* **653**: 462. arXiv:astro-ph/0603773. Bibcode:2006ApJ...653..462G.doi:10.1086/5

- Guetta, D. and Piran, T. (2006). "The BATSE-Swift luminosity and redshift distributions of short-duration GRBs". *Astronomy and Astrophysics* **453** (3): 823–828. arXiv:astro-ph/0511239. Bibcode:2006A&A...453..823G. doi:10. 6361:20054498.

- Hakkila, J.; et al. (2003). "How Sample Completeness Affects Gamma-Ray Burst Classification". *Astrophysical Journal* **582**: 320. arXiv:astro-ph/0209073. Bibcode:2003ApJ...582..320H. doi:10.1086/344568.

- Horvath, I. (1998). "A Third Class of Gamma-Ray Bursts?". *Astrophysical Journal* **508** (2): 757. arXiv:astro-ph/9803077. Bibcode:1998ApJ...508..757H. doi:10.1086/306416.

- Hjorth, J.; et al. (2005). "GRB 050509B: Constraints on Short Gamma-Ray Burst Models". *Astrophysical Journal Letters* **630** (2): L117–L120. arXiv:astro-ph/0506123. Bibcode:2005ApJ...630L.117H. doi:10.1086/491733.

- Hurley, K., Cline, T. and Epstein, R. (1986). "Error Boxes and Spatial Distribution". In Liang, E.P. and Petrosian, V. *AIP Conference Proceedings*. Gamma-Ray Bursts **141**. American Institute of Physics. pp. 33–38. ISBN 0-88318-340-4.

- Hurley, K. (1992). "Gamma-Ray Bursts – Receding from Our Grasp". *Nature* **357** (6374): 112. Bibcode:1992Natur .357..112H. doi:10.1038/357112a0.

- Hurley, K. (2003). "A Gamma-Ray Burst Bibliography, 1973–2001" (PDF). In Ricker, G.R. and Vanderspek, R.K. *Gamma-Ray Burst and Afterglow Astronomy, 2001: A Workshop Celebrating the First Year of the HETE Mission*. American Institute of Physics. pp. 153–155. ISBN 0-7354-0122-5.

- Hurley, K.; et al. (2005). "An exceptionally bright flare from SGR 1806–20 and the origins of short-duration gamma-ray bursts". *Nature* **434** (7037): 1098–1103. arXiv:astro-ph/0502329. Bibcode:2005Natur.434.1098H. doi:10.1038/nature03519. PMID 15858565.

- Katz, J.I. (2002). *The Biggest Bangs*. Oxford University Press. ISBN 0-19-514570-4.

- Klebesadel, R.; et al. (1973). "Observations of Gamma-Ray Bursts of Cosmic Origin". *Astrophysical Journal Letters* **182**: L85. Bibcode:1973ApJ...182L..85K. doi:10.1086/181225.

- Kochanek, C.S. and Piran, T. (1993). "Gravitational Waves and Gamma-Ray Bursts". *Astrophysical Journal Letters* **417**: L17–L23. arXiv:astro-ph/9305015. Bibcode:1993ApJ...417L..17K. doi:10.1086/187083.

- Kouveliotou, C.; et al. (1993). "Identification of two classes of gamma-ray bursts". *Astrophysical Journal Letters* **413**: L101. Bibcode:1993ApJ...413L.101K. doi:10.1086/186969.

- Lamb, D.Q. (1995). "The Distance Scale to Gamma-Ray Bursts". *Publications of the Astronomical Society of the Pacific* **107**: 1152. Bibcode:1995PASP..107.1152L. doi:10.1086/133673.

- Lazzati, D. (2005). "Precursor activity in bright, long BATSE gamma-ray bursts". *Monthly Notices of the Royal Astronomical Society* **357** (2): 722–731. arXiv:astro-ph/0411753. Bibcode:2005MNRAS.357..722L. doi:10.1111/j.1 -2966.2005.08687.x.

- Krolik J. and Piran T.. (2011). "Swift J1644+57: A White Dwarf Tidally Disrupted by a $10^4$ M_{odot} Black Hole?". *The Astrophysical Journal* **743** (2): 134. arXiv:1106.0923. Bibcode:2011ApJ...743..134K. doi:10.1088/00 637x/743/2/134.

- Levan, A. J.; et al. (2011). "An Extremely Luminous Panchromatic Outburst from the Nucleus of a Distant Galaxy". *Science* **332** (6039): 199. arXiv:1104.3356. Bibcode:2011Sci...333..199L. doi:10.1126/science.1207143.

- MacFadyen, A.I. and Woosley, S. (1999). "Collapsars: Gamma-Ray Bursts and Explosions in "Failed Supernovae"".*Astrophysical Journal* **524**: 262–289. arXiv:astro-ph/9810274. Bibcode:1999ApJ...524..262M.doi:10.1

- MacFadyen, A.I. (2006). "Late flares from GRBs — Clues about the Central Engine". *AIP Conference Proceedings* **836**: 48–53. Bibcode:2006AIPC..836...48M. doi:10.1063/1.2207856.

- Marani, G.F.; et al. (1997). "On Similarities among GRBs". *Bulletin of the American Astronomical Society* **29**: 839. Bibcode:1997AAS...190.4311M.

- Mazzali, P.A.; et al. (2005). "An Asymmetric Energetic Type Ic Supernova Viewed Off-Axis, and a Link to Gamma Ray Bursts". *Science* **308** (5726): 1284–1287. arXiv:astro-ph/0505199. Bibcode:2005Sci...308.1284M. doi:10.1126/science.1111384. PMID 15919986.

- "The Annihilating Effects of Space Travel". The University of Sydney. 2012. arXiv:1202.5708v1.

- Meegan, C.A.; et al. (1992). "Spatial distribution of gamma-ray bursts observed by BATSE". *Nature* **355** (6356): 143. Bibcode:1992Natur.355..143M. doi:10.1038/355143a0.

- Melott, A.L.; et al. (2004). "Did a gamma-ray burst initiate the late Ordovician mass extinction?". *International Journal of Astrobiology* **3**: 55–61. arXiv:astro-ph/0309415. Bibcode:2004IJAsB...3...55M.doi:10.1017/S1473550

- Meszaros, P. and Rees, M.J. (1997). "Optical and Long-Wavelength Afterglow from Gamma-Ray Bursts".*Astrophy Journal* **476**: 232. arXiv:astro-ph/9606043. Bibcode:1997ApJ...476..232M. doi:10.1086/303625.

- Metzger, B.; et al. (2007). "Proto-Neutron Star Winds, Magnetar Birth, and Gamma-Ray Bursts". *AIP Conference Proceedings SUPERNOVA 1987A: 20 YEARS AFTER: Supernovae and Gamma-Ray Bursters* **937**. pp. 521–525. Bibcode:2007AIPC..937..521M. doi:10.1063/1.2803618.

- Mukherjee, S.; et al. (1998). "Three Types of Gamma-Ray Bursts". *Astrophysical Journal* **508**: 314. arXiv:astro-ph/9802085. Bibcode:1998ApJ...508..314M. doi:10.1086/306386.

- Nakar, E. (2007). "Short-hard gamma-ray bursts". *Physics Reports* **442**: 166–236. arXiv:astro-ph/0701748. Bibcode:2007PhR...442..166N. doi:10.1016/j.physrep.2007.02.005.

- McCray, Richard; et al. "Report of the 2008 Senior Review of the Astrophysics Division Operating Missions" (PDF).

- "Very Large Array Detects Radio Emission From Gamma-Ray Burst" (Press release). National Radio Astronomy Observatory. 15 May 1997. Retrieved 2009-04-04.

- Nousek, J.A.; et al. (2006). "Evidence for a Canonical Gamma-Ray Burst Afterglow Light Curve in the Swift XRT Data". *Astrophysical Journal* **642**: 389–400. arXiv:astro-ph/0508332. Bibcode:2006ApJ...642..389N. doi:10.1086/500724.

- Paczyński, B. and Rhoads, J.E. (1993). "Radio Transients from Gamma-Ray Bursters". *The Astrophysical Journal* **418**: 5. arXiv:astro-ph/9307024. Bibcode:1993ApJ...418L...5P. doi:10.1086/187102.

- Paczyński, B. (1995). "How Far Away Are Gamma-Ray Bursters?". *Publications of the Astronomical Society of the Pacific* **107**: 1167. arXiv:astro-ph/9505096. Bibcode:1995PASP..107.1167P. doi:10.1086/133674.

- Paczyński, B. (1999). "Gamma-Ray Burst–Supernova relation". In M. Livio, N. Panagia, K. Sahu. *Supernovae and Gamma-Ray Bursts: The Greatest Explosions Since the Big Bang*. Space Telescope Science Institute. pp. 1–8. ISBN 0-521-79141-3.

- Pedersen, H.; et al. (1986). "Deep Searches for Burster Counterparts". In Liang, Edison P.; Petrosian, Vahé. *AIP Conference Proceedings*. Gamma-Ray Bursts **141**. American Institute of Physics. pp. 39–46. ISBN 0-88318-340-4.

- Plait, Phil (2 March 2008). "WR 104: A nearby gamma-ray burst?". *Bad Astronomy*. Retrieved 2009-01-07.

- Piran, T. (1992). "The implications of the Compton (GRO) observations for cosmological gamma-ray bursts". *Astrophysical Journal Letters* **389**: L45. Bibcode:1992ApJ...389L..45P. doi:10.1086/186345.

- Piran, T. (1997). "Toward understanding gamma-ray bursts". In Bahcall, J.N. and Ostriker, J. *Unsolved Problems in Astrophysics*. p. 343. Bibcode:1997upa..conf..343P.

- Podsiadlowski, Ph.; et al. (2004). "The Rates of Hypernovae and Gamma-Ray Bursts: Implications for Their Progenitors". *Astrophysical Journal Letters* **607**: L17. arXiv:astro-ph/0403399. Bibcode:2004ApJ...607L..17P. doi:10.1086/421347.

- Pontzen, A.; et al. (2010). "The nature of HI absorbers in GRB afterglows: clues from hydrodynamic simulations". *MNRAS* **402** (3): 1523. arXiv:0909.1321. Bibcode:2010MNRAS.402.1523P. doi:10.1111/j.1365-2966.2009.16017.x.

- Prochaska, J.X.; et al. (2006). "The Galaxy Hosts and Large-Scale Environments of Short-Hard Gamma-Ray Bursts".*Astrophysical Journal***641**(2): 989. arXiv:astro-ph/0510022. Bibcode:2006ApJ...642..989P.doi:10.1086

- Racusin, J.L.; et al. (2008). "Broadband observations of the naked-eye gamma-ray burst GRB080319B". *Nature* **455**(7210): 183–188. arXiv:0805.1557. Bibcode:2008Natur.455..183R.doi:10.1038/nature07270. PMID

- Reddy, F. (28 April 2009). "New Gamma-Ray Burst Smashes Cosmic Distance Record" (Press release). NASA. Retrieved 2009-05-16.

- Ricker, G.R. and Vanderspek, R.K. (2003). "The High Energy Transient Explorer (HETE): Mission and Science Overview". In Ricker, G.R. and Vanderspek, R.K. *Gamma-Ray Burst and Afterglow Astronomy 2001: A Workshop Celebrating the First Year of the HETE Mission*. American Institute of Physics Conference Series **662**. pp. 3–16. Bibcode:2003AIPC..662....3R. doi:10.1063/1.1579291.

- Reichart, Daniel E. (1998). "The Redshift of GRB 970508". *Astrophysical Journal Letters* **495** (2): L99. arXiv:astro-ph/9712100. Bibcode:1998ApJ...495L..99R. doi:10.1086/311222.

- Rykoff, E.; et al. (2009). "Looking Into the Fireball: ROTSE-III and Swift Observations of Early GRB Afterglows". *Astrophysical Journal* **702**: 489. arXiv:0904.0261. Bibcode:2009ApJ...702..489R. doi:10.1088/0004-637X/702/1/489.

- Sari, R; Piran, T; Narayan, R (1998). "Spectra and Light Curves of Gamma-Ray Burst Afterglows". *Astrophysical Journal Letters* **497** (5): L17. arXiv:astro-ph/9712005. Bibcode:1998ApJ...497L..17S. doi:10.1086/311269.

- Sari, R; Piran, T; Halpern, J.P (1999). "Jets in Gamma-Ray Bursts". *Astrophysical Journal Letters* **519**: L17–L20. arXiv:astro-ph/9903339. Bibcode:1999ApJ...519L..17S. doi:10.1086/312109.

- Schilling, Govert (2002). *Flash! The hunt for the biggest explosions in the universe*. Cambridge University Press. ISBN 0-521-80053-6.

- "Gamma-Ray Flash Came from Star Being Eaten by Massive Black Hole". *Science Daily web site*. ScienceDaily LLC. 2011-06-16. Retrieved 2011-06-19. External link in |work= (help)

- Simić, S.; et al. (2005). "A model for temporal variability of the GRB light curve". In Bulik, T., Rudak, B, and Madejski, G. *Astrophysical Sources of High Energy Particles and Radiation*. American Institute of Physics Conference Series **801**. pp. 139–140. Bibcode:2005AIPC..801..139S. doi:10.1063/1.2141849.

- Stanek, K.Z.; et al. (2006). "Protecting Life in the Milky Way: Metals Keep the GRBs Away" (PDF). *Acta Astronomica* **56**: 333. arXiv:astro-ph/0604113. Bibcode:2006AcA....56..333S.

- Stern, Boris E. and Poutanen, Juri (2004). "Gamma-ray bursts from synchrotron self-Compton emission". *Monthly Notices of the Royal Astronomical Society* **352** (3): L35–L39. arXiv:astro-ph/0405488. Bibcode:2004MNRAS.352 doi:10.1111/j.1365-2966.2004.08163.x.

- Thorsett, S.E. (1995). "Terrestrial implications of cosmological gamma-ray burst models". *Astrophysical Journal Letters* **444**: L53. arXiv:astro-ph/9501019. Bibcode:1995ApJ...444L..53T. doi:10.1086/187858.

- "TNG caught the farthest GRB observed ever". Fundación Galileo Galilei. 24 April 2009. Retrieved 2009-04-25.

- van Paradijs, J.; et al. (1997). "Transient optical emission from the error box of the gamma-ray burst of 28 February 1997". *Nature* **386** (6626): 686. Bibcode:1997Natur.386..686V. doi:10.1038/386686a0.

- Vedrenne, G and Atteia, J.-L. (2009). *Gamma-Ray Bursts: The brightest explosions in the Universe*. Springer. ISBN 978-3-540-39085-5.

- Vietri, M. and Stella, L. (1998). "A Gamma-Ray Burst Model with Small Baryon Contamination". *Astrophysical Journal Letters* **507**: L45–L48. arXiv:astro-ph/9808355. Bibcode:1998ApJ...507L..45V. doi:10.1086/311674.

- Virgili, F.J., Liang, E.-W. and Zhang, B. (2009). "Low-luminosity gamma-ray bursts as a distinct GRB population: a firmer case from multiple criteria constraints". *Monthly Notices of the Royal Astronomical Society* **392**: 91–103. arXiv:0801.4751. Bibcode:2009MNRAS.392...91V. doi:10.1111/j.1365-2966.2008.14063.x.

- Wanjek, Christopher (4 June 2005). "Explosions in Space May Have Initiated Ancient Extinction on Earth". NASA. Retrieved 2007-09-15.

- Watson, D.; et al. (2006). "Are short γ-ray bursts collimated? GRB 050709, a flare but no break". *Astronomy and Astrophysics* **454** (3): L123–L126. arXiv:astro-ph/0604153. Bibcode:2006A&A...454L.123W. doi:10.1051/0004-6361:20065380.

- Woosley, S.E. and Bloom, J.S. (2006). "The Supernova Gamma-Ray Burst Connection". *Annual Review of Astronomy and Astrophysics* **44**: 507–556. arXiv:astro-ph/0609142. Bibcode:2006ARA&A..44..507W. doi:10.1146/03.

- Wozniak, P.R.; et al. (2009). "Gamma-Ray Burst at the Extreme: The Naked-Eye Burst GRB 080319B". *Astrophysical Journal* **691**: 495–502. arXiv:0810.2481. Bibcode:2009ApJ...691..495W. doi:10.1088/0004-637X/

- Zhang, B.; et al. (2009). "Discerning the physical origins of cosmological gamma-ray bursts based on multiple observational criteria: the cases of z = 6.7 GRB 080913, z = 8.2 GRB 090423, and some short/hard GRBs". *Astrophysical Journal* **703** (2): 1696–1724. arXiv:0902.2419. Bibcode:2009ApJ...703.1696Z. doi:10.1088/0004-637X/703/2/1696.

## 31.12 External links

### GRB Mission Sites

- Swift Gamma-Ray Burst Mission:
    - Official NASA Swift Homepage
    - UK Swift Science Data Centre
    - Swift Mission Operations Center at Penn State

- HETE-2: High Energy Transient Explorer (Wiki entry)

- INTEGRAL: INTErnational Gamma-Ray Astrophysics Laboratory (Wiki entry)

- BATSE: Burst and Transient Source Explorer

- Fermi Gamma-ray Space Telescope (Wiki entry)

- AGILE: Astro-rivelatore Gamma a Immagini Leggero (Wiki entry)

- EXIST: Energetic X-ray Survey Telescope

### GRB Follow-up Programs

- The Gamma-ray bursts Coordinates Network (GCN) (Wiki entry)

- BOOTES: Burst Observer and Optical Transient Exploring System (Wiki entry)

- GROND: Gamma-Ray Burst Optical Near-infrared Detector (Wiki entry)

- KAIT: The Katzman Automatic Imaging Telescope (Wiki entry)

- MASTER: Mobile Astronomical System of the Telescope-Robots

- PAIRITEL: Peters Automated Infrared Imaging Telescope

- PROMPT: Panchromatic Robotic Optical Monitoring and Polarimetry Telescopes (Wiki entry)

- RAPTOR: Rapid Telescopes for Optical Response

- ROTSE: Robotic Optical Transient Search Experiment (Wiki entry)

- REM: Rapid Eye Mount

# Chapter 32

# Hypernova

For the Iranian band, see Hypernova (band). For the The Browning album, see Hypernova (album).

A **hypernova** (pl. **hypernovae** or **hypernovas**) is a type of star explosion with an energy substantially higher than that of standard supernovae. An alternative term for most hypernova is "superluminous supernova" (SLSN). Such explosions are thought to be the origin of long-duration gamma-ray bursts.[1]

Just like supernovae in general, hypernovae are produced by several different types of stellar explosion: some well modelled and observed in recent years, some still tentatively suggested for observed hypernovae, and some entirely theoretical. Numerous hypernovae have been observed corresponding to supernovae type Ic and type IIn, and possibly also at least one of type IIb.[2]

The word *collapsar*, short for *collapsed star*, was formerly used to refer to the end product of stellar gravitational collapse, a stellar-mass black hole. The word is now sometimes used to refer to a specific model for the collapse of a fast-rotating star, as discussed below.

## 32.1   History of the term

Before the 1990s, the term "hypernova" was used sporadically to describe the theoretical extremely energetic explosions of extremely massive population III stars. It has also been used to describe other extreme energy events, such as mergers of supermassive black holes.

In 1998, a paper suggesting a link between gamma-ray bursts and young massive stars[3] formally proposed to use the term "hypernova" for the visible after-glow from those gamma-ray bursts. The energy of such events was speculated to be up to several hundred times that of known supernovae.

Almost simultaneously, various over-luminous supernovae were being discovered and investigated.[4][5][6][7] These events were described as hypernovae and varied from less than five to around 50 times as energetic as other supernovae and up to 20 times as luminous as a standard type Ia supernova at its peak. This definition has become standard for the term "hypernova", although not all of them are associated with gamma-ray bursts.

Investigation of these types of luminous supernovae suggests that some of them are due to explosions of extremely massive low metallicity stars by the pair instability mechanism, although not with the energies that were speculated for them decades earlier.[8][9]

## 32.2   Gamma-ray bursts

Main article: Gamma-ray burst

Gamma-ray bursts are some of the most energetic events observed in the universe, but their origin was entirely speculative[10]

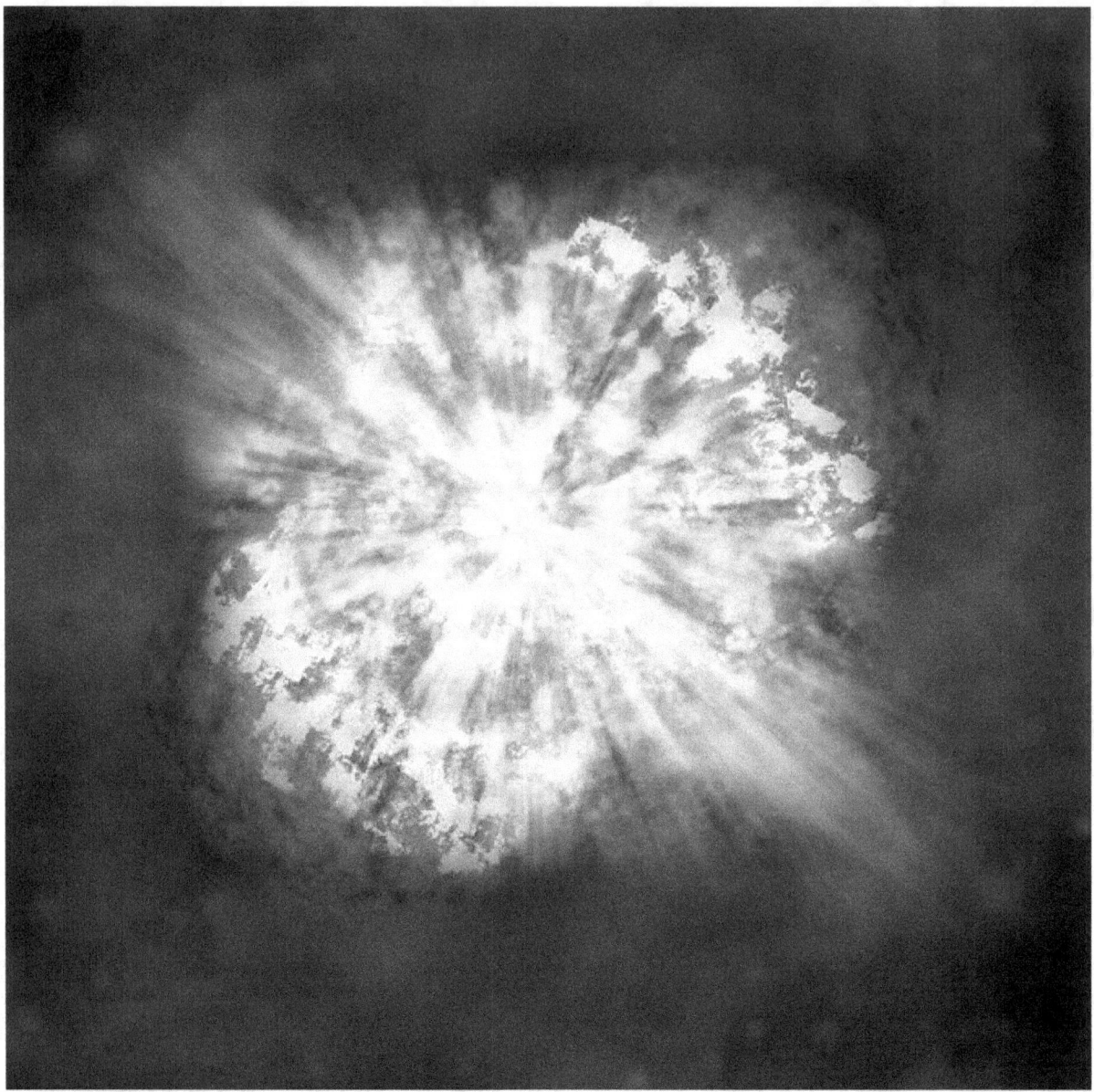

*NASA's artist impression of SN 2006gy, one of the most luminous hypernovae seen*

until around the year 2000. Now, supernova explosions are known to cause at least some gamma-ray bursts, although some gamma-ray bursts are likely from completely different events and not all supernovae are necessarily associated with gamma-ray bursts.

A nearby gamma-ray burst could destroy life on Earth; however, no likely candidate progenitors are close enough to be a danger. Some have suggested that a gamma-ray burst may have caused the Ordovician–Silurian mass extinction on Earth 440 million years ago, but no categorical evidence for this hypothesis exists.[11]

## 32.3   Causes of hypernovae

A wide variety of models have been proposed to explain events an order of magnitude or more greater than standard supernovae. The collapsar and CSM models are widely accepted and a number of events are well-observed. Other models are still only tentatively observed or entirely theoretical.

### 32.3.1   Collapsar model

For a completely collapsed star, see stellar black hole.

The collapsar model is a type of hypernova that produces a gravitationally collapsed object, or black hole. When core

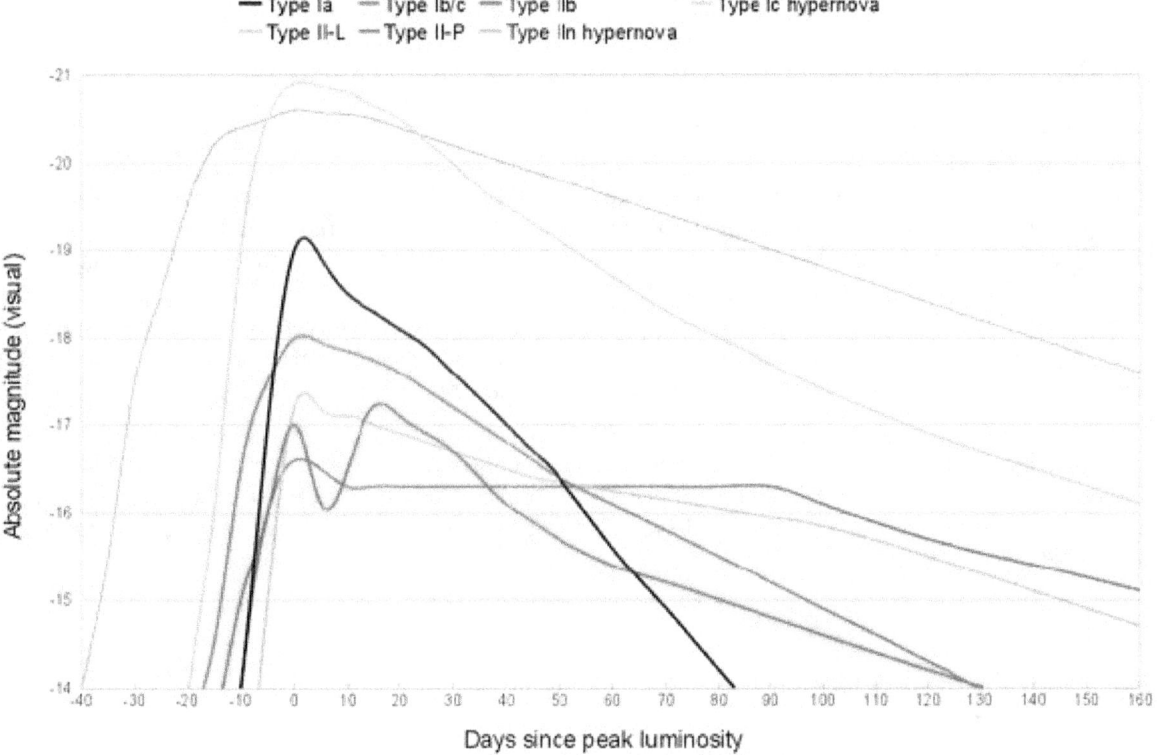

*Light curves compared to normal supernovae*

collapse occurs in a star with a core at least around fifteen times the sun's mass ($M\odot$)—though chemical composition and rotational rate are also significant—the explosion energy is insufficient to expel the outer layers of the star, and it will collapse into a black hole without producing a visible supernova outburst.

A star with a core mass slightly below this level—in the range of 5–15 $M\odot$—will undergo a supernova explosion, but so much of the ejected mass falls back onto the core remnant that it still collapses into a black hole. If such a star is rotating slowly, then it will produce a faint supernova, but if the star is rotating quickly enough, then the fallback to the black hole will produce relativistic jets. The energy that these jets transfer into the ejected shell renders the visible outburst substantially more luminous than a standard supernova. The jets also beam high energy particles and gamma rays directly outward and thereby produce x-ray or gamma-ray bursts; the jets can last for several seconds or longer and correspond to long-duration gamma-ray bursts, but they do not appear to explain short-duration gamma-ray bursts.

A star with a 5–15 $M\odot$ core has an approximate total mass of 25–90 $M\odot$ if the star has not undergone mass loss. Such a star will still have a hydrogen envelope and will explode as a type II supernova. Faint type II supernovae have been observed, but no definite candidates for a type II hypernova (except type IIn, which are not thought to be jet supernovae). Only the very lowest metallicity population III stars will reach this stage of their life with little mass loss. Other stars, including most of those visible to us, will have had most of their outer layers blown away by their high luminosity to become a Wolf–Rayet star and will explode as type Ib or type Ic supernovae. Many observed hypernovae are type Ic and those associated with gamma-ray bursts are almost all type Ic, and these are very good candidates for having relativistic jets produced by fallback to a black hole. Not all type Ic hypernovae correspond to observed gamma-ray bursts but the burst would only be visible if one of the jets were aimed towards us.

In recent years a great deal of observational data on long-duration gamma-ray bursts has significantly increased our understanding of these events and made clear that the collapsar model produces explosions that differ only in detail from more or less ordinary supernovae and have energy ranges from approximately normal to around 100 times larger. Nevertheless, they continue sometimes to be referred to in the literature as hypernovae. The word *hypernova* itself was coined by S.E. Woosley.[12]

A good example of a collapsar hypernova is Sn1998bw,[13] which was associated with the gamma-ray burst GRB 980425. It is classified as a type Ic supernova due to its distinctive spectral properties in the radio spectrum, indicating the presence of relativistic matter.

### 32.3.2   CSM model (circumstellar material)

Almost all observed hypernovae have had spectra similar to either a type Ic or type IIn supernova. The type Ic hypernovae are thought to be produced by jets from fallback to a black hole, but type IIn hypernovae have significantly different light curves and are not associated with gamma-ray bursts. Type IIn supernovae are all embedded in a dense nebula probably expelled from the progenitor star itself, and this circumstellar material (CSM) is thought to be the cause of the extra luminosity.[14] When material expelled in an initial normal supernova explosion meets dense nebular or material or dust close to the star, the shockwave converts kinetic energy efficiently into visible radiation. Thus we see an extremely luminous supernova of extended duration even though the initial explosion energy was the same as that of a normal supernova.

Although any supernova type could potentially produce a type IIn hypernova, given suitable surrounding CSM, the constraints on the size and density of the CSM mean that it will almost always be produced from the star itself immediately prior to the supernova explosion. Such stars are hypergiants and LBVs undergoing substantial mass loss due to Eddington instability, for example SN2005gl.[15]

### 32.3.3   Pair-instability supernova

Main article: Pair-instability supernova

Another type of hypernova is a pair-instability supernova, of which SN 2006gy[16] may possibly be the first observed example. This supernova event was observed in a galaxy about 238 million light years (73 megaparsecs) from Earth.

The theoretical basis for pair-instability collapse has been known for many decades[17] and was suggested as a dominant source of higher mass elements in the early universe as super-massive population III stars exploded. In a pair-instability supernova, the pair production effect causes a sudden pressure drop in the star's core, leading to a rapid partial collapse. Gravitational potential energy from the collapse causes runaway fusion of the core which entirely destroys the star, leaving no remnant.

Current models show that this phenomenon only happens in stars with extreme low metallicity and masses between about 140 and 260 times the Sun, making observing them in the local universe extremely unlikely. Although originally expected to produce hypernova explosions hundreds of times greater than a supernova, they actually produce luminosities ranging from about the same as a normal core collapse supernova to perhaps 50 times brighter, although remaining bright for much longer.[18]

### 32.3.4   Magnetar energy release

Models of the creation and subsequent spin down of a magnetar yield much higher luminosities than regular supernova[19][20] events and match the observed properties[21][22] of at least some hypernovae. In cases where pair-instability supernova may not be a good fit for explaining a hypernova,[23] a magnetar explanation is more plausible.

### 32.3.5 Other models

There are still models for hypernova explosions produced from binary systems, white dwarf or neutron stars in unusual arrangements or undergoing mergers, and some of these are proposed to account for gamma-ray bursts.

## 32.4 See also

- Gamma-ray burst progenitors

- Quark star

## 32.5 References

[1] "A Hypernova: The Super-charged Supernova and its link to Gamma-Ray Bursts". *Imagine the Universe!*. NASA. Retrieved 9 December 2011.

[2] Hamuy, M.; Deng, J.; Mazzali, P. A.; Morrell, N. I.; Phillips, M. M.; Roth, M.; Gonzalez, S.; Thomas-Osip, J.; Krzeminski, W.; Contreras, C.; Maza, J.; González, L.; Huerta, L.; Folatelli, G. N.; Chornock, R.; Filippenko, A. V.; Persson, S. E.; Freedman, W. L.; Koviak, K.; Suntzeff, N. B.; Krisciunas, K. (2009). "Supernova 2003bg: The First Type IIb Hypernova" (pdf). *The Astrophysical Journal* **703** (2): 1612–1623. arXiv:0908.1783. Bibcode:2009ApJ...703.1612H. doi:10.1088/0004-637X/703/2/1612.

[3] Paczyński, B. (1998). "Are Gamma-Ray Bursts in Star-Forming Regions?" (pdf). *The Astrophysical Journal* **494** (1): L45–L48. arXiv:astro-ph/9710086. Bibcode:1998ApJ...494L..45P. doi:10.1086/311148.

[4] Iwamoto, K.; Nakamura, T.; Nomoto, K. I.; Mazzali, P. A.; Danziger, I. J.; Garnavich, P.; Kirshner, R.; Jha, S.; Balam, D.; Thorstensen, J. (2000). "The Peculiar Type Ic Supernova 1997ef: Another Hypernova" (pdf). *The Astrophysical Journal* **534** (2): 660–669. Bibcode:2000ApJ...534..660I. doi:10.1086/308761.

[5] Nomoto, K.; Iwamoto, K.; Mazzali, P. A.; Umeda, H.; Nakamura, T.; Patat, F.; Danziger, I. J.; Young, T. R.; Suzuki, T.; Shigeyama, T.; Augusteijn, T.; Doublier, V.; Gonzalez, J. -F.; Boehnhardt, H.; Brewer, J.; Hainaut, O. R.; Lidman, C.; Leibundgut, B.; Cappellaro, E.; Turatto, M.; Galama, T. J.; Vreeswijk, P. M.; Kouveliotou, C.; Van Paradijs, J.; Pian, E.; Palazzi, E.; Frontera, F. (1998). "A Hypernova Model for the Supernova Associated with the Big γ-Ray Burst of 25 April 1998". *Nature* **395** (6703): 672–674. doi:10.1038/27155.

[6] Mazzali, P. A.; Deng, J.; Maeda, K.; Nomoto, K.; Umeda, H.; Hatano, K.; Iwamoto, K.; Yoshii, Y.; Kobayashi, Y.; Minezaki, T.; Doi, M.; Enya, K.; Tomita, H.; Smartt, S. J.; Kinugasa, K.; Kawakita, H.; Ayani, K.; Kawabata, T.; Yamaoka, H.; Qiu, Y. L.; Motohara, K.; Gerardy, C. L.; Fesen, R.; Kawabata, K. S.; Iye, M.; Kashikawa, N.; Kosugi, G.; Ohyama, Y.; Takada-Hidai, M.; Zhao, G. (2002). "The Type Ic Hypernova SN 2002ap" (pdf). *The Astrophysical Journal* **572** (1): L61–L65. Bibcode:2002ApJ...572L..61M. doi:10.1086/341504.

[7] Mazzali, P. A.; Deng, J.; Pian, E.; Malesani, D.; Tominaga, N.; Maeda, K.; Nomoto, K. I.; Chincarini, G.; Covino, S.; Della Valle, M.; Fugazza, D.; Tagliaferri, G.; Gal-Yam, A. (2006). "Models for the Type Ic Hypernova SN 2003lw associated with GRB 031203" (pdf). *The Astrophysical Journal* **645** (2): 1323–1330. arXiv:astro-ph/0603516. Bibcode:2006ApJ...645.1323M. doi:10.1086/504415.

[8] Gal-Yam, A.; Mazzali, P.; Ofek, E. O.; Nugent, P. E.; Kulkarni, S. R.; Kasliwal, M. M.; Quimby, R. M.; Filippenko, A. V.; Cenko, S. B.; Chornock, R.; Waldman, R.; Kasen, D.; Sullivan, M.; Beshore, E. C.; Drake, A. J.; Thomas, R. C.; Bloom, J. S.; Poznanski, D.; Miller, A. A.; Foley, R. J.; Silverman, J. M.; Arcavi, I.; Ellis, R. S.; Deng, J. (2009). "Supernova 2007bi as a pair-instability explosion". *Nature* **462** (7273): 624–627. doi:10.1038/nature08579. PMID 19956255.

[9] Kasen, D.; Woosley, S. E.; Heger, A. (2011). "Pair Instability Supernovae: Light Curves, Spectra, and Shock Breakout" (pdf). *The Astrophysical Journal* **734** (2): 102. arXiv:1101.3336. Bibcode:2011ApJ...734..102K. doi:10.1088/0004-637X/734/2/102.

[10] Higdon, J. C.; Lingenfelter, R. E. (1990). "Gamma-Ray Bursts". *Annual Review of Astronomy and Astrophysics* **28**: 401. Bibcode:1990ARA&A..28..401H. doi:10.1146/annurev.aa.28.090190.002153.

[11] Minard, Anne (April 3, 2009). "Gamma-Ray Burst Caused Mass Extinction?". National Geographic News. Retrieved 16 April 2010.

[12] Woosley, S. E.; Weaver, T. A. (1982). "Theoretical Models for Supernovae". In Rees, M. J.; Stoneham, R. J. *Supernovae: A Survey of Current Research*. NATO ASI Series **C90**. Dordrecht: D. Reidel Publishing. p. 79. Bibcode:1982sscr.conf...79W. ISBN 9789027714428.

[13] Fujimoto, S. I.; Nishimura, N.; Hashimoto, M. A. (2008). "Nucleosynthesis in Magnetically Driven Jets from Collapsars" (pdf). *The Astrophysical Journal* **680** (2): 1350–1358. arXiv:0804.0969. Bibcode:2008ApJ...680.1350F. doi:10.1086/529416.

[14] Smith, N.; Chornock, R.; Li, W.; Ganeshalingam, M.; Silverman, J. M.; Foley, R. J.; Filippenko, A. V.; Barth, A. J. (2008). "SN 2006tf: Precursor Eruptions and the Optically Thick Regime of Extremely Luminous Type IIn Supernovae" (pdf). *The Astrophysical Journal* **686** (1): 467–484. arXiv:0804.0042. Bibcode:2008ApJ...686..467S. doi:10.1086/591021.

[15] Gal-Yam, A.; Leonard, D. C. (2009). "A Massive Hypergiant Star as the Progenitor of the Supernova SN 2005gl" (pdf). *Nature* **458** (7240): 865–867. doi:10.1038/nature07934. PMID 19305392.

[16] Smith, N.; Chornock, R.; Silverman, J. M.; Filippenko, A. V.; Foley, R. J. (2010). "Spectral Evolution of the Extraordinary Type IIn Supernova 2006gy" (pdf). *The Astrophysical Journal* **709** (2): 856–883. arXiv:0906.2200. Bibcode:2010ApJ...709..856S. doi:10.1088/0004-637X/709/2/856.

[17] Fraley, G. S. (1968). "Supernovae Explosions Induced by Pair-Production Instability" (pdf). *Astrophysics and Space Science* **2** (1): 96–114. Bibcode:1968Ap&SS...2...96F. doi:10.1007/BF00651498.

[18] Kasen, D.; Woosley, S. E.; Heger, A. (2011). "Pair Instability Supernovae: Light Curves, Spectra, and Shock Breakout" (pdf). *The Astrophysical Journal* **734** (2): 102. arXiv:1101.3336. Bibcode:2011ApJ...734..102K. doi:10.1088/0004-637X/734/2/102.

[19] Woosley, S.E. (August 2010). "Bright Supernovae From Magnetar Birth". *Astrophysical Journal Letters* **719** (2): L204. arXiv:0911.0698. Bibcode:2010ApJ...719L.204W. doi:10.1088/2041-8205/719/2/L204. Retrieved 27 December 2013.

[20] Kasen, Daniel; Bildsten, Lars (2010). "Supernova Light Curves Powered by Young Magnetars". *Astrophysical Journal* **717**: 245–249. arXiv:0911.0680. Bibcode:2010ApJ...717..245K. doi:10.1088/0004-637X/717/1/245. Retrieved 27 December 2013.

[21] C.Inserra, C.; et al. (June 2013). "Super Luminous Ic Supernovae: catching a magnetar by the tail". *The Astrophysical Journal* **770** (2): 128. arXiv:1304.3320. Bibcode:2013ApJ...770..128I. doi:10.1088/0004-637X/770/2/128. Retrieved 27 December 2013.

[22] D.A. Howell, D. A.; et al. (October 2013). "Two superluminous supernovae from the early universe discovered by the Supernova Legacy Survey". *Astrophysical Journal* **779** (2): 98. arXiv:1310.0470. Bibcode:2013ApJ...779...98H. doi:10.1088/0004-637X/779/2/98. Retrieved 27 December 2013.

[23] M. Nicholl, et. al, M.; Smartt, S. J.; Jerkstrand, A.; Inserra, C.; McCrum, M.; Kotak, R.; Fraser, M.; Wright, D.; Chen, T.-W.; Smith, K.; Young, D. R.; Sim, S. A.; Valenti, S.; Howell, D. A.; Bresolin, F.; Kudritzki, R. P.; Tonry, J. L.; Huber, M. E.; Rest, A.; Pastorello, A.; Tomasella, L.; Cappellaro, E.; Benetti, S.; Mattila, S.; Kankare, E.; Kangas, T.; Leloudas, G.; Sollerman, J.; Taddia, F.; et al. (October 2013). "Slowly fading super-luminous supernovae that are not pair-instability explosions". *Nature* **502** (7471): 346. arXiv:1310.4446. Bibcode:2013Natur.502..346N. doi:10.1038/nature12569. PMID 24132291. Retrieved 27 December 2013.

## 32.6   Further reading

- MacFadyen, A. I.; Woosley, S. E. (1999). "Collapsars: Gamma-Ray Bursts and Explosions in 'Failed Supernovae'" (PDF). *Astrophysical Journal* **524** (1): 262–289. arXiv:astro-ph/9810274. Bibcode:1999ApJ...524..262M. doi:10.1086/307790.

- Woosley, S. E. (1993). "Gamma-ray bursts from stellar mass accretion disks around black holes" (PDF). *Astrophysical Journal* **405** (1): 273–277. Bibcode:1993ApJ...405..273W. doi:10.1086/172359.

- Piran, T. (2004). "The Physics of Gamma-Ray Bursts" (PDF). *Reviews of Modern Physics* **76** (4): 1143–1210. arXiv:astro-ph/0405503v1. Bibcode:2004RvMP...76.1143P. doi:10.1103/RevModPhys.76.1143.

- "Cosmological Gamma-Ray Bursts and Hypernovae Conclusively Linked (SN 2003dh and GRB 030329)". European Southern Observatory (ESO). 2003.

- Hjorth, J.; Sollerman, J.; Møller, P.; Fynbo, J. P. U.; Woosley, S. E.; Kouveliotou, C.; Tanvir, N. R.; Greiner, J. (2003). "A very energetic supernova associated with the γ-ray burst of 29 March 2003" (PDF). *Nature* **423** (6942): 847–850. arXiv:astro-ph/0306347. Bibcode:2003Natur.423..847H. doi:10.1038/nature01750. PMID 12815425.

# Chapter 33

# Astrophysical jet

An **astrophysical jet** (hereafter 'jet') is a phenomenon often seen in astronomy, where streams of matter are emitted along the axis of rotation of a compact object. It is usually caused by the dynamic interactions within an accretion disc. When matter is emitted at speeds approaching the speed of light, these jets are called relativistic jets, because the effects of special relativity become important. The largest jets are those seen in active galaxies such as quasars and radio galaxies. Other systems which often contain jets include cataclysmic variable stars, X-ray binaries and T Tauri stars. Herbig–Haro objects are caused by the interaction of jets with the interstellar medium. Bipolar outflows or jets may also be associated with protostars (young, forming stars),[1] or with evolved post-AGB stars (often in the form of bipolar nebulae).

While it is still the subject of ongoing research to understand how jets are formed and powered, the two most often proposed origins of this power are the central object (such as a black hole), and the accretion disc. Accretion discs around many stellar objects are able to produce jets, although those around a black hole are the fastest and most active. This is because the speed of the jet is around the same speed as the escape velocity of the central object. This makes the speed of a jet from an accreting black hole near the speed of light, while protostellar jets are much slower. While it is not known exactly how accretion discs manage to produce jets, they are thought to generate tangled magnetic fields that cause the jets to collimate. The hydrodynamics of a de Laval nozzle may also give a hint to the mechanisms involved.

One of the best ways of exploring how jets are produced is to determine the composition of the jets at a radius where they can be directly observed. For example, if a jet originates in the accretion disc, its plasma is likely to have electron-ion composition, whereas if it originates in the black hole it will likely be electron-positron in nature. Also, the plasma emits various forms of radiation such as X-rays and radio waves, which aid diagnosis.

## 33.1 Relativistic jet

**Relativistic jets** are extremely powerful jets[2] of plasma, with speeds close to the speed of light, that are emitted near the central massive objects of some active galaxies, notably radio galaxies and quasars. Their lengths can reach several thousand[3] or even hundreds of thousands of light years.[4] Because the jet speed is close to the speed of light, the effects of the Special Theory of Relativity are important; in particular, relativistic beaming will change the apparent brightness. The mechanics behind both the creation of the jets[5][6] and the composition of the jets[7] are still a matter of much debate in the scientific community. In terms of composition, some studies favour a model in which the jets are composed of an electrically neutral mixture of electrons, positrons, and protons in some proportion, while others are consistent with a jet of electron-positron plasma.[8]

Similar jets, though on a much smaller scale, can develop around the accretion disks of neutron stars and stellar black holes. These systems are often called microquasars. An example is SS433, whose well-observed jet has a velocity of 0.23c, although other microquasars appear to have much higher (but less well measured) jet velocities. Even weaker and less relativistic jets may be associated with many binary systems; the acceleration mechanism for these jets may be similar to the magnetic reconnection processes observed in the Earth's magnetosphere and the solar wind.

The general hypothesis among astrophysicists is that the formation of relativistic jets is the key to explaining the production

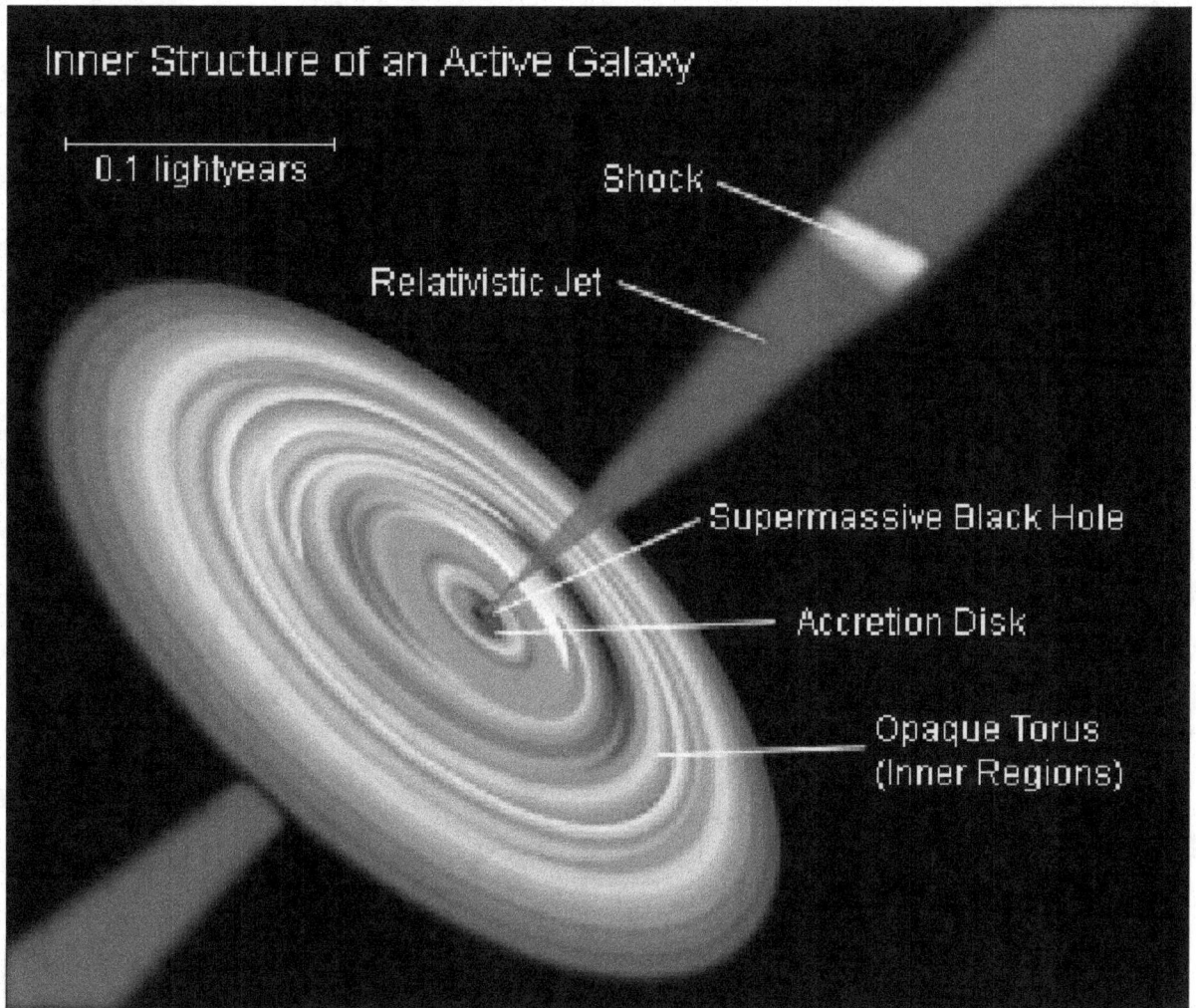

*Relativistic jet. The environment around the AGN where the relativistic plasma is collimated into jets which escape along the pole of the supermassive black hole.*

of gamma-ray bursts. These jets have Lorentz factors of ~100 (that is, speeds of roughly 0.99995c), making them some of the swiftest celestial objects currently known.

### 33.1.1 Rotating black hole as energy source

Because of the enormous amount of energy needed to launch a relativistic jet, some jets are thought to be powered by spinning black holes. There are two well known theories for how the energy is transferred from the black hole to the jet.

- *Blandford–Znajek process*.[9] This is the most popular theory for the extraction of energy from the central black hole. The magnetic fields around the accretion disk are dragged by the spin of the black hole. The relativistic material is possibly launched by the tightening of the field lines.

- *Penrose mechanism*.[10] This extracts energy from a rotating black hole by frame dragging. This theory was later proven to be able to extract relativistic particle energy and momentum,[11] and subsequently shown to be a possible mechanism for the formation of jets.[12]

A Hubble Space Telescope survey indicated that relativistic jets may be more likely to form from supermassive black holes resulting from the merger of two galaxies and their galaxy centre's black holes. Not all galaxy mergers create relativistic

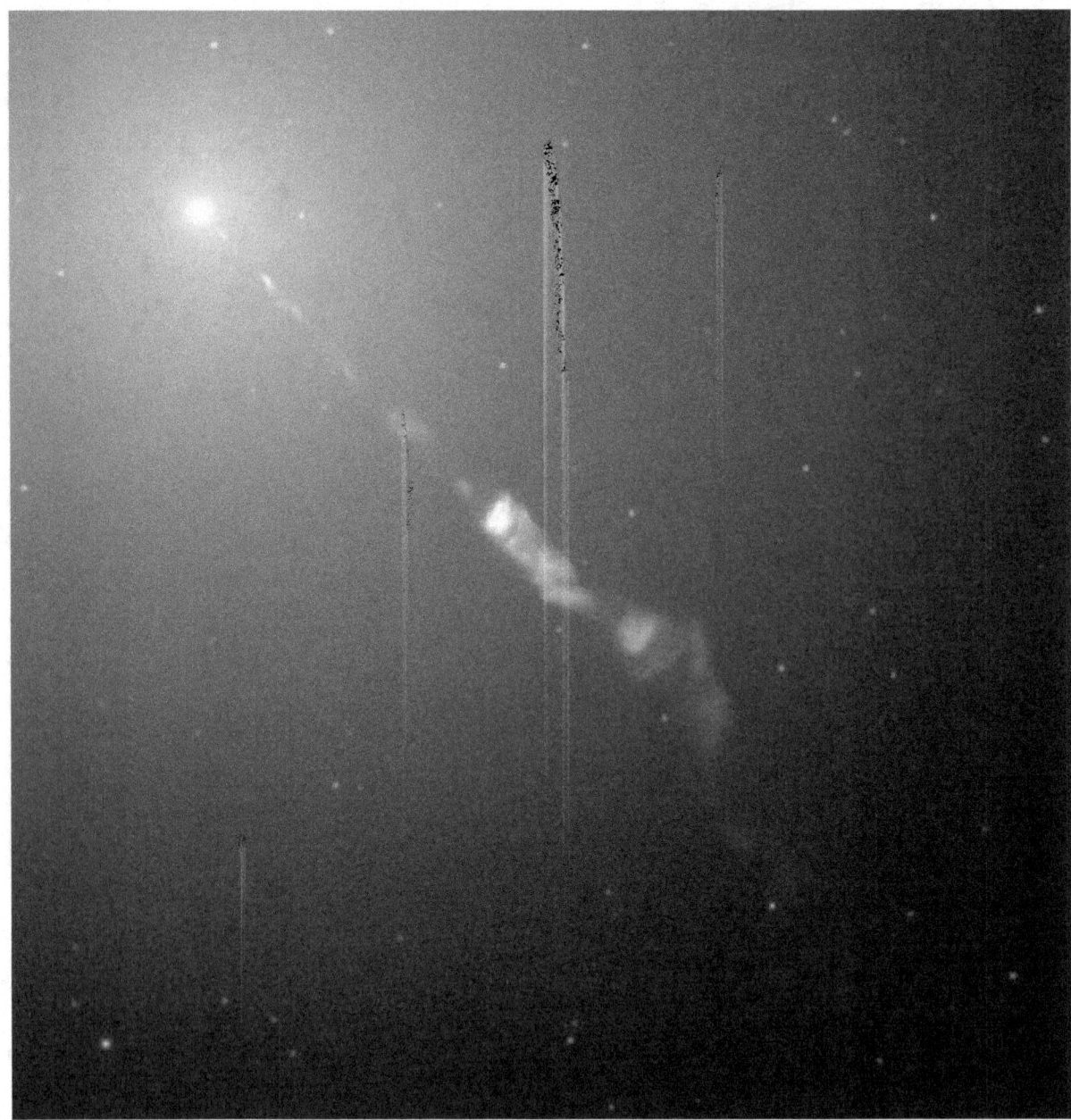

*Elliptical galaxy M87 emitting a relativistic jet, as seen by the Hubble Space Telescope.*

jets.[13][14] NASA/ESA Video

## 33.2   Other images

- Centaurus A in x-rays showing the relativistic jet

- The M87 jet seen by the Very Large Array in radio frequency (the viewing field is larger and rotated with respect to the above image.)

- Hubble Legacy Archive Near-UV image of the relativistic jet in 3C 66B

- Galaxy NGC 3862, an extragalactic jet of material moving at nearly the speed of light can be seen at the three o'clock position.[1]

1. ^ "Hubble Video Shows Shock Collision Inside Black Hole Jet". 27 May 2015.

## 33.3 See also

- Accretion disc
- Bipolar outflow
- Blandford–Znajek process
- List of plasma (physics) articles

## 33.4 References

[1] "Star sheds via reverse whirlpool". Astronomy.com. 27 December 2007. Retrieved 26 May 2015.

[2] Wehrle, A.E.; Zacharias, N.; Johnston, K.; et al. (11 Feb 2009). "What is the structure of Relativistic Jets in AGN on Scales of Light Days?" (PDF). *Astro2010: the Astronomy and Astrophysics Decadal Survey* **2010**: 310. Bibcode:2009astro2010S.310W.

[3] Biretta, J. (6 Jan 1999). "Hubble Detects Faster-Than-Light Motion in Galaxy M87".

[4] "Evidence for Ultra-Energetic Particles in Jet from Black Hole". *Yale University - Office of Public Affairs*. 20 June 2006. Archived from the original on 2008.

[5] Meier, David L (2003). "The theory and simulation of relativistic jet formation: Towards a unified model for micro- and macro-quasars". *New Astronomy Reviews* **47** (6–7): 667. arXiv:astro-ph/0312048. Bibcode:2003NewAR..47..667M. doi:10.1016/S1387-6473(03)00120-9.

[6] Semenov, V.; Dyadechkin, Sergey; Punsly, Brian (2004). "Simulations of Jets Driven by Black Hole Rotation". *Science* **305** (5686): 978–980. arXiv:astro-ph/0408371. Bibcode:2004Sci...305..978S. doi:10.1126/science.1100638. PMID 15310894.

[7] Georganopoulos, Markos; Kazanas, Demosthenes; Perlman, Eric; Stecker, Floyd W. (2005). "Bulk Comptonization of the Cosmic Microwave Background by Extragalactic Jets as a Probe of Their Matter Content". *The Astrophysical Journal* **625** (2): 656. arXiv:astro-ph/0502201. Bibcode:2005ApJ...625..656G. doi:10.1086/429558.

[8] Wardle, J.F.C. "Electron–positron jets associated with the quasar 3C279". *Nature* **395** (1 October 1998): 457–461. Bibcode:. doi:10.1038/26675.

[9] Blandford, R. D.; Znajek, R. L. (1977). "Electromagnetic extraction of energy from Kerr black holes". *Monthly Notices of the Royal Astronomical Society* **179** (3): 433. Bibcode:1977MNRAS.179..433B. doi:10.1093/mnras/179.3.433.

[10] Penrose, Roger (1969). "Gravitational Collapse: The Role of General Relativity". *Rivista del Nuovo Cimento* **1**: 252–276. Bibcode:1969NCimR...1..252P. Reprinted in: Penrose, R. (2002), ""Golden Oldie": Gravitational Collapse: The Role of General Relativity", *General Relativity and Gravitation* **34** (7): 1141, Bibcode:2002GReGr..34.1141P, doi:10.1023/A:10165784

[11] R.K. Williams (1995). "Extracting x rays, Ɣ rays, and relativistic e⁻e⁺ pairs from supermassive Kerr black holes using the Penrose mechanism". *Physical Review* **51** (10): 5387–5427. Bibcode:1995PhRvD..51.5387W. doi:10.1103/PhysRevD.51.5387.

[12] Williams, Reva Kay (2004). "Collimated Escaping Vortical Polar e⁻e⁺ Jets Intrinsically Produced by Rotating Black Holes and Penrose Processes". *The Astrophysical Journal* **611** (2): 952. arXiv:astro-ph/0404135. Bibcode:2004ApJ...611..952W. doi:10.1086/422304.

[13] "Galaxy Crashes May Give Birth to Powerful Space Jets". Retrieved 2015-05-29.

[14] "Merging galaxies break radio silence - Large Hubble survey confirms link between mergers and supermassive black holes with relativistic jets". *www.spacetelescope.org*. Retrieved 2015-05-29. |first1= missing |last1= in Authors list (help)

## 33.5   External links

- NASA - Ask an Astrophysicist: Black Hole Bipolar Jets
- SPACE.com - Twisted Physics: How Black Holes Spout Off
- Compact Objects and Accretion Disks

## 33.6   Videos

- Hubble Video Shows Shock Collision inside Black Hole Jet (Article)

# Chapter 34

# Binary star

For the hip hop group, see Binary Star (band).

A **binary star** is a star system consisting of two stars orbiting around their common center of mass. Systems of two,

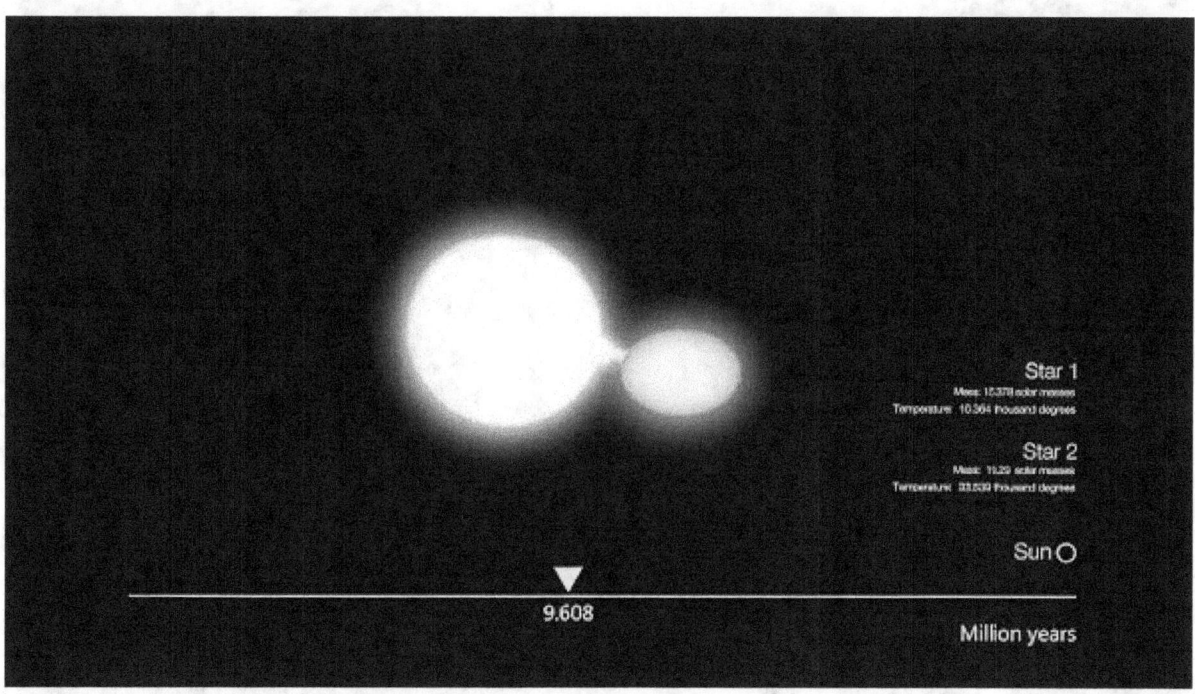

*Artist's impression of the evolution of a hot high-mass binary star.*

three, four, or even more stars are called *multiple star systems*. These systems, especially when more distant, often appear to the unaided eye as a single point of light, and are then revealed as double (or more) by other means. Research over the last two centuries suggests that half or more of visible stars are part of multiple star systems.[1]

The term *double star* is often used synonymously with *binary star*; however, double star can also mean *optical double star*. Optical doubles are so called because the two stars appear close together in the sky as seen from the Earth; they are almost on the same line of sight. Nevertheless, their "doubleness" depends only on this optical effect; the stars themselves are distant from one another and share no physical connection. A double star can be revealed as optical by means of differences in their parallax measurements, proper motions, or radial velocities. Most known double stars have not been studied sufficiently closely to determine whether they are optical doubles or they are doubles physically bound through gravitation into a multiple star system.

Binary star systems are very important in astrophysics because calculations of their orbits allow the masses of their com-

*Hubble image of the Sirius binary system, in which Sirius B can be clearly distinguished (lower left)*

ponent stars to be directly determined, which in turn allows other stellar parameters, such as radius and density, to be indirectly estimated. This also determines an empirical mass-luminosity relationship (MLR) from which the masses of single stars can be estimated.

Binary stars are often detected optically, in which case they are called *visual binaries*. Many visual binaries have long orbital periods of several centuries or millennia and therefore have orbits which are uncertain or poorly known. They may also be detected by indirect techniques, such as spectroscopy (*spectroscopic binaries*) or astrometry (*astrometric binaries*). If a binary star happens to orbit in a plane along our line of sight, its components will eclipse and transit each other; these pairs are called *eclipsing binaries*, or, as they are detected by their changes in brightness during eclipses and transits, *photometric binaries*.

If components in binary star systems are close enough they can gravitationally distort their mutual outer stellar atmospheres. In some cases, these *close binary systems* can exchange mass, which may bring their evolution to stages that single stars cannot attain. Examples of binaries are Sirius and Cygnus X-1 (Cygnus X-1 being a well known black hole). Binary stars are also common as the nuclei of many planetary nebulae, and are the progenitors of both novae and type Ia supernovae.

## 34.1 Discovery

The term *binary* was first used in this context by Sir William Herschel in 1802,[2] when he wrote:[3]

> "If, on the contrary, two stars should really be situated very near each other, and at the same time so far insulated as not to be materially affected by the attractions of neighbouring stars, they will then compose a separate system, and remain united by the bond of their own mutual gravitation towards each other. This should be called a real double star; and any two stars that are thus mutually connected, form the binary sidereal system which we are now to consider."

By the modern definition, the term *binary star* is generally restricted to pairs of stars which revolve around a common center of mass. Binary stars which can be resolved with a telescope or interferometric methods are known as *visual binaries*.[4][5] For most of the known visual binary stars one whole revolution has not been observed yet, they are observed to have travelled along a curved path or a partial arc.[6]

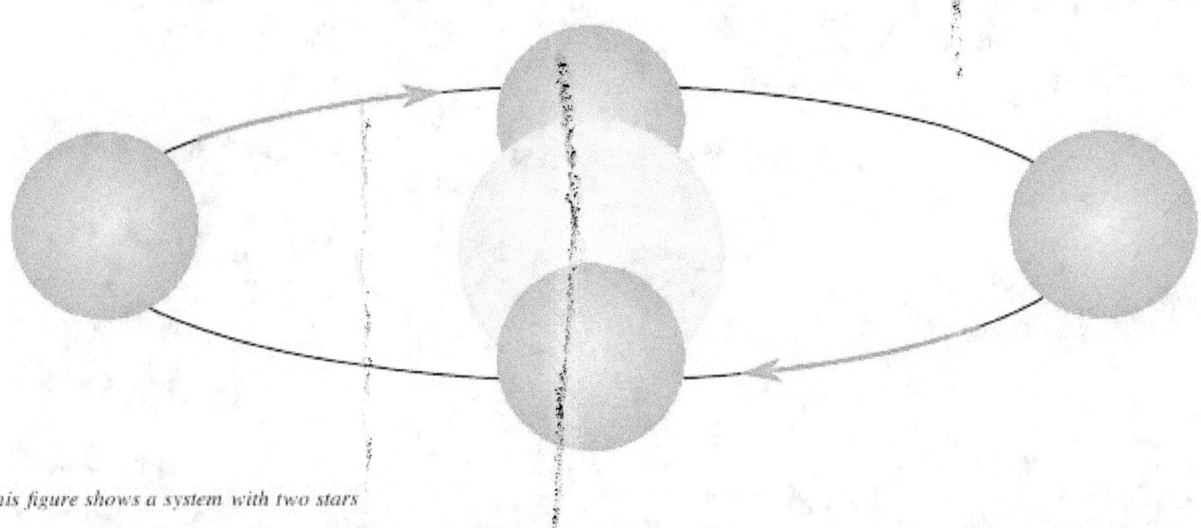

*This figure shows a system with two stars*

The more general term *double star* is used for pairs of stars which are seen to be close together in the sky.[2] This distinction is rarely made in languages other than English.[4] Double stars may be binary systems or may be merely two stars that appear to be close together in the sky but have vastly different true distances from the Sun. The latter are termed *optical doubles* or *optical pairs*.[7]

Since the invention of the telescope, many pairs of double stars have been found. Early examples include Mizar and Acrux. Mizar, in the Big Dipper (Ursa Major), was observed to be double by Giovanni Battista Riccioli in 1650[8][9] (and probably earlier by Benedetto Castelli and Galileo).[10] The bright southern star Acrux, in the Southern Cross, was discovered to be double by Father Fontenay in 1685.[8]

John Michell was the first to suggest that double stars might be physically attached to each other when he argued in 1767 that the probability that a double star was due to a chance alignment was small.[11][12] William Herschel began observing double stars in 1779 and soon thereafter published catalogs of about 700 double stars.[13] By 1803, he had observed changes in the relative positions in a number of double stars over the course of 25 years, and concluded that they must be binary systems;[14] the first orbit of a binary star, however, was not computed until 1827, when Félix Savary computed the orbit of Xi Ursae Majoris.[15] Since this time, many more double stars have been catalogued and measured. The

Washington Double Star Catalog, a database of visual double stars compiled by the United States Naval Observatory, contains over 100,000 pairs of double stars,[16] including optical doubles as well as binary stars. Orbits are known for only a few thousand of these double stars,[17] and most have not been ascertained to be either true binaries or optical double stars.[18] This can be determined by observing the relative motion of the pairs. If the motion is part of an orbit, or if the stars have similar radial velocities and the difference in their proper motions is small compared to their common proper motion, the pair is probably physical.[19] One of the tasks that remains for visual observers of double stars is to obtain sufficient observations to prove or disprove gravitational connection.

## 34.2    Classifications

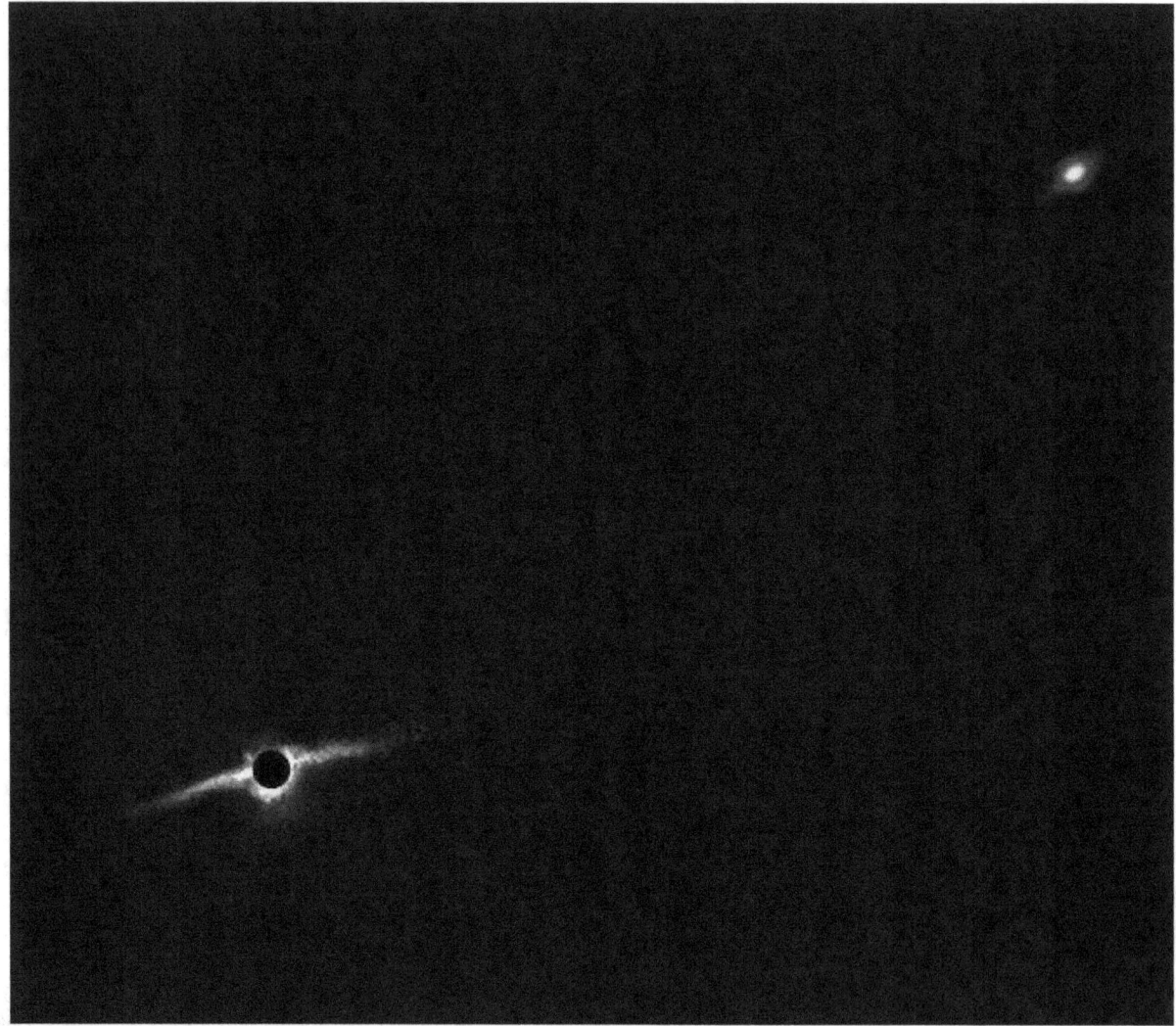

*Edge-on disc of gas and dust present around the binary star system HD 106906AB.*[20]

### 34.2.1    Methods of observation

Binary stars are classified into four types according to the way in which they are observed: visually, by observation; spectroscopically, by periodic changes in spectral lines; photometrically, by changes in brightness caused by an eclipse;

or astrometrically, by measuring a deviation in a star's position caused by an unseen companion.[4][21] Any binary star can belong to several of these classes; for example, several spectroscopic binaries are also eclipsing binaries.

### Visual binaries

A *visual binary* star is a binary star for which the angular separation between the two components is great enough to permit them to be observed as a double star in a telescope, or even high-powered binoculars. The angular resolution of the telescope is an important factor in the detection of visual binaries, and as better angular resolutions are applied to binary star observations increasing number of visual binaries will be detected. The relative brightness of the two stars is also an important factor, as glare from a bright star may make it difficult to detect the presence of a fainter component.

The brighter star of a visual binary is the *primary* star, and the dimmer is considered the *secondary*. In some publications (especially older ones), a faint secondary is called the *comes* (plural *comites*; companion). If the stars are the same brightness, the discoverer designation for the primary is customarily accepted.[22]

The position angle of the secondary with respect to the primary is measured, together with the angular distance between the two stars. The time of observation is also recorded. After a sufficient number of observations are recorded over a period of time, they are plotted in polar coordinates with the primary star at the origin, and the most probable ellipse is drawn through these points such that the Keplerian law of areas is satisfied. This ellipse is known as the *apparent ellipse*, and is the projection of the actual elliptical orbit of the secondary with respect to the primary on the plane of the sky. From this projected ellipse the complete elements of the orbit may be computed, where the semi-major axis can only be expressed in angular units unless the stellar parallax, and hence the distance, of the system is known.[5]

### Spectroscopic binaries

Sometimes, the only evidence of a binary star comes from the Doppler effect on its emitted light. In these cases, the binary consists of a pair of stars where the spectral lines in the light emitted from each star shifts first toward the blue, then toward the red, as each moves first toward us, and then away from us, during its motion about their common center of mass, with the period of their common orbit.

In these systems, the separation between the stars is usually very small, and the orbital velocity very high. Unless the plane of the orbit happens to be perpendicular to the line of sight, the orbital velocities will have components in the line of sight and the observed radial velocity of the system will vary periodically. Since radial velocity can be measured with a spectrometer by observing the Doppler shift of the stars' spectral lines, the binaries detected in this manner are known as *spectroscopic binaries*. Most of these cannot be resolved as a visual binary, even with telescopes of the highest existing resolving power.

In some spectroscopic binaries, spectral lines from both stars are visible and the lines are alternately double and single. Such a system is known as a double-lined spectroscopic binary (often denoted "SB2"). In other systems, the spectrum of only one of the stars is seen and the lines in the spectrum shift periodically towards the blue, then towards red and back again. Such stars are known as single-lined spectroscopic binaries ("SB1").

The orbit of a spectroscopic binary is determined by making a long series of observations of the radial velocity of one or both components of the system. The observations are plotted against time, and from the resulting curve a period is determined. If the orbit is circular then the curve will be a sine curve. If the orbit is elliptical, the shape of the curve will depend on the eccentricity of the ellipse and the orientation of the major axis with reference to the line of sight.

It is impossible to determine individually the semi-major axis $a$ and the inclination of the orbit plane $i$. However, the product of the semi-major axis and the sine of the inclination (i.e. $a \sin i$) may be determined directly in linear units (e.g. kilometres). If either $a$ or $i$ can be determined by other means, as in the case of eclipsing binaries, a complete solution for the orbit can be found.[23]

Binary stars that are both visual and spectroscopic binaries are rare, and are a precious source of valuable information when found. Visual binary stars often have large true separations, with periods measured in decades to centuries; consequently, they usually have orbital speeds too small to be measured spectroscopically. Conversely, spectroscopic binary stars move fast in their orbits because they are close together, usually too close to be detected as visual binaries. Binaries that are both visual and spectroscopic thus must be relatively close to Earth.

**Eclipsing binaries**

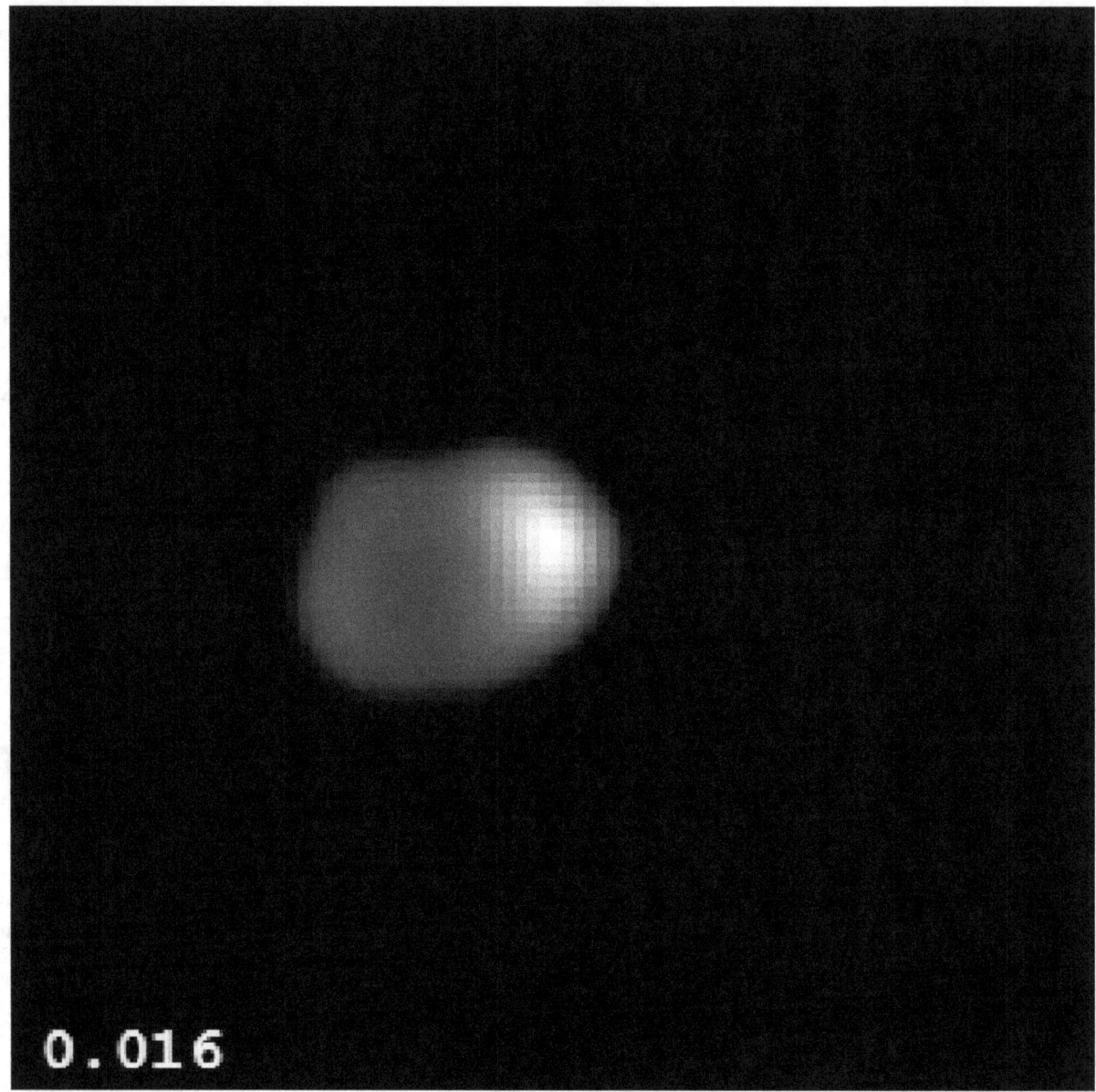

0.016

*Algol B orbits Algol A. This animation was assembled from 55 images of the CHARA interferometer in the near-infrared H-band, sorted according to orbital phase.*

An *eclipsing binary star* is a binary star in which the orbit plane of the two stars lies so nearly in the line of sight of the observer that the components undergo mutual eclipses. In the case where the binary is also a spectroscopic binary and the parallax of the system is known, the binary is quite valuable for stellar analysis.[24] Algol is the best-known example of an eclipsing binary.[25]

In the last decade, measurement of extragalactic eclipsing binaries' fundamental parameters has become possible with 8 meter class telescopes. This makes it feasible to use them to directly measure the distances to external galaxies, a process that is more accurate than using standard candles.[26] Recently, they have been used to give direct distance estimates to the LMC, SMC, Andromeda Galaxy and Triangulum Galaxy. Eclipsing binaries offer a direct method to gauge the distance to galaxies to a new improved 5% level of accuracy.[27]

Eclipsing binaries are variable stars, not because the light of the individual components vary but because of the eclipses.

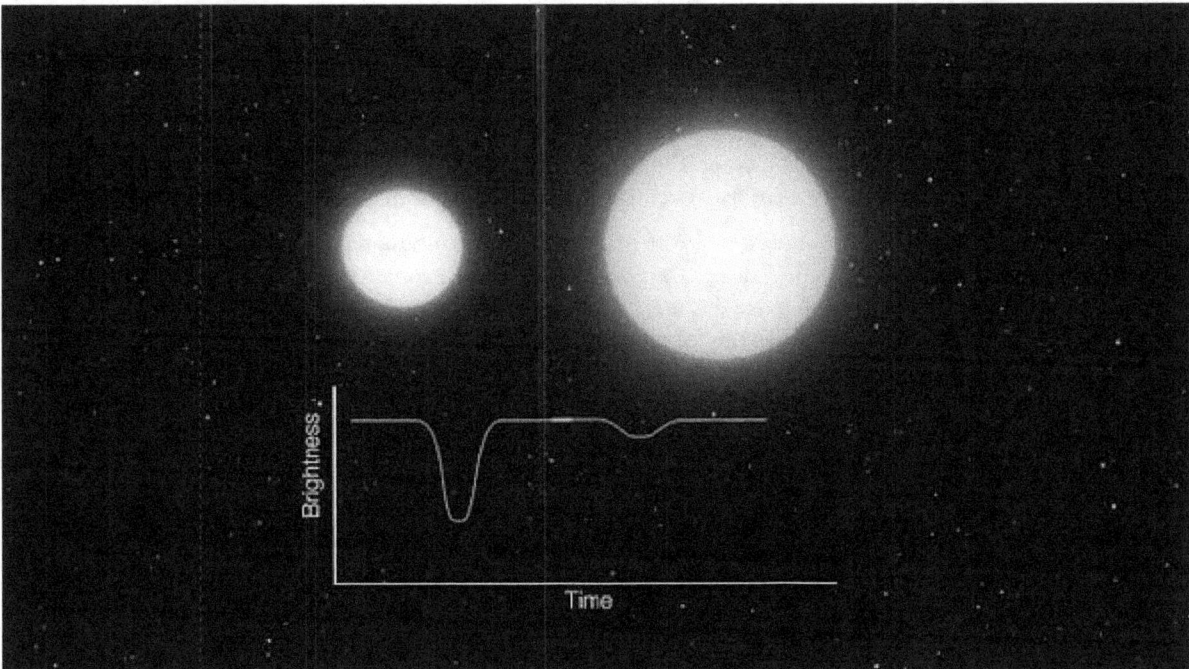

*This video shows an artist's impression of an eclipsing binary star system. As the two stars orbit each other they pass in front of one another and their combined brightness, seen from a distance, decreases.*

The light curve of an eclipsing binary is characterized by periods of practically constant light, with periodic drops in intensity. If one of the stars is larger than the other, one will be obscured by a total eclipse while the other will be obscured by an annular eclipse.

The period of the orbit of an eclipsing binary may be determined from a study of the light curve, and the relative sizes of the individual stars can be determined in terms of the radius of the orbit by observing how quickly the brightness changes as the disc of the near star slides over the disc of the distant star. If it is also a spectroscopic binary the orbital elements can also be determined, and the mass of the stars can be determined relatively easily, which means that the relative densities of the stars can be determined in this case.[28]

### Non-eclipsing binaries that can be detected through photometry

Close non-eclipsing binaries can also be photometrically detected by observing how the stars affect each other. First is by observing extra light which the stars reflect from their companion. Second is by observing ellipsoidal light variations which are caused by deformation of the star's shape by their companions. Third effect is by looking at how relativistic beaming affects the apparent magnitude of the stars. Detecting binaries with these methods requires accurate photometry and has to be done with space telescopes.[29]

### Astrometric binaries

Astronomers have discovered some stars that seemingly orbit around an empty space. *Astrometric binaries* are relatively nearby stars which can be seen to wobble around a point in space, with no visible companion. The same mathematics used for ordinary binaries can be applied to infer the mass of the missing companion. The companion could be very dim, so that it is currently undetectable or masked by the glare of its primary, or it could be an object that emits little or no electromagnetic radiation, for example a neutron star.[30]

The visible star's position is carefully measured and detected to vary, due to the gravitational influence from its counterpart. The position of the star is repeatedly measured relative to more distant stars, and then checked for periodic shifts in position. Typically this type of measurement can only be performed on nearby stars, such as those within 10 parsecs.

Nearby stars often have a relatively high proper motion, so astrometric binaries will appear to follow a *wobbly* path across the sky.

If the companion is sufficiently massive to cause an observable shift in position of the star, then its presence can be deduced. From precise astrometric measurements of the movement of the visible star over a sufficiently long period of time, information about the mass of the companion and its orbital period can be determined.[31] Even though the companion is not visible, the characteristics of the system can be determined from the observations using Kepler's laws.[32]

This method of detecting binaries is also used to locate extrasolar planets orbiting a star. However, the requirements to perform this measurement are very exacting, due to the great difference in the mass ratio, and the typically long period of the planet's orbit. Detection of position shifts of a star is a very exacting science, and it is difficult to achieve the necessary precision. Space telescopes can avoid the blurring effect of Earth's atmosphere, resulting in more precise resolution.

## 34.2.2 Configuration of the system

*Artist's conception of a cataclysmic variable system*

Another classification is based on the distance of the stars, relative to their sizes:[33]

*Detached binaries* are binary stars where each component is within its Roche lobe, i.e. the area where the gravitational pull of the star itself is larger than that of the other component. The stars have no major effect on each other, and essentially evolve separately. Most binaries belong to this class.

*Semidetached binary stars* are binary stars where one of the components fills the binary star's Roche lobe and the other does not. Gas from the surface of the Roche-lobe-filling component (donor) is transferred to the other, accreting star. The mass transfer dominates the evolution of the system. In many cases, the inflowing gas forms an accretion disc around the accretor.

A *contact binary* is a type of binary star in which both components of the binary fill their Roche lobes. The uppermost part of the stellar atmospheres forms a *common envelope* that surrounds both stars. As the friction of the envelope brakes the orbital motion, the stars may eventually merge.[34]

### 34.2.3 Cataclysmic variables and X-ray binaries

When a binary system contains a compact object such as a white dwarf, neutron star or black hole, gas from the other (donor) star can accrete onto the compact object. This releases gravitational potential energy, causing the gas to become hotter and emit radiation. Cataclysmic variable stars, where the compact object is a white dwarf, are examples of such systems.[35] In X-ray binaries, the compact object can be either a neutron star or a black hole. These binaries are classified as low-mass or high-mass according to the mass of the donor star. High-mass X-ray binaries contain a young, early-type, high-mass donor star which transfers mass by its stellar wind, while low-mass X-ray binaries are semidetached binaries in which gas from a late-type donor star or a white dwarf overflows the Roche lobe and falls towards the neutron star or black hole.[36] Probably the best known example of an X-ray binary is the high-mass X-ray binary Cygnus X-1. In Cygnus X-1, the mass of the unseen companion is estimated to be about nine times that of the Sun,[37] far exceeding the Tolman–Oppenheimer–Volkoff limit for the maximum theoretical mass of a neutron star. It is therefore believed to be a black hole; it was the first object for which this was widely believed.[38]

## 34.3 Orbital period

Orbital periods can be less than an hour (for AM CVn stars), or a few days (components of Beta Lyrae), but also hundreds of thousands of years (Proxima Centauri around Alpha Centauri AB).

### 34.3.1 Variations in period

Main article: Applegate mechanism

The Applegate mechanism explains long term orbital period variations seen in certain eclipsing binaries. As a main-sequence star goes through an activity cycle, the outer layers of the star are subject to a magnetic torque changing the distribution of angular momentum, resulting in a change in the star's oblateness. The orbit of the stars in the binary pair is gravitationally coupled to their shape changes, so that the period shows modulations (typically on the order of $\Delta P/P \sim 10^{-5}$) on the same time scale as the activity cycles (typically on the order of decades).[39]

Another phenomenon observed in some Algol binaries has been monotonic period increases. This is quite distinct from the far more common observations of alternating period increases and decreases explained by the Applegate mechanism. Monotonic period increases have been attributed to mass transfer, usually (but not always) from the less massive to the more massive star[40]

## 34.4 Designations

### 34.4.1 A and B

The components of binary stars are denoted by the suffixes *A* and *B* appended to the system's designation, *A* denoting the primary and *B* the secondary. The suffix *AB* may be used to denote the pair (for example, the binary star α Centauri AB consists of the stars α Centauri A and α Centauri B.) Additional letters, such as *C*, *D*, etc., may be used for systems with more than two stars.[41] In cases where the binary star has a Bayer designation and is widely separated, it is possible that the members of the pair will be designated with superscripts; an example is Zeta Reticuli, whose components are $\zeta^1$ Reticuli and $\zeta^2$ Reticuli.[42]

### 34.4.2   Discoverer designations

Double stars are also designated by an abbreviation giving the discoverer together with an index number.[43] α Centauri, for example, was found to be double by Father Richaud in 1689, and so is designated *RHD 1*.[8][44] These discoverer codes can be found in the Washington Double Star Catalog.[45]

### 34.4.3   Hot and cold

The components of a binary star system may be designated by their relative temperatures as the *hot companion* and *cool companion*.

Examples:

- Antares (Alpha Scorpii) is a red supergiant star in a binary system with a hotter blue main-sequence star Antares B. Antares B can therefore be termed a hot companion of the cool supergiant.[46]

- Symbiotic stars are binary star systems composed of a late-type giant star and a hotter companion object. Since the nature of the companion is not well-established in all cases, it may be termed a "hot companion".[47]

- The luminous blue variable Eta Carinae has recently been determined to be a binary star system. The secondary appears to have a higher temperature than the primary and has therefore been described as being the "hot companion" star. It may be a Wolf–Rayet star.[48]

- R Aquarii shows a spectrum which simultaneously displays both a cool and hot signature. This combination is the result of a cool red supergiant accompanied by a smaller, hotter companion. Matter flows from the supergiant to the smaller, denser companion.[49]

- NASA's Kepler mission has discovered examples of eclipsing binary stars where the secondary is the hotter component. KOI-74b is a 12,000 K white dwarf companion of KOI-74 (KIC 6889235), a 9,400 K early A-type main-sequence star.[50][51][52] KOI-81b is a 13,000 K white dwarf companion of KOI-81 (KIC 8823868), a 10,000 K late B-type main-sequence star.[50][51][52]

## 34.5   Evolution

### 34.5.1   Formation

While it is not impossible that some binaries might be created through gravitational capture between two single stars, given the very low likelihood of such an event (three objects are actually required, as conservation of energy rules out a single gravitating body capturing another) and the high number of binaries, this cannot be the primary formation process. Also, the observation of binaries consisting of pre main-sequence stars, supports the theory that binaries are already formed during star formation. Fragmentation of the molecular cloud during the formation of protostars is an acceptable explanation for the formation of a binary or multiple star system.[53][54]

The outcome of the three-body problem, where the three stars are of comparable mass, is that eventually one of the three stars will be ejected from the system and, assuming no significant further perturbations, the remaining two will form a stable binary system.

### 34.5.2   Mass transfer and accretion

As a main-sequence star increases in size during its evolution, it may at some point exceed its Roche lobe, meaning that some of its matter ventures into a region where the gravitational pull of its companion star is larger than its own.[55] The result is that matter will transfer from one star to another through a process known as Roche lobe overflow (RLOF), either being absorbed by direct impact or through an accretion disc. The mathematical point through which this transfer happens is called the first Lagrangian point.[56] It is not uncommon that the accretion disc is the brightest (and thus sometimes the only visible) element of a binary star.

If a star grows outside of its Roche lobe too fast for all abundant matter to be transferred to the other component, it is also possible that matter will leave the system through other Lagrange points or as stellar wind, thus being effectively lost to both components.[57] Since the evolution of a star is determined by its mass, the process influences the evolution of both companions, and creates stages that cannot be attained by single stars.[58][59]

Studies of the eclipsing ternary Algol led to the *Algol paradox* in the theory of stellar evolution: although components of a binary star form at the same time, and massive stars evolve much faster than the less massive ones, it was observed that the more massive component Algol A is still in the main sequence, while the less massive Algol B is a subgiant at a later evolutionary stage. The paradox can be solved by mass transfer: when the more massive star became a subgiant, it filled its Roche lobe, and most of the mass was transferred to the other star, which is still in the main sequence. In some binaries similar to Algol, a gas flow can actually be seen.[60]

### 34.5.3   Runaways and novae

It is also possible for widely separated binaries to lose gravitational contact with each other during their lifetime, as a result of external perturbations. The components will then move on to evolve as single stars. A close encounter between two binary systems can also result in the gravitational disruption of both systems, with some of the stars being ejected at high velocities, leading to runaway stars.[61]

If a white dwarf has a close companion star that overflows its Roche lobe, the white dwarf will steadily accrete gases from the star's outer atmosphere. These are compacted on the white dwarf's surface by its intense gravity, compressed and heated to very high temperatures as additional material is drawn in. The white dwarf consists of degenerate matter, and so is largely unresponsive to heat, while the accreted hydrogen is not. Hydrogen fusion can occur in a stable manner on the surface through the CNO cycle, causing the enormous amount of energy liberated by this process to blow the remaining gases away from the white dwarf's surface. The result is an extremely bright outburst of light, known as a nova.[62]

In extreme cases this event can cause the white dwarf to exceed the Chandrasekhar limit and trigger a supernova that destroys the entire star, and is another possible cause for runaways.[63][64] An example of such an event is the supernova SN 1572, which was observed by Tycho Brahe. The Hubble Space Telescope recently took a picture of the remnants of this event.

## 34.6   Astrophysics

Binaries provide the best method for astronomers to determine the mass of a distant star. The gravitational pull between them causes them to orbit around their common center of mass. From the orbital pattern of a visual binary, or the time variation of the spectrum of a spectroscopic binary, the mass of its stars can be determined. In this way, the relation between a star's appearance (temperature and radius) and its mass can be found, which allows for the determination of the mass of non-binaries.

Because a large proportion of stars exist in binary systems, binaries are particularly important to our understanding of the processes by which stars form. In particular, the period and masses of the binary tell us about the amount of angular momentum in the system. Because this is a conserved quantity in physics, binaries give us important clues about the conditions under which the stars were formed.

### 34.6.1   Calculating the center of mass in binary stars

In a simple binary case, $r_1$, the distance from the center of the first star to the center of mass, is given by:

$$r_1 = a \cdot \frac{m_2}{m_1 + m_2} = \frac{a}{1 + m_1/m_2}$$

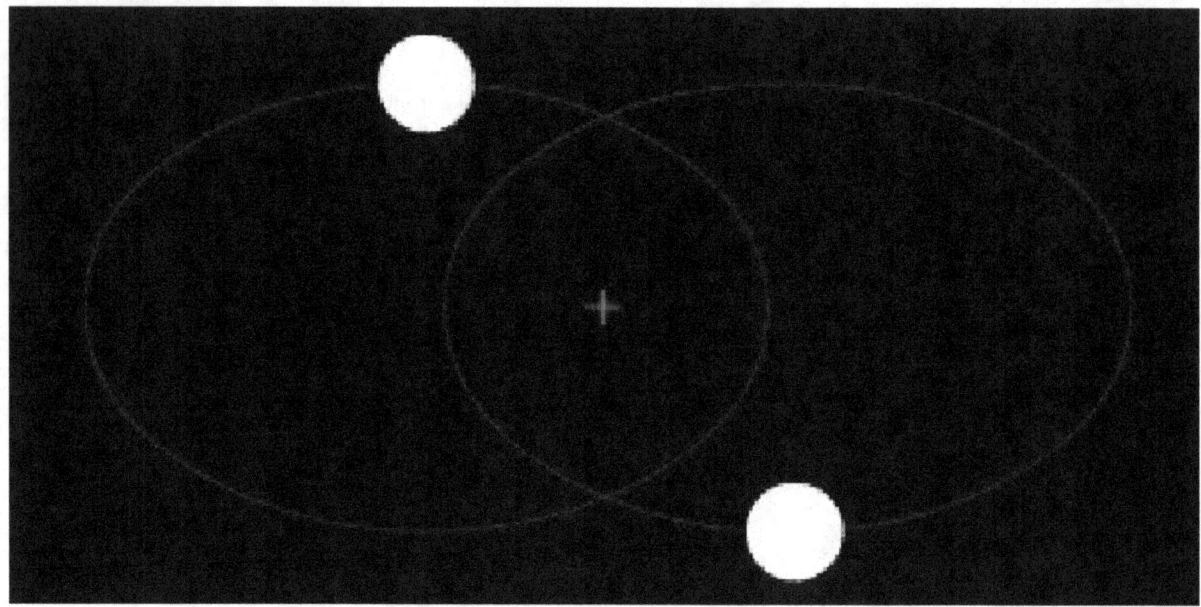

*A simulated example of a binary star, where two bodies with similar mass orbit around a common barycenter in elliptic orbits*

where:

> $a$ is the distance between the two stellar centers and
>
> $m_1$ and $m_2$ are the masses of the two stars.

If $a$ is taken to be the semi-major axis of the orbit of one body around the other, then $r_1$ will be the semimajor axis of the first body's orbit around the center of mass or *barycenter*, and $r_2 = a - r_1$ will be the semimajor axis of the second body's orbit. When the center of mass is located within the more massive body, that body will appear to wobble rather than following a discernible orbit.

### 34.6.2   Center of mass animations

Images are representative, not simulated. The position of the red cross indicates the center of mass of the system.

### 34.6.3   Research findings

It is estimated that approximately 1/3 of the star systems in the Milky Way are binary or multiple, with the remaining 2/3 consisting of single stars.[65]

There is a direct correlation between the period of revolution of a binary star and the eccentricity of its orbit, with systems of short period having smaller eccentricity. Binary stars may be found with any conceivable separation, from pairs orbiting so closely that they are practically in contact with each other, to pairs so distantly separated that their connection is indicated only by their common proper motion through space. Among gravitationally bound binary star systems, there exists a so-called log normal distribution of periods, with the majority of these systems orbiting with a period of about 100 years. This is supporting evidence for the theory that binary systems are formed during star formation.[66]

In pairs where the two stars are of equal brightness, they are also of the same spectral type. In systems where the brightnesses are different, the fainter star is bluer if the brighter star is a giant star, and redder if the brighter star belongs to the main sequence.[67]

The mass of a star can be directly determined only from its gravitational attraction. Apart from the Sun and stars which act as gravitational lenses, this can be done only in binary and multiple star systems, making the binary stars an important

*Artist's impression of the sight from a (hypothetical) moon of planet HD 188753 Ab (upper left), which orbits a triple star system. The brightest companion is just below the horizon.*

class of stars. In the case of a visual binary star, after the orbit and the stellar parallax of the system has been determined, the combined mass of the two stars may be obtained by a direct application of the Keplerian harmonic law.[68]

Unfortunately, it is impossible to obtain the complete orbit of a spectroscopic binary unless it is also a visual or an eclipsing binary, so from these objects only a determination of the joint product of mass and the sine of the angle of inclination relative to the line of sight is possible. In the case of eclipsing binaries which are also spectroscopic binaries, it is possible to find a complete solution for the specifications (mass, density, size, luminosity, and approximate shape) of both members of the system.

### Planets

Main article: Habitability of binary star systems

Science fiction has often featured planets of binary or ternary stars as a setting, for example George Lucas' Tatooine from *Star Wars*, and one notable story, "Nightfall", even takes this to a six-star system. In reality, some orbital ranges are impossible for dynamical reasons (the planet would be expelled from its orbit relatively quickly, being either ejected from the system altogether or transferred to a more inner or outer orbital range), whilst other orbits present serious challenges for eventual biospheres because of likely extreme variations in surface temperature during different parts of the orbit. Planets that orbit just one star in a binary system are said to have "S-type" orbits, whereas those that orbit around both stars have "P-type" or "circumbinary" orbits. It is estimated that 50–60% of binary systems are capable of supporting habitable terrestrial planets within stable orbital ranges.[69]

Simulations have shown that the presence of a binary companion can actually improve the rate of planet formation within stable orbital zones by "stirring up" the protoplanetary disk, increasing the accretion rate of the protoplanets within.[69]

Detecting planets in multiple star systems introduces additional technical difficulties, which may be why they are only rarely found.[70] Examples include the white dwarf-pulsar binary PSR B1620-26, the subgiant-red dwarf binary Gamma Cephei, and the white dwarf-red dwarf binary NN Serpentis. More planets around binaries are listed in: [Muterspaugh; Lane; Kulkarni; Maciej Konacki; Burke; Colavita; Shao; Hartkopf; Boss (2010). "The PHASES Differential Astrometry Data Archive. V. Candidate Substellar Companions to Binary Systems". *The Astronomical Journal* **140** (6): 1657. arXiv:1010.4048. doi:10.1088/0004-6256/140/6/1657.].

A study of fourteen previously known planetary systems found three of these systems to be binary systems. All planets were found to be in S-type orbits around the primary star. In these three cases the secondary star was much dimmer than the primary and so was not previously detected. This discovery resulted in a recalculation of parameters for both the planet and the primary star.[71]

## 34.7  Examples

The large distance between the components, as well as their difference in color, make Albireo one of the easiest observable visual binaries. The brightest member, which is the third brightest star in the constellation Cygnus, is actually a close binary itself. Also in the Cygnus constellation is Cygnus X-1, an X-ray source considered to be a black hole. It is a high-mass X-ray binary, with the optical counterpart being a variable star.[72] Sirius is another binary and the brightest star in the night time sky, with a visual apparent magnitude of −1.46. It is located in the constellation Canis Major. In 1844 Friedrich Bessel deduced that Sirius was a binary. In 1862 Alvan Graham Clark discovered the companion (Sirius B; the visible star is Sirius A). In 1915 astronomers at the Mount Wilson Observatory determined that Sirius B was a white dwarf, the first to be discovered. In 2005, using the Hubble Space Telescope, astronomers determined Sirius B to be 12,000 km (7,456 mi) in diameter, with a mass that is 98% of the Sun.[73]

An example of an eclipsing binary is Epsilon Aurigae in the constellation Auriga. The visible component belongs to the spectral class F0, the other (eclipsing) component is not visible. The last such eclipse occurred from 2009–2011, and it is hoped that the extensive observations that will likely be carried out may yield further insights into the nature of this system. Another eclipsing binary is Beta Lyrae, which is a semi-detached binary star system in the constellation of Lyra.

Other interesting binaries include 61 Cygni (a binary in the constellation Cygnus, composed of two K class (orange) main-sequence stars, 61 Cygni A and 61 Cygni B, which is known for its large proper motion), Procyon (the brightest star in the constellation Canis Minor and the eighth brightest star in the night time sky, which is a binary consisting of the main star with a faint white dwarf companion), SS Lacertae (an eclipsing binary which stopped eclipsing), V907 Sco (an eclipsing binary which stopped, restarted, then stopped again) and BG Geminorum (an eclipsing binary which is thought to contain a black hole with a K0 star in orbit around it).

## 34.8  Multiple star examples

Systems with more than two stars are termed multiple stars. Algol is the most noted ternary (long thought to be a binary), located in the constellation Perseus. Two components of the system eclipse each other, the variation in the intensity of Algol first being recorded in 1670 by Geminiano Montanari. The name Algol means "demon star" (from Arabic: الغول al-ghūl), which was probably given due to its peculiar behavior. Another visible ternary is Alpha Centauri, in the southern constellation of Centaurus, which contains the fourth brightest star in the night sky, with an apparent visual magnitude of −0.01. This system also underscores the fact that binaries need not be discounted in the search for habitable planets. Alpha Centauri A and B have an 11 AU distance at closest approach, and both should have stable habitable zones.[74]

There are also examples of systems beyond ternaries: Castor is a sextuple star system, which is the second brightest star in the constellation Gemini and one of the brightest stars in the nighttime sky. Astronomically, Castor was discovered to be a visual binary in 1719. Each of the components of Castor is itself a spectroscopic binary. Castor also has a faint and widely separated companion, which is also a spectroscopic binary. The Alcor–Mizar visual binary in Ursa Majoris also consists of six stars, four comprising Mizar and two comprising Alcor.

*The two visibly distinguishable components of Albireo.*

## 34.9   See also

- 104 Aquarii

- 107 Aquarii

- Beta Centauri

- Binary stars in fiction

- Binary system

- HD 30453

*Artist's impression of the double-star system GG Tauri-A.*

- Rotational Brownian motion (astronomy)

- Two-body problem in general relativity

## 34.10   Notes and references

[1] Filippenko, Alex, *Understanding the Universe* (of *The Great Courses* on DVD), Lecture 46, time 1:17, The Teaching Company, Chantilly, VA, USA, 2007

[2] *The Binary Stars*, Robert Grant Aitken, New York: Dover, 1964, p. ix.

[3] Herschel, William (1802). "Catalogue of 500 New Nebulae, Nebulous Stars, Planetary Nebulae, and Clusters of Stars; With Remarks on the Construction of the Heavens". *Philosophical Transactions of the Royal Society of London* **92**: 477–528 [481]. Bibcode:1802RSPT...92..477H. doi:10.1098/rstl.1802.0021. JSTOR 107131.

[4] Heintz, W. D. (1978). *Double Stars*. Dordrecht: D. Reidel Publishing Company. pp. 1–2. ISBN 90-277-0885-1.

[5] "Visual Binaries". University of Tennessee.

[6] Heintz, W. D. (1978). *Double Stars*. Dordrecht: D. Reidel Publishing Company. p. 5. ISBN 90-277-0885-1.

[7] Heintz, W. D. (1978). *Double Stars*. D. Reidel Publishing Company, Dordrecht. p. 17. ISBN 90-277-0885-1.

[8] *The Binary Stars*, Robert Grant Aitken, New York: Dover, 1964, p. 1.

[9] Vol. 1, part 1, p. 422, *Almagestum Novum*, Giovanni Battista Riccioli, Bononiae: Ex typographia haeredis Victorij Benatij, 1651.

[10] A New View of Mizar, Leos Ondra, accessed on line May 26, 2007.

[11] pp. 10–11, *Observing and Measuring Double Stars*, Bob Argyle, ed., London: Springer, 2004, ISBN 1-85233-558-0.

[12] pp. 249–250, An Inquiry into the Probable Parallax, and Magnitude of the Fixed Stars, from the Quantity of Light Which They Afford us, and the Particular Circumstances of Their Situation, John Michell, *Philosophical Transactions (1683–1775)* **57** (1767), pp. 234–264.

[13] Heintz, W. D. (1978). *Double Stars*. Dordrecht: D. Reidel Publishing Company. p. 4. ISBN 90-277-0885-1.

[14] Account of the Changes That Have Happened, during the Last Twenty-Five Years, in the Relative Situation of Double-Stars: With an Investigation of the Cause to Which They Are Owing, William Herschel, *Philosophical Transactions of the Royal Society of London* **93** (1803), pp. 339–382.

[15] p. 291, French astronomers, visual double stars and the double stars working group of the Société Astronomique de France, E. Soulié, *The Third Pacific Rim Conference on Recent Development of Binary Star Research*, proceedings of a conference sponsored by Chiang Mai University, Thai Astronomical Society and the University of Nebraska-Lincoln held in Chiang Mai, Thailand, 26 October-1 November 1995, *ASP Conference Series* **130** (1997), ed. Kam-Ching Leung, pp. 291–294, Bibcode: 1997ASPC..130..291S.

[16] "Introduction and Growth of the WDS", The Washington Double Star Catalog, Brian D. Mason, Gary L. Wycoff, and William I. Hartkopf, Astrometry Department, United States Naval Observatory, accessed on line August 20, 2008.

[17] Sixth Catalog of Orbits of Visual Binary Stars, William I. Hartkopf and Brian D. Mason, United States Naval Observatory, accessed on line August 20, 2008.

[18] The Washington Double Star Catalog, Brian D. Mason, Gary L. Wycoff, and William I. Hartkopf, United States Naval Observatory. Accessed on line December 20, 2008.

[19] Heintz, W. D. (1978). *Double Stars*. Dordrecht: D. Reidel Publishing Company. pp. 17–18. ISBN 90-277-0885-1.

[20] "Planet-hunting SPHERE Images First Circumbinary Planet System with Disc". Retrieved 26 October 2015.

[21] "Binary Stars". Cornell Astronomy.

[22] *The Binary Stars*, Robert Grant Aitken, New York: Dover, 1964, p. 41.

[23] Herter, T. "Stellar Masses". Cornell University. Archived from the original on June 17, 2012.

[24] Bruton, D. "Eclipsing Binary Stars". Stephen F. Austin State University.

[25] Bruton, D. "Eclipsing Binary Stars". Stephen F. Austin State University.

[26] Wilson, R. E. (1 January 2008). "Eclipsing Binary Solutions in Physical Units and Direct Distance Estimation". *The Astrophysical Journal* **672** (1). Bibcode:2008ApJ...672..575W. doi:10.1086/523634.

[27] Bonanos, Alceste Z. (2006). "Eclipsing Binaries: Tools for Calibrating the Extragalactic Distance Scale". *Proceedings of the International Astronomical Union* **2**. arXiv:astro-ph/0610923. doi:10.1017/S1743921307003845.

[28] Worth, M. "Binary Stars" (PowerPoint). Stephen F. Austin State University.

[29] Lev Tal-Or, Simchon Faigler, Tsevi Mazeh (2014). "Seventy-two new non-eclipsing BEER binaries discovered in CoRoT lightcurves and confirmed by RVs from AAOmega". arXiv:1410.3074.

[30] Bock, D. "Binary Neutron Star Collision". NCSA.

[31] Asada, H.; T. Akasaka; M. Kasai (27 September 2004). "Inversion formula for determining parameters of an astrometric binary". *Publ.Astron.Soc.Jap* **56**: L35–L38. arXiv:astro-ph/0409613. Bibcode:2004PASJ...56L..35A. doi:10.1093/pasj/56.6.L35.

[32] "Astrometric Binaries". University of Tennessee.

[33] Nguyen, Q. "Roche model". San Diego State University.

[34] Voss, R.; T.M. Tauris (2003). "Galactic distribution of merging neutron stars and black holes". *Monthly Notices of the Royal Astronomical Society* **342** (4): 1169–1184. arXiv:0705.3444. Bibcode:2003MNRAS.342.1169V. doi:10.1046/j.1365-8711.2003.06616.x.

[35] Robert Connon Smith (November 2006). "Cataclysmic Variables". *Contemporary Physics* **47** (6): 363–386. arXiv:astro-ph/0701654. Bibcode:2007astro.ph...1654C. doi:10.1080/00107510601181175.

[36] Neutron Star X-ray binaries, *A Systematic Search of New X-ray Pulsators in ROSAT Fields*, Gian Luca Israel, Ph. D. thesis, Trieste, October 1996.

[37]  Iorio, Lorenzo (July 24, 2007). "On the orbital and physical parameters of the HDE 226868/Cygnus X-1 binary system". *E-print* **315** (1–4): 335. arXiv:0707.3525. Bibcode:2008Ap&SS.315..335I. doi:10.1007/s10509-008-9839-y.

[38]  Black Holes, Imagine the Universe!, NASA. Accessed on line August 22, 2008.

[39]  Applegate, James H. (1992). "A mechanism for orbital period modulation in close binaries". *Astrophysical Journal, Part 1* **385**: 621–629. Bibcode:1992ApJ...385..621A. doi:10.1086/170967.

[40]  Hall, Douglas S. (1989). "The relation between RS CVn and Algol". *Space Science Reviews* **50**: 219–233. Bibcode:1989SSRv... doi:10.1007/BF00215932.

[41]  Heintz, W. D. (1978). *Double Stars*. Dordrecht: D. Reidel Publishing Company. p. 19. ISBN 90-277-0885-1.

[42]  "Binary and Multiple Star Systems". Lawrence Hall of Science at the University of California.

[43]  pp. 307–308, *Observing and Measuring Double Stars*, Bob Argyle, ed., London: Springer, 2004, ISBN 1-85233-558-0.

[44]  Entry 14396-6050, discoverer code RHD 1AB.The Washington Double Star Catalog, United States Naval Observatory. Accessed on line August 20, 2008.

[45]  References and discoverer codes, The Washington Double Star Catalog, United States Naval Observatory. Accessed on line August 20, 2008.

[46]  – see essential notes: "Hot companion to Antares at 2.9arcsec; estimated period: 678yr."

[47]  Kenyon, S. J.; Webbink, R. F. (1984). "The nature of symbiotic stars". *Astrophysical Journal* **279**: 252–283. Bibcode:1984ApJ doi:10.1086/161888.

[48]  Iping, Rosina C.; Sonneborn, George; Gull, Theodore R.; Massa, Derck L.; Hillier, D. John (2005). "Detection of a Hot Binary Companion of η Carinae". *The Astrophysical Journal* **633** (1): L37–L40. arXiv:astro-ph/0510581. Bibcode:2005ApJ...633L..37I. doi:10.1086/498268.

[49]  Nigel Henbest; Heather Couper (1994). *The guide to the galaxy*. ISBN 978-0-521-45882-5.

[50]  Rowe, Jason F.; Borucki, William J.; Koch, David; Howell, Steve B.; Basri, Gibor; Batalha, Natalie; Brown, Timothy M.; Caldwell, Douglas; Cochran, William D.; Dunham, Edward; Dupree, Andrea K.; Fortney, Jonathan J.; Gautier, Thomas N.; Gilliland, Ronald L.; Jenkins, Jon; Latham, David W.; Lissauer, Jack J.; Marcy, Geoff; Monet, David G.; Sasselov, Dimitar; Welsh, William F. (2010). "Kepler Observations of Transiting Hot Compact Objects". *The Astrophysical Journal Letters* **713** (2): L150–L154. arXiv:1001.3420. Bibcode:2010ApJ...713L.150R. doi:10.1088/2041-8205/713/2/L150.

[51]  van Kerkwijk, Marten H.; Rappaport, Saul A.; Breton, René P.; Justham, Stephen; Podsiadlowski, Philipp; Han, Zhanwen (2010). "Observations of Doppler Boosting in Kepler Light Curves". *The Astrophysical Journal* **715** (1): 51–58. arXiv:1001.4539. Bibcode:2010ApJ...715...51V. doi:10.1088/0004-637X/715/1/51.

[52]  Borenstein, Seth (4 January 2010). "Planet-hunting telescope unearths hot mysteries" (6:29 pm EST).

[53]  Boss, A. P. (1992). "Formation of Binary Stars". In J. Sahade; G. E. McCluskey; Yoji Kondo. *The Realm of Interacting Binary Stars*. Dordrecht: Kluwer Academic. p. 355. ISBN 0-7923-1675-4.

[54]  Tohline, J. E.; J. E. Cazes; H. S. Cohl. "The Formation of Common-Envelope, Pre-Main-Sequence Binary Stars". Louisiana State University.

[55]  Kopal, Z. (1989). *The Roche Problem*. Kluwer Academic. ISBN 0-7923-0129-3.

[56]  "Contact Binary Star Envelopes" by Jeff Bryant, Wolfram Demonstrations Project.

[57]  "Mass Transfer in Binary Star Systems" by Jeff Bryant with Waylena McCully, Wolfram Demonstrations Project.

[58]  Boyle, C.B. (1984). "Mass transfer and accretion in close binaries – A review". *Vistas in Astronomy* **27**: 149–169. Bibcode:1984 doi:10.1016/0083-6656(84)90007-2.

[59]  Vanbeveren, D.; W. van Rensbergen; C. de Loore (2001). *The Brightest Binaries*. Springer. ISBN 0-7923-5155-X.

[60]  Blondin, J. M.; M. T. Richards; M. L. Malinowski. "Mass Transfer in the Binary Star Algol". American Museum of Natural History.

[61] Hoogerwerf, R.; de Bruijne, J.H.J.; de Zeeuw, P.T. (December 2000). "The Origin of Runaway Stars". *Astrophysical Journal* **544** (2): L133. arXiv:astro-ph/0007436. Bibcode:2000ApJ...544L.133H. doi:10.1086/317315.

[62] Prialnik, D. (2001). "Novae". *Encyclopaedia of Astronomy and Astrophysics*. pp. 1846–1856.

[63] Icko, I. (1986). "Binary Star Evolution and Type I Supernovae". *Cosmogonical Processes*. p. 155.

[64] Fender, R. (2002). "Relativistic outflows from X-ray binaries (a.k.a. 'Microquasars')". *Lect. Notes Phys.* Lecture Notes in Physics **589** (101): 101. arXiv:astro-ph/0109502. Bibcode:2002LNP...589..101F. doi:10.1007/3-540-46025-X_6. ISBN 978-3-540-43518-1.

[65] Most Milky Way Stars Are Single. Harvard-Smithsonian Center for Astrophysics

[66] Hubber, D. A.; A.P. Whitworth (2005). "Binary Star Formation from Ring Fragmentation". *Astronomy and Physics* **437**: 113–125. arXiv:astro-ph/0503412. Bibcode:2005A&A...437..113H. doi:10.1051/0004-6361:20042428.

[67] Schombert, J. "Birth and Death of Stars". University of Oregon.

[68] "Binary Star Motions". Cornell Astronomy.

[69] Elisa V. Quintana; Jack J. Lissauer (2007). "Terrestrial Planet Formation in Binary Star Systems". arXiv:0705.3444 [astro-ph].

[70] Schirber, M (17 May 2005). "Planets with Two Suns Likely Common". Space.com.

[71] Daemgen, S.; Hormuth, F.; Brandner, W.; Bergfors, C.; Janson, M.; Hippler, S.; Henning, T. (2009). "Binarity of transit host stars – Implications for planetary parameters" (PDF). *Astronomy and Astrophysics* **498** (2): 567–574. arXiv:0902.2179. Bibcode:2009A&A...498..567D. doi:10.1051/0004-6361/200810988.

[72] See sources at Cygnus X-1

[73] McGourty, C. (2005-12-14). "Hubble finds mass of white dwarf". BBC News. Retrieved 2010-01-01.

[74] Elisa V. Quintana; Fred C. Adams; Jack J. Lissauer & John E. Chambers (2007). "Terrestrial Planet Formation around Individual Stars within Binary Star Systems". *Astrophysical Journal* **660**: 807. arXiv:astro-ph/0701266. Bibcode:2007ApJ...660..807Q. doi:10.1086/512542.

## 34.11 External links

- The Double Star Library, at the U.S. Naval Observatory

- ianridpath.com: List of the best visual binaries, for amateurs, with orbital elements

- Pictures of binaries at Hubblesite.org

- Chandra X-ray Observatory

- Binary Stars at DMOZ

- An extensive simulation for the Algol system by North Carolina State University

- Selected visual double stars and their relative position as a function of time

- Artistic representations of binary stars by Mark A. Garlick

- Orbits and Velocity Curves of Spectroscopic Binaries, J. Miller Barr (1908)

- Eclipsing Binaries in the 21st Century—Opportunities for Amateur Astronomers

# Chapter 35

# Quark-nova

A **quark-nova** is the theorized violent explosion resulting from the conversion of a neutron star to a quark star. Analogous to a supernova heralding the birth of a neutron star, a quark nova signals the creation of a quark star. The concept of quark-novae was suggested by Dr. Rachid Ouyed[1] (University of Calgary, Canada) and Drs. Dey and Dey (Calcutta University, India).[2]

When a neutron star spins down, it may convert to a quark star through a process known as quark deconfinement. The resultant star would have quark matter in its interior. The process would release immense amounts of energy, perhaps explaining the most energetic explosions in the universe; calculations have estimated that as much as $10^{47}$ J could be released from the phase transition inside a neutron star.[3] Quark-novae may be one cause of gamma ray bursts. According to Jaikumar et al.,[4] they may also be involved in producing heavy elements such as platinum through r-process nucleosynthesis.

Rapidly spinning neutron stars with masses between 1.5 and 1.8 solar masses are theoretically the best candidates for conversion due to spin down of the star within a Hubble time. This amounts to a small fraction of the projected neutron star population. A conservative estimate based on this indicates that up to two quark-novae may occur in the observable universe each day.

Theoretically, quark stars would be radio-quiet, so radio-quiet neutron stars may be quark stars.

Direct evidence for quark-novae is scant; however, recent observations of supernovae SN 2006gy, SN 2005gj and SN 2005ap may point to their existence.[5][6]

## 35.1    See also

- Quark matter

- Quark-degenerate matter

- SN 2006gy

- SN 2005gj

## 35.2    References

[1] "Quark Nova Project". Retrieved 7 Sep 2012.

[2] R. Ouyed; J. Dey; M. Dey (2002). "Quark-Nova". *Astronomy and Astrophysics* **390**: L39–L42. arXiv:astro-ph/0105109. Bibcode:2002A&A...390L..39O. doi:10.1051/0004-6361:20020982.

[3] "Theories of Quark-novae". Retrieved 29 June 2008.

[4] Prashanth Jaikumar; Meyer; Kaori Otsuki; Rachid Ouyed (2007). "Nucleosynthesis in neutron-rich ejecta from Quark-Novae". *Astronomy and Astrophysics* **471**: 227–236. arXiv:nucl-th/0610013. Bibcode:2007A&A...471..227J.doi:10.1051/0004-636

[5] Astronomy Now Online - Second Supernovae Point to Quark Stars

[6] Leahy, Denis; Ouyed, Rachid (2008). "Supernova SN2006gy as a first ever Quark Nova?". *Monthly Notices of the Royal Astronomical Society* **387** (3): 1193. arXiv:0708.1787. Bibcode:2008MNRAS.387.1193L. doi:10.1111/j.1365-2966.2008.13312.x.

## 35.3 External links

- Quark-novae produce neutrino bursts, which can be detected by neutrino observatories

- Quark Stars Could Produce Biggest Bang (SpaceDaily) June 7, 2006

- Quark Nova Project animations (University of Calgary)

# Chapter 36

# Galaxy morphological classification

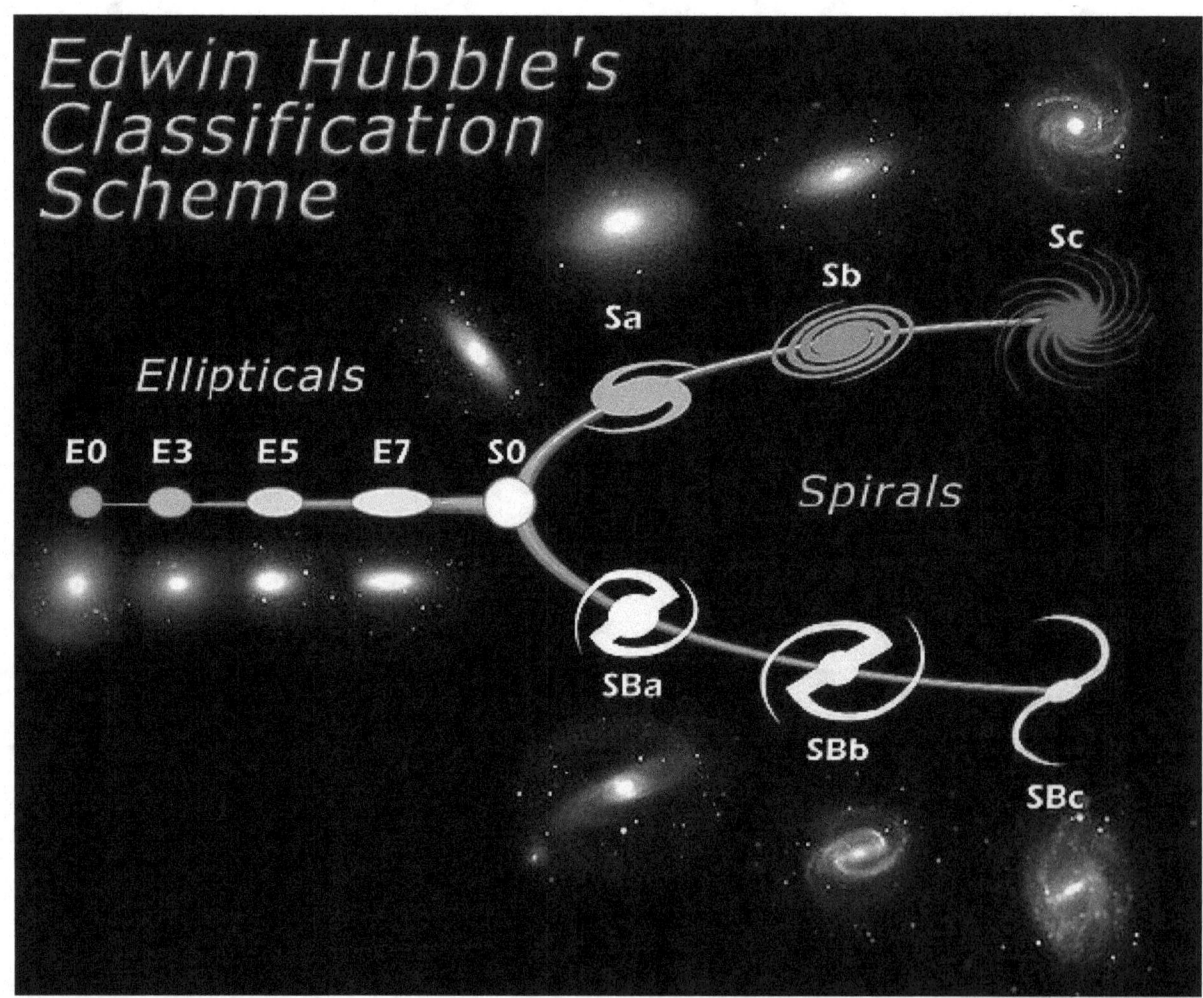

*Tuning-fork style diagram of the Hubble sequence*

**Galaxy morphological classification** is a system used by astronomers to divide galaxies into groups based on their visual appearance. There are several schemes in use by which galaxies can be classified according to their morphologies, the most famous being the Hubble sequence, devised by Edwin Hubble and later expanded by Gérard de Vaucouleurs and Allan Sandage.

# 36.1   Hubble sequence

Main article: Hubble sequence

The Hubble sequence is a morphological classification scheme for galaxies invented by Edwin Hubble in 1926.[1][2] It is often known colloquially as the "Hubble tuning-fork" because of the shape in which it is traditionally represented. Hubble's scheme divides galaxies into three broad classes based on their visual appearance (originally on photographic plates):

- **Elliptical galaxies** have smooth, featureless light distributions and appear as ellipses in images. They are denoted by the letter $E$, followed by an integer $n$ representing their degree of ellipticity on the sky.

- **Spiral galaxies** consist of a flattened disk, with stars forming a (usually two-armed) spiral structure, and a central concentration of stars known as the bulge, which is similar in appearance to an elliptical galaxy. They are given the symbol "S". Roughly half of all spirals are also observed to have a bar-like structure, extending from the central bulge. These barred spirals are given the symbol "S.B.".

- **Lenticular galaxies** (designated S0) also consist of a bright central bulge surrounded by an extended, disk-like structure but, unlike spiral galaxies, the disks of lenticular galaxies have no visible spiral structure and are not actively forming stars in any significant quantity.

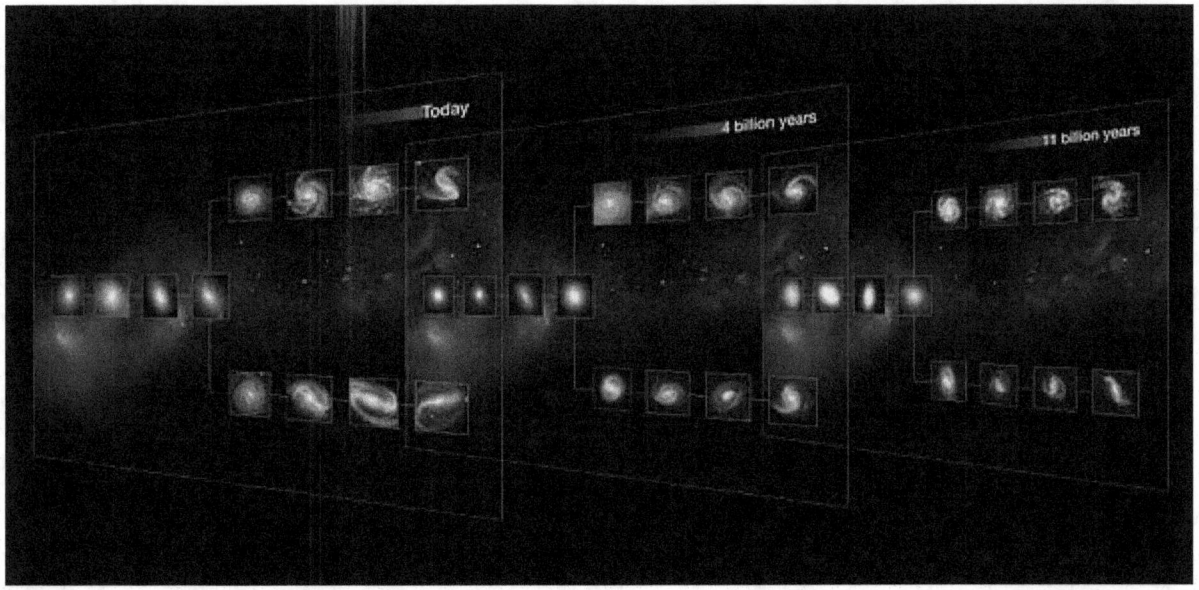

*The Hubble sequence throughout the universe's history.*[3]

These broad classes can be extended to enable finer distinctions of appearance and to encompass other types of galaxies, such as irregular galaxies, which have no obvious regular structure (either disk-like or ellipsoidal).

The Hubble sequence is often represented in the form of a two-pronged fork, with the ellipticals on the left (with the degree of ellipticity increasing from left to right) and the barred and unbarred spirals forming the two parallel prongs of the fork. Lenticular galaxies are placed between the ellipticals and the spirals, at the point where the two prongs meet the "handle".

To this day, the Hubble sequence is the most commonly used system for classifying galaxies, both in professional astronomical research and in amateur astronomy.

## 36.2    De Vaucouleurs system

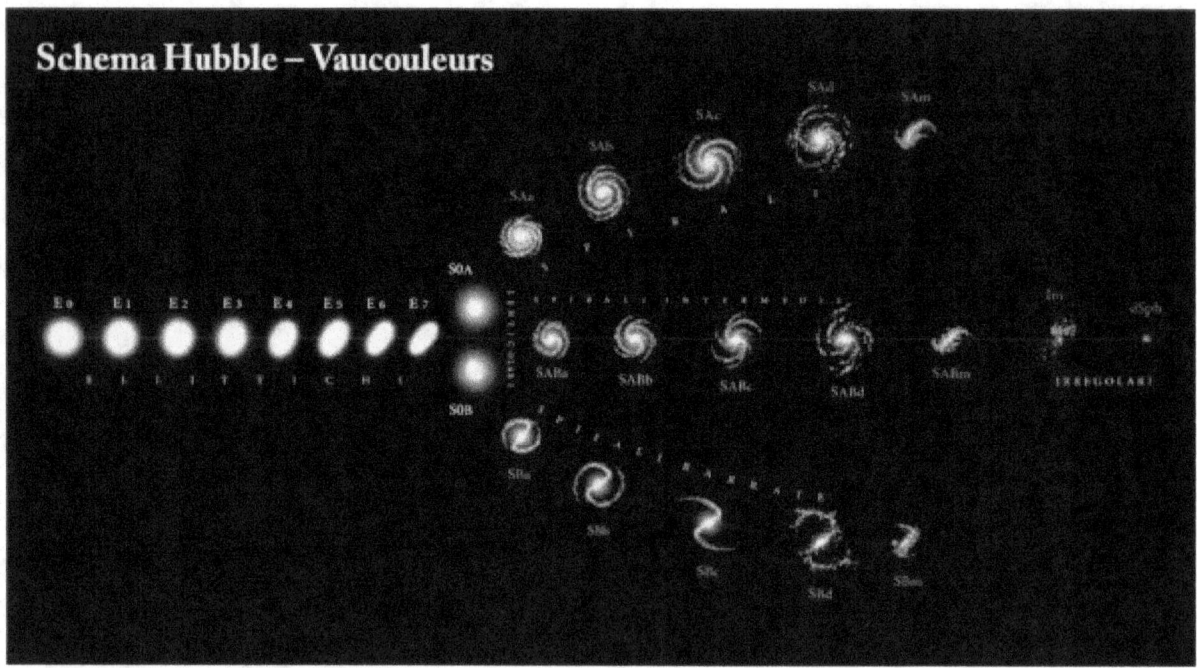

The de Vaucouleurs system for classifying galaxies is a widely used extension to the Hubble sequence, first described by Gérard de Vaucouleurs in 1959.[4] De Vaucouleurs argued that Hubble's two-dimensional classification of spiral galaxies—based on the tightness of the spiral arms and the presence or absence of a bar—did not adequately describe the full range of observed galaxy morphologies. In particular, he argued that rings and lenses are important structural components of spiral galaxies.[5]

The de Vaucouleurs system retains Hubble's basic division of galaxies into ellipticals, lenticulars, spirals and irregulars. To complement Hubble's scheme, de Vaucouleurs introduced a more elaborate classification system for spiral galaxies, based on three morphological characteristics:

- **Bars.** Galaxies are divided on the basis of the presence or absence of a nuclear bar. De Vaucouleurs introduced the notation SA to denote spiral galaxies without bars, complementing Hubble's use of SB for barred spirals. He also allowed for an intermediate class, denoted SAB, containing weakly barred spirals.[6] Lenticular galaxies are also classified as unbarred (SA0) or barred (SB0), with the notation S0 reserved for those galaxies for which it is impossible to tell if a bar is present or not (usually because they are edge-on to the line-of-sight).

- **Rings.** Galaxies are divided into those possessing ring-like structures (denoted '(r)') and those without rings (denoted '(s)'). So-called 'transition' galaxies are given the symbol (rs).[6]

- **Spiral arms.** As in Hubble's original scheme, spiral galaxies are assigned to a class based primarily on the tightness of their spiral arms. The de Vaucouleurs scheme extends the arms of Hubble's tuning fork to include several additional spiral classes:

  - Sd (SBd) - diffuse, broken arms made up of individual stellar clusters and nebulae; very faint central bulge
  - Sm (SBm) - irregular in appearance; no bulge component
  - Im - highly irregular galaxy

Most galaxies in these three classes were classified as Irr I in Hubble's original scheme. In addition, the Sd class contains some galaxies from Hubble's Sc class. Galaxies in the classes Sm and Im are termed the "Magellanic"

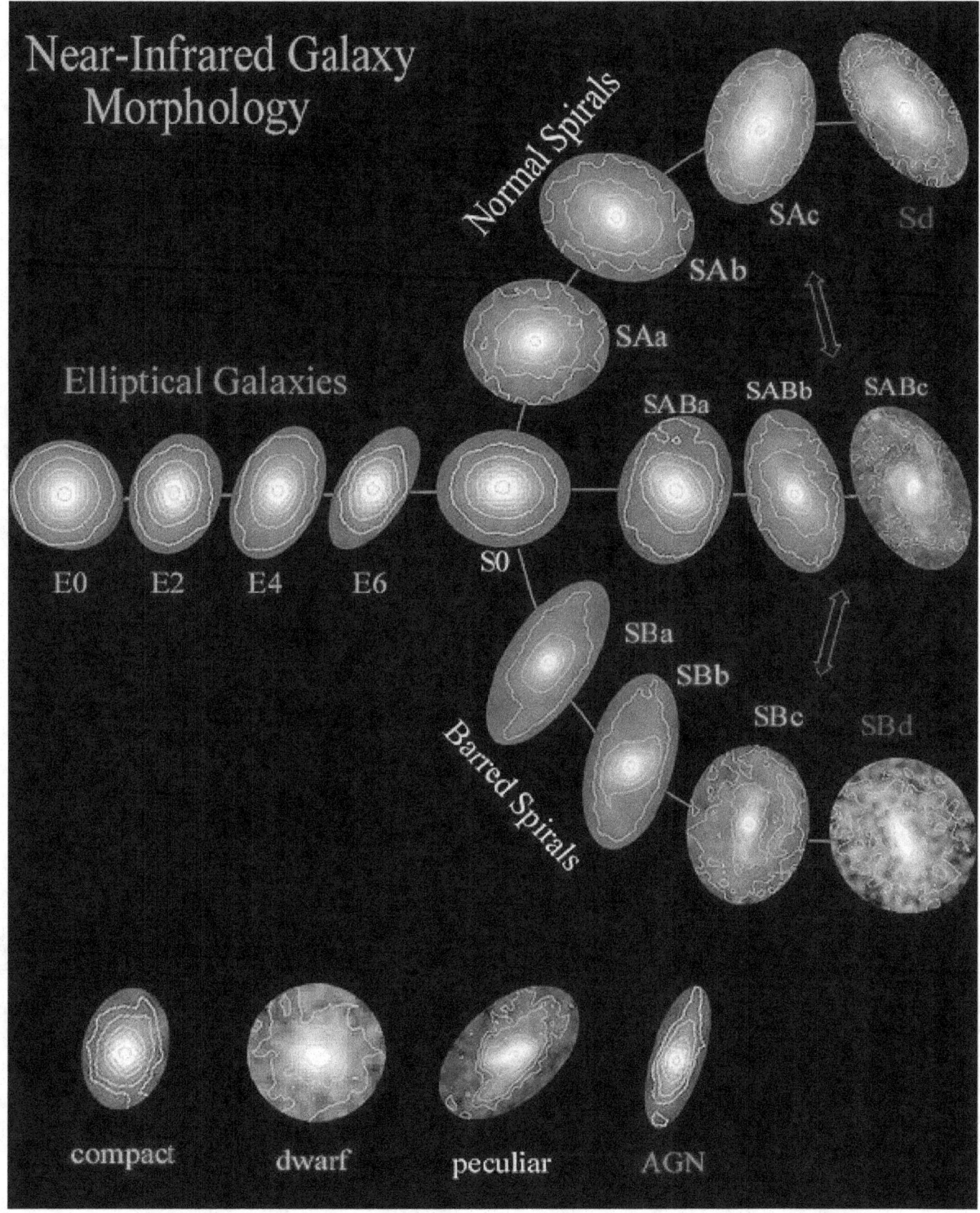

Near-Infrared Galaxy Morphology

Normal Spirals

Elliptical Galaxies

Barred Spirals

SAc   Sd

SAb

SAa

E0   E2   E4   E6

SABa   SABb   SABc

S0

SBa

SBb

SBc   SBd

compact   dwarf   peculiar   AGN

spirals and irregulars, respectively, after the Magellanic Clouds. The Large Magellanic Cloud is of type SBm, while the Small Magellanic Cloud is an irregular (Im).

The different elements of the classification scheme are combined — in the order in which they are listed — to give the complete classification of a galaxy. For example, a weakly barred spiral galaxy with loosely wound arms and a ring is

*NGC 6782: a spiral galaxy (type SB(r)0/a) with three rings of different radii, as well as a bar.*

denoted SAB(r)c.

Visually, the de Vaucouleurs system can be represented as a three-dimensional version of Hubble's tuning fork, with stage (spiralness) on the x-axis, family (barredness) on the y-axis, and variety (ringedness) on the z-axis.[7]

### 36.2.1   Numerical Hubble stage

De Vaucouleurs also assigned numerical values to each class of galaxy in his scheme. Values of the numerical Hubble stage T run from −6 to +10, with negative numbers corresponding to early-type galaxies (ellipticals and lenticulars) and positive numbers to late types (spirals and irregulars). Elliptical galaxies are divided into three 'stages': compact ellipticals (cE), normal ellipticals (E) and late types (E$^+$). Lenticulars are similarly subdivided into early (S$^-$), intermediate (S$^0$) and late (S$^+$) types. Irregular galaxies can be of type magellanic irregulars (T = 10) or 'compact' (T = 11).

The use of numerical stages allows for more quantitative studies of galaxy morphology.

*NGC 7793: a spiral galaxy of type SA(s)d.*

## 36.3 Yerkes (or Morgan) scheme

Created by American astronomer William Wilson Morgan. Together with Philip Keenan, Morgan developed the MK system for the classification of stars through their spectra. The Yerkes scheme uses the spectra of stars in the galaxy; the shape, real and apparent; and the degree of the central concentration to classify galaxies.

Thus, for example, the Andromeda Galaxy is classified as kS5.

## 36.4 See also

- Morphological Catalogue of Galaxies
- Galaxy color–magnitude diagram

*The Large Magellanic Cloud: a type SBm galaxy.*

- Galaxy Zoo

- William Wilson Morgan

- Fritz Zwicky

## 36.5   References

[1]  Hubble, E. P. (1926). "Extra-galactic nebulae". *Contributions from the Mount Wilson Observatory / Carnegie Institution of Washington* **324**: 1–49. Bibcode:1926CMWCL.324....1H.

[2]  Hubble, E. P. (1936). *The Realm of the Nebulae*. New Haven: Yale University Press. LCCN 36018182.

[3]  "Hubble explores the origins of modern galaxies". *ESA/Hubble Press Release*. Retrieved 20 August 2013.

[4]  De Vaucouleurs, G.(1959). "Classification and Morphology of External Galaxies".*Handbuch der Physik***53**: 275. Bibcode:1959

[5]  Binney, J.; Merrifield, M. (1998). *Galactic Astronomy*. Princeton: Princeton University Press. ISBN 978-0-691-02565-0.

[6]  de Vaucouleurs, Gérard (April 1963). "Revised Classification of 1500 Bright Galaxies". *Astrophysical Journal Supplement* **8**: 31. Bibcode:1963ApJS....8...31D. doi:10.1086/190084.

[7]  De Vaucouleurs, G. (1994). "Global Physical Parameters of Galaxies" (PostScript). Retrieved 2008-01-02.

[8]  Binney, J.; Merrifield, M. (1998). *Galactic Astronomy*. Princeton: Princeton University Press. ISBN 978-0-691-02565-0.

## 36.6  External links

- Galaxies and the Universe - an introduction to galaxy classification

- Near-Infrared Galaxy Morphology Atlas, T.H. Jarrett

- The Spitzer Infrared Nearby Galaxies Survey (SINGS) Hubble Tuning-Fork, SINGS Spitzer Space Telescope Legacy Science Project

- Go to GalaxyZoo.org to try your hand at classifying galaxies as part of an Oxford University open community project

# Chapter 37

# Near-Earth supernova

Main article: History of supernova observation

A **near-Earth supernova** is an explosion resulting from the death of a star that occurs close enough to the Earth (roughly less than 100 light-years away) to have noticeable effects on its biosphere.

## 37.1  Effects on Earth

On average, a supernova explosion occurs within 10 parsecs (33 light-years) of the Earth every 240 million years.[lower-alpha 1] Gamma rays are responsible for most of the adverse effects a supernova can have on a living terrestrial planet. In Earth's case, gamma rays induce a chemical reaction in the upper atmosphere, converting molecular nitrogen into nitrogen oxides, depleting the ozone layer enough to expose the surface to harmful solar and cosmic radiation. Phytoplankton and reef communities would be particularly affected, which could severely deplete the base of the marine food chain.[2][3]

## 37.2  Risk by supernova type

Speculation as to the effects of a nearby supernova on Earth often focuses on large stars as Type II supernova candidates. Several prominent stars within a few hundred light years from the Sun are candidates for becoming supernovae in as little as a millennium. Although they would be spectacular to look at, were these "predictable" supernovae to occur, they are thought to have little potential to affect Earth.

Recent estimates predict that a Type II supernova would have to be closer than eight parsecs (26 light-years) to destroy half of the Earth's ozone layer.[4] Such estimates are mostly concerned with atmospheric modeling and considered only the known radiation flux from SN 1987A, a Type II supernova in the Large Magellanic Cloud. Estimates of the rate of supernova occurrence within 10 parsecs of the Earth vary from 0.05-0.5 per Ga[5] to 10 per Ga.[6] Several authors have based their estimates on the idea that supernovae are concentrated in the spiral arms of the galaxy, and that supernova explosions near the Sun usually occur during the ~10 million years that the Sun takes to pass through one of these regions (we are now in or entering the Orion Arm). The relatively recent paper by Gehrels *et al.* uses a value of 3 supernovae less than 10 parsecs away per Ga.[4] The frequency within a distance D is proportional to $D^3$ for small values of D, but for larger values is proportional to $D^2$ because of the finite thickness of the galactic disk (at intergalactic distances $D^3$ is again appropriate). Examples of relatively near supernovae are the Vela Supernova Remnant (~800 ly, ~12,000 years ago) and Geminga (~550 ly, ~300,000 years ago).

Type Ia supernovae are thought to be potentially the most dangerous if they occur close enough to the Earth. Because Type Ia supernovae arise from dim, common white dwarf stars, it is likely that a supernova that could affect the Earth will occur unpredictably and take place in a star system that is not well studied. The closest known candidate is IK Pegasi.[7] It is currently estimated, however, that by the time it could become a threat, its velocity in relation to the Solar System would have carried IK Pegasi to a safe distance.[4]

*The Crab Nebula is a pulsar wind nebula associated with the 1054 supernova. It is located about 6,500 light-years from the Earth.*[1]

## 37.3   Past events

Evidence from daughter products of short-lived radioactive isotopes shows that a nearby supernova helped determine the composition of the Solar System 4.5 billion years ago, and may even have triggered the formation of this system.[8] Supernova production of heavy elements over astronomic periods of time ultimately made the chemistry of life on Earth possible.

In 1996, astronomers at the University of Illinois at Urbana-Champaign theorized that traces of past supernovae might be detectable on Earth in the form of metal isotope signatures in rock strata. Subsequently, iron-60 enrichment has been reported in deep-sea rock of the Pacific Ocean by researchers from the Technical University of Munich.[9][10][11] 23 atoms of this iron isotope were found in the top 2 cm of crust, and these date from the last 13 million years or so. It is estimated that the supernova must have occurred in the last 5 million years or else it would have had to happen very close to the solar system to account for so much iron-60 still being here. A supernova occurring so close would have probably caused a mass extinction, which didn't happen in that time frame.[12] The quantity of iron seems to indicate that the supernova was less than 30 parsecs away. On the other hand, the authors estimate the frequency of supernovae at a distance less than $D$ (for reasonably small $D$) as around $(D/10 \text{ pc})^3$ per Ga, which gives a probability of only around 5% for a supernova within 30 pc in the last 5 million years. They point out that the probability may be higher because our solar system is entering the Orion Arm of the Milky Way.

Adrian L. Melott *et al.* estimated that gamma ray bursts from "dangerously close" supernova explosions occur two or more times per billion years, and this has been proposed as the cause of the end Ordovician extinction, which resulted in the death of nearly 60% of the oceanic life on Earth.[13]

In 1998 a supernova remnant, RX J0852.0-4622, was found in front (apparently) of the larger Vela Supernova Remnant.[14] Gamma rays from the decay of titanium-44 (half-life about 60 years) were independently discovered emanating from it,[15] showing that it must have exploded fairly recently (perhaps around 1200 CE), but there is no historical record of it. The flux of gamma rays and x-rays indicates that the supernova was relatively close to us (perhaps 200 parsecs or 660 ly). If so, this is a surprising event because supernovae less than 200 parsecs away are estimated to occur less than once per 100,000 years.[11]

In 2009, researchers have found nitrates in ice cores from Antarctica at depths corresponding to the known supernovae of 1006 and 1054 CE, as well as from around 1060 CE. The nitrates were apparently formed from nitrogen oxides created by gamma rays from the supernovae. This technique should be able to detect supernovae going back several thousand years.[16]

## 37.4   See also

- List of supernova candidates

## 37.5   Footnotes

[1]   Since a radius of 100 light years contains approximately 27.8 times as much volume as one of 33 light years, a supernova should occur within a radius of 100 light years from Earth approximately once every 8.6 million years. A supernova would occur within a radius of 200 light years approximately once every million years, within 500 light years every 69,000 years, and within 1,000 light years roughly every 8,625 years.

## 37.6   References

[1]   Kaplan, D. L.; Chatterjee, S.; Gaensler, B. M.; Anderson, J. (2008). "A Precise Proper Motion for the Crab Pulsar, and the Difficulty of Testing Spin-Kick Alignment for Young Neutron Stars". *Astrophysical Journal* **677** (2): 1201. arXiv:0801.1142. Bibcode:2008ApJ...677.1201K. doi:10.1086/529026.

[2]   Ellis, John; Schramm, David N. (March 1993). "Could a nearby supernova explosion have caused a mass extinction?". ARXIV. arXiv:hep-ph/9303206. Bibcode:1993hep.ph....3206E.

[3]   Whitten, R. C.; Borucki, W. J.; Wolfe, J. H.; Cuzzi, J. (September 30, 1976). "Effect of nearby supernova explosions on atmospheric ozone". *Nature* **263** (5576): 398–400. Bibcode:1976Natur.263..398W. doi:10.1038/263398a0.

[4]   Gehrels, Neil; Laird, Claude M.; et al. (2003-03-10). "Ozone Depletion from Nearby Supernovae". *Astrophysical Journal* **585** (2): 1169–1176. arXiv:astro-ph/0211361. Bibcode:2003ApJ...585.1169G. doi:10.1086/346127.

[5]   Whitten, R. C.; Cuzzi, J.; Borucki W. J.; Wolfe, J. H. (1976). "Effect of nearby supernova explosions on atmospheric ozone". *Nature* **263** (5576): 263. Bibcode:1976Natur.263..398W. doi:10.1038/263398a0. Retrieved 2007-02-01.

[6]   Clark, D. H.; McCrea, W. H.; Stephenson, F. R. (1977). "Frequency of nearby supernovae and climactic and biological catastrophes". *Nature* **265** (5592): 318–319. Bibcode:1977Natur.265..318C. doi:10.1038/265318a0. Retrieved 2007-02-01.

[7]   Garlick, Mark (March 2007). "The Supernova Menace". *Sky & Telescope*.

[8] Taylor, G. Jeffrey (2003-05-21). "Triggering the Formation of the Solar System". Planetary Science Research. Retrieved 2006-10-20.

[9] Staff (Fall/Winter 2005–2006). "Researchers Detect 'Near Miss' Supernova Explosion". University of Illinois College of Liberal Arts and Sciences. p. 17. Archived from the original on 2006-09-01. Retrieved 2007-02-01. Check date values in: |date= (help)

[10] Knie, K.; et al. (2004). "$^{60}$Fe Anomaly in a Deep-Sea Manganese Crust and Implications for a Nearby Supernova Source". *Physical Review Letters* **93** (17): 171103–171106. Bibcode:2004PhRvL..93q1103K. doi:10.1103/PhysRevLett.93.171103.

[11] Fields, B. D.; Ellis, J. (1999). "On Deep-Ocean Fe-60 as a Fossil of a Near-Earth Supernova". *New Astronomy* **4** (6): 419–430. arXiv:astro-ph/9811457. Bibcode:1999NewA....4..419F. doi:10.1016/S1384-1076(99)00034-2.

[12] Fields & Ellis, p. 10

[13] Melott, A.; et al. (2004). "Did a gamma-ray burst initiate the late Ordovician mass extinction?". *International Journal of Astrobiology* **3** (2): 55–61. arXiv:astro-ph/0309415. Bibcode:2004IJAsB...3...55M. doi:10.1017/S1473550404001910.

[14] Aschenbach, Bernd (1998-11-12). "Discovery of a young nearby supernova remnant". *Letters to Nature* **396** (6707): 141–142. Bibcode:1998Natur.396..141A. doi:10.1038/24103.

[15] Iyudin, A. F.; et al. (November 1998). "Emission from $^{44}$Ti associated with a previously unknown Galactic supernova". *Nature* **396** (6707): 142–144. Bibcode:1998Natur.396..142I. doi:10.1038/24106.

[16] "Ancient supernovae found written into the Antarctic ice". *New Scientist* (2698). 2009-03-04. Retrieved 2009-03-09. Refers to .

# Chapter 38

# List of supernova candidates

This is a **list of supernova candidates**, or stars that astronomers have suggested are supernova progenitors. Type II supernova progenitors include stars with at least 10 solar masses that are in the final stages of their evolution. (Prominent examples of stars in this mass range include Antares, Spica,[1] Gamma Velorum,[2] Mu Cephei, and members of the Quintuplet Cluster.[3]) Type Ia supernova progenitors are white dwarf stars that are close to the Chandrasekhar limit of about 1.44 solar masses and are accreting matter from a binary companion star. The list includes massive Wolf–Rayet stars, which may become Type Ib/Ic supernovae.

*This is an incomplete list that may never be able to satisfy particular standards for completeness. You can help by expanding it with reliably sourced entries.*

## 38.1   Notes

[1] The Kitt Peak Downes star.

## 38.2   References

[1] *Supernova Remnants and Neutron Stars*, Harvard-Smithsonian Center for Astrophysics, 2005-08-02, retrieved 2006-06-08

[2] Kaler, James B., "Regor", *Stars* (University of Illinois), retrieved 2007-01-08

[3] Lloyd, Robin (2006-09-04), *Strange Space Pinwheels Spotted*, space.com, retrieved 2007-01-08

[4] Samuel, Eugenie (2002-05-23), *Supernova poised to go off near Earth*, New Scientist, retrieved 2007-01-12

[5] Tzekova, S. Y.; et al. (2004), *IK Pegasi (HR 8210)*, ESO, retrieved 2007-01-12

[6] "Big and Giant Stars: Alpha Lupi". Jumk.de. Retrieved 2013-01-03.

[7] "What will happen when Antares explodes?". Disassociated.com. 2009-01-13. Retrieved 2013-01-03.

[8] Smith, Nathan; Hinkle, Kenneth H.; Ryde, Nils (March 2009). "Red Supergiants as Potential Type IIn Supernova Progenitors: Spatially Resolved 4.6 μm CO Emission Around VY CMa and Betelgeuse". *The Astrophysical Journal* **137** (3): 3558–3573. arXiv:0811.3037. Bibcode:2009AJ....137.3558S. doi:10.1088/0004-6256/137/3/3558.

[9] Beech, Martin (December 2011), "The past, present and future supernova threat to Earth's biosphere", *Astrophysics and Space Science* **336** (2): 287–302. Bibcode:2011Ap&SS.336..287B, doi:10.1007/s10509-011-0873-9

[10] Moravveji, Ehsan; Guinan, Edward F.; Shultz, Matt; Williamson, Michael H.; Moya, Andres (March 2012). "Asteroseismology of the nearby SN-II Progenitor: Rigel. Part I. The *MOST* High-precision Photometry and Radial Velocity Monitoring", *The Astrophysical Journal* **747** (1): 108–115. arXiv:1201.0843. Bibcode:2012ApJ...747..108M. doi:10.1088/0004-637X/747/2/108

[11] Than, Ker (2006-07-19). *Mystery of Explosive Star Solved*. space.com. retrieved 2007-01-08

[12] Staff (2006-07-25). *Astronomers See Future Supernova Developing*. SpaceDaily. retrieved 2006-12-01

[13] "Recurrent Novae as a Progenitor System of Type Ia Supernovae. I. RS Ophiuchi Subclass: Systems with a Red Giant Companion - Abstract - The Astrophysical Journal - IOPscience". Iopscience.iop.org. 2008-12-19. Retrieved 2013-01-03.

[14] Lloyd, Robin (2006-09-04). "Strange Space Pinwheels Spotted". space.com. Retrieved 2007-01-08.

[15] Schaefer, Bradley E.; Pagnotta, Ashley; Shara, Michael M. (January 2010). "The Nova Shell and Evolution of the Recurrent Nova T Pyxidis". *The Astrophysical Journal* **708**: 381–402. arXiv:0906.0933. Bibcode:2010ApJ...708..381S. doi:10.1088/0004-637X/708/1/381.

[16] Edwards, Lin (2010). "Massive white dwarf in our galaxy may go supernova". *PhysOrg*. Retrieved 2010-02-26.

[17] Weaver, D.; Humphreys, R. (2007-01-08). "Astronomers Map a Hypergiant Star's Massive Outbursts". HubbleSite NewsCenter. Retrieved 2007-01-16.

[18] Monnier, J. D.; Millan-Gabet, R.; Tuthill, P. G.; Traub, W. A.; Carleton, N. P.; Coude Du Foresto, V.; Danchi, W. C.; Lacasse, M. G.; Morel, S.; Perrin, G.; Porro, I. L.; Schloerb, F. P.; Townes, C. H. (2004). "High-Resolution Imaging of Dust Shells by Using Keck Aperture Masking and the IOTA Interferometer". *The Astrophysical Journal* **605**: 436. arXiv:astro-ph/0401363. Bibcode:2004ApJ...605..436M. doi:10.1086/382218.

[19] van Boekel, R.; Schöller, M.; Herbst, T. (2003-11-18). "Biggest Star in Our Galaxy Sits within a Rugby-Ball Shaped Cocoon". European Organisation for Astronomical Research in the Southern Hemisphere (ESO). Retrieved 2007-01-08.

[20] Milan, Wil (2000-03-07). "Possible Hypernova Could Affect Earth". space.com. Retrieved 2007-01-06.

[21] Smith, Nathan (March 2007). "Discovery of a Nearby Twin of SN 1987A's Nebula around the Luminous Blue Variable HD 168625: Was Sk −69 202 an LBV?". *The Astronomical Journal* **133** (3): 1034–1040. arXiv:astro-ph/0611544. Bibcode:2007AJ. doi:10.1086/510838.

[22] Tuthill, Peter G.; et al. (March 2008). "The Prototype Colliding-Wind Pinwheel WR 104". *The Astrophysical Journal* **675** (1): 698–710. arXiv:0712.2111. Bibcode:2008ApJ...675..698T. doi:10.1086/527286.

[23] Kaler, Jim (1999-04-09). "WR 104: Pinwheel Star". Astronomy Picture of the Day. Retrieved 2007-01-08.

[24] Jones, Terry Jay; et al. (July 1993). "IRC +10420 - A cool hypergiant near the top of the H-R diagram". *Astrophysical Journal, Part I* **411** (1): 323–335. Bibcode:1993ApJ...411..323J. doi:10.1086/172832.

[25] Than, Ker (2004-10-04). "Astronomers Demonstrate a Global Internet Telescope". University of Manchester. Retrieved 2007-01-08.

[26] Staff (2003-01-31). "The William Herschel telescope finds the best candidate for a supernova explosion". Particle Physics and Astronomy Research Council. Retrieved 2007-01-05.

[27] Staff (2011-09-28). "'Fried Egg' Nebula Cracks Open Rare Hypergiant Star". Space.com. Retrieved 2015-06-13.

[28] Smartt, S. J.; Lennon, D. J.; Kudritzki, R. P.; Rosales, F.; Ryans, R. S. I.; Wright, N. (September 2002). "The evolutionary status of Sher 25 - implications for blue supergiants and the progenitor of SN 1987A". *Astronomy and Astrophysics* **391** (3): 979–991. arXiv:astro-ph/0205242. Bibcode:2002A&A...391..979S. doi:10.1051/0004-6361:20020829.

[29] Jura, M.; Velusamy, T.; Werner, M. W. (2001-06-05). "What next for the Likely Pre-Supernova, HD 179821?". American Astronomical Society. Retrieved 2007-01-08.

[30] Josselin, E.; Lèbre, A. (2001). "Probing the post-AGB nature of HD 179821". *Astronomy and Astrophysics* **367** (3): 826–830. Bibcode:2001A&A...367..826J. doi:10.1051/0004-6361:20000496.

[31] Woudt, P. A.; et al. (November 2009). "The Expanding Bipolar Shell of the Helium Nova V445 Puppis". *The Astrophysical Journal* **706** (1): 738–746. arXiv:0910.1069. Bibcode:2009ApJ...706..738W. doi:10.1088/0004-637X/706/1/738.

[32] Thoroughgood, T. D.; Dhillon, V. S.; Littlefair, S. P.; Marsh, T. R.; Smith, D. A. (2002). "The recurrent nova U Scorpii -- A type Ia supernova progenitor". *The Physics of Cataclysmic Variables and Related Objects* **261**. San Francisco, CA: Astronomical Society of the Pacific. Bibcode:2002ASPC..261...77T. Retrieved 2009-01-24.

[33] Maxted, P. F. L.; Marsh, T. R.; North, R. C. (September 2000). "KPD 1930+2752: a candidate Type Ia supernova progenitor". *Monthly Notices of the Royal Astronomical Society* **317** (3): L41–L44. arXiv:astro-ph/0007257. Bibcode:2000MNRAS.317L... doi:10.1046/j.1365-8711.2000.03856.x.

[34] Kanipe, Jeff (2000-08-05). "Skywatch—Watch This Space!". space.com. Retrieved 2007-01-06.

# Chapter 39

# Timeline of white dwarfs, neutron stars, and supernovae

**Timeline of white dwarfs, neutron stars, and supernovae**

Note that this list is mainly about the development of knowledge, but also about some supernovae taking place. For a separate list of the latter, see the article List of supernovae. All dates refer to when the supernova was observed on Earth or would have been observed on Earth had powerful enough telescopes existed at the time.

- 185 – Chinese astronomers become the first to record observations of a supernova, the SN 185.

- 1006 – Ali ibn Ridwan and Chinese astronomers observe the brightest (magnitude −7.5) recorded supernova, SN 1006, which is observed in the constellation of Lupus.

- 1054 – Chinese, American Indian and Arab astronomers observe the SN 1054, the Crab Nebula supernova explosion.

- 1181 – Chinese astronomers observe the SN 1181 supernova.

- 1572 – Tycho Brahe discovers a supernova (SN 1572) in the constellation Cassiopeia.

- 1604 – Johannes Kepler's supernova, SN 1604, in Serpens is observed.

- 1862 – Alvan Graham Clark observes Sirius B.

- 1866 – William Huggins studies the spectrum of a nova and discovers that it is surrounded by a cloud of hydrogen.

- 1885 – A supernova, S Andromedae, is observed in the Andromeda Galaxy leading to recognition of supernovae as a distinct class of novae.

- 1910 – the spectrum of 40 Eridani B is observed, making it the first confirmed white dwarf.

- 1914 – Walter Sydney Adams determines an incredibly high density for Sirius B.

- 1926 – Ralph Fowler uses Fermi–Dirac statistics to explain white dwarf stars.

- 1930 – Subrahmanyan Chandrasekhar discovers the white dwarf maximum mass limit.

- 1933 – Fritz Zwicky and Walter Baade propose the neutron star idea and suggest that supernovae might be created by the collapse of normal stars to neutron stars—they also point out that such events can explain the cosmic ray background.

- 1939 – Robert Oppenheimer and George Volkoff calculate the first neutron star models.

- 1942 – J.J.L. Duyvendak, Nicholas Mayall, and Jan Oort deduce that the Crab Nebula is a remnant of the 1054 supernova observed by Chinese astronomers.

- 1958 – Evry Schatzman, Kent Harrison, Masami Wakano, and John Wheeler show that white dwarfs are unstable to inverse beta decay.

- 1962 – Riccardo Giacconi, Herbert Gursky, Frank Paolini, and Bruno Rossi discover Scorpius X-1.

- 1967 – Jocelyn Bell and Antony Hewish discover radio pulses from a pulsar.

- 1967 – J.R. Harries, Kenneth G. McCracken, R.J. Francey, and A.G. Fenton discover the first X-ray transient (Cen X-2).

- 1968 – Thomas Gold proposes that pulsars are rotating neutron stars.

- 1969 – David Staelin, E.C. Reifenstein, William Cocke, Mike Disney, and Donald Taylor discover the Crab Nebula pulsar thus connecting supernovae, neutron stars, and pulsars.

- 1971 – Riccardo Giacconi, Herbert Gursky, Ed Kellogg, R. Levinson, E. Schreier, and H. Tananbaum discover 4.8 second X-ray pulsations from Centaurus X-3.

- 1972 - Charles Kowal discovers the Type Ia supernova SN 1972e in NGC 5253, which would be observed for more than a year and become the basis case for the type.

- 1974 – Russell Hulse and Joseph Taylor discover the binary pulsar PSR B1913+16.

- 1977 – Kip Thorne and Anna Żytkow present a detailed analysis of Thorne–Żytkow objects.

- 1982 – Donald Backer, Shrinivas Kulkarni, Carl Heiles, Michael Davis, and Miller Goss discover the millisecond pulsar PSR B1937+214.

- 1985 – Michiel van der Klis discovers 30 Hz quasi-periodic oscillations in GX 5-1.

- 1987 – Ian Shelton discovers SN 1987A in the Large Magellanic Cloud ...

- 2006 – Robert Quimby and P. Mondol discover SN 2006gy (a possible hypernova) in NGC 1260.

# 39.1 Text and image sources, contributors, and licenses

## 39.1.1 Text

- **Supernova** *Source:* https://en.wikipedia.org/wiki/Supernova?oldid=687855539 *Contributors:* AxelBoldt, Chenyu, Bryan Derksen, AstroNome, Malcolm Farmer, Alex.tan, Roadrunner, DavidLevinson, Ark~enwiki, Montrealais, Hfastedge, Lorenzarius, Lir, Gdarin, Gabbe, Wwwwolf, Wapcaplet, lxfd64, Minesweeper, Alfio, Kosebamse, Ahoerstemeier, Nanobug, BigFatBuddha, Mihai~enwiki, Julesd, Glenn, Lancevortex, Pizza Puzzle, Hike395, Lenaic, Emperorbma, Timwi, Dcoetzee, Wikiborg, Daniel Quinlan, Wik, Tpbradbury, Furrykef, Nv8200pa, SEWilco, Nickshanks, Pilaf~enwiki, Francs2000, Robbot, Altenmann, Peak, Lowellian, Babbage, Yosri, Academic Challenger, Rursus, Meelar, Rtfisher, Jeroen, Wereon, Borislav, Kairos, Peter L, Adam78, Cbane, Stirling Newberry, Decumanus, Giftlite, DocWatson42, Christopher Parham, Achurch, Laudaka, Anarchiuswiki, Herbee, Peruvianllama, Anville, Curps, Guanaco, Pascal666, Nayuki, Gzornenplatz, Python eggs, SWAdair, DÂ.ugosz, Jrdioko, Wmahan, Decoy, Zeimusu, Slowking Man, ConradPino, Antandrus, Tom the Goober, HorsePunchKid, OverlordQ, Karol Langner, Rdsmith4, DragonflySixtyseven, Bodnotbod, GeoGreg, Bk0, Nickptar, Iantresman, Urhixidur, Wheresmysocks, TJSwoboda, Jh51681, Adashiel, Mike Rosoft, DanielCD, RossPatterson, Diagonalfish, Discospinster, Rich Farmbrough, FT2, Gadykozma, Rama, Schuetzm, Vsmith, Wk muriithi, ArnoldReinhold, Mani1, Paul August, Bender235, Rubicon, ESkog, Janderk, StupendousMan, Brian0918, RJHall, Livajo, El C, Nonpareility, Shanes, Tom, Frankenschulz, RoyBoy, Afed, Dralwik, Martey, Harley peters, DrYak, Shenme, ParticleMan, I9Q79oL78KiL0QTFHgyc, La goutte de pluie, Cherlin, Sam Korn, Fleurot~enwiki, Jumbuck, Alansohn, Gary, Tablizer, Anthony Appleyard, Eric Kvaalen, Arthena, Andrewpmk, Plumbago, Iothiania, Keflavich, Hu, Bootstoots, Snowolf, Rick Sidwell, Deathphoenix, DV8 2XL, Gene Nygaard, Dan East, Netkinetic, Nick Mks, HenryLi, Ceyockey, Adrian.benko, Flying fish, Siafu, WilliamKF, Nuno Tavares, OwenX, Jacen Aratan, GVOLTT, Nuggetboy, Pol098, Peter Beard, Hurricane Angel, JeremyA, The Wordsmith, Dave7331, Kgrr, Mangojuice, Kralizec!, Smartech~enwiki, Chrkl, Christopher Thomas, MrSomeone, RichardWeiss, Rnt20, Ashmoo, Magister Mathematicae, Keeves, Terryn3, Jclemens, Drbogdan, Sjakkalle, Rjwilmsi, Coemgenus, WCFrancis, Commander, Strait, John187, Mike Peel, Brighterorange, Krash, The wub, Nandesuka, Mordecai, SanGatiche, Cassowary, Yamamoto Ichiro, X1987x, Diablo-D3, FlaBot, SchuminWeb, RobertG, Nihiltres, Nivix, Anurag Garg, DannyZ, R Lee E, Physchim62, Smithbrenon, Chobot, DVdm, Shopsinc, Gwernol, SamusAran997, YurikBot, Wavelength, Spacepotato, RobotE, Brandmeister (old), Pip2andahalf, Jtkiefer, Diliff, Sillylizard, SpuriousQ, Chaser, Stian, Gaius Cornelius, Rsrikanth05, Wimt, GeeJo, Tavilis, NawlinWiki, Wiki alf, Pagrashtak, R.e.s., D-Katana, Anetode, Inhighspeed, Rmky87, JPMcGrath, Dbfirs, Morgan Leigh, DeadEyeArrow, Bota47, Perry Middlemiss, Uber nemo, Djdaedalus, Smkolins, Georgewilliamherbert, Mythobeast, Zzuuzz, Chesnok, Nikkimaria, Closedmouth, Errabee, Eddie tejeda, HereToHelp, Anclation~enwiki, Jaranda, Banus, GrinBot~enwiki, DVD R W, Algae, WesleyDodds, The Yeti, Marquez~enwiki, Dragon of the Pants, AndrewWTaylor, SG, KnightRider~enwiki, Zfervent, SmackBot, Aciampolini, MaskedSlacker, Prodego, KnowledgeOfSelf, Ma8thew, Jagged 85, RedSpruce, Renesis, Jipcy, Veesicle, Edgar181, Yamaguchi先生, Cachedio, Gilliam, Ohnoitsjamie, Hmains, GwydionM, Ewok Slayer, Gorman, Saros136, Chris the speller, Bidgee, Keegan, Audacity, Josh215, Trebor, MalafayaBot, Hibernian, Solargroovy, RomaC, Sbharris, Colonies Chris, Rcbutcher, Darth Panda, Ekrenor, Scwlong, NYKevin, TheK-Man, Addshore, Kcordina, Memming, Cybercobra, Blake-, Dreadstar, DMPalmer, PsychoJosh, The Smilodon, Pkeets, Pilotguy, Kukini, Ventiac, Esrever, Chronopsis~enwiki, Beasterline, Erimus, Swatjester, Doug Bell, Harryboyles, Zearin, JzG, Kuru, Scientizzle, J 1982, Calum MacÙisdean, Bydand, Lazylaces, BlisteringFreakachu, IronGargoyle, Heliogabulus, Zzzzzzzzzzz, FrostyBytes, Chrisch, Slakr, Muadd, George The Dragon, Rglovejoy, SandyGeorgia, Mmiller0712, RHB, WilliamJE, Iridescent, Spebudmak, Kencf0618, Kostya30~enwiki, TwistOfCain, Grblomerth, Judgesurreal777, Xaagkx, Newone, Dsspiegel, RekishiEJ, Silent reverie86, Tim lalala, Coffee Atoms, Mssgill, Tawkerbot2, Hoagssculptor, Hammer Raccoon, CalebNoble, Fvasconcellos, Joniscool98, Tj2, JForget, Friendly Neighbour, Deon, Myrrhlin, Ecthros, Iced Kola, Big Jock Knew, Basawala, Ruslik0, Michaelbarreto, Karenjc, No1lakersfan, Myasuda, King Hildebrand, Chris83, Vectro, Abeg92, Gogo Dodo, Chasingsol, Rracecarr, Skittleys, Michael C Price, Clovis Sangrail, Chrislk02, Robertinventor, Mattjball, Kozuch, Editor at Large, Zalgo, Janoside, Satori Son, Casliber, Thijs!bot, Meol, Epbr123, Knakts, Gaijin42, Pajz, Qwyrxian, Mbell, Daniel, Keraunos, Headbomb, Marek69, John254, SGGH, Lars Lindberg Christensen, Ideogram, Dfrg.msc, Uruiamme, Natalie Erin, SpellingD, Porqin, AntiVandalBot, Konman72, Roflbater, Luna Santin, Seaphoto, Uvaphdman, Phyzixchica, Prolog, Doc Tropics, Ericleimbach, Jj137, Fayenatic london, Ryugin, Dr. Submillimeter, Darklilac, Jahrsper, Canadian-Bacon, Kariteh, Deflective, Nthep, M.C., QuantumEngineer, Dricherby, Acroterion, WolfmanSF, Murgh, Bongwarrior, VoABot II, Jamie L, SineWave, Nyttend, Brusegadi, Wormcast, BrianGV, Fallschirmjäger, 28421u2232nfenfcenc, Allstarecho, Schumi555, Cpl Syx, Mollwollfumble, Shijualex, DerHexer, JaGa, Edward321, Indianstar, Lelkesa, Partymetroid, Cocytus, NatureA16, Otvaltak, Hdt83, MartinBot, Observer 144, Pagw, Rettetast, Anaxial, Jay Litman, Tubbs334, Mschel, AlexiusHoratius, RWyn, PrestonH, Arsenijette, Outlaw640, Watch37264, J.delanoy, Pharaoh of the Wizards, DrKay, Trusilver, Rrostrom, Catmoongirl, Uncle Dick, Meltro, Iadar, Nsande01, Jerry, Softcurls55, Tandy1989, DanielEng, TheChrisD, It Is Me Here, Thomas Larsen, PocklingtonDan, Tahlei bon, AntiSpamBot, NewEnglandYankee, 83d40m, Ontarioboy, Brian Pearson, Pundit, Juliancolton, Aen Tan, Hydramus, King Toadsworth, DMCer, JudahNielsen, MishaPan, Pdcook, Jarry1250, Dorftrottel, S, CJackman, CardinalDan, JeffreyRMiles, Spellcast, Nivekx, Vedran8080, Lights, Kevsredsox, RingtailedFox, TheOtherJesse, Philip Trueman, TXiKiBoT, Oshwah, Hqb, Dj thegreat, Daniel adjaye, Anonymous Dissident, Hububletelescope, KT00100, Crohnie, Frebeque, Piperh, Cwf1138, Rich Janis, Dendodge, Fbs. 13, Markp93, Raymondwinn, Wingedsubmariner, Saturn star, Hablahablaha, Madhero88, Vladsinger, RandomXYZb, Milkbreath, Andy Dingley, Gloverspud, Jpeeling, Az99, Synthebot, James McBride, ColemanJ, Falcon8765, Richtom80, Enviroboy, RaseaC, Newsaholic, Michael Frind, The Mad Genius, Mpmagi, Projectstar9, Dedication, RedRabbit1983, D. Recorder, Linelor, Gaelen S., Bob freeman1, SieBot, Monkeyboi, Sonicology, Tresiden, Jauerback, KGyST, Plinkit, Parhamr, Skylark42, Caltas, Acorah, Matthew Yeager, Nathan, Wing gundam, The way, the truth, and the light, Gravitan, RadicalOne, Exert, Oda Mari, Arbor to SJ, Hxhbot, Vbond, Je-SOCO, Oxymoron83, Antonio Lopez, Faradayplank, Star28, AnonGuy, Lightmouse, Bjwilli2, Wlegro, Hashte, Jocelynisgreat, The-G-Unit-Boss, J3nn r0ckz y0ur s0x, Qgain1, Gloombringer, Gunde123456789, Sir~enwiki, Anyeverybody, Mtaylor848, Dust Filter, Nn123645, Pinkadelica, Nergaal, Ratinglink, André Oliva, Velvetron, Troy 07, SallyForth123, Mr. Granger, Twinsday, Egdcltd, Preds1998, Sfan00 IMG, GarbagEcol, ClueBot, Coastercrazy10, Kevjumba, Strider12233, MileyWorld, Rassisch, BlacksheepPAUL, D'Acorah, Tobyklynsmith, Ladnav1989, The Thing That Should Not Be, Rodhullandemu, Herakles01, Vtr555, Drmies, Neverquick, Luke4545, DragonBot, Robert Skyhawk, Excirial, Sgmuse, Lartoven, Barbarinaz, Sun Creator, Alexcosta7, Tyler, IllusionFREAK, Ember of Light, Razorflame, Mikaey, Kakofonous, La Pianista, Thingg, Aitias, Chasecarter, Versus22, Burner0718, SoxBot III, HumphreyW, ברוקולי, Motto25, Drewtaylor1978, NellieBly, PL290, Bit Lordy, Ikihi123, MystBot, NonvocalScream, Addbot, Domminator36, Roentgenium111, AVand, DOI bot, Jojhutton, Metagraph, Cmking99, Fieldday-sunday, Mr. Wheely Guy, Andycat bossman, Bte99, CanadianLinuxUser, Fluffernutter, Ederiel, Sn2007Cb, SoSaysChappy, AndersBot, Debresser,

LinkFA-Bot, 5 albert square, Chem-MTFC, Numbo3-bot, Tyim419, Tide rolls, Gameseeker, Elbow94, Luckas-bot, Yobot, 2D, Andreasmperu, Newportm, Donfbreed, ArchonMagnus, Mmxx, THEN WHO WAS PHONE?, Ncda, 73lackluck, AnomieBOT, SieteDeMayo, Jim1138, AdjustShift, Kingpin13, Waterden, KPackard, Bejesus, Yale.2001, Materialscientist, Hunnjazal, Argondo, The High Fin Sperm Whale, Citation bot, Icosmology, Maxis ftw, Roux-HG, Xqbot, Joinpep, Sionus, Sriram.aeropsn, The Evil IP address, Maddie!, AFDHKJ, Lithopsian, Papalew, GrouchoBot, Armbrust, Goatseatgrass, Greeneggsandcheese, Omnipaedista, Rusty10990, Sam10443, Doulos Christos, Moxy, Habou97, Abdul hadi 750, Kylethompson91, A. di M., Fotaun, Griffinofwales, Wolficus, FrescoBot, Originalwana, Tobby72, Dondelara07, Recognizance, Ionutzmovie, Braidanbash, EmilTyf, Citation bot 1, Awsomezacste, Anywhere road, Pinethicket, I dream of horses, HRoestBot, Jonesey95, Tom.Reding, Pmokeefe, Calham106, RedBot, Pikiwyn, Midnight Comet, Bjdabomb777, Merlion444, Double sharp, Trappist the monk, Belchman, فليق عقيف, Vrenator, Natethehawk, Extra999, Daniellecoley88, Xiaomao123, JLincoln, Jeffrd10, Diannaa, Gwyneth99, TheMesquito, UnitedWashclothExpress, Cluemaster25452, DARTH SIDIOUS 2, Mean as custard, Aadiljaleel, Yaush, Stewartwebster, Gold Claw3, Bitil Guilderstrone, EmausBot, John of Reading, Dewritech, RA0808, Ğaaw, Pcorty, TreacherousWays, Jmencisom, Slightsmile, Cplpike, Savh, Hhhippo, Chemicalcuke, ZéroBot, Brilbri, PS., Josve05a, StringTheory11, Dffgd, Magasjukur, H3llBot, Gomphothere, AManWithNoPlan, Imlaging, Brandmeister, Coasterlover1994, L Kensington, MercWithMouth, Zukaz, Targaryen, Czeror, Planet photometry, DASHBotAV, Petrb, Fjörgynn, ClueBot NG, Coverman6, This lousy T-shirt, Satellizer, ArchieOof, Danceroid, OperaJoeGreen, Ventti013, Kenisaim, Soupguy1234, GoldenGlory84, GeorgeYee, Moneya, Widr, Helpful Pixie Bot, Anxiousswift, Happykt, Gob Lofa, Bibcode Bot, BG19bot, Krenair, StevenBjerke, Allypop1, MusikAnimal, AvocatoBot, B wik, Mark Arsten, Ninney, Landnanners, Daisysalas, Aardwolf first, Youcantrustme420, Vractomorph1, Zedshort, Nitrobutane, RiseUpAgain, Cheeseyum21, Jason from nyc, BattyBot, Harryzuoz, Pratyya Ghosh, SUPERUNDEAD, Sbrucetree, Swathi.gadde, Dexbot, Crazedpostdoc, CuriousMind01, Dave Bowman - Discovery Won, Reatlas, Iamjohn102, Praemonitus, Babitaarora, Haminoon, Thehumanpuddle, Foura, The Herald, Prokaryotes, Califragic, Ginsuloft, Kahtar, Zourcilleux, Seabuckthorn, Tumse matlub, BlueAzulon, Qwerty12330, Monkbot, LujViton, BethNaught, FOLLOWYOURDREAMSTODAY, Wtangg, Neeraj Bhakta, MuseumGeek, KH-1, ChamithN, ZachoRockz, Meshak Nath Ghoshal, Dretler, Tetra quark, Vilica1, Isambard Kingdom, Mr.ahunnit, Nicholas123hi, Qwertyuiopasdfghjkl;asdfgcvfhdtgvbvdxssweeeed, KasparBot, Buechman256, DragonRage21, ThatOneScientist, Grassik and Anonymous: 1115

- **Supernova remnant** *Source:* https://en.wikipedia.org/wiki/Supernova_remnant?oldid=656762635 *Contributors:* CYD, Vicki Rosenzweig, Mav, Bryan Derksen, The Anome, Berek, AstroNomer~enwiki, -- April, Alan Peakall, Minesweeper, Alfio, Nickshanks, April~enwiki, Jai, Jyril, Marcika, JamesHoadley, Iantresman, Rich Farmbrough, Janderk, RJHall, La goutte de pluie, Jcrocker, Kitch, RyanGerbil10, Richard Arthur Norton (1958- ), Etacar11, GregorB, Funhistory, Bebenko, Graham87, Drbogdan, Rjwilmsi, Quale, Marasama, Krash, Mordecai, Wrightbus, Bgwhite, YurikBot, Hairy Dude, Lofty, Pagrashtak, Zzuuzz, SmackBot, Eskimbot, DHN-bot~enwiki, Andrei Stroe, Robofish, NongBot~enwiki, DabMachine, Tawkerbot2, George100, The Mark, Rleiton, Marek69, Kauczuk, IanOsgood, .anacondabot, Tubbs334, CommonsDelinker, Nono64, Rod57, STBotD, Bináris, Idioma-bot, VolkovBot, 1981willy, AlleborgoBot, Thunderbird2, Sakkura, KGyST, Nergaal, Martarius, ClueBot, Guy34565, Abdelqader, Alexbot, Estirabot, Addbot, DOI bot, BrainMarble, LaaknorBot, Saavek47, Zorrobot, Luckas-bot, Yobot, Amirobot, Amble, Vini 17bot5, Dreamer08, Citation bot, ArthurBot, Xqbot, WingedSkiCap, GrouchoBot, Unbeatable0, FrescoBot, Originalwana, Citation bot 1, Tom.Reding, RedBot, دیرانی محامد عبـاد, Orenburg1, Vrenator, Ripchip Bot, Jmencisom, ZéroBot, Medeis, Parusaro, ClueBot NG, Bibcode Bot, BG19bot, Negativecharge, AvocatoBot, Xlicolts613, Koiravva, SkyFlubbler, Spideratseds and Anonymous: 53

- **Supernova nucleosynthesis** *Source:* https://en.wikipedia.org/wiki/Supernova_nucleosynthesis?oldid=687010207 *Contributors:* Tedernst, Ken Arromdee, Rursus, Crimson30, Mateuszica, Karol Langner, Mrtrey99, Vsmith, RJHall, CDN99, Bradkittenbrink, Tycho, AlexTiefling, Scott-Davis, Drbogdan, Rjwilmsi, Strait, MarSch, Srleffler, Chobot, Beanyk, Uber nemo, Cadillac, Georgewilliamherbert, Modify, Cmglee, MacsBug, SmackBot, Eskimbot, Gilliam, Bluebot, Darth Panda, Rogermw, Voyajer, Mystman666, Jmnbatista, Flyguy649, Friendly Neighbour, CmdrObot, Fokion, King Hildebrand, James E B, Underpants, Headbomb, Orionus, Seddon, Magioladitis, Rhadamante, DinoBot, DAID, Sheliak, Claydonald, Oshwah, UnitedStatesian, BotKung, Wasted Sapience, Dpeinador~enwiki, Peterckw, Nergaal, Masterblooregard, Roberto Mura, Addbot, Roentgenium111, DOI bot, Lightbot, Ibmua, Tom.Reding, Jvandonsel, Jusses2, Arbnos, ClueBot NG, Helpful Pixie Bot, Calabe1992, Bibcode Bot, Kristaoz, Zedshort, Андрей Бондарь, Mfuerte, TheJJJunk, Coconutporkpie and Anonymous: 45

- **Interstellar medium** *Source:* https://en.wikipedia.org/wiki/Interstellar_medium?oldid=687801188 *Contributors:* Bryan Derksen, Tarquin, -- April, XJaM, Heron, Patrick, Boud, Dante Alighieri, Alfio, Looxix~enwiki, J'raxis, Mark Foskey, Glenn, Doradus, Fvw, Wetman, Shantavira, Rursus, Rtfisher, Xanzzibar, Paisa, Bfinn, No Guru, Home Row Keysplurge, Waltpohl, Eequor, Foobar, Christopherlin, Beland, Karol Langner, Iantresman, Esperant, Noisy, Rich Farmbrough, RJHall, Mr. Billion, Pjf, El C, Hayabusa future, I9Q79oL78KiL0QTFHgyc, Giraffedata, Nk, Kamyar~enwiki, Eric Kvaalen, Bruce Bebow, Erik, Gene Nygaard, Siafu, WilliamKF, Amara, Isnow, Eyreland, Marky1981, Aarghdvaark, Ashmoo, Drbogdan, Rjwilmsi, Nightscream, Uxh, Krash, DoubleBlue, Mordecai, FlaBot, Ian Pitchford, Jmkprime, Chobot, Bgwhite, Roberto de Ajvol, YurikBot, Hairy Dude, Conscious, Van der Hoorn, Howcheng, The Obfuscator, Gadget850, Smkolins, Ketsuekigata, Reyk, Jolt76, Caco de vidro, Mejor Los Indios, Mhardcastle, KnightRider~enwiki, SmackBot, Fireworks, Ashill, MHD, Sir Spike, Bluebot, Miquonranger03, Rediahs, Baronnet, Colonies Chris, Rkinch, Modest Genius, Njál, Asgekar, T-borg, SpacemanAfrica, Just plain Bill, Henning Makholm, A5b, Jiminy pop, J 1982, JorisvS, Concept2, Mgiganteus1, Ocatecir, Ckatz, Inglorion, 16@r, Danilot, Etafly, Joseph Solis in Australia, UncleDouggie, CapitalR, CmdrObot, Wikifried, CuriousEric, Boulderinionian, Boardhead, Orionus, Acasp, Dr. Submillimeter, Salgueiro~enwiki, JAnDbot, Spacehippy, Rothorpe, Magioladitis, Eldumpo, Mollwollfumble, Vssun, R'n'B, 2012Olympian, Googolian, Tarotcards, Bramblez, Kvdveer, Idioma-bot, TXiKiBoT, The Original Wildbear, TheresJamInTheHills, Fbs. 13, BotKung, James McBride, Hellothere17, The Mad Genius, SieBot, JD554, Jdaloner, Lightmouse, Nancy, Martarius, ClueBot, Deedub1983, Tprayx, Alexis Brooke M, Brews ohare, Scog, Jnprather, ברקקווי, Camboxer, Rror, Rreagan007, Alexius08, Addbot, DOI bot, Tide rolls, Lightbot, Zorrobot, Legobot, Luckas-bot, Yobot, Andreasmperu, Ptbotgourou, TaBOT-zerem, Citation bot, Roux-HG, LilHelpa, Marshallsumter, Xqbot, Nickkid5, Tyrol5, CRea80, Brett Lally, Geek12597, Fotaun, FrescoBot, Originalwana, Physdragon, Citation bot 1, Tom.Reding, FoxBot, EmausBot, WikitanvirBot, Jmencisom, ZéroBot, Dondervogel 2, Thatfield977, ChuispastonBot, Bonnie Engles, ClueBot NG, Nebulosus, Sleddog116, Frietjes, Lincoln Josh, Helpful Pixie Bot, Bibcode Bot, BG19bot, Northamerica1000, Shawn Worthington Laser Plasma, Cliff12345, Filiosus's Saga, BattyBot, Khazar2, Zeeyanwiki, Justashuman960, AKYF, The Herald, Kogge, Elenceq, D.bcota, EckartHb, KasparBot and Anonymous: 100

- **Nuclear fusion** *Source:* https://en.wikipedia.org/wiki/Nuclear_fusion?oldid=686726342 *Contributors:* AxelBoldt, Magnus Manske, Chenyu, Trelvis, Mav, Bryan Derksen, The Anome, AstroNomer~enwiki, Taw, Malcolm Farmer, Verloren, Andre Engels, Ted Longstaffe, Jkominek, Youssefsan, XJaM, Peterlin~enwiki, Ben-Zin~enwiki, Maury Markowitz, Heron, Tobin Richard, Stevertigo, Patrick, JohnOwens, Tim Starling, Ixfd64, Looxix~enwiki, Ellywa, Ahoerstemeier, William M. Connolley, Angela, Andrewa, Aarchiba, Glenn, Kaihsu, Jedidan747, Ghewgill,

Hello162626, Thom801, Drclaptop, Ddude1969, Pffeifer, Promptjump, Rub117, Cyberbot II, The Illusive Man, Tarafauss, ChrisGualtieri, Lesvesla, Angelagibson11, SD5bot, Aschuess, Sapce Cowboy, Mahesh gandikota, Lsclear, Dexbot, Cyro43, Erjablow, Teleohapsis, Ancdefg, PeerRevision, Lugia2453, Agrrules, Reatlas, Joeinwiki, Akihabarabankinya, Kalyanpadmandar, Aj7s6, Hellogj, Teraminato, Samhg, Momnkey1997, Drewvillines, Morg00, Brett6781, WikiHelper2134, Hevatroid, LieutenantLatvia, HAKANYASARKAYA, Bill theuser88, Crow, Lette Sgo, JaconaFrere, Epic Failure, Sethep, Wyn.junior, Clubclubclub, Tysonf3, Heathflugruger, PaulZapata, Meinneger, Dilipkumardk, CharlieFerry, Aguy77, PedroGodoyP, DewDewey, Borgieporgie, Oiyarbepsy, Dgasparri, Strongjam, Wijowa, Xelevationzx, Udontknowmynamerandom, Hater lov hotdogs, Cratter the matter, Chickennuggets886, KasparBot, Yeddyeggel, Vul Vokun Ah and Anonymous: 1158

- **Compact star** *Source:* https://en.wikipedia.org/wiki/Compact_star?oldid=682873366 *Contributors:* Bryan Derksen, Tarquin, XJaM, Caid Raspa, Schneelocke, Robbot, BitwiseMan, Sverdrup, MarchHare, Rursus, Jheise, Tobias Bergemann, Eequor, Isidore, Icairns, Rich Farmbrough, RJHall, Eldar, Cmdrjameson, JW1805, Mr. Brownstone, Eras-mus, Tevatron~enwiki, Rjwilmsi, Koavf, Marasama, YurikBot, Wavelength, Spacepotato, Neitherday, Hairy Dude, Phmer, Hellbus, Ilmari Karonen, The Yeti, SmackBot, Ashill, Incnis Mrsi, Kmarinas86, DHN-bot~enwiki, Lchiarav, Mohsenmxmx, Edwy, JorisvS, Stelio, Zzzzzzzzzz, J7n, Kurtan~enwiki, Vaughan Pratt, Pegasubot, Friendlystar, Thijs!bot, Headbomb, Mattva01, Dr. Submillimeter, Rothorpe, CptEpoch, Spoxjox, Warut, VolkovBot, SieBot, Hamiltondaniel, Sfan00 IMG, Alexbot, Muro Bot, CorpITGuy, TimothyRias, IngerAlHaosului, MystBot, SkyLined, Addbot, Lightbot, Zorrobot, Luckas-bot, Yobot, AnomieBOT, Citation bot, Xqbot, Ataleh, Charvest, FrescoBot, Tom.Reding, Nacen, Bkdd, WikitanvirBot, ZéroBot, StringTheory11, Alex Nico, Snotbot, Braincricket, Bibcode Bot, ChrisGualtieri, MichaelMedford, RhinoMind, Astredita, Monkbot and Anonymous: 41

- **Gravitational collapse** *Source:* https://en.wikipedia.org/wiki/Gravitational_collapse?oldid=681203060 *Contributors:* Bryan Derksen, Aarchiba, AnthonyQBachler, Korath, Sverdrup, Robinh, Jheise, M-Falcon, Eequor, Antandrus, FT2, RJHall, Myria, IMeowbot, Yansa, Mpatel, Eteq, Durin, Itinerant1, Lynxara, Chobot, Bgwhite, YurikBot, BOT-Superzerocool, Theda, E Wing, SmackBot, Bluebot, Snori, Hibernian, Baa, Tsca.bot, FelisLeo, JorisvS, Ckatz, Newone, Tawkerbot2, Hiroshi-br, Thijs!bot, Nick Number, Luna Santin, Widefox, JAnDbot, WolfmanSF, Catgut, SwedishPsycho, Extransit, Mathglot, Jeepday, Potatoswatter, DorganBot, Idioma-bot, VolkovBot, TXiKiBoT, Aajacksoniv, Sonicology, Calusarul, Flyer22 Reborn, Breny47, PhySusie, Awikiwikiwik!!!, Thingg, 1ForTheMoney, Addbot, LaaknorBot, Margin1522, Luckasbot, Yobot, AnomieBOT, MauritsBot, Capricorn42, Ataleh, Dngnta, Mnmngb, Vbrcat, Louperibot, Dhlab, Murv21, TeigeRyan, Dinamikbot, Ripchip Bot, EmausBot, John of Reading, WikitanvirBot, GoingBatty, Brandmeister, Davidaedwards, ClueBot NG, Schmidto, Gmeredith15855, Galactic Rishabh, Phleg1, Astronaut Willis, KasparBot, Brandon Defrise Carter and Anonymous: 46

- **White dwarf** *Source:* https://en.wikipedia.org/wiki/White_dwarf?oldid=687426193 *Contributors:* Bryan Derksen, Robert Merkel, The Anome, AstroNomer~enwiki, Wayne Hardman, DavidLevinson, Ben-Zin~enwiki, Spiff~enwiki, Frecklefoot, TeunSpaans, Alan Peakall, Natbat, Minesweeper, Alfio, Looxix~enwiki, Ahoerstemeier, Glenn, Andres, Ghewgill, Pizza Puzzle, Schneelocke, Hashar, Agtx, Timwi, Stone, Fuzheado, IceKarma, Shizhao, Lumos3, Jni, Twang, Donarreiskoffer, Robbot, Fredrik, Nilmerg, Rursus, DHN, Borislav, Jheise, GreatWhiteNortherner, Sushi~enwiki, Tobias Bergemann, DarkHorizon, Pabouk, Giftlite, Ryanrs, Jyril, Xerxes314, Wwoods, Jacob1207, P.T. Aufrette, Curps, Leonard G., Yekrats, Eequor, CryptoDerk, Antandrus, 1297, Icairns, GeoGreg, Joyous!, Waza, Adashiel, DanielCD, Jkl, Discospinster, Solitude, Rich Farm-brough, Vsmith, StephanKetz, Aris Katsaris, Mani1, Bender235, Evice, RJHall, Project2501a, El C, Worldtraveller, Tom, Art LaPella, Sim-fish, Causa sui, Bobo 192, Stesmo, Twilo, Pmbrennan, Nyenyec, Tronno, .:Ajvol:., Malafaya, Nk, PiccoloNamek, Jumbuck, Alansohn, Foant, Coma28, MattBowen, QVanillaQ, Keenan Pepper, Deathphoenix, Allen McC.~enwiki, Gene Nygaard, Blaxthos, WilliamKF, Angr, Grouchy-Dan, GVOLTT, Etacar11, Camw, Rocastelo, BillC, Valkyrian Einherjar, GregorB, Palica, RichardWeiss, Rnt20, Ashmoo, Graham87, Ahsen, Rjwilmsi, Marasama, TAS, Brighterorange, Old Redneck Jokes, FlaBot, Patrick1982, Skyfiler, RobertG, Gnostic804, Algri, Carrionluggage, Chobot, Moocha, DVdm, Mhking, Coonhunter, YurikBot, Wavelength, Spacepotato, JabberWok, Jengelh, Stephenb, NawlinWiki, Misza13, Tony1, Haemo, Zzuuzz, Lt-wiki-bot, Reyk, Timothyarnold85, One, The Yeti, SmackBot, Sonoma-rich, Incnis Mrsi, Tom Lougheed, Knowl-edgeOfSelf, Nickst, Ozone77, Kadaveri, ERcheck, Grokmoo, Kmarinas86, Hraefen, Guess Who, TimBentley, MalafayaBot, Colonies Chris, Hgrosser, Modest Genius, Phaedriel, Ratel, Thegraham, Ohconfucius, Doug Bell, Freewol, Gobonobo, JoshuaZ, JorisvS, Ckatz, SandyGe-orgia, Interlingua, Dekaels~enwiki, LeyteWolfer, Newone, Freelance Intellectual, Mssgill, Tawkerbot2, Friendly Neighbour, JohnCD, Elec-tricmic, MrFish, Cydebot, Wahying, Ez5698, Jameboy, ChrisIk02, Robertinventor, Optimist on the run, Kozuch, Casliber, Blobpic, JamesAM, Headbomb, Hcobb, AntiVandalBot, Majorly, Martyn Smith, Morngnstar, Kaini, Gökhan, JAnDbot, IanOsgood, East718, WolfmanSF, Wal-loper69, Murgh, Bongwarrior, Mlindroo, Chris G, DerHexer, Kayau, SquidSK, NatureA16, MartinBot, RP88, Tgeairn, J.delanoy, DrKay, Tikiwont, Uncle Dick, Jonpro, Extransit, SU Linguist, Physicsboy, Ryan Postlethwaite, NewEnglandYankee, Mdmahir, 2help, KylieTastic, Cometstyles, SirBob42, Gemini1980, CardinalDan, Vedran8080, B0b is a b0b, VolkovBot, John Darrow, RingtailedFox, Almazi, Mocirne, Semilemon, Maghnus, Pparazorback, Philip Trueman, TXiKiBoT, Salvar, Eeron80, Henrykus, Wingedsubmariner, BigDunc, Falcon8765, Anton Gutsunaev, Jobberone, Onceonthisisland, Logan, Parthava, SieBot, BotMultichill, DNJH, Gerakibot, Dawn Bard, Mimihitam, Wle-gro, StaticGull, Starcluster, Explicit, ClueBot, Satkomuni, Fyyer, Drmies, Piledhigheranddeeper, Jiltedmatriarch, Brewcrewer, DragonBot, Excirial, TheUnknown285, NuclearWarfare, Arjayay, Njardarlogar, Kaiba, Warmbloods123, Dekisugi, BOTarate, Tauris, Kevdav63, Wnt, Vanished user uih38riiw4hjlsd, NERIC-Security, Little Mountain 5, Avoided, HarlandQPitt, Good Olfactory, Aritate, Piasoff, Thatguyflint, Addbot, Mortense, Willking1979, DOI bot, 11341134b, Professorastronomy, Ronhjones, Ironholds, NjardarBot, SoSaysChappy, AndersBot, LinkFA-Bot, 5 albert square, Bloodyninja, Lineface, Somerandomjackass, Safjdfdsvg, Theking17825, VASANTH S.N., Tide rolls, Mira7, Luckas-bot, Yobot, Tohd8 BohaithuGh1, IW.HG, Robert Treat, AnomieBOT, Astroman617, Jim1138, Tom87020, Vil Adanrath, Kingpin13, Materialscientist, RobertEves 92, The High Fin Sperm Whale, Citation bot, Maxis ftw, EugeneForrester, Obersachsebot, Xqbot, Nickkid5, WingedSkiCap, Lithopsian, Moxy, SchnitzelMannGreek, Interstellar Man, Dailycare, FrescoBot, Originalwana, Recognizance, DivineAlpha, Cannolis, Careful With That Axe, Eugene, Intelligentsium, Pinethicket, Jonesey95, Tom.Reding, Phearson, BlackHades, Bostonfanforealz, Rrrarrr, Cococake, Marissa alagna, Footwarrior, Sentra246, FoxBot, Calle Cool, Trappist the monk, DixonDBot, Vrenator, Extra999, Dian-naa, Taylor2323, Brakoholic, TjBot, Yaush, Hajatvrc, Newty23125, DASHBot, Orphan Wiki, Heracles31, Sadalsuud, Racerx11, Jmencisom, HarDNox, Gibbie144, Italia2006, Hhhippo, ZéroBot, Cogiati, Yiosie2356, Mitch123mitch21, Donner60, Yaffley, Whoop whoop pull up, ClueBot NG, MelbourneStar, Chester Markel, Helpful Pixie Bot, Bibcode Bot, BG19bot, Mynameisnoted, Wiki13, Dodshe, Sokepler, MarkArsten, Cadiomals, Zedshort, BattyBot, Geofreeman32, U-95, ChrisGualtieri, LordElektroWIKI, KeroCarGT, Dexbot, Stas1995, Curious-Mind01, Faizan, Eyesnore, Praemonitus, RandomMan3000, AnthonyJ Lock, Ispek13, Nestrs, A85Spiike, Pietro13 Jwratner1, Toroloco4925, JustBerry, Astredita, Vieque, BethNaught, DSCrowned, Flower f5a9b8, Turtlemancho, Tetra quark, KasparBot, White Anunnaki, Votlickiller, Davidf6720, ThatOneScientist, Dp99 and Anonymous: 434

- **Stellar evolution** *Source:* https://en.wikipedia.org/wiki/Stellar_evolution?oldid=686512458 *Contributors:* Mav, Bryan Derksen, Zundark, The

Anome, Wayne Hardman, Andre Engels, Xaonon, Arvindn, Roadrunner, DavidLevinson, Heron, Edward, Lorenzarius, D. Boud, Tim Starling, Alan Peakall, Trevor H., Alfio, Looxix–enwiki, G–enwiki, Aarchiba, Jedidan747, Smack, Pizza Puzzle, Rednblu, Jakenelson, Maximus Rex, Mathus–enwiki, SirJective, Cvaneg, Jni, Dale Arnett, Fredrik, Tlogmer, Arkuat, Thunderbolt16, Academic Challenger, Rursus, Jimduck, Fuelbottle, Jheise, Stirling Newberry, Ancheta Wis, Giftlite, Graeme Bartlett, Gene Ward Smith, Harp, Herbee, Everyking, Jacob1207, Niteowlneils, Mboverload, Jackol, Beland, Icairns, ELApro, Rfl, N328KF, Moverton, Rich Farmbrough, Sladen, Vsmith, Eric Forste, RJHall, El C, Worldtraveller, Tom, Bobo192, GattoRandagio, La goutte de pluie, Cherlin, Haham hanuka, Conny, Mote, Alansohn, Tablizer, Ungtss, Tezeti, ATG, Mysdaao, Sciurinæ, Gene Nygaard, Vadim Makarov, Anarchimede, WilliamKF, Etacar11, Palica, RichardWeiss, Graham87, Drbogdan, Rjwilmsi, Angusmclellan, Oblivious, S Schaffter, THE KING, Skyfiler, CJLL Wright, Chobot, Spacepotato, Xihr, Gravecat, NawlinWiki, Wiki alf, Pagrashtak, Gadget850, Dan337, Lt-wiki-bot, Reyk, CharlesHBennett, Sitegod, Kungfuadam, Banus, Serendipodous, Cmglee, Luk, Chrisw428, KnightRider–enwiki, SmackBot, Ashill, Vald, Ltepper, Eskimbot, Gilliam, Ppntori, Saros136, TheDarkArchon, Wellspring, Whispering, Colonies Chris, Rogermw, Can't sleep, clown will eat me, Lchiarav, Pnkrockr, Aldaron, Hgilbert, Sbluen, WayKurat, Mathiasrex, J 1982, JorisvS, Fig wright, Fuzzy510, Casull, Gorniac, Rhetth, Achoo5000, Tawkerbot2, Kurtan–enwiki, CmdrObot, Olaf Davis, Ruslik0, NE Ent, Funnyfarmofdoom, Michael C Price, Epbr123, Sagaciousuk, DaveJ7, Big Bird, Mentifisto, Majorly, Chubbles, Seaphoto, Orionus, Bakabaka, Dr. Submillimeter, G Rose, MagiMaster, Leuko, WolfmanSF, Bongwarrior, VoABot II, BSVulturis, Mbc362, The Enlightened, Sodabottle, Stiip, Xtifr, Geboy, CommonsDelinker, PrestonH, J.delanoy, Pharaoh of the Wizards, Mattsmasher0000, Rod57, Janus Shadowsong, RTBoyce, Rominandreu, MEMcNeil, Funandtrvl, Xenonice, Lights, AlnoktaBOT, Ryan032, TXiKiBoT, Walor, Someguy1221, BobM, Duncanjwh, Martin451, Mzmadmike, LeaveSleaves, Gbaor, Psyche825, BotKung, Ilyushka88, 1981willy, Blurpeace, FKmailliW, Bzotter, SieBot, Timb66, PlanetStar, YonaBot, Ajayfermi, BotMultichill, Caltas, Wing gundam, Keilana, Bentogoa, Flyer22 Reborn, Tiptoety, CaelumArisen, Pika ten10, Tombomp, Poindexter Propellerhead, Anchor Link Bot, Hamiltondaniel, Freewayguy, ClueBot, The Thing That Should Not Be, Lawrence Cohen, EknellerWIKI, Agge1000, ChandlerMapBot, Ktr101, Verwoerd, MacedonianBoy, Scog, BOTarate, Mlaffs, Aitias, Dohgon Immortal, DumZiBoT, Arianewiki1, Skarebo, Drewtaylor1978, Ich42, Hobbema, Addbot, Xp54321, DOI bot, LaaknorBot, LinkFA-Bot, 84user, Wizziwizard, NoNonsenseHumJock, Bigzteve, Tide rolls, Dmartin25, WuBot, Dulcineaven, Luckas-bot, Yobot, Senator Palpatine, Robert Treat, Backslash Forwardslash, AnomieBOT, JackieBot, Materialscientist, The High Fin Sperm Whale, Citation bot, Klingon83, Xqbot, Capricorn42, Br77rino, Lithopsian, Little billy 123, Trafford09, Moxy, Mjasfca, FrescoBot, Originalwana, EmilTyf, Citation bot 1, Citation bot 4, Gaba p, Pinethicket, Tom.Reding, SaturdayNightSpecial, Steve2011, IVAN3MAN, Elena Emanova, Trappist the monk, Lotje, Begoon, Vidia234, Diannaa, Stroppolo, Yaush, EmausBot, Gfoley4, MrRandomPerson, Yowife, Passionless, Dcirovic, ZéroBot, Claudio M Souza, Fæ, StringTheory11, Rcsprinter123, Tiiliskivi, L Kensington, Quantumor, Donner60, NTox, LikeLakers2, Whoop whoop pull up, ClueBot NG, Widr, Helpful Pixie Bot, SzMithrandir, Bibcode Bot, J824h, Kai Ojima, Wiki13, MusikAnimal, Sonasonic, Cadiomals, GGShinobi, Fluidphysics, Wer900, Szczureq, Nitrobutane, Jason from nyc, BattyBot, Cyberbot II, ChrisGualtieri, Soulbust, Dexbot, Garuda0001, BenMorbeck, Epicgenius, NottNott, Kogge, Astredita, Ithinkicahn, Kkosman, Monkbot, Hithere1998, Neeraj Bhakta, Jim W Davis, The Original Bob, Quackduck39 and Anonymous: 348

- **Gravitational energy** *Source:* https://en.wikipedia.org/wiki/Gravitational_energy?oldid=681437711 *Contributors:* SebastianHelm, Tobias Bergemann, Mporter, OverlordQ, Mike Rosoft, Larry V, Musiphil, Bobrayner, FlaBot, DVdm, Enormousdude, Closedmouth, Josh3580, SmackBot, Eskimbot, Andy M. Wang, Chris the speller, Bluebot, Colonies Chris, Javalenok, Decltype, Aleenf1, CmdrObot, Meno25, Michael C Price, Thijs!bot, Pajz, Headbomb, Elmoosecapitan, Steelpillow, Golgofrinchian, Bongwarrior, BatteryIncluded, Krushdiva, Tgeairn, J.delanoy, Uncle Dick, Lantonov, VolkovBot, Skingski, Keilana, KoenDelaere, DaDrought3, Denisarona, ClueBot, Excirial, Bagworm, Vanished user uih38riiw4hjlsd, Addbot, Bwr6, Ronhjones, PV=nRT, LuK3, Modbear, Rubinbot, Oracelau, Materialscientist, Tomdo08, Paine Ellsworth, Tom.Reding, Vrenator, Clarkcj12, Alph Bot, EmausBot, ZéroBot, ElationAviation, Access Denied, Bamyers99, Crown Prince, ClueBot NG, Jack Greenmaven, Movses-bot, Widr, MerllwBot, NuclearEnergy, Roberticus, Snaevar-bot, Lugia2453, BioticPixels, OnlyWikki, HMSLavender, Kanishkkumar123, Anyrurhrgn and Anonymous: 73

- **History of supernova observation** *Source:* https://en.wikipedia.org/wiki/History_of_supernova_observation?oldid=680639264 *Contributors:* Rtfisher, Sonjaaa, Thorwald, Rich Farmbrough, RJHall, Eric Kvaalen, Falcorian, WilliamKF, Billhpike, Rjwilmsi, Mike Peel, Wavelength, Brandmeister (old), RussBot, LiamE, Tony1, Wknight94, Bamse, SmackBot, Jagged 85, Liaocyed, ComaDivine, HalfShadow, Thanatosimii, JH-man, JoshuaZ, Joshua Scott, Mr Stephen, Rock4arolla, Nehrams2020, WilliamJE, GiantSnowman, Headbomb, Auke Slotegraaf, Dekimasu, BSVulturis, بارس, Tonicthebrown, Trilobitealive, MishaPan, TXiKiBoT, KGyST, Maelgwnbot, Nergaal, ClueBot, Meisterkoch, Atlytle, Rsellis, PSimeon, Addbot, Roentgenium111, DOI bot, Canon.vs.nikon, Kjaer, Tucoxn, Materialscientist, Citation bot, DSisyphBot, Winged-SkiCap, Prezbo, Jc3s5h, Dnessett, Citation bot 1, Tom.Reding, Bigd666, Extra999, Polylepsis, John of Reading, Jmencisom, RCThomas, H3llBot, Brandmeister, ClueBot NG, Satellizer, RndmPrsn13, Helpful Pixie Bot, Bibcode Bot, ElphiBot, Dexbot, Monkbot, MuseumGeek, Spideratseds, Dretler and Anonymous: 29

- **Type Ia supernova** *Source:* https://en.wikipedia.org/wiki/Type_Ia_supernova?oldid=686215256 *Contributors:* Roadrunner, Nv8200pa, Topbanana, Rursus, Melikamp, Cyclopia, Neko-chan, RJHall, Shenme, Alansohn, Ceyockey, Falcorian, Etacar11, Chrkl, Drbogdan, Rjwilmsi, Mike Peel, Hairy Dude, Bjf, Długosz, Ilmari Karonen, Algae, SmackBot, Gilliam, Modest Genius, Soulkeeper, George The Dragon, Rock4arolla, Nehrams2020, Newone, Nutster, Robertinventor, Keraunos, Headbomb, Nick Number, Orionus, Igodard, WolfmanSF, R'n'B, Fconaway, Ihutchesson, Tambora1815, KylieTastic, DorganBot, Idioma-bot, Vestboy Myst, Rei-bot, UnitedStatesian, BotKung, AlleborgoBot, KGyST, LeadSongDog, Reuqr, Lightmouse, Nergaal, Fedhere, ClueBot, GambitNC, StigBot, Noca2plus, Sun Creator, Crowsnest, Drewtaylor1978, SkyLined, Hobbema, Addbot, DOI bot, Sn2007Cb, Proxima Centauri, LinkFA-Bot, Somerandomjackass, Wikbot, Wisconsinsurfer, Yobot, Piano non troppo, Citation bot, Maxis ftw, Xqbot, WingedSkiCap, Lithopsian, Mnmngb, Fotaun, FrescoBot, LucienBOT, Originalwana, JohnMenninghaus7, Citation bot 1, Emaus, Jonesey95, Tom.Reding, Jim Fitzgerald, Fartherred, IVAN3MAN, Ukmaxi, JLincoln, Louiselives, TGCP, EmausBot, WikitanvirBot, Primefac, GoingBatty, Sp33dyphil, Luiscalcada, EMurciano, Solarflare100, ZéroBot, Quondum, Daniel.kassl, ChuispastonBot, ClueBot NG, ReecyBoy42, Helpful Pixie Bot, Calabe1992, Bibcode Bot, BG19bot, Narayan89, Vagobot, AvocatoBot, Mark Arsten, Gallina3795, MythosMagic, Wer900, BattyBot, Khazar2, Forty Seven Nine, Dexbot, Rfassbind, Devious3144, YiFeiBot, Kogge, Snkoysean, Craneis, Monkbot and Anonymous: 44

- **Type Ib and Ic supernovae** *Source:* https://en.wikipedia.org/wiki/Type_Ib_and_Ic_supernovae?oldid=664618566 *Contributors:* Jordi Burguet Castell, Rich Farmbrough, Pie4all88, RJHall, Rjwilmsi, Koavf, Mike s, Spacepotato, JustAddPeter, SmackBot, Chris the speller, Modest Genius, Daniel.o.jenkins, CmdrObot, Thijs!bot, Headbomb, WolfmanSF, SchumiUCD, Lightmouse, Nergaal, Artichoker, Alexbot, Dana boomer, Trulystand700, Parejkoj, Addbot, DOI bot, Edoberger, Citation bot, ArthurBot, LilHelpa, TechBot, Lithopsian, GrouchoBot, Mnmngb, Citation bot 1, Tom.Reding, Full-date unlinking bot, IVAN3MAN, Calle Cool, RjwilmsiBot, Aircorn, EmausBot, Bibcode Bot, Grizzly-Pear, Monkbot and Anonymous: 9

- **Type II supernova** *Source:* https://en.wikipedia.org/wiki/Type_II_supernova?oldid=683963057 *Contributors:* Nealmcb, Trisweb, Rursus, Rich Farmbrough, FT2, RJHall, Tom, Duk, Axl, Gene Nygaard, WadeSimMiser, Sjö, Rjwilmsi, Koavf, Mike s, Physchim62, Smithbrenon, Hellbus, Ashill, Andrei Stroe, Smith609, Rickington, Nehrams2020, Esurnir, Danrok, Casliber, Headbomb, Eltanin, AstroPaul, IanOsgood, Maias, WolfmanSF, Redian, John Darrow, Rei-bot, UnitedStatesian, SwordSmurf, AlleborgoBot, Maelgwnbot, Nergaal, HACubs, EoGuy, Tarlneustaedter, SkyLined, Addbot, Mortense, DOI bot, LinkFA-Bot, Lightbot, Yobot, Kilom691, Sallenmd, Mintrick, Citation bot, Xqbot, Gap9551, Srich32977, Lithopsian, Mnmngb, LucienBOT, Citation bot 2, Citation bot 1, Tom.Reding, Pmokeefe, Trappist the monk, Ifly6, Aircorn, EmausBot, Immunize, Winner 42, ZéroBot, LelandHerder, Ego White Tray, Gary Dee, Aleazar84, ClueBot NG, TRauscher, Danim, Helpful Pixie Bot, Bibcode Bot, Narayan89, Krenair, Trevayne08, Aardwolf first, Khazar2, Dexbot, Anderson, Tsiolkovsky, Kogge, Monkbot, Bannanamanbanana, Tonathan100 and Anonymous: 37

- **Light curve** *Source:* https://en.wikipedia.org/wiki/Light_curve?oldid=680274442 *Contributors:* AstroNomer~enwiki, Heron, Alfio, Poor Yorick ,Hike395, Charles Matthews, Sverdrup, RJHall, Huntster, Evil Monkey, Search4Lancer, Rjwilmsi, Quuxplusone, MalafayaBot, Jan.Kamenicek, 81120906713, Aaronp808, Spebudmak, Krylonblue83, Julia Rossi, Alphachimpbot, Meodudlye, RP88, Colincbn, VolkovBot, TXiKiBoT,Aua, Alexbot, Roosvelte, Addbot, LaaknorBot, Luckas-bot, Citation bot, Xqbot, Hauganm, Citation bot 1, Tom.Reding, Full-date unlinkingbot, Iferrinv, RjwilmsiBot, EmausBot, ZéroBot, A2soup, H3llBot, ClueBot NG, Bibcode Bot, IluvatarBot, Wer900, OakRunner, Rfassbind,Acw113, Monkbot, Dretler and Anonymous: 7

- **Apparent magnitude** *Source:* https://en.wikipedia.org/wiki/Apparent_magnitude?oldid=684129050 *Contributors:* Mav, AstroNomer~enwiki, Wayne Hardman, Andre Engels, Eclecticology, XJaM, Roadrunner, Patrick, Michael Hardy, Imcool123, Alfio, Looxix~enwiki, Alvaro, Susurrus, Andres, Cherkash, HPA, Pizza Puzzle, Dino, Denni, Timc, Jakenelson, Dragons flight, Xyb, Fvw, Pakaran, BenRG, Hajor, Robbot, Fredrik, RedWolf, Naddy, Merovingian, Rursus, Hemanshu, Modeha, Wikibot, Carnildo, Giftlite, Jyril, Sj, Harp, Marcika, Monedula, Curps, Niteowl-neils, Pne, Jrdioko, Utcursch, Antandrus, Karol Langner, Tomruen, Bosmon, Icairns, Deleting Unnecessary Words, Urhixidur, Mike Rosoft, Geof, Jkl, Discospinster, Rich Farmbrough, Guanabot, Spundun, Bender235, RJHall, Circeus, Geoff.green, SpeedyGonsales, Ben77, Haham hanuka, Alphonsus, Wendell, Jhertel, Mlm42, Wtshymanski, Evil Monkey, Dirac1933, Lerdsuwa, Skatebiker, Pcd72, Firsfron, JeremyA, Knuckles, BartBenjamin, Akavel~enwiki, Smartech~enwiki, Bebenko, Rnt20, Ketiltrout, Rjwilmsi, Commander, Marasama, Mike s, Mau-rog, Maxim Razin, Javalizard, Vuong Ngan Ha, FlaBot, Patrick1982, JYOuyang, Jobeus, Thecurran, Glenn L. Chobot, Bgwhite, Wavelength, RobotE, Shawn81, Gaius Cornelius, CambridgeBayWeather, Johnny Pez, Test-tools~enwiki, Prickus, Zephalis, Light current, Chesnok, Cara-binieri, Tim R, Garion96, John Broughton, Cmglee, Kalsermar, SmackBot, MattieTK, Elonka, NorthernFire, C.Fred, Phaldo, AtilimGunes-Baydin, B00P, MalafayaBot, Tianxiaozhang~enwiki, Chtit draco, Tsca.bot, WinstonSmith, Pulu, Dougmc, Hgilbert, Maelnuneb, Kotjze, ALK, JorisvS, Accurizer, Frango com Nata, Interlingua, Novangelis, Beefyt, Joseph Solis in Australia, Newone, DOMICH, Robsdot, Chrislk02, Coelacan, Fournax, Najro, Marek69, Ufwuct, Orionus, Lklundin, Res2216firestar, James39, JAnDbot, Husond, Rothorpe, Magioladitis, Linkinpark342, Shijualex, Kheider, Ztobor, MartinBot, Nono64, J.delanoy, Dinoguy1000, Richontaban, IdLoveOne, Nwbeeson, MKoltnow, William pascoe, Ahtih, VolkovBot, RingtailedFox, BBence, TXiKiBoT, Sagittarian Milky Way, Java7837, A4bot, Martin451, ARUNKUMAR P.R, Synthebot, James McBride, Sardonicone, Legoktm, SieBot, Hertz1888, Byrialbot, Faradayplank, Robertcurrey, Lightmouse, Nn123645, Velvetron, Twinsday, Martarius, Sfan00 IMG, ClueBot, Bobathon71, Pomona17, Punk1234, Mumiemonstret, Excirial, Jusdafax, Tomeasy, Jot-terbot, Njardarlogar, Belarius Slade, Thingg, Rangiras, Crowsnest, Godfather71190, D0nj03, Arianewiki1, Interferometrist, WikiDao, Addbot, Roentgenium111, Tanhabot, Adrian 1001, Jwsinclair, 84user, Lightbot, Abosaleh911, Legobot, Luckas-bot, Yobot, Ptbotgourou, Rrokkedd, THEN WHO WAS PHONE?, TestEditBot, AnomieBOT, Icarusmonkey, Materialscientist, Stefansquintet, Hunnjazal, DirlBot, Xqbot, The-godofbigthings, Ubcule, Nasa-verve, GrouchoBot, Appple, RibotBOT, CRea80, Tanze1, Constructive editor, Зоэрь Ə, HDWK, AllCluesKey, 10metreh, Tom.Reding, FoxBot, Dinamik-bot, Clintondiy, Reaper Eternal, Andrea105, Ripchip Bot, 123Mike456Winston789, Newty23125, DASHBot, WikitanvirBot, Sadalsuud, Cpobrech, Jmencisom, Spiritrong, Jenks24, Dondervogel 2, Hevron1998, Bamyers99, H3llBot, Cline92, Brandmeister, Donner60, PeterDReed, Fjörgynn, ClueBot NG, LeastCommonAncestor, Joefromrandb, CorniceBowlSkier, Helpful Pixie Bot, Gob Lofa, Bibcode Bot, Tycho-Is-Great, Zedshort, ChrisGualtieri, Dillydillysilly, Stas1995, Makecat-bot, Tony Mach, SteenthIWbot, Rfass-bind, Winguse, Eyesnore, Sss14740, Terrill Shepard Soules, Murzy.d.jhabvala, Jianhui67, Tushar Shrotriya, Spideratseds, Fleivium, Calcula2, I am. furhan. and Anonymous: 194

- **Chandrasekhar limit** *Source:* https://en.wikipedia.org/wiki/Chandrasekhar_limit?oldid=687386102 *Contributors:* Wesley, Bryan Derksen, The Anome, Tarquin, AstroNomer~enwiki, Ap, Verloren, Wayne Hardman, Andre Engels, Josh Grosse, Montrealais, Stevertigo, Alan Peakall, Alfio, Suisui, Marco Krohn, Lenaic, Phr, The Anomebot, Jnc, Mathus~enwiki, Robbot, Arkuat, Mirv, Davidcannon, Xyzzyva, ShaunMacPher-son, Harp, Wolfkeeper, Wyss, Duncharris, Sundar, DefLog~enwiki, Eranb, Sam Hocevar, B.d.mills, Jayjg, Rich Farmbrough, Vsmith, Chow-ells, MuDavid, Bender235, ZeroOne, RJHall, El C, Olve Utne, La goutte de pluie, Tezeti, Burn, Allen McC.~enwiki, Adrian.benko, David Haslam, Miaow Miaow, Joke137, Christopher Thomas, Rnt20, Search4Lancer, Avia, Pradeepsinghhbti, Jehochman, Gparker, Lithiyum, Lynxara, Physchim62, Chobot, Spacepotato, RobotE, Hairy Dude, Gaius Cornelius, David R. Ingham, Leutha, The Yeti, SmackBot, Inc-nis Mrsi, Unyoyega, Richard B. Slaniel, Ohnoitsjamie, Kevin Ryde, DHN-bot~enwiki, Colonies Chris, Salmar, SundarBot, T-borg, Alcuin, Andrei Stroe, SashatoBot, Malixsys, Fimus, JorisvS, Kyoko, Antonio Prates, PhoenixSeraph, Friendly Neighbour, CRGreathouse, Wegge-Bot, Myasuda, Cydebot, Asymptote, Thijs!bot, CharlotteWebb, Deflective, Quentar~enwiki, Epinheiro, Coolhandscot, WolfmanSF, Justin-Green, MartinBot, STBot, Fconaway, BigrTex, Numbo3, Peter Chastain, IdLoveOne, DorganBot, Idioma-bot, JCMP, John Darrow, Philap Trueman, TXiKiBoT, Ask123, Markp93, SieBot, Sahilm, OsamaBinLogin, Arthur Smart, OKBot, Martarius, ClueBot, Garyzx, Agge1000, Jotterbot, BOTarate, Techroach, Daffydd, MystBot, Mortense, DOI bot, Colinb007, Lightbot, Mira7, Zorrobot, Legobot, Yobot, Donfbreed, Amirobot, PianoDan, Materialscientist, Citation bot, Teleprinter Sleuth, ArthurBot, Xqbot, Nfr-Maat, Tomwsulcer, Andrestand, Jonesey95, A412, Rosa67, Σ, Puzl bustr, DARTH SIDIOUS 2, Andrea105, Ripchip Bot, EmausBot, John of Reading, WikitanvirBot, Solomonfromfin-land, Hhhippo, I. Kensington, Llightex, Whoop whoop pull up, ClueBot NG, Braincricket, Helpful Pixie Bot, Bibcode Bot, Majee chinmay, Mynameisnoted, Dodshe, RickV88, Dexbot, Frosty, Monkbot, Lince meladath, RKhehehe and Anonymous: 111

- **Electron degeneracy pressure** *Source:* https://en.wikipedia.org/wiki/Electron_degeneracy_pressure?oldid=653831428 *Contributors:* The Anome, Jni, Bearcat, Owain, JDoolin, M1ss1ontomars2k4, La goutte de pluie, Computerjoe, Rjwilmsi, Chobot, Hairy Dude, Jengelh, GeeJo, Charles Randles, Timothyarnold85, SmackBot, Grokmoo, Wikipedia brown, Pilotguy, JorisvS, JRSpriggs, Cydebot, Lutskovp, WolfmanSF, TheSlyFox, Freelance Physicist, LorenzoB, Pete okell, R'n'B, Cuzkatzimhut, Calwiki, Nxavar, Fizzackerly, Agge1000, RP459, Addbot,

Luckas-bot, Yobot, Captain Quirk, BCROCE, FrescoBot, Citation bot 1, Bigdok, EngineerFromVega, RjwilmsiBot, Newty23125, John of Reading, JasonSaulG, Itspragee, AvicAWB, Quondum, JaneStillman, Bibcode Bot, Trevayne08, Ariomhooks, Monkbot and Anonymous: 37

- **Carbon-burning process** *Source:* https://en.wikipedia.org/wiki/Carbon-burning_process?oldid=678525658 *Contributors:* Bryan Derksen, Edward, Xavic69, Looxix~enwiki, Andrewa, Rursus, Cdheald, RJHall, Kwamikagami, Shenme, I9Q79oL78KiL0QTFHgyc, Nik42, RJFJR, Christopher Thomas, Rjwilmsi, Strait, Salix alba, JWB, Poppy, Nekura, SmackBot, Bluebot, AB, Artman40, Saxbryn, CmdrObot, Ruslik0, Andkore, Myasuda, Underpants, Jono4174, Headbomb, Escarbot, Orionus, WolfmanSF, JamesBWatson, Rominandreu, Sheliak, QuackGuru, TXiKiBoT, Venny85, Betalph, Hamiltondaniel, Sfan00 IMG, MystBot, SkyLined, Addbot, Yobot, Robert Treat, Materialscientist, Citation bot, GrouchoBot, FrescoBot, Citation bot 1, IVAN3MAN, TobeBot, Trappist the monk, Puzl bustr, RjwilmsiBot, Slon02, EmausBot, Slightsmile, ZéroBot, Arbnos, Andewbuggy, Whoop whoop pull up, Helpful Pixie Bot, Bibcode Bot, BattyBot, U-95, Monkbot and Anonymous: 19

- **Main sequence** *Source:* https://en.wikipedia.org/wiki/Main_sequence?oldid=682885899 *Contributors:* Mav, Bryan Derksen, Zundark, Malcolm Farmer, XJaM, William Avery, Roadrunner, Patrick, Michael Hardy, Alfio, Looxix~enwiki, Rboatright, Angela, Mark Foskey, Susurrus, Samw, Pizza Puzzle, WhisperToMe, Robbot, Babbage, Rursus, Timrollpickering, Tobycat, Modeha, Aetheling, Pmcray, David Gerard, Stirling Newberry, Lestatdelc, Kaldari, Thincat, Icairns, Moverton, Rich Farmbrough, Vsmith, Mani1, Wadewitz, RJHall, Kwamikagami, Art LaPella, Jon the Geek, Grick, Bobo192, Wayfarer, Eric Kvaalen, ATG, Stillnotelf, GeorgeStepanek, Jun-Dai, Blaxthos, WilliamKF, Nuno Tavares, Pol098, Arrkhal, Bebenko, RichardWeiss, Lawrence King, Graham87, BD2412, Rjwilmsi, Raddick, Mike s, Brighterorange, Maxim Razin, SiriusB, Chobot, Sharkface217, Gwernol, Kdehl, The Rambling Man, YurikBot, Wavelength, Spacepotato, Sceptre, Hydrargyrum, Ksyrie, Kvn8907, Ragesoss, Bozoid, Gadget850, Perry Middlemiss, Chaos syndrome, Junglecat, Hal peridol, SmackBot, NickyMcLean, Moeron, Tom Lougheed, Unyoyega, C.Fred, Saros136, Hibernian, Adun12, Jerome Charles Potts, Modest Genius, Tlusťa, ConMan, Salamurai, Erimus, JorisvS, Like tears in rain, Dan Gluck, Dekaels~enwiki, Røed, Mssgill, JForget, David s graff, Ruslik0, Funnyfarmofdoom, Mato, Jayen466, Michael C Price, Thijs!bot, Epbr123, Oerjan, Marek69, Ialsoagree, AntiVandalBot, Orionus, Spartaz, JAnDbot, Plantsurfer, East718, Acroterion, Penubag, WolfmanSF, Secret Squirrel, Bongwarrior, VoABot II, Ronstew, Kokin, Catgut, DerHexer, Stargazer360, Geboy, Leyo, Fellwalker57, El0i, J.delanoy, Pharaoh of the Wizards, DrKay, Rrostrom, All Is One, Gzkn, Dividing, Nwbeeson, Bcp67, Idioma-bot, X!, Larryisgood, Philip Trueman, GimmeBot, BotKung, Jmac1962, James McBride, FKmailliW, Junkinbomb, Timb66, Rfts, Gerakibot, Snideology, Radzewicz, Techman224, Jruderman, Ethan1701, Freewayguy, Loren.wilton, ClueBot, Ukabia, DanielDeibler, Boing! said Zebedee, Blanchardb, Rotational, Piledhigheranddeeper, Auntof6, DragonBot, Estirabot, Maradona01, Thingg, The Sock of Maelgwn, Stefanole, Spitfire, LoneStar77, Addbot, DOI bot, Landon1980, CanadianLinuxUser, Jaydec, Lightbot, CountryBot, Legobot, Luckas-bot, Aldebaran66, Rrokkedd, KamikazeBot, Kristen Eriksen, Galoubet, Citation bot, Eumolpo, Ulysse2000, LovesMacs, The Fiddly Leprechaun, DSisyphBot, Thatfeel, Lithopsian, TopHyatt, Moxy, Matteobachetti, FrescoBot, Pepper, 28club, Millslogle, Tom.Reding, IVAN3MAN, RjwilmsiBot, Limequat, Bhawani Gautam, Regancy42, TGCP, EmausBot, John of Reading, Orphan Wiki, Mhdkandil, Dewritech, RenamedUser01302013, Jmencisom, StringTheory11, Davido44, H3llBot, Tolly4bolly, Michaelandsandy, Whoop whoop pull up, ClueBot NG, CocuBot, A520, Mgribov, Alex Nico, Jj1236, Kevin Gorman, Ryan Vesey, MerllwBot, Bibcode Bot, Solomon7968, Snow Blizzard, Ggreybeard, Oznitecki, StarryGrandma, Cyberbot II, Rhlozier, Frosty, Malerooster, Acetotyce, Melonkelon, Nouvelle Planète, Dillon128, Kogge, Monkbot, Yangch17, Tetra quark, PlaQu, Sdgedfegw and Anonymous: 207

- **Black hole** *Source:* https://en.wikipedia.org/wiki/Black_hole?oldid=687289359 *Contributors:* Chenyu, CYD, Ansible, Bryan Derksen, Zundark, Timo Honkasalo, The Anome, Tarquin, AstroNomer~enwiki, Gareth Owen, Ed Poor, Wayne Hardman, Eclecticology, Graham Chapman, XJaM, Arvindn, Roadrunner, SimonP, Ben-Zin~enwiki, Apollia, Modemac, Chris Q, Dbundy, Stevertigo, Frecklefoot, Edward, Patrick, RTC, Boud, JohnOwens, PhilipMW, Ken Arromdee, Michael Hardy, Tim Starling, EddEdmondson, Kwertii, Isomorphic, Fuzzie, Jketola, Sam Francis, Ixfd64, Bcrowell, Iluveapra, AquaRichy, Minesweeper, Alfio, Kosebamse, Stw, Looxix~enwiki, Ahoerstemeier, Cyp, Anders Feder, Ronz, William M. Connolley, Theresa knott, Snoyes, Suisui, Angela, Den fjättrade ankan~enwiki, Jebba, Glenn, Susurrus, Evercat, Samuel~enwiki, Mxn, Schneelocke, Hike395, Emperorbma, Frieda, Fry-kun, Vanished user 5zariu3jisj0j4irj, Wikiborg, Paul Stansifer, Jwrosenzweig, The Anomebot, Doradus, Tpbradbury, Marshman, Maximus Rex, Furrykef, Morwen, Saltine, Taxman, Rei, Ed g2s, Rnbc, Thue, Lord Emsworth, Joy, Raul654, BenRG, Banno, Jhobson1, Jeffq, Owen, RadicalBender, Mrdice, Northgrove, SD6-Agent, Phil Boswell, Vt-aoe, AlexPlank, Robbot, Sander123, Craig Stuntz, TomPhil, Alrasheedan, RedWolf, Donreed, Altenmann, Romanm, Lowellian, Merovingian, Sverdrup, Meelar, Auric, JB82, DHN, Davodd, Hadal, Quincy, JesseW, Wikibot, Wereon, Borislav, Reid, Jheise, JerryFriedman, Diberri, Jholman, Dina, Tobias Bergemann, Alan Liefting, Enochlau, Giftlite, DocWatson42, Christopher Parham, MPF, Gtrmp, Awolf002, Andy, Barbara Shack, Castaa, Cobaltbluetony, Lethe, Tom harrison, Art Carlson, Lupin, Herbee, SheikYerBooty, Xerxes314, Paul Pogonyshev, Peruvianllama, Everyking, Plautus satire, Anville, NASA~enwiki, Curps, David Johnson, Home Row Keysplurge, Sriehl, Joe Kress, Cantus, Rpyle731, Andris, Guanaco, Avsa, Jorge Stolfi, Sundar, Eequor, Solipsist, Nathan Hamblen, Foobar, Dan Gardner, PlatinumX, SWAdair, DÅ‚ugosz, Becomm, Bobblewik, Joseph Dwayne, RcktScientistX, Stevietheman, StuartH, Chowbok, Geni, Gdr, Fpahl, Antandrus, HorsePunchKid, Beland, Onco p53, MadIce, Noirum, Rdsmith4, Maximaximax, Jokestress, Aranoff, Jobrober, Variant, Kevin B12, Bosmon, Satori, GeoGreg, Sam Hocevar, Tzarius, Gscshoyru, Iantresman, Neutrality, Joyous!, Quota, TJSwoboda, Jewbacca, Degir6328, Temujin9, Jwlidtnet, Ayager, Mike Rosoft, Ouro, Simonides, Freakofnurture, Spiffy sperry, Poccil, Bactram, Indosauros, زرـزل, Discospinster, Solitude, Rich Farmbrough, Guanabot, Yuval madar, Igorivanov~enwiki, FT2, Pjacobi, Vsmith, Jpk, Ponder, Antaeus Feldspar, Mani1, Olau, Paul August, Xjaymanx, Dmr2, MJSS, Bender235, ESkog, Zaslav, Kjoonlee, Cucumberslumber, Kalel, Nabla, Brian0918, RJHall, Livajo, El C, Lankiveil, Parklandspanaway, Edward Z. Yang, Shanes, Arete~enwiki, Spearhead, Susvolans, Rsmelt, Art LaPella, RoyBoy, Dalf, Jpgordon, Iridia, Causa sui, Shoujun, Bobo192, 23skidoo, Billymac00, Flxmghvgvk, Draco2, Reuben, Jguk 2, JW1805, Redquark, I9Q79oL78KiL0QTFHgyc, Timl, Chbarts, Toh, La goutte de pluie, Nk, BM, NickSchweitzer, Doozer, Hi3322110, Tuskey, Cherlin, Apostrophe, Haham hanuka, Hagerman, Tms, Wayfarer, Solocommand, MetalMilitia, Papeschr, Knucmo2, Jumbuck, Jérôme, Alansohn, Jamyskis, Tek022, Keenan Pepper, Andrewpmk, Tezeti, Ricky81682, Andrew Gray, Lord Pistachio, Punarbhava, Riana, Wikidea, AzaToth, Keflavich, Lectonar, Axl, R Calvete, Mac Davis, Cdc, Grobertson, Transcend~enwiki, Wtmitchell, BanyanTree, BRW, QuixoticKate, Almafeta, Cecil, Yuckfoo, Jheald, Count Iblis, H2g2bob, ThomasWinwood, Gortu, Computerjoe, Jchillerup, Freyr, DV8 2XL, Mordero, Gene Nygaard, Axeman89, Nick Mks, Kazvorpal, Markaci, Njk, Dmitry Brant, Bobrayner, ChrisJMoor, Richard Arthur Norton (1958- ), Rorschach, OwenX, JarlaxleArtemis, Camw, LOL, Pinball22, Merlinme, Prophile, Orchew, BillC, Mazca, HFarmer, JeremyA, Direwolf, Mpatel, Tabletop, Ianweller, Schzmo, Nirmalya, Terence, MFH, PhoenixPinion, GregorB, Macaddct1984, El Suizo, CharlesC, Jon Harald Søby, Joke137, Christopher Thomas, Rufous, Rgbea, Bebenko, GSlicer, Rnt20, Graham87, Magister Mathematicae, GoldRingChip, BD2412, Qwertyus, Chun-hian, Kbdank71, Eteq, RxS, BorgHunter, Drbogdan, Akubhai, Coneslayer, Sjakkalle, Rjwilmsi, Zbxgscqf, Phileas, Arie~enwiki, WCFrancis, Wikibofh, Vary, Strait, Marasama, Strake, Sdornan, Captain Disdain, HandyAndy, Mike s, Nick R, Elkester, Ligulem, Jehochman, SeanMack, Ems57fcva, AndyKali, Bhadani, Jackdriscoll, Maurog,

GregAsche, Sango123, Yamamoto Ichiro, KaiMartin, W00d, Lionelbrits, FayssalF, FlaBot, Patrick1982, Ian Pitchford, SchuminWeb, Vegardw, RobertG, Musical Linguist, SiriusB, Harmil, Nivix, Chanting Fox, Itinerant1, RexNL, Gurch, Schumps, Hansamurai, Algri, Gmz1023, Poderis, Fresheneesz, Pete.Hurd, Jesse0986, Alphachimp, Diza, Tedder, Kri, Imnotminkus, Ahsankhan, GringoCroco, Essaregee, Chobot, Fourdee, DVdm, RashBold, GreyedOut, Bgwhite, Ahpook, Cactus.man, Eric B, NSR, Tone, Amaurea, CaseKid, Mike5904, Wiserd911, Siddhant, McGinnis, YurikBot, Wavelength, Extraordinary Machine, Splintercellguy, Sceptre, Hairy Dude, Deeptrivia, Rt66lt, Jimp, Hillman, Brandmeister (old), StuffOfInterest, Tznkai, Phantomsteve, RussBot, Arado, Gunblade-enwiki, TheDoober, Xihr, Splash, Chris Capoccia, SnoopY-enwiki, JabberWok, Jengelh, Anomaly1, SpuriousQ, Stephenb, Argentino, Gaius Cornelius, CambridgeBayWeather, Lavenderbunny, Morphh, Salsb, Tavilis, Anomalocaris, NawlinWiki, Wiki alf, Joshdboz, ErkDemon, ThunderE6, John Newbury, Joelr31, Thiseye, JocK, SCZenz, Irishguy, Nick, Aaron Brenneman, ArmadniGeneral, Jpowell, Ravedave, Eipipuz, Schmock, EverettColdwell, Raven4x4x, Moe Epsilon, Farmanesh, Swen, Lomn, Semperf, Beanyk, Raskolnikov The Penguin, Tony1, Bucketsofg, Linkofazeroth, Gadget850, DeadEyeArrow, Rjrawlings-enwiki, .marc., RyanJones, Mistercow, Wknight94, JECompton, SamuelRiv, Richardcavell, BazookaJoe, WAS 4.250, Light current, Albus Dumbledore-enwiki, Enormousdude, TheKoG, Chesnok, Ageekgal, Oysteinp, Chase me ladies, I'm the Cavalry, Theda, Ketsuekigata, Fang Aili, Brz7, Aeon1006, Alias Flood, CWenger, Alain r, LeonardoRob0t, Fram, HereToHelp, Emc2, JLaTondre, ArielGold, PhS, Caco de vidro, Stuhacking, Nsevs, Banus, RG2, Benandorsqueaks, Infinity0, GrinBot-enwiki, Serendipodous, DVD R W, DocendoDiscimus, Sardanaphalus, Snottily, MacsBug, SmackBot, Aim Here, Jo marie, Terrancommander, JoeCollver, Imz, Kurochka, Varunbhalerao, EinsteinIV, Brianyoumans, Tom Lougheed, Herostratus, Stellea, Prodego, Melchoir, MJMyers2-enwiki, Brokenfrog, Unyoyega, CyclePat, Jim62sch, Prototime, WilyD, KocjoBot-enwiki, Davewild, Silpion, Evanhatesspam, Canthusus, Hbackman, Jpvinall, Man with two legs, HalfShadow, Bb1, CorvinZahn, Aksi great, Gilliam, Ohnoitsjamie, Oscarthecat, GwydionM, Cabe6403, Cowman109, Saros136, Izehar, Scaife, Kurykh, Keegan, Basejumper123-enwiki, Temiree, Njerseyguy, Persian Poet Gal, Omghgomg, HubHikari, Jfsamper, Kungming2, DHN-bot-enwiki, Cassivs, Colonies Chris, Firetrap9254, Scwlong, Audriusa, Brainblaster52, Can't sleep, clown will eat me, Rludlow, Scott3, Jefffire, Oscar Bravo, Skidude9950, D roc16, Sephirothrr, God of War, TheKMan, Starexplorer, Andy120290, LeContexte, Kcordina, Ishanz, Grover cleveland, CamXV, Portcho, Khoikhoi, Jmlk17, Sloverlord, Flyguy649, Fuhghettaboutit, Iapetus, Mwmoretti, Tiki2099, Nakon, Jiddisch-enwiki, Brithackemack, Mustanglover, John D. Croft, Hoof Hearted, "alyosha", Aidepolcyene, Dream out loud, Pwjb, Richard001, Eran of Arcadia, Invincible Ninja, Uriel-238, Kellyprice, Maximum bobby, DMacks, Doooook, Daniel.Cardenas, Tangsyde, Pilotguy, Yevgeny Kats, Mega-Hasher, The undertow, SashatoBot, Nishkid64, Rory096, Beasterline, Robomaeyhem, Richard L. Peterson, John, Thedoj, Swlenz, Sfuerst, Philosophus, Dog Eat Dog World, Filthish, Cronholm144, Kipala, Alex Arnold, The Infidel, Rijkbenik, Soumyasch, Dhesi, Shadowlynk, AstroChemist, JorisvS, Mancroft, Mgigantus1, CredoFromStart, Shawdow, Berrick, Ben Moore, Zzzzzzzzzzz, Stratadrake, Slakr, TheHYPO, George.howitt, Optimale, Aeluwas, Childzy, Hypnosifl, InedibleHulk, Waggers, Mets501, Funnybunny, Ryulong, Serlin, Citicat, EEPROM Eagle, MTSbot-enwiki, SmokeyJoe, John F. Amitch, KJS77, Hetar, T boyd, TaggedJC, ILovePlankton, Buntykawale, Michaelbusch, TerryE, Clarityfiend, Abel Cavaşi, D Hill, Dreftymac, Joseph Solis in Australia, Chyko, JoeBot, T.O. Rainy Day, Newone, Icefox2k, Turbokoala, MOBle, UncleDouggie, Solipse, Fsotrain09, Tony Fox, CapitalR, Domitori, Tuttt, Humanperson0, Anger22, Dpeters11, Laplace's Demon, Rwst, Tawkerbot2, JRSpriggs, Filelakeshoe, Chetvorno, Cryptic C62, Flubeca, Hammer Raccoon, IronChris, Orangutan, Hsjawanda, Fvasconcellos, Kotepho, SkyWalker, Firehawk1717, JForget, Danras, CRGreathouse, Calmargulis, Nityann, Lenky, Crescentnebula, Capefeather, Wikifried, D.N.Parrish, ClovisHopman, Syphondu, Lmcelhiney, Benwildeboer, Green caterpillar, Crabnebula, ShelfSkewed, S.Bowen, Some P. Erson, Moreschi, Kjknohw, Rotiro, Terre, Cydebot, Natasha2006, Kanags, Zima65, Reywas92, Ramitmahajan, Gogo Dodo, Gagueci, JFreeman, Boardhead, Bazzargh, Wikipediarules2221, Difluoroethene, Jlmorgan, Dancter, Codingmasters, Michael C Price, Tdvance, Tawkerbot4, BMG-enwiki, DumbBOT, Chrislk02, Jay32183, Blm22, Narcosa, Omicronpersei8, Sharkbait784, Ephyon, Quophnix, Lo2u, Gimmetrow, Graham21kidd, Blobpic, BetacommandBot, CieloEstrellado, Thergvk, Thijs!bot, Lord Hawk, Crockspot, Mercury-enwiki, Qwyrxian, Waynesun, Markus Pössel, Kablammo, N5iln, Wahlin, Oerjan, Berria, MrXow, Headbomb, Newton2, Simeon H, Bobblehead, Kathovo, Tellyaddict, Cool Blue, Gvbn, Dfrg.msc, Infophile, Dgies, CharlotteWebb, Greg L, Srose, FreeKresge, Sam42, Wikidenizen, Anarchopedia, Dawnseeker2000, Elert, Escarbot, KrakatoaKatie, WikiSlasher, AntiVandalBot, Ais523, Macmanui, User's name, Majorly, Yonatan, Luna Santin, CodeWeasel, StantheGarbageMan, Yomangani, Voortle, Doc Tropics, Edokter, Phil.a, Messiah21, Dr. Submillimeter, Rsocol, LibLord, Danger, Science History, Glennwells, Spartaz, Daniels 9212, Archmagusrm, Byrgenwulf, Elaragirl, OGGVOB, Myanw, Lklundin, Uusitunnus, Kigali1, Bobvila2, Komponisto, MER-C, Never been to spain, Instinct, IanOsgood, Tonyrocks922, Davidpage, Andonic, 100110100, Cameron.walsh, Kirrages, Denimadept, Bigresearcher, LittleOldMe, DavidLaurenson, Acroterion, Yahel Guhan, Pervect, Gtation, Magioladitis, Mikemill, Gekedo, WolfmanSF, VoABot II, Raduberinde, Myopic, Antientropic, Praveenp, Farquaadhnchmn, Xeddy, Michele123, Lonewolf79 04, Loqi, Niele2006, Matt Bartlett74, SparrowsWing, BrianGV, JaKoBay, Crunchy Numbers, Giggy, Tuncrypt, Jeroje, Dyert, Mlsquad, Fluffy snowey, Disney freak!, Ceolwulf-enwiki, Chris G, DerHexer, Irishchieftain, Jomom, Patstuart, Jman73, Olsonist, Robin S, Joshua Davis, NatureA16, Otvaltak, DancingPenguin, Dr. Morbius, MartinBot, Mogus0226, Shentino, Vanessaezekowitz, Arjun01, NAHID, Gigaknight, UnfriendlyFire, John okell, Waynephinney, John Millikin, InnerJustice, Rettetast, Jay Litman, Filksinger, Loof1, Mschel, CommonsDelinker, 4.18GB, AlexiusHoratius, Pbroks13, MapleTree, Popeye Doyle, Siliconov, Sheila Rogers, LedgendGamer, Nuclearfusion, Ssolbergj, Natsirtguy, RockMFR, J.delanoy, Trusilver, Ledzep3012, Allbraves08, Svetovid, Rgoodermote, Philcha, JamesR, Bogey97, Wa3frp, Melamed katz, BillWSmithJr, Catmoongirl, Uncle Dick, Mike Winters, Jonpro, Qatter, Jreferee, Tomgibbons, Bumblebee55555, Lantonov, Dargaud, St.daniel, Turtlebean2, Mozzley, LordAnubisBOT, Whilding87, Jimbothechicken, Jerggp, Crakkpot, Zedmelon, Adam Snapp, Territory, ReekRend, AntiSpamBot, Lordaal, Plasticup, Mcaigjt, Anton1234, Glens userspace watcher, Goingstuckey, NewEnglandYankee, Ryan858, Charmander trainer, Cobi, Touch Of Light, Seanskusindinmamma, Fui fui moi moi2, Minesweeper.007, Han Solar de Harmonics, Angular, Brancron, BrettAllen, Reversepolarity, SBKT, Natl1, Gtg204y, TWCarlson, Andy Marchbanks, H1voltage, Lseixas, TKM625, Billebrooks, Idioma-bot, Sheliak, Azuriteking, Signalhead, Scunnane, ACSE, Cactus Guru, Vranak, Ironrooster, VolkovBot, Parker2010, Mocirne, Milenita-enwiki, WOSlinker, Wolfnix, Philip Trueman, Fran Rogers, DarkShroom, Canopus27, TXiKiBoT, Rollo44, Jreut, Mrkwtrs, Z.E.R.O., Anonymous Dissident, Ryan shell, Oplek, Italiandevil0505, Ask123, JayC, Vanished user ikijeirw34iuaeolaseriffic, Lradrama, Melsaran, Gekritzl, Corvus cornix, Mzmadmike, Abdullais4u, Driski555, Cremepuff222, PouponOnToast, Maxim, Maksdo, Zvbxrpl, Tfmmushroom, ViresetHonestas, Happycore3, Rex Imperator, Brittadudette, Andy Dingley, MP 12, Jon1992, Lamro, Rouhiheki, Miko3k, James McBride, Ridow, Falcon8765, Spinningspark, WatermelonPotion, Bigevan1, Brianga, Mike4ty4, Gunnville, Bobo The Ninja, Bk2001050, Wisamzaqoot, The Mad Genius, Radical Robert, San Diablo, Iamnotastarwarsfan, Bufrost, LuigiManiac, DarthBotto, Harshil8, Cowlinator, Sfmammamia, Callix, NHRHS2010, Nogood202, Kvncrtr, Steven Weston, D. Recorder, Bigev1, Brattbratt, Kbrose, Sureshonsearch, Tutszilla, Gaelen S., Bob freeman1, Lylefor, SieBot, BalanceRestored, Netgem21, Nabiki87, Timb66, Taftgod, NonChalance, JamesA, Work permit, Euryalus, I Like Cheeseburgers, Clissold07, Paradoctor, Bengal fan13, Joncam, Viskonsas, Caltas, ConfuciusOrnis, Poopstix, Smenge32, SolusX, Wayne317, Andersmusician, SiegeLord, Yulu, Ujjwol, Iames, Likebox, RadicalOne, Oysterguitarist, Poopypoopypiepie, Emperorfurkan,

JetLover, Hello4719, Stilkver, Godfinger, DevOhm, Oxymoron83, Bfesta14, Cmac16, Nuttycoconut, Canadianboyjd, Zharradan.angelfire, JBauer24, John fromer, Gangsterls, Lightmouse, Mydoggcoco, Poindexter Propellerhead, The Great Attractor, The-G-Unit-Boss, Benoni-Bot~enwiki, A kaldenhoven, DivineBurner, SteakNotShake, Dsmith7707, Coldcreation, Soulofdarknes01, David xie, Forser5, Jeroen888, Cosmo0, Randomblue, Hamiltondaniel, Movieguru2006, Vanished User 8902317830, Dust Filter, Thekingofspain, Payno, Gantuya eng, Phantomkaiser, Monmnom, Colin012, Dstebbins, Saltwell1986, Aidan180495, ArepoEn, Martarius, Beeblebrox, ClueBot, Stevekirst7, Robwalsh, Ander549, Suli1000, Andrew Nutter, PipepBot, Snigbrook, Scribble07, Patrickfongfong106, The Thing That Should Not Be, ArdClose, Kapohenry, Supersonicstars, Techdawg667, Vikasatkin, Wwheaton, CyrilThePig4, Arakunem, Andr0o, Drmies, Jimbo jones9, Control-alt-delete, Russ143, Chewlett, BlackJunebug, Mcnurse, Tyguth123, Boing! said Zebedee, Rotational, Agge1000, Phenylalanine, *blissfully ignorant*BETCHES, Aua, DragonBot, Snaxalotl, Ktr101, CohesionBot, Three-quarter-ten, GoldenGoose100, Carninia, Eujin16, Timsdad, Jemxia, Leonard^Bloom, Gwguffey, Josephmd, Cenarium, Nmoo, Bracton, Jotterbot, PhySusie, Scog, Dleiter, Jwaits12, C628, 3CUTiE--PiE, Thingg, Pisceesumsprecan, Lx 121, AC+79 3888, Trulystand700, Armhouse, DumZiBoT, TimothyRias, BarretB, Baron von HoopleDoople, Oldnoah, Psycholian, WikHead, Holoeconomics, Benjamnjoel2, SilvonenBot, Zetsubo666, Sweetpoet, Jd027, JinJian, ZooFari, MaizeAndBlue86, Fiskbil, ElMeBot, Lemmey, Parejkoj, Whtrz, Supermonkey443, Pogozelski123, Addbot, Lkvlamen, Crissyman, TheNightRyder, 11341134a, Uruk2008, DOI bot, JJ606, Snakeboy144, Gnatbuzz, Crazysane, Artie bristles, Jugbug2, Bte99, Groundsquirrel13, WFPM, Haasfelix, Proxima Centauri, Delaszk, Syber Sid, Debresser, AnnaFrance, LinkFA-Bot, Elen of the Roads, Prim Ethics, Harvardstudent, Blmichel, Ryttaren, Tide rolls, Whatintheworldisthat, OlEnglish, Potekhin, Samuel Pepys, ScienceApe, Snookerman, Krukouski, Legobot, Luckas-bot, Yobot, LoneRubberDragon, Bunnyhop11, Tohd8BohaithuGh1, Legobot II, Lolchanges, VZ9, Gum Stuck on Bottom of Shoe, Pigetrational, KamikazeBot, Rubin16, Ayrton Prost, Szajci, AnomieBOT, AndrooUK, Archon 2488, Grey Fox-9589, Message From Xenu, AdjustShift, Ornamentalone, Asoer, Powerzilla, Materialscientist, RobertEves92, Citation bot, Eumolpo, Palitzsch250, Xqbot, Meewam, Emerydora, DSisyphBot, Hanberke, Tad Lincoln, NASCAR Nathan, Runaway9995, UlmPhysiker, GrouchoBot, Mpe.mpg.de, ProtectionTaggingBot, Nlilovic, Omnipaedista, Kurtdriver, RibotBOT, Seeleschneider, Der Falke, JediMaster362, Moxy, WillMall, Imperators II, A. di M., Interstellar Man, Sesu Prime, FrescoBot, Feneeth of Borg, Akuvar, Originalwana, Goodbye Galaxy, Worrycharm, THENEWMIKON8ER, Europi3n, Mfwitten, Steve Quinn, Citation bot 2, Skull33, Robo37, Fruit.Smoothie, Citation bot 1, Javert, Careful With That Axe, Eugene, Gil987, Jonesey95, Tom.Reding, Achim1999, Concernedresident's butler, SpaceFlight89, Xaviertan, An elite, OldManNlck, Savemaxim, MertyWiki, Tempk, SanDiego7, Ashishg1984, TheInforment, Revenge12345678, Seattle Jörg, Nora lives, IVAN3MAN, Lemmiwinks2, Thames Aldwych W. Mines, Rajeev Goutam, Meier99, FoxBot, TobeBot, Belchman, Randomlogan, D climacus, Jordgette, Lolcakes1414, Williame3, Jamie s w, Extra999, EventHorizon5488, Spikescape, RjwilmsiBot, Mifield, Mrfencey, Hardikvasa, NamelsRon, Chriss.2, Mchcopl, Burmiester, Newty23125, Salvio giuliano, Billare, EmausBot, John of Reading, WikitanvirBot, JCRules, AlexUT, Grrow, Racerx11, Joseph507357, PoeticVerse, Dangoerman, Jmencisom, Challisrussia, Cpl-pike, Chricho, Italia2006, Hhhippo, Ida Shaw, Stanford96, Socioj, StringTheory11, Xabier Armendaritz, Nicolas Eynaud, H3llBot, Brandmeister, Y-barton, Crux007, ChuispastonBot, RockMagnetist, One.Ouch.Zero, Herk1955, ClueBot NG, Gilderien, Iloveandrea, Nijilravipp, Jj1236, Tabletrack, Garlikguy2, Rezabot, JoetheMoe25, Danim, Pluma, Helpful Pixie Bot, Asdfjkl1235, Bibcode Bot, BG19bot, Pine, Furkhaocean, Badon, Pascal yuiop, Cadiomals, Blaspie55, BattyBot, U-95, ChrisGualtieri, Khazar2, Ducknish, Dexbot, Webclient101, Mogism, Stas1995, Cerabot~enwiki, CuriousMind01, SFK2, Graphium, Cserez, Max14182000, Corn cheese, Among Men, Reatlas, Joeinwiki, Anastronomer, Rfassbind, Donfbreed2, Greengreengreenred, MatthewJ00, Smortypi, Light Peak, Ryenocerous, Jakec, Rolf h nelson, SuicideRider003, Space core192, Blakethecake333, Comp.arch, Kharkiv07, Baconfry, Ritviksaharan, Kogge, Mark Matthew Dalton, Anrnusna, Sudoiusudo, Signoredexter, Elenceq, Monkbot, Paul Masson, Garfield Garfield, SkyFlubbler, ChamithN, DangerousJXD, Freshness For Lettuce, Tetra quark, JuanLT2045, DN-boards1, Jerodlycett, Fogbannana, KasparBot, Ceannlann gorm, EternalNomad, CheeseStick1, Brandon Defrise Carter and Anonymous: 1420

- **Neutron star** *Source:* https://en.wikipedia.org/wiki/Neutron_star?oldid=687132683 *Contributors:* AxelBoldt, Tobias Hoevekamp, Ansible, Bryan Derksen, AstroNomer~enwiki, Manning Bartlett, Andre Engels, Roadrunner, DavidLevinson, Heron, Bdesham, JohnOwens, Tim Starling, Lquilter, Alfio, Todd, Looxix~enwiki, Caid Raspa, Jebba, Julesd, Glenn, Poor Yorick, Andres, Jonik, Timwi, Stismail, Doradus, Maximus Rex, Furrykef, Omegatron, Pakaran, Jni, Northgrove, Vt-aoe, Robbot, Mirv, Rebrane, Wikibot, Borislav, Jimduck, Jheise, David Gerard, Giftlite, Herbee, Xerxes314, Wwoods, Jacob1207, Alison, Wikibob, Niteowlneils, Leonard G., WalkinDownThirtyThree, Pascal666, Eequor, Lawfulhippo, Pne, Golbez, Physicist, Yath, Antandrus, HorsePunchKid, Russell E, Karol Langner, Balcer, Satori, Icairns, Sam Hocevar, Peter bertok, JohnArmagh, Scovetta, M1ss1ontomars2k4, Mike Rosoft, Rich Farmbrough, Cacycle, Vsmith, RJHall, Art LaPella, Mike Schwartz, Feitclub, Smalljim, Draco2, Foobaz, ParticleMan, La goutte de pluie, Cherlin, Haham hanuka, Quaoar, Alansohn, Andrew Gray, Axl, Malber, Amorymeltzer, Vuo, ShawnVW, Gene Nygaard, Ringbang, Feezo, Richard Arthur Norton (1958- ), Firsfron, Mindmatrix, StradivariusTV, BillC, Robert K S, Daniel Vollmer, Arzachel, Fxer, Gimboid13, Wisq, Christopher Thomas, Chinacat, Palica, NeonGeniuses, RichardWeiss, Ashmoo, Nanite, Drbogdan, Rjwilmsi, Zbxgscqf, Marasama, Mike Peel, Ligulem, Oo64eva, Titoxd, FlaBot, Patrick1982, Latka, SiriusB, Gark, RexNL, Gurch, KFP, Algri, Smithbrenon, DVdm, Guliolopez, Hall Monitor, Debivort, YurikBot, Wavelength, Spacepotato, Sceptre, Hairy Dude, AVM, Jengelh, Hellbus, Akamad, Gaius Cornelius, Eleassar, Salsb, Grafen, ZacBowling, Ondenc, FFLaguna, Jamesmcguigan, Brandon, Coderzombie, Raven4x4x, Lomn, Gadget850, Bota47, Djdaedalus, Mikespoff, FF2010, Lt-wiki-bot, Closedmouth, Arthur Rubin, Modify, JDC, Alain r, Sambe, Alemily, Fram, Curpsbot-unicodify, Garion96, Endymi0n, Darrel francis, JDspeeder1, GrinBot~enwiki, Airconswitch, The Yeti, That Guy, From That Show!, SmackBot, Doktor~enwiki, Pavlović, Elfsareus, Delldot, Bobzchemist, Hongshi, Onebravemonkey, Shai-kun, Yamaguchi⬚⬚, Pretendo, Peter Isotalo, Gilliam, GwydionM, Dauto, Blairck, Kmarinas86, Wikicali00, Chris the speller, Xpi6, Anchoress, SchfiftyThree, DHN-bot~enwiki, Darth Panda, Can't sleep, clown will eat me, Richardmilgate, Mwinog2777, Rrburke, Downwards, Nibuod, Kendrick7, Acdx, Spam Max, Andrei Stroe, PXE-M0F, IgWannA, Doug Bell, Titus III, Richard L. Peterson, FrozenMan, J 1982, Vampus, JorisvS, Mgiganteus1, Zzzzzzzzzz, Randomtime, Geologyguy, Ryulong, MTSbot~enwiki, Stephen B Streater, Olivierd, Alan.ca, Iridescent, Dekaels~enwiki, GDallimore, Zero sharp, Igoldste, Longlivefolkmusic, Tawkerbot2, Xcentaur, The Prince of Darkness, Pegasusbot, JPilborough, Myrrhlin, Van helsing, Eric, Ruslik0, GHe, NickW557, Megahmad, Skybon, Myasuda, Fl, DumbBOT, Mattjball, Ebyabe, Zalgo, Thijs!bot, Epbr123, Jman16984, Rotate, Markus Pössel, Oerjan, Headbomb, Marek69, "", Dfrg.msc, Greg L, Dawnseeker2000, Noclevername, Escarbot, AntiVandalBot, Luna Santin, Seaphoto, KP Botany, Bob 123456, JAnDbot, MER-C, Db099221, Fourchannel, Acroterion, WolfmanSF, Bongwarrior, VoABot II, Nyttend, Hekerui, LorenzoB, DerHexer, Patstuart, Wikianon, Lost tourist, Kheider, Gwern, MartinBot, Prgrmr@wrk, Anaxial, Mausy5043, J.delanoy, Maurice Carbonaro, Katalaveno, NewEnglandYankee, Supernova87a, Nwbeeson, Potatoswatter, Railwayfan2005, KylieTastic, Rlclark3, Dorftrottel, Steel1943, Lights, Maniaphobic, Vooz, Craigheinke, AlnoktaBOT, Traitor de, Dfpawlowski, TheOtherJesse, Philip Trueman, TXiKiBoT, OverSS, Red Act, MattieN, Anonymous Dissident, Erikev, BelsKr, Martin451, Jackfork, LeaveSleaves, Dirkbb, Redyoshi49q, Enviroboy, Anton Gutsunaev, Bryansworld, Edman007, Gspinoza, Nibios, Ceranthor, EmxBot, SieBot, PlanetStar, Scarian, Weeliljimmy, Bobbuilder3, Jsc83, Keilana, RadicalOne, Tiptoety, PieSnipa85, Jdaloner, Tombomp, J3nn r0ckz y0ur s0x, AMackenzie, Eouw0o83hf, Svick, Sevenman, Sean.hoyland, Hamiltondaniel, Tifaret, ImageRemovalBot, SpectrumAnalyser, Martarius, ClueBot, Trojancowboy, Cigarshaped, Nootron44, Eugene chan, RonBeeCNC, Foxj, Rjd0060, Beyonda, Adrianwn, Mild Bill Hiccup, Polyamorph, Wikisteff, Rrsilbar, Cirt, PixelBot, Vandalz0rs, Thebeast373, Brews ohare, NuclearWarfare, Millionsandbillions, Jotterbot, Njardarlogar, Dmattscott, Unmerklich, Tauris, MelonBot, SoxBot III, DumZiBoT, IngerAlHaosului, Birbikram, WikiDao, Jojo96, Addbot, Huntered, Denali134, AVand, Sygnett, DOI bot, Hda3ku, Elaak, Fieldday-sunday, NjardarBot, Cst17, J1138fleming, Favonian, 84user, Tide rolls, Nicholas Crestone, Lightbot, OlEnglish, Qemist, Potekhin, Gail, Legobot, Clay Juicer, Luckas-bot, Yobot, Crispmuncher, Nallimbot, Robert Treat, Synchronism, AnomieBOT, Jim1138, JackieBot, RBM 72, BCROCE, Materialscientist, USConsLib, Cababunga, The High Fin Sperm Whale, Citation bot, E2eamon, Marshallsumter, Xqbot, Alexa7890, Jnatar, What!?Why?Who?, Mouagip, Dayman2, Armbrust, Ribot-BOT, NoRad, Moxy, Hakunamenta, DeNoel, FalconL, Originalwana, Amilnerwhite, Ilovecatsnicole, Zero Thrust, Cannolis, Citation bot 1, Lanulos, Biker Biker, Adlerbot, Tom.Reding, Nacen, MondalorBot, SpaceFlight89, Elentirno, Xeworlebi, Jauhienij, IVAN3MAN, Trappist the monk, Zonafan39, Mono, Michael9422, Extra999, HawkE65, Ioan Wynne-Jones, Zucchini7, Рулин, TjBot, Jpatros, EmausBot, Vanished user zq46pw21, Tommy2010, K6ka, Hhhippo, ZéroBot, Josve05a, MithrandirAgain, StringTheory11, Druzhnik, Medeis, Cline92, Lone St4lk3r, David J Johnson, Coolbob2422, KKPie, Inka 888, Zueignung, Mainy1996, Carmichael, JavinComi, ChuispastonBot, Robertschulze, Zeta ζ, Czeror, Mhvk, Whoop whoop pull up, Madibootay, ClueBot NG, Gilderien, Frietjes, Jlattimer, Doctree, Helpful Pixie Bot, Calabe1992, Bibcode Bot, Regulov, IzackN, BG19bot, Albert instine, Vicky.singh092, MusikAnimal, Exobiologist, FiveColourMap, Robert the Devil, Cadiomals, Shulse123, GregorDS, Thisdick1300794318, MrJohnnyMorales, Chowder98100, Zedshort, Doctorwhofan2013, Danijm, BattyBot, Cyberbot II, EuroCarGT, Wtrebla, G.Kiruthikan, Zacattack147, Reatlas, Joeinwiki, Randompoopcake, Rfassbind, Chris90nz, Sageattorney, Tentinator, Brobof, Nestrs, Raphael.concorde, Jwratner1, Johndric Valdez, Mfb, Fuckyourmomma2, Signoredexter, Monkbot, Raichu234352, Christometh, Sebgod, Steampunk09, AntHerder, SkyFlubbler, Poiuytrewqvtaatv123321, Tetra quark, Des2020, Rmh2020, Corvus-TAU, Znbn, Pulkitmidha, KasparBot, Ilikchese, Mosovon64 and Anonymous: 543

- **Electron capture** *Source:* https://en.wikipedia.org/wiki/Electron_capture?oldid=686919015 *Contributors:* Mav, Andre Engels, Imran, GaryW, Pstudier, Twang, Donarreiskoffer, Gentgeen, Romanm, SpellBott, Mikez, Art Carlson, Dratman, Icairns, B.d.mills, Hax0rw4ng, Newhoggy, Discospinster, Vsmith, Sunborn, Joanjoc~enwiki, Brim, Foobaz, Riana, DV8 2XL, Forteblast, Richard Arthur Norton (1958- ), Benbest, Rjwilmsi, Chobot, DVdm, Tone, YurikBot, Spacepotato, Hairy Dude, Shawn81, Shaddack, Anomalocaris, Dna-webmaster, Modify, LeonardoRob0t, Incnis Mrsi, Jagged 85, Betacommand, Sbharris, Vladis1av, BIL, Drphilharmonic, Daniel.Cardenas, Untitleduser, C.jeynes, Diverman, Magere Hein, Icek~enwiki, Michael C Price, Headbomb, Hcobb, Roches, Dirac66, LorenzoB, Vinograd19, AstroHurricane001, Howa0082, Yonidebot, Jutiphan, Vatie7, Sheliak, VolkovBot, TXiKiBoT, Pamputt, SieBot, YonaBot, Flyer22 Reborn, ClueBot, Cmj91uk, SchreiberBike, Oldnoah, NellieBly, SkyLined, Debzer, Addbot, LaaknorBot, Zorrobot, Skippy le Grand Gourou, Luckas-bot, AnomieBOT, ArthurBot, Xqbot, GrouchoBot, FrescoBot, PigFlu Oink, Minivip, Miracle Pen, AndyHe829, MartinThoma, Bibcode Bot, Snow Rise, Eio, Zedshort, Pani Slepičková, JPBrod, Maysens, BsGTeo, Spyglasses, Meteor sandwich yum, Monkbot, Haveasweater, Jsaur, Wqwt, Alma.f.r, KasparBot and Anonymous: 53

- **Pair-instability supernova** *Source:* https://en.wikipedia.org/wiki/Pair-instability_supernova?oldid=678404418 *Contributors:* Roadrunner, Arkuat, FT2, Bender235, Tom, Keflavich, BRW, Mazca, Pol098, Rjwilmsi, Strait, McGinnis, Hellbus, Georgewilliamherbert, 2over0, Modify, MacsBug, Sewlong, Rogermw, Henning Makholm, OhioFred, JorisvS, Ruslik0, Dtgriscom, CosineKitty, WolfmanSF, Paulshikleejr, Warrickball, Dorftrottel, TXiKiBoT, HarryAlffa, Wjhudson, Nergaal, ClueBot, Tarlneustaedter, Drolz09, Addbot, Luckas-bot, KamikazeBot, Robert Treat, Xqbot, Lithopsian, Rainald62, FrescoBot, Cs32en, Tom.Reding, RjwilmsiBot, EmausBot, Hanavy, Bibcode Bot, BattyBot, Dexbot, Makecat-bot, Parkus.aurelius, Femkemilene, Monkbot, RasverixX82 and Anonymous: 29

- **Photodisintegration** *Source:* https://en.wikipedia.org/wiki/Photodisintegration?oldid=678821109 *Contributors:* Julesd, Gamma~enwiki, Stone, Dratman, Julianonions, Rich Farmbrough, Strait, Lockesdonkey, Georgewilliamherbert, Dan337, Modify, King Hildebrand, A876, Headbomb, Orionus, PloniAlmoni, Morngnstar, LorenzoB, Pekaje, DAID, Sheliak, Addbot, ماني, Luckas-bot, Robert Treat, AnomieBOT, Citation bot, Trewal, Tom.Reding, EmausBot, WikitanvirBot, Helpful Pixie Bot, Pooh12550, Bibcode Bot, Zedshort, Nitrobutane, Lucquessoy and Anonymous: 15

- **Metallicity** *Source:* https://en.wikipedia.org/wiki/Metallicity?oldid=687816047 *Contributors:* Chrislintott, Roadrunner, Ahoerstemeier, Cherkash, Reddi, Donreed, Rorro, Rursus, Angilbas, Antandrus, Karol Langner, RetiredUser2, B.d.mills, FT2, RJHall, Worldtraveller, Mpvdm, Quaoar, Gene Nygaard, Zanaq, FeanorStar7, Pol098, Jeff3000, Mpatel, RichardWeiss, Drbogdan, Rjwilmsi, Angusmclellan, Tar-Palantir, Adaj, Joffan, YurikBot, Hairy Dude, Red Slash, Chris Capoccia, Gaius Cornelius, Ospalh, Dbfirs, Serendipodous, SmackBot, Saravask, Dav2008, Unyoyega, Eskimbot, Saros136, Someonesdad363616, Rogermw, Xwinger, Weregerbil, SashatoBot, Erimus, JorisvS, Zzzzzzzzzzz, Richard Nowell, IvanLanin, Mssgill, Mmmpotatoes, Dia^, Hilmarz, CRGreathouse, Michael C Price, Christian75, Crum375, Thijs!bot, Keraunos, Headbomb, Peter Gulutzan, Orionus, Paul from Michigan, JAnDbot, BSVulturis, Soulbot, Hekerui, PsyMar, John Millikin, Nando.sm, TechnoFaye, CommonsDelinker, Fcsuper, GravityFong, Hans Dunkelberg, Nigholith, Ian.thomson, Rod57, Janus Shadowsong, NewEnglandYankee, Knulclunk, Ohms law, Quominus, Spiral Wave, VoidLurker, DorganBot, Steel1943, Sheliak, VolkovBot, Larryisgood, Harfarhs, Arnd Klotz, A4bot, Mzmadmike, Markp93, UnitedStatesian, DrSpiff, Oliepedia, Trefusius, SieBot, Timb66, Phe-bot, Dead Fish Jr, Moletrouser, Michael.Urban, ChandlerMapBot, Scog, MelonBot, Arianewiki1, Coder11235, SilvonenBot, Hobbema, CosmologyProfessor, Addbot, Mr0t1633, DOI bot, AstroDave, Bigzteve, Lightbot, OlEnglish, Zorrobot, Tosca23, Yobot, Robert Treat, AnomieBOT, Citation bot, Maxis ftw, LilHelpa, Xqbot, Gap9551, Almabot, Mnmngb, Moxy, FrescoBot, AllCluesKey, Gaba p, I dream of horses, Tom.Reding, Kfk13, TjBot, Yaush, WildBot, Steve03Mills, WikitanvirBot, Dewritech, Primefac, Hidrofago, Hhhippo, Claudio M Souza, StringTheory11, SporkBot, Brownie Charles, Whoop whoop pull up, TheRealLanceManly, ClueBot NG, Alex Nico, Tideflat, Sephirohq, Bibcode Bot, DBigXray, Zedshort, BattyBot, Mooratov, Reatlas, Epicgenius, Praemonitus, Robevans123, Zorrible, KimPG, Tetra quark, Robj37 and Anonymous: 98

- **Supergiant** *Source:* https://en.wikipedia.org/wiki/Supergiant?oldid=682981306 *Contributors:* Roadrunner, Peterlin~enwiki, Zoe, Zimriel, Ixfd64, Alfio, Looxix~enwiki, Ahoerstemeier, Julesd, Netsnipe, Pizza Puzzle, Schneelocke, Nickshanks, Imesj, RedWolf, Rursus, Jyril, Cam, Icairns, Discospinster, Nk, Jeodesie, Dillee1, Alansohn, Redfarmer, Pauli133, Saxifrage, Jacen Aratan, Etacar11, Terence, Steinbach, Palica, Ketiltrout, FlaBot, Patrick1982, Margosbot~enwiki, Algri, Chobot, Spacepotato, Gaius Cornelius, Jim Apple, Argo Navis, Luk, Attilios, KnightRider~enwiki, SmackBot, Ashill, Moeron, Eskimbot, Commander Keane bot, John Hyams, Nakon, JephSullivan, OhioFred, JorisvS, Hans van Deukeren, Feureau, Doczilla, Tawkerbot2, JForget, Preetikapoor0, Thijs!bot, Epbr123, Al Lemos, Natalie Erin, Insulanus, Scepia, Smartse, Spencer, JAnDbot, VoABot II, Captain panda, Jeepday, (jarbarf), Tygrrr, Idioma-bot, VolkovBot, Toddles29, Philip Trueman, TXiKiBoT, Mawkernewek, Ferengi, BotKung, Happysailor, Escape Orbit, Martarius, ClueBot, Fyyer, Mild Bill Hiccup, BlueAmethyst, Ktr101,

Waldorfgx, Dabomb87, Grbsokk, Twinsday, Martarius, Sfan00 IMG, ClueBot, Artichoker, The Thing That Should Not Be, Rodhullandemu, Wwheaton, Jappalang, Jsbloom1, Hceline, Pkubanek, Otolemur crassicaudatus, Piledhigheranddeeper, Cirt, House13, Excirial, Alexbot, BobKawanaka, Estirabot, Torsmo, Taranet, DumZiBoT, Adacore, Jehochman2, Sstankowitz, Ikihi123, Grandkalyan, MaizeAndBlue86, Kbdankbot, Addbot, DOI bot, Khamosh, Ronhjones, Vishnava, Mohamed Magdy, Glane23, LinkFA-Bot, Tyw7, Zorrobot, Luckas-bot, Yobot, Ptbotgourou, Legobot II, Buddy431, Baxxterr, AnomieBOT, Rubinbot, Grburster, Materialscientist, GammaRayBurst, Citation bot, ArthurBot, LilHelpa, Xqbot, Tripodian, Addhockey10, DSisyphBot, Manojtd, Gap9551, Lithopsian, Papalew, Astro Reeves, GrouchoBot, Doulos Christos, Rabsmith, Ngbingsheng, Captain Cheeks, CANCER ONE, ايلى2010, D'ohBot, Ysyoon, AstaBOTh15, Hallucegenia, Dogaru Florin, Pinethicket, Jonesey95, Tom.Reding, MJ94, Σ, Fartherred, SkyMachine, IVAN3MAN, JMMuller, Vprashanth87, Michael9422, Pbrower2a, AndrewvdBK, Earthandmoon, RjwilmsiBot, Mathewsyriac, Antioch07, EmausBot, WikitanvirBot, Look4light, RA0808, Jmencisom, Passionless, Tommy2010, Wikipelli, Thecheesykid, John Cline, Illegitimate Barrister, Jenks24, NicatronTg, NinjaFishy, Quondum, MikeBlockCPA, AManWithNoPlan, Makecat, Ihardlythinkso, One.Ouch.Zero, Rocketrod1960, IanAbel, ClueBot NG, Gsmyth99, Ilovefatkids, Dragodinobub, HarryBowman, Euty, Danim, Crazymonkey1123, Bibcode Bot, BG19bot, ASDosar, KateWishing, Scottyhoohow, Dr meetsingh, Wikih101, Mimzy89, Quickcrazy78, Tsvipiran, Zedshort, BattyBot, Mdann52, BrightStarSky, Dexbot, Acoma Magic, Reatlas, Harrybraviner, Loganfalco, Madreterra, Rolf h nelson, Brucegendre, Jwratner1, Impsswoon, Owllord97, Monkbot, Bbzhang, I am. furhan., KasparBot and Anonymous: 329

- **Hypernova** *Source:* https://en.wikipedia.org/wiki/Hypernova?oldid=687565329 *Contributors:* Bryan Derksen, SimonP, Maury Markowitz, Stevertigo, Glenn, Wikiborg, Morwen, Geraki, Owen, Twang, Robbot, 1984, Xanzzibar, Timvasquez, MPF, Laudaka, Mooquackwooftweetmeow, ConradPino, Beland, GeoGreg, Jafro, Deadlock, Urvabara, Gadykozma, Vsmith, Warpflyght, Zaslav, Kbh3rd, Aranel, Tom, 23skidoo, Cwolfsheep, MetalMilitia, Dillee1, 578, Anthony Appleyard, Slugmaster, GeorgeStepanek, Computerjoe, Gene Nygaard, Zereshk, Siafu, FeanorStar7, Etacar11, Jason Palpatine, Kamal3, Pol098, JHBledsoe, Aristotle Pagaltzis, Jugger90, Christopher Thomas, Palica, Emerson7, RuM, Eteq, Mike Peel, Jehochman, Krash, Mordecai, Wildespace, Patrick1982, Charliemouse, TeaDrinker, Jameswilmot2000, Wrightbus, Zotel, YurikBot, Xihr, Shawn81, Trovatore, Mysid, Robost, Smkolins, Georgewilliamherbert, Geoffrey.landis, Groyolo, The Yeti, Attilios, SmackBot, Ashill, RedSpruce, Declare, GreggHB, Hibernian, Sbharris, Modest Genius, Mrwuggs, Addshore, Elendil's Heir, Wen D House, Akriasas, Weregerbil, Victor Lopes, Roger.lee, Lambiam, Doug Bell, Petsoukos-enwiki, Zearin, ML5, Svartkell, JorisvS, Zzzzzzzzzz, Agent gruer, D S H, Muadd, Rock4arolla, PRRfan, Iridescent, Exander, Cryptic C62, CmdrObot, Jom-enwiki, Azndragonemperor, Cydebot, Gogo Dodo, Omicronpersei8, Sharkbait784, Thijs!bot, Keraunos, Nonagonal Spider, Dtgriscom, Oosh, Gioto, Jj137, Dr. Submillimeter, Grant Gussie, E Pluribus Americanus, Magioladitis, VoABot II, Pyromancer102, Vanished user qjewfin34thsdjskjwjh423e, DGG, NatureA16, STBot, Tuganax-enwiki, RTBoyce, NewEnglandYankee, Vranak, VolkovBot, Cygfrydd Llewellyn, TXiKiBoT, Malljaja, Mzmadmike, The Mad Genius, Gbawden, Sonicology, Coati123, Andrewjlockley, Oda Mari, Twinsday, Martarius, Elsweyn, Ktr101, AssegaiAli, Hypercott, DumZiBoT, Heironymous Rowe, Lycanthrope321, PSimeon, Stickee, WikHead, SilvonenBot, MystBot, MaizeAndBlue86, Skeletor 0, Addbot, Roentgenium111, Auspex1729, Vikiçizer, Jasper Deng, Bfigura's puppy, Lightbot, Zorrobot, Luckas-bot, Yobot, Zoe17, The Vector Kid, AnomieBOT, Shockna, Icalanise, Materialscientist, Citation bot, ArthurBot, GnawnBot, Xqbot, Gap9551, Lithopsian, Papalew, Armbrust, Coosbane, Bez, VS6507, Hallucegenia, Adlerbot, Abductive, Tom.Reding, RedBot, Piandcompany, IVAN3MAN, عباد ديوان مجاهد ة, Trappist the monk, Miracle Pen, Ink Falls, Mean as custard, Slightsmile, Wikipelli, Hhhippo, ZéroBot, StringTheory11, TheSusy, Whoop whoop pull up, ClueBot NG, Gob Lofa, Bibcode Bot, BG19bot, Wikimaniac458161, Mynameisnoted, Shah jay vipul, Trevayne08, Vractomorph1, Duxwing, EricEnfermero, Dexbot, Webclient101, Effy Shaf, Reatlas, AmaryllisGardener, Smellygarb, GianXXIV, Varkman, Aubreybardo, Ajbilan, JHobson3, Monkbot, Blackninja23, PyroRodgers123, AbHiSHARMA143, SkyFlubbler, C.inserra, Blank25007, Porthdude and Anonymous: 157

- **Astrophysical jet** *Source:* https://en.wikipedia.org/wiki/Astrophysical_jet?oldid=685905941 *Contributors:* Ixfd64, AWhiteC, JackofOz, PBP, Curps, Worldtraveller, Axeman89, Christopher Thomas, Chobot, StuRat, Mhardcastle, SmackBot, Jrockley, Gilliam, Kurtan-enwiki, Cydebot, Alaibot, RogierBrussee, Leyo, Mathglot, GlassCobra, Martarius, Mild Bill Hiccup, Silent Key, Langphysics, Feline Hymnic, PauloHelene, Addbot, Basilicofresco, Ka Faraq Gatri, Protonk, Chzz, Zorrobot, Yobot, AnomieBOT, Citation bot, Gap9551, Racerx11, AManWithNoPlan, Bibcode Bot, StarryGrandma, CuriousMind01, Comp.arch, Kogge, Elenceq, Monkbot, Unome101, Heinerj and Anonymous: 10

- **Binary star** *Source:* https://en.wikipedia.org/wiki/Binary_star?oldid=687555772 *Contributors:* Mav, Bryan Derksen, Jeronimo, PierreAbbat, Heron, Patrick, Ixfd64, Alfio, Looxix-enwiki, Nikai, Alaric, Smack, Vanished user 5zariu3jisj0j4irj, Timwi, Jallan, Rvolz, Tpbradbury, Jeffrey Smith, Jose Ramos, Robbot, Hankwang, Rursus, Auric, Rtfisher, Hadal, Nerval, Giftlite, MSGJ, Anville, Curps, Henry Flower, Radius, Bobblewik, Edcolins, Rrw, Antandrus, The Singing Badger, Madmagic, Quarl, Kusunose, Karol Langner, Icairns, SamSim, Urhixidur, Janneok-enwiki, Canterbury Tail, ChrisRuvolo, Qutezuce, Bender235, RJHall, El C, Walden, Kwamikagami, Skeppy, Pilatus, PhilHibbs, Giraffedata, La goutte de pluie, Sasquatch, BillCook, Jumbuck, Jim Spinner, Alansohn, Smegpt86, Keenan Pepper, Andrewpmk, Ricky81682, FDJf, Denniss, MrBudgens, Stephen Hodge, Suruena, Simon Dodd, Zxcvbnm, Zoohouse, Computerjoe, Achievist, Nick Mks, WilliamKF, Woohookitty, FeanorStar7, Miaow Miaow, BillC, Pol098, Taragui, Ianweller, Fxer, Palica, Marskell, BD2412, Yurik, Rjwilmsi, Susan Davis, Zbxgscqf, Quiddity, BlueMoonlet, Mike s, Mike Peel, Ligulem, Brighterorange, Fish and karate, RobertG, Margosbot-enwiki, Alfred Centauri, Gurch, Algri, Windharp, Chobot, Jared Preston, Gplefka, YurikBot, Wavelength, Spacepotato, Stephenb, Gaius Cornelius, CambridgeBayWeather, NawlinWiki, Irishguy, Doctorindy, Zwobot, Varano, Tirerim, Jcrook1987, Maximusveritas, Phgao, Lt-wiki-bot, Ageekgal, Chaos syndrome, Spliffy, Raveled, Ilmari Karonen, Katieh5584, DVD R W, WindFish, SmackBot, RDBury, Bigbluefish, Drorbn, David Shear, AndyZ, Nickst, Jrockley, Cuddlyopedia, Sloman, Gilliam, Hmains, B00P, DHN-bot-enwiki, Colonies Chris, Verrai, Modest Genius, John Hyams, Fred Rasio, OrphanBot, Binarystarmusic, Bowlhover, Red1-enwiki, Zawthet, IRua, Iridescence, Just plain Bill, Andrei Stroe, SashatoBot, CFLeon, Robomaeyhem, Doug Bell, Srikeit, Sophia, J 1982, JorisvS, Mgiganteus1, Caviare, Zzzzzzzzzz, 81120906713, Segurador, Artman40, Bay Flam, Wjejskenewr, Courcelles, DKqwerty, LessHeard vanU, Olaf Davis, Jokes Free4Me, Sax Russell, Moreschi, Laura S, Stebbins, Kanags, Tawkerbot4, Brad101, Omicronpersei8, Robertsteadman, Thijs!bot, Epbr123, Lord Hawk, Headbomb, Tellyaddict, Davidhorman, TarkusAB, Vala M, AntiVandalBot, The Obento Musubi, Uvaphdman, Fru1tbat, Dr. Submillimeter, Kevin Nelson, Praxle1, ZIPZAPZOOM, Raúl G, Jordan8800, Blackpeoplewilldielol, JAnDbot, Deflective, Murgh, VoABot II, Soulbot, SwiftBot, Pyromancer102, Xtifr, Kheider, Geboy, MartinBot, Schmloof, CarlFeynman, Creol, J.delanoy, DrKay, Skeptic2, Peter Chastain, Hans Dunkelberg, Mellowship, Noelle65, KylieTastic, Treisijs, Xiahou, MadMathematician, Jeter751, VolkovBot, CWii, Pleasantville, RingtailedFox, Dirkterrell, AlnoktaBOT, Philip Trueman, TXiKiBoT, A4bot, Ryan shell, Jonasmike, Martin451, BotKung, Cwilliamsdog, RobertFritzius, Fotek, SieBot, Timb66, Euryalus, WereSpielChequers, ToePeu.bot, VVVBot, Evleos-enwiki, Lightmouse, The Stickler, Janggeom, Anchor Link Bot, Denisarona, Freewayguy, ClueBot, GorillaWarfare, WurmWoode, Mattjones16, Wysprgr2005, Mild Bill Hiccup, Bbb2007, NuclearWarfare, An-

## 39.1.2 Images

- **File:Flag_of_Pakistan.svg** *Source:* https://upload.wikimedia.org/wikipedia/commons/3/32/Flag_of_Pakistan.svg *License:* Public domain *Contributors:* The drawing and the colors were based from flagspot.net. *Original artist:* User:Zscout370

- **File:Flag_of_Portugal.svg** *Source:* https://upload.wikimedia.org/wikipedia/commons/5/5c/Flag_of_Portugal.svg *License:* Public domain *Contributors:* http://jorgesampaio.arquivo.presidencia.pt/pt/republica/simbolos/bandeiras/index.html#imgs *Original artist:* Columbano Bordalo Pinheiro (1910; generic design); Vítor Luís Rodrigues; António Martins-Tuválkin (2004; this specific vector set: see sources)

- **File:Flag_of_Russia.svg** *Source:* https://upload.wikimedia.org/wikipedia/en/f/f3/Flag_of_Russia.svg *License:* PD *Contributors:* ? *Original artist:* ?

- **File:Flag_of_South_Korea.svg** *Source:* https://upload.wikimedia.org/wikipedia/commons/0/09/Flag_of_South_Korea.svg *License:* Public domain *Contributors:* Ordinance Act of the Law concerning the National Flag of the Republic of Korea, Construction and color guidelines (Russian/English) ← This site is not exist now.(2012.06.05) *Original artist:* Various

- **File:Flag_of_Spain.svg** *Source:* https://upload.wikimedia.org/wikipedia/en/9/9a/Flag_of_Spain.svg *License:* PD *Contributors:* ? *Original artist:* ?

- **File:Flag_of_Sweden.svg** *Source:* https://upload.wikimedia.org/wikipedia/en/4/4c/Flag_of_Sweden.svg *License:* PD *Contributors:* ? *Original artist:* ?

- **File:Flag_of_Switzerland.svg** *Source:* https://upload.wikimedia.org/wikipedia/commons/f/f3/Flag_of_Switzerland.svg *License:* Public domain *Contributors:* PDF Colors Construction sheet *Original artist:* User:Marc Mongenet

  Credits:

- **File:Flag_of_Ukraine.svg** *Source:* https://upload.wikimedia.org/wikipedia/commons/4/49/Flag_of_Ukraine.svg *License:* Public domain *Contributors:* ДСТУ 4512:2006 - Державний прапор України. Загальні технічні умови

  SVG: 2010

  *Original artist:* України

- **File:Flag_of_the_Czech_Republic.svg** *Source:* https://upload.wikimedia.org/wikipedia/commons/c/cb/Flag_of_the_Czech_Republic.svg *License:* Public domain *Contributors:*

  - -xfi-'s file
  - -xfi-'s code
  - Zirland's codes of colors

  *Original artist:*
  (of code): SVG version by cs:-xfi-.

- **File:Flag_of_the_Netherlands.svg** *Source:* https://upload.wikimedia.org/wikipedia/commons/2/20/Flag_of_the_Netherlands.svg *License:* Public domain *Contributors:* Own work *Original artist:* Zscout370

- **File:Flag_of_the_People'{}s_Republic_of_China.svg** *Source:* https://upload.wikimedia.org/wikipedia/commons/f/fa/Flag_of_the_People%27s_Republic_of_China.svg *License:* Public domain *Contributors:* Own work, http://www.protocol.gov.hk/flags/eng/n_flag/design.html *Original artist:* Drawn by User:SKopp, redrawn by User:Denelson83 and User:Zscout370

- **File:Flag_of_the_United_Kingdom.svg** *Source:* https://upload.wikimedia.org/wikipedia/en/a/ae/Flag_of_the_United_Kingdom.svg *License:* PD *Contributors:* ? *Original artist:* ?

- **File:Flag_of_the_United_States.svg** *Source:* https://upload.wikimedia.org/wikipedia/en/a/a4/Flag_of_the_United_States.svg *License:* PD *Contributors:* ? *Original artist:* ?

- **File:Folder_Hexagonal_Icon.svg** *Source:* https://upload.wikimedia.org/wikipedia/en/4/48/Folder_Hexagonal_Icon.svg *License:* Cc-by-sa-3.0 *Contributors:* ? *Original artist:* ?

- **File:Fusion_rxnrate.svg** *Source:* https://upload.wikimedia.org/wikipedia/commons/d/d0/Fusion_rxnrate.svg *License:* CC BY 2.5 *Contributors:* Own work *Original artist:* Dstrozzi

- **File:FusionintheSun.svg** *Source:* https://upload.wikimedia.org/wikipedia/commons/7/78/FusionintheSun.svg *License:* CC BY-SA 3.0 *Contributors:* Own work *Original artist:* Borb

- **File:G299-Remnants-SuperNova-Type1a-20150218.jpg** *Source:* https://upload.wikimedia.org/wikipedia/commons/b/bd/G299-Remnants jpg *License:* Public domain *Contributors:* http://www.nasa.gov/sites/default/files/thumbnails/image/g299.jpg *Original artist:* NASA/CXC/U.T

- **File:GKPersei-MiniSuperNova-20150316.jpg** *Source:* https://upload.wikimedia.org/wikipedia/commons/1/11/GKPersei-MiniSuperNova-jpg *License:* Public domain *Contributors:* http://www.nasa.gov/sites/default/files/thumbnails/image/gkper.jpg *Original artist:* X-ray: NASA/ et al;Optical: NASA/STScI;Radio: NRAO/VLA

- **File:GRB080319B_illustration_NASA.jpg** *Source:* https://upload.wikimedia.org/wikipedia/commons/8/83/GRB080319B_illustration_NASA.jpg *License:* Public domain *Contributors:* http://imagine.gsfc.nasa.gov/docs/features/news/10sep08.html[1] *Original artist:* NASA/Swift /MaryPat Hrybyk-Keith and John Jones

- **File:GRB_BATSE_12lightcurves.png** *Source:* https://upload.wikimedia.org/wikipedia/commons/e/ef/GRB_BATSE_12lightcurves.png *License:* Public domain *Contributors:* Own work *Original artist:* Daniel Perley

- **File:Galaxies_AGN_Inner-Structure-of.jpg** *Source:* https://upload.wikimedia.org/wikipedia/commons/4/40/Galaxies_AGN_Inner-jpg *License:* CC-BY-SA-3.0 *Contributors:* en.wikipedia *Original artist:* Mrbrak

- **File:Galaxy_morphology.jpg** *Source:* https://upload.wikimedia.org/wikipedia/commons/8/85/Galaxy_morphology.jpg *License:* CC-BY-SA-3.0 *Contributors:* http://www.ipac.caltech.edu/2mass/gallery/galmorph/ *Original artist:* Dr. T.H. Jarrett (Caltech)

- **File:Gamma_ray_burst.jpg** *Source:* https://upload.wikimedia.org/wikipedia/commons/6/63/Gamma_ray_burst.jpg *License:* Attribution *Contributors:* National Science Foundation Press Release 05-156: Gamma-Ray Burst Smashes a Record *Original artist:* Nicolle Rager Fuller of the NSF

- **File:Gwiazda_podwójna_zaćmieniowa_schemat.svg** *Source:* https://upload.wikimedia.org/wikipedia/commons/3/3d/Gwiazda_podw%C3%B3jna_za%C4%87mieniowa_schemat.svg *License:* CC-BY-SA-3.0 *Contributors:* Own work *Original artist:* MesserWoland

- **File:HR-diag-no-text-2.svg** *Source:* https://upload.wikimedia.org/wikipedia/commons/6/67/HR-diag-no-text-2.svg *License:* CC-BY-SA-3.0 *Contributors:* Modified version of Image:HR-diag-no-text.svg, written by User:Rursus *Original artist:* User:Spacepotato

- **File:HRDiagram.png** *Source:* https://upload.wikimedia.org/wikipedia/commons/6/6b/HRDiagram.png *License:* CC BY-SA 2.5 *Contributors:* The Hertzsprung Russell Diagram *Original artist:* Richard Powell

- **File:HST_SN_1987A_20th_anniversary.jpg** *Source:* https://upload.wikimedia.org/wikipedia/commons/5/50/HST_SN_1987A_20th_anniversary.jpg *License:* Public domain *Contributors:* http://hubblesite.org/newscenter/archive/releases/2007/10/image/a/(direct link) *Original artist:* NASA,ESA,P. Challis,and R. Kirshner(Harvard-Smithsonian Center for Astrophysics)

- **File:He1523a.jpg** *Source:* https://upload.wikimedia.org/wikipedia/commons/5/5f/He1523a.jpg *License:* CC BY 4.0 *Contributors:* http://www.solstation.com/x-objects/he1523.htm *Original artist:* ESO, European Southern Observatory

- **File:Herbig-Haro_object_HH_110.jpeg** *Source:* https://upload.wikimedia.org/wikipedia/commons/b/bd/Herbig-Haro_110_%28captured_by_the_Hubble_Space_Telescope%29.tif *License:* Public domain *Contributors:* http://hubblesite.org/newscenter/archive/releases/2012/30/image/a/ (direct link) *Original artist:* NASA, ESA, and the Hubble Heritage Team (STScI/AURA)

- **File:Host_Galaxies_of_Calcium-Rich_Supernovae.jpg** *Source:* https://upload.wikimedia.org/wikipedia/commons/5/5e/Host_Galaxies_of_Calcium-Rich_Supernovae.jpg *License:* CC BY 3.0 *Contributors:* http://www.spacetelescope.org/images/opo1528a/ *Original artist:* NASA, ESA, and R. Foley (University of Illinois)

- **File:Hot_and_brilliant_O_stars_in_star-forming_regions.jpg** *Source:* https://upload.wikimedia.org/wikipedia/commons/6/6d/Hot_and_brilliant_O_stars_in_star-forming_regions.jpg *License:* CC BY 3.0 *Contributors:* http://www.eso.org/public/images/eso1230b/ *Original artist:* ESO

- **File:Hubble-Vaucouleurs.png** *Source:* https://upload.wikimedia.org/wikipedia/commons/5/5d/Hubble-Vaucouleurs.png *License:* CC BY-SA 3.0 *Contributors:* Own work *Original artist:* Antonio Ciccolella

- **File:HubbleTuningFork.jpg** *Source:* https://upload.wikimedia.org/wikipedia/commons/2/21/HubbleTuningFork.jpg *License:* Public domain *Contributors:* Transferred from en.wikipedia to Commons. *Original artist:* The original uploader was Cosmo0 at English Wikipedia

- **File:Hubble_Spies_Vast_Gas_Disk_around_Unique_Massive_Star.jpg** *Source:* https://upload.wikimedia.org/wikipedia/commons/0/00/Hubble_Spies_Vast_Gas_Disk_around_Unique_Massive_Star.jpg *License:* CC BY 3.0 *Contributors:* http://www.spacetelescope.org/images/opo1521a/ *Original artist:* NASA, ESA, and G. Bacon (STScI) Science Credit: NASA, ESA, and J. Mauerhan (University of California, Berkeley)

- **File:Hubble_captures_infrared_glow_of_a_kilonova_blast.jpg** *Source:* https://upload.wikimedia.org/wikipedia/commons/7/7d/Hubble_captures_infrared_glow_of_a_kilonova_blast.jpg *License:* Public domain *Contributors:* http://www.spacetelescope.org/images/opo1329a/ *Original artist:* NASA, ESA, N. Tanvir (University of Leicester), A. Fruchter (STScI), and A. Levan (University of Warwick)

- **File:Hubble_sees_a_cosmic_caterpillar.jpg** *Source:* https://upload.wikimedia.org/wikipedia/commons/c/cf/Hubble_sees_a_cosmic_caterpillar.jpg *License:* CC BY 3.0 *Contributors:* http://www.spacetelescope.org/images/opo1335a/ *Original artist:* NASA, ESA, the Hubble HeritageTeam(STScI/AURA),and IPHAS

- **File:IC_755_HST.jpg** *Source:* https://upload.wikimedia.org/wikipedia/commons/a/ae/IC_755_HST.jpg *License:* CC BY 3.0 *Contributors:* http://www.spacetelescope.org/images/potw1129a/ *Original artist:* ESA/Hubble & NASA

- **File:Ilc_9yr_moll4096.png** *Source:* https://upload.wikimedia.org/wikipedia/commons/3/3c/Ilc_9yr_moll4096.png *License:* Public domain *Contributors:* http://map.gsfc.nasa.gov/media/121238/ilc_9yr_moll4096.png *Original artist:* NASA / WMAP Science Team

- **File:Images_of_gas_cloud_being_ripped_apart_by_the_black_hole_at_the_centre_of_the_Milky_Way_ESO.jpg** *Source:* https://upload.wikimedia.org/wikipedia/commons/1/15/Images_of_gas_cloud_being_ripped_apart_by_the_black_hole_at_the_centre_of_the_Milky_Way_ESO.jpg *License:* CC BY 4.0 *Contributors:* http://www.eso.org/public/images/eso1332a/ *Original artist:* ESO/S. Gillessen

- **File:IonringBlackhole.jpeg** *Source:* https://upload.wikimedia.org/wikipedia/commons/2/25/IonringBlackhole.jpeg *License:* CC0 *Contributors:* No machine-readable source provided. Own work assumed (based on copyright claims). *Original artist:* No machine-readable author provided. Brandon Defrise Carter assumed (based on copyright claims).

- **File:Isochrone_ZAMS_Z2pct.png** *Source:* https://upload.wikimedia.org/wikipedia/en/0/0f/Isochrone_ZAMS_Z2pct.png *License:* PD *Contributors:* ? *Original artist:* ?

- **File:IsolatedNeutronStar.jpg** *Source:* https://upload.wikimedia.org/wikipedia/commons/0/01/IsolatedNeutronStar.jpg *License:* Public domain *Contributors:* Transferred from en.wikipedia *Original artist:* Original uploader was Northgrove at en.wikipedia

- **File:IvyMike2.jpg** *Source:* https://upload.wikimedia.org/wikipedia/commons/3/3d/IvyMike2.jpg *License:* Public domain *Contributors:* This image is available from the National Nuclear Security Administration Nevada Site Office Photo Library under number XX-11. *Original artist:* Federal Government of the United States

- **File:Keplers_supernova.jpg** *Source:* https://upload.wikimedia.org/wikipedia/commons/d/d4/Keplers_supernova.jpg *License:* Public domain *Contributors:* http://www.nasa.gov/multimedia/imagegallery/image_feature_219.html Larger version uploaded from http://chandra.harvard.edu/photo/printgallery/2004/ a NASA-sponsored site. Per *Bridgeman Art Library v. Corel Corp.,* no new copyright should apply anyway. *Original artist:* NASA/ESA/JHU/R.Sankrit & W.Blair

- **File:Orbit3.gif** *Source:* https://upload.wikimedia.org/wikipedia/commons/5/59/Orbit3.gif *License:* Public domain *Contributors:* Own work *Original artist:* User:Zhatt

- **File:Orbit4.gif** *Source:* https://upload.wikimedia.org/wikipedia/commons/5/5a/Orbit4.gif *License:* Public domain *Contributors:* Own work *Original artist:* User:Zhatt

- **File:Orbit5.gif** *Source:* https://upload.wikimedia.org/wikipedia/commons/0/0e/Orbit5.gif *License:* Public domain *Contributors:* Own work *Original artist:* User:Zhatt

- **File:PIA18467-NuSTAR-Plot-BlackHole-BlursLight-20140812.png***Source:* https://upload.wikimedia.org/wikipedia/commons/d/d9/PI png *License:* Public domain *Contributors:* http://photojournal.jpl.nasa.gov/jpeg/PIA18467.jpg *Original artist:* NASA/JPL-Caltech/Institute for Astronomy, Cambridge

- **File:PIA18848-PSRB1509-58-ChandraXRay-WiseIR-20141023.jpg***Source:* https://upload.wikimedia.org/wikipedia/commons/a/a6/PId jpg *License:* Public domain *Contributors:* http://www.nasa.gov/sites/default/files/pia18848-wisefacepalm.jpg *Original artist:* NASA/CXC/SAO (X-Ray), NASA/JPL-Caltech (Infrared)

- **File:PIA19822-MagneticBlackHoleWaves-AlfvenS-waves-20150709.jpg** *Source:* https://upload.wikimedia.org/wikipedia/commons/f/fe/ PIA19822-MagneticBlackHoleWaves-AlfvenS-waves-20150709.jpg *License:* Public domain *Contributors:* http://photojournal.jpl.nasa.gov/ jpeg/PIA19822.jpg *Original artist:* NASA/JPL-Caltech

- **File:Planet-hunting_SPHERE_Images_First_Circumbinary_Planet_System_with_Disc.jpg** *Source:* https://upload.wikimedia.org/wiki commons/9/9c/Planet-hunting_SPHERE_Images_First_Circumbinary_Planet_System_with_Disc.jpg*License:* CC BY4.0*Contributors:* http:/ /www.eso.org/public/images/potw1543a/*Original artist:* ESO.A. M.Lagrange(Université Grenoble Alpes)

- **File:Portal-puzzle.svg** *Source:* https://upload.wikimedia.org/wikipedia/en/f/fd/Portal-puzzle.svg *License:* Public domain *Contributors:* ? *Original artist:* ?

- **File:Potw1508a.tif** *Source:* https://upload.wikimedia.org/wikipedia/commons/5/54/Potw1508a.tif *License:* CC BY 3.0 *Contributors:* http:// www.spacetelescope.org/images/potw1508a/ http://www.spacetelescope.org/static/archives/images/original/potw1508a.tif *Original artist:* Credit: ESA/Hubble & NASA Acknowledgement: Gilles Chapdelaine

- **File:Progenitor_IA_supernova.svg** *Source:* https://upload.wikimedia.org/wikipedia/commons/c/ce/Progenitor_IA_supernova.svg *License:* CC BY 3.0 *Contributors:* http://hubblesite.org/newscenter/archive/releases/star/supernova/2004/34/image/d/ *Original artist:* NASA. ESA and A. Feild (STScI); vectorisation by chris 🤖

- **File:Question_book-new.svg** *Source:* https://upload.wikimedia.org/wikipedia/en/9/99/Question_book-new.svg *License:* Cc-by-sa-3.0 *Contributors:*

  Created from scratch in Adobe Illustrator. Based on Image:Question book.png created by User:Equazcion *Original artist:* Tkgd2007

- **File:RXTE_Detects_Heartbeat_Of_Smallest_Black_Hole_Candidate.ogv** *Source:* https://upload.wikimedia.org/wikipedia/commons/a/ a7/RXTE_Detects_Heartbeat_Of_Smallest_Black_Hole_Candidate.ogv *License:* Public domain *Contributors:* Goddard Multimedia *Original artist:* NASA/Goddard Space Flight Center

- **File:R_66_and_R_126_disc_illustration.png** *Source:* https://upload.wikimedia.org/wikipedia/commons/4/49/R_66_and_R_126_disc_ illustration.png*License:* Public domain*Contributors:* http://www.spitzer.caltech.edu/images/1571-ssc2006-05b-Supersized-Disk*Original artist:* NASA/JPL-Caltech/R.Hurt(SSC)

- **File:Radioactive.svg** *Source:* https://upload.wikimedia.org/wikipedia/commons/b/b5/Radioactive.svg *License:* Public domain *Contributors:* Created by Cary Bass using Adobe Illustrator on January 19, 2006. *Original artist:* Cary Bass

- **File:Remnants_of_single_massive_stars.svg** *Source:* https://upload.wikimedia.org/wikipedia/commons/1/18/Remnants_of_single_massive_ stars.svg *License:* CC BY-SA 3.0 *Contributors:* Own work based on A. Heger, C. L. Fryer, S. E. Woosley, N. Langer, D. H. Hartmann: How massive single stars end their life. The Astrophysical Journal, 591:288–300, 2003 July 1 *Original artist:* Fulvio314

- **File:Representative_lifetimes_of_stars_as_a_function_of_their_masses.svg** *Source:* https://upload.wikimedia.org/wikipedia/commons/ a/a9/Representative_lifetimes_of_stars_as_a_function_of_their_masses.svg *License:* CC BY-SA 3.0 *Contributors:* http://www.worldscientific. com/worldscibooks/10.1142/8573 *Original artist:* Carlos A. Bertulani

- **File:Rho_Cassiopeiae_Sol_VY_Canis_Majoris.png** *Source:* https://upload.wikimedia.org/wikipedia/commons/a/aa/Rho_Cassiopeiae_Sol_ VY_Canis_Majoris.png *License:* CC BY-SA 3.0 *Contributors:* Own work by uploader. *Original artist:* Anynobody

- **File:SN1994D.jpg** *Source:* https://upload.wikimedia.org/wikipedia/commons/a/a2/SN1994D.jpg *License:* CC BY 3.0 *Contributors:* http:// www.spacetelescope.org/images/html/opo9919i.html *Original artist:* NASA/ESA, The Hubble Key Project Team and The High-Z Supernova Search Team

- **File:SN1998aq_max_spectra.svg** *Source:* https://upload.wikimedia.org/wikipedia/commons/2/25/SN1998aq_max_spectra.svg *License:* CC BY-SA 3.0 *Contributors:* self-made, data from Matheson et al. 2008. *Original artist:* Falcorian

- **File:SNIIcurva.png** *Source:* https://upload.wikimedia.org/wikipedia/commons/a/a9/SNIIcurva.png *License:* CC-BY-SA-3.0 *Contributors:* ? *Original artist:* ?

- **File:SNIIcurva.svg** *Source:* https://upload.wikimedia.org/wikipedia/commons/0/0e/SNIIcurva.svg *License:* Public domain *Contributors:* Own work *Original artist:* Paulsmith99

- **File:SNIacurva.png** *Source:* https://upload.wikimedia.org/wikipedia/commons/8/88/SNIacurva.png *License:* CC-BY-SA-3.0 *Contributors:* No machine-readable source provided. Own work assumed (based on copyright claims). *Original artist:* No machine-readable author provided. Xenoforme~commonswiki assumed (based on copyright claims).

- **File:Supernovae_as_initial_mass-metallicity.svg** *Source:* https://upload.wikimedia.org/wikipedia/commons/1/14/Supernovae_as_initial_mass-metallicity.svg *License:* CC BY-SA 3.0 *Contributors:* Own work based on A. Heger, C. L. Fryer, S. E. Woosley, N. Langer, D. H. Hartmann: How massive single stars end their life. The Astrophysical Journal, 591:288–300, 2003 July 1 *Original artist:* Fulvio314

- **File:Swift_spacecraft.jpg** *Source:* https://upload.wikimedia.org/wikipedia/commons/6/64/Swift_spacecraft.jpg *License:* Public domain *Contributors:* Obtained from Swift web site. *Original artist:* NASA

- **File:Symbol_book_class2.svg** *Source:* https://upload.wikimedia.org/wikipedia/commons/8/89/Symbol_book_class2.svg *License:* CC BY-SA 2.5 *Contributors:* Mad by Lokal_Profil by combining: *Original artist:* Lokal_Profil

- **File:Symbol_list_class.svg** *Source:* https://upload.wikimedia.org/wikipedia/en/d/db/Symbol_list_class.svg *License:* Public domain *Contributors:* ? *Original artist:* ?

- **File:TCV_vue_gen.jpg** *Source:* https://upload.wikimedia.org/wikipedia/commons/1/14/TCV_vue_gen.jpg *License:* CC BY-SA 2.5 *Contributors:* ? *Original artist:* ?

- **File:The_Hubble_Sequence_throughout_the_Universe'{}s_history.jpg** *Source:* https://upload.wikimedia.org/wikipedia/commons/4/44/The_Hubble_Sequence_throughout_the_Universe%27s_history.jpg *License:* Public domain *Contributors:* http://www.spacetelescope.org/images/heic1315a/ *Original artist:* NASA, ESA, M. Kornmesser

- **File:The_Rise_and_Fall_of_a_Supernova.jpg** *Source:* https://upload.wikimedia.org/wikipedia/commons/b/b3/The_Rise_and_Fall_of_a_Supernova.jpg *License:* CC BY 4.0 *Contributors:* http://www.eso.org/public/images/potw1323a/ *Original artist:* ESO/IRAP-CNRS-UPS/A.Klotz

- **File:The_Sun_by_the_Atmospheric_Imaging_Assembly_of_NASA'{}s_Solar_Dynamics_Observatory_-_20100819.jpg** *Source:* https://upload.wikimedia.org/wikipedia/commons/b/b4/The_Sun_by_the_Atmospheric_Imaging_Assembly_of_NASA%27s_Solar_Dynamics_Observatory_-_20100819.jpg *License:* Public domain *Contributors:* http://sdo.gsfc.nasa.gov/assets/img/browse/2010/08/19/20100819_003221_4096_0304.jpg *Original artist:* NASA/SDO (AIA)

- **File:The_field_around_yellow_hypergiant_star_HR_5171.jpg** *Source:* https://upload.wikimedia.org/wikipedia/commons/c/ca/The_field_around_yellow_hypergiant_star_HR_5171.jpg *License:* CC BY 4.0 *Contributors:* http://www.eso.org/public/images/eso1409a/ *Original artist:* ESO/Digitized Sky Survey 2

- **File:The_life_cycle_of_a_Sun-like_star.ogg** *Source:* https://upload.wikimedia.org/wikipedia/commons/6/67/The_life_cycle_of_a_Sun-like_star.ogg *License:* CC BY 4.0 *Contributors:* ESO *Original artist:* ESO/M. Kornmesser

- **File:The_life_of_Sun-like_stars.jpg** *Source:* https://upload.wikimedia.org/wikipedia/commons/e/eb/The_life_of_Sun-like_stars.jpg *License:* CC BY 4.0 *Contributors:* ESO *Original artist:* ESO/S. Steinhöfel

- **File:Three-dim-pillars-creation.jpg** *Source:* https://upload.wikimedia.org/wikipedia/commons/e/e7/Three-dim-pillars-creation.jpg *License:* CC BY 4.0 *Contributors:* http://www.eso.org/public/archives/images/large/eso1518a.jpg *Original artist:* ESO/M. Kornmesser

- **File:Triple-star_sunset.jpg** *Source:* https://upload.wikimedia.org/wikipedia/commons/9/90/Triple-star_sunset.jpg *License:* Public domain *Contributors:* http://photojournal.jpl.nasa.gov/catalog/PIA03520 *Original artist:*

- The original uploader was SnoopY at English Wikipedia

- **File:Type_Ia_supernova_simulation_-_Argonne_National_Laboratory.jpg** *Source:* https://upload.wikimedia.org/wikipedia/commons/b/b7/Type_Ia_supernova_simulation_-_Argonne_National_Laboratory.jpg *License:* CC BY 2.0 *Contributors:* http://www.flickr.com/photos/argonne/5352540236/ *Original artist:* Argonne National Laboratory / U.S. Department of Energy

- **File:UY_Scuti_zoomed_in,_Rutherford_Observatory,_07_September_2014.jpeg** *Source:* https://upload.wikimedia.org/wikipedia/commons/4/48/UY_Scuti_zoomed_in%2C_Rutherford_Observatory%2C_07_September_2014.jpeg *License:* CC BY-SA 3.0 *Contributors:* I captured this picture from the Rutherford Observatory as I was granted access there to use the primary telescope in2011. *Original artist:* Haktarfone

- **File:VISTA_Magellanic_Cloud_Survey_view_of_the_Tarantula_Nebula.jpg** *Source:* https://upload.wikimedia.org/wikipedia/commons/3/30/VISTA_Magellanic_Cloud_Survey_view_of_the_Tarantula_Nebula.jpg *License:* CC BY 4.0 *Contributors:* http://www.eso.org/public/images/eso1033a/ *Original artist:* ESO/M.-R. Cioni/VISTA Magellanic Cloud survey

- **File:Voyager.jpg** *Source:* https://upload.wikimedia.org/wikipedia/commons/d/d2/Voyager.jpg *License:* Public domain *Contributors:* NASA website *Original artist:* NASA

- **File:Vraagteken.svg** *Source:* https://upload.wikimedia.org/wikipedia/commons/0/02/Vraagteken.svg *License:* Public domain *Contributors:* ? *Original artist:* ?

- **File:WHAM_survey.png** *Source:* https://upload.wikimedia.org/wikipedia/commons/f/ff/WHAM_survey.png *License:* Attribution *Contributors:* Transferred from en.wikipedia to Commons by User:Kauczuk using CommonsHelper. *Original artist:* Ashill at en.wikipedia

- **File:WR124.jpg** *Source:* https://upload.wikimedia.org/wikipedia/commons/8/8f/WR124.jpg *License:* CC BY 4.0 *Contributors:* http://www.eso.org/public/images/wr124/ *Original artist:* ESO

- **File:WhiteDwarf_mass-radius_en.svg** *Source:* https://upload.wikimedia.org/wikipedia/commons/1/14/WhiteDwarf_mass-radius_en.svg *License:* CC BY 3.0 *Contributors:* File:WhiteDwarf mass-radius fr.svg *Original artist:*

- user:AllenMcC. (original jpg file)

- **File:White_Dwarf_Ages.ogv** *Source:* https://upload.wikimedia.org/wikipedia/commons/6/6b/White_Dwarf_Ages.ogv *License:* Public domain *Contributors:* HubbleSite *Original artist:* NASA and G. Bacon (STScI)

- **File:White_dwarfs_circling_each_other_and_then_colliding.gif** *Source:* https://upload.wikimedia.org/wikipedia/commons/c/cc/White_dwarfs_circling_each_other_and_then_colliding.gif *License:* Public domain *Contributors:* http://www.nasa.gov/vision/universe/starsgalaxies/collide_whitedwarf.html *Original artist:* Dana Berry

- **File:Wikibooks-logo-en-noslogan.svg** *Source:* https://upload.wikimedia.org/wikipedia/commons/d/df/Wikibooks-logo-en-noslogan.svg *License:* CC BY-SA 3.0 *Contributors:* Own work *Original artist:* User:Bastique, User:Ramac et al.

### 39.1.3 Content license

# Chapter 40

## Editors Comments

**Supernova 1997ff and its Impact on Fundamental Cosmology Theory**

By Paul F. Kisak

**" I take the positivist viewpoint that a physical theory is just a
mathematical model, and that it is meaningless to ask whether it corresponds to reality.  All that one can ask
is that its predictions should be in agreement with observation "**

Stephen Hawking (1942 - )
" The Nature of Space and Time 1996 p 3-4 "

Section Outline:

1. A Perspective of Sn1997ff
2. SN1997 ff Pertinent Observational Data
3. Executive Summary
4. Method of Discovery
5. Why is SN1997ff so important and how will it impact cosmology theory?
6. Constants of Proportionality
7. In the Beginning There Was Hubble
8. Some Fundamental Background on ' The Big Bang Theory '
9. Einstein's Theory of Gravity
10. Cosmological Observations
11. Conclusion
12. Bibliography

If the reader is well versed in math and physics, especially as they impact cosmology theory, then sections 1-5 deal with the overview.  If the reader would like to learn some details pertaining to the fundamentals of cosmology and "Big Bang Theory " in particular, sections 6-10 pertain.

Variables that are listed in bold represent vectors, and those that are not, represent scalar quantities.  Some text is reprinted from NASA press releases and NASA photos.  The author wishes to thank NASA, The Hubble Space Science Institute, Lawrence Berkeley National Laboratory and his colleagues for their contributions to this article.

## 1. A Perspective of Sn1997ff :

11.3 (+/- .2) billion years ago a star exploded. It is framed by the constellation Ursa Major. 3.5 years ago (1997) it was photographed and 4 weeks ago it's impact on cosmology theory was being contemplated by a group of individuals at the Hubble Space Telescope Institute, one of whom had never even seen the photograph. Tomorrow, as a result, revisions will start being made to the standard model of the Universe.

This exploding star is the Supernova 1997ff type Ia (SN1997ff).

The " type Ia " classification is based on the fact that the spectral images show no hydrogen lines but do show spectral lines for silica. The " ff " designator in the title "SN 1997ff " is an indicator of which period, in the photographic imaging cycle, the photos were taken.

**Distant Supernova in the Hubble Deep Field**      HST • WFPC2
NASA and A. Riess (STScI) • STScI-PRC01-09

In addition, type Ia supernovae are believed to arise from an accreting white dwarf which is a very old, dense, relatively low mass star that is not undergoing thermonuclear reactions in its core. The white dwarf becomes a supernova when detonation of its supersonic burning front or a deflagration by subsonic burning occurs. These events are ultimately dependent on the critical mass of the star. Every element found in nature, except for hydrogen and helium, was made in either a star or supernovae explosion according to current science.

Supernovae reveal the last stages of stellar evolution. They also transmit energy into the interstellar gas, and play a crucial role in the calculation of extragalactic distance, scale and cosmology. There have been slightly over 2000 supernova discovered since 1885. (1,2,3,4)

" What is your substance, whereof are you made ..."

- William Shakespeare (1564-1616) -
" Sonnet 53 "

## 2. SN1997ff Pertinent Observational Data

The following represent the latest data concerning SN1997ff as of 20 April 2001.

**Name:**
> SN1997ff

**Object Description:**
> Supernova Type Ia in Hubble Deep Field North Galaxy 4-403.0

**Position (J2000):**
> R.A. 12h 36m 44.18s Dec. +62° 12' 44.8"

**Constellation:**
> Ursa Major

**Distance:**
> 3 billion parsecs (10 billion light-years)
> {updated to 11.3 (+/- .2) billion light years on 26 April 2001)

**Redshift:**
> $z = \sim 1.7$

**Magnitude:**
> 26.8

**Instrument:**
> NICMOS; WFPC2

**Exposure Dates:**
> December 23, 1997 - June 22, 1998

**Total Exposure Time:**
> ~ 6 days

**Filters:**
> F814W (I), F110W (J), F160W (H)

**Principal Astronomers:**

A.G. Riess (STScI), R. Thompson (U Arizona), R. Gilliland (STScI), P. Nugent (LBNL), B. Schmidt (MSSSO), J.Tonry (U Hawaii), M. Dickinson (STScI), T. Budavari (JHU), M. Livio (STScI), H. Spinrad (UC Berkeley), D. Stern (UC Berkeley), D. Sanders (U Hawaii), and S. Veilleux (U Maryland) .

**Image Credit:**
NASA and A. Riess (STScI)

**Release Date:**
April 2, 2001 1:00p.m. EDT (1,2,3)

> **" It is a capital mistake to theorize before one has data.**
> **Insensibly one begins to twist facts to suit theories,**
> **instead of theories to suit facts. "**

- Arthur Conan Doyle  (1859-1930) -
" A Scandal in Bohemia "

## 3.  Executive Summary :

SN1997ff reveals that the measurement of its distance as a function of luminosity (26.8), is significantly different than that shown by its red shift (1.7), according to current models.  SN1997ff is farther from the earth than current big bang theory and nucleosynthesis would predict.  Supernovae, of evolution " type Ia ", are predictably stable and used as a measuring stick for the universe specifically because of their stability.  The universe is expanding faster than it was approximately 5-7 billion years ago.  Nobody knows why.

There are approximately 26 ways of measuring the distance of stellar objects.  These techniques are now being used on SN1997ff in a more aggressive manner to reduce the error in measuring these great distances.  Astronomers, however, prefer to speak of light years rather than distance due to the anomalies under consideration and the way that the models handle luminosity and magnitude.

This observation/anomaly has surfaced before but there was not enough data, especially from the deep field, to create a consensus on whether or not this was an actual phenomenon.  SN1997ff, being the most distant supernovae ever discovered along with the fact that it is a type Ia, seems to verify that the universe is accelerating when you look beyond the 5-7 billion light year range.

By measuring the predictable light output of supernovae, astronomers can estimate how far they are from Earth.  But supernovas blaze so brightly that they can be seen far across space.  That's why some astronomers also call them "cosmic mile markers."  Their light provides important information about the universe's behavior.  Supernovas illuminate the dark corners of space, allowing astronomers to map the history of the universe's expansion.

Supernovae occur when a relatively massive star (greater than approximately 7.6 solar masses) no longer has enough fuel for the fusion process that takes place in the core of the star to create the outward thermodynamic pressure which combats the inward gravitational pull of the star's great mass.  When this occurs, the star will swell into what is termed a red supergiant.  However the core of the star eventually yields to gravity and begins shrinking. As it shrinks, it grows progressively hotter and denser.

The first fusion stage converts hydrogen into helium.  The fusion continues converting helium into carbon, carbon into neon, neon into oxygen, oxygen into magnesium, magnesium into silicon, silicon into sulphur & phosphorus. Silicon is finally converted into iron, and a dense iron core is formed at the center of the star. All these differing reactions take increasingly shorter periods of time, and release progressively less amounts of energy. At this point, the structure of the star is layered like an onion - there are shells of progressively less heavy chemical elements surrounding the core.

This is where the curve of binding energy requires a change in the process occur.  The abovementioned reactions create energy (an exothermic reaction). But to convert Iron into other elements (e.g. cobalt & nickel) requires energy (an endothermic reaction). Thus, fusion halts. At the very high temperatures now present in the core of the star (much greater than 109 K), a process known as photodisintegration occurs. As a result of photodisintegration, the core starts to rapidly collapse.

Different parts of the core collapse at different rates, with the result that the inner core uncouples from the outer core, leaving it behind. During the collapse, speeds can reach 70,000 km/s in the outer core, and within about one second, a volume the size of Earth has been compressed down to a radius of 50 kilometers. As a result the rest of the star is left in the precarious position of being almost suspended above the catastrophically collapsing core. This collapse of the inner iron core continues until the density there exceeds approximately $8 \times 10^{14}$ grams/cm-3. At this point, the material that now makes up the inner core stiffens (as a result of the nuclei of the atoms present becoming repelled by each other), with the result that the inner core now rebounds somewhat, sending pressure waves outwards into the infalling material of the outer core. These pressure waves, when they reach the local speed of sound, form a shock wave that starts moving outwards.

As the shock wave propagates outwards, it encounters the falling inner iron core. The extremely high temperatures that occur as a result of this cause further photodisintegration, robbing the shock of most of its energy. If what's left of the iron core is not too massive (less than 1.2 solar masses), the shock will fight its way through the rest of the outer core - which takes about 20 milliseconds, and collide with the remainder of the outer layers of the star. On the other hand, if the iron core is massive enough, the shock stalls, becoming nearly stationary, with infalling material now accreting onto it. At this point, the neutrinos now streaming from the core superheat the material beneath the shock wave; the resulting plumes of hot material push the shock wave outwards and allow it to continue its march towards the surface, driving all before it (Janka 2001).

As the shock encounters material in the star's outer layers, the material is heated, fusing to form new elements and radioactive isotopes. The shock then propels the various outer layers of the star out into space, leaving the inner core behind. The total energy in the expanding material is in the order of 1051 ergs (or less), but 1053 ergs of energy (100 times more) is radiated away in the form of neutrinos. Finally, vast amounts of photons are released, resulting in a spectacular optical display, equivalent of one billion Suns, giving an absolute magnitude of about -18. Due to the radioactive decay of heavy elements produced in the explosion (Mochizuki & Kumagai 1988; Wanajo et al. 2001), it then starts to slowly fade, at rate of approximately 6 to 8 magnitudes each year. Type II supernovae are not as luminous as Type Ia supernovae, by a factor of at least three. Bethe (1993), Wallerstein et al. (1997), Mezzacappa (2000) and Liebendoerfer et al. (2001) deal with the mechanics of this type of supernovae in detail.

Astronomer Adam G. Reiss of The Hubble Space Telescope Science Institute in Baltimore, MD has stated that "the universe is actually expanding faster now than it was 5 to 7 billion years ago. It actually seems to be accelerating so it will be expanding forever." The supernova's velocity and distance from Earth suggests "dark energy " may be accelerating the universe's expansion. Dark energy is a mysterious force originally proposed by Einstein.

Much of the pertinent data and calculations for SN1997ff, that show the expected distance compared to the accelerated distance, are still unpublished. I have been informed that an article is expected to be published in several astrophysics journals in the coming months. I am indebted to Paul Preuss of Lawrence Berkeley National Laboratory for his time in helping to clarify some of the issues that surround the SN1997ff anomalies.

" Not from the stars do I judgement pluck,
and yet methinks I have astronomy ... "

- William Shakespeare (1564-1616) -
" Sonnet 14 "

## 4. Method of Discovery :

The Hubble Space Telescope Science Institute in Baltimore, MD photographed SN1997ff during the period December 23, 1997 - June 22, 1998 while doing a routine 'Deep Field Study.'

The deep field studies are done to analyze stellar objects, which represent the oldest objects in the universe. One prerequisite for being the oldest is that these objects must also be some of the furthest objects from the earth; hence the name deep field or far field.

The usual method of finding 'deep field' supernova such as SN1997ff involves the computer comparison of 'deep field' photographs that are taken at separate times. The first image of this particular 'deep field' area was taken in 1995. The next image used was taken in 1997. A computer is then used to scan the many images and identify any changes in brightness in overlapping areas.

The computer identified the SN1997ff discrepancy in brightness amongst two images on 2 April 2001. The lapse in time between the taking of the photograph and the computer detection is due to the voluminous backlog of photographs and the complex method of analysis using different spectra such as the infrared. The light from SN1997ff did not appear in the 1995 photograph of the same 'deep field' area. The fact is that when SN1997ff exploded, 11.3 +/- .2 billion years ago, the light from that explosion did not reach earth until sometime after 1995. This method is similar to the technique that is used to detect comets.

The fact that SN1997ff is a "type Ia" supernova allows the astronomers to use the "type Ia" supernovae expansion and nucleosynthesis models. These models can predict SN1997ff luminosities as a function of time and help corroborate the distance measurements that are gathered from Doppler studies done on other stars in the same galaxy.

**" The end is where we start from. "**

- T. S. Eliot (1888-1965) -
" Four Quartets "

### 5.  Why is SN1997ff so important and how will it impact cosmology theory ?

SN1997ff is important because it represents an anomaly according to the presiding ' Big Bang Theory.' SN1997ff is also the most distant supernova ever observed.  The observational data of SN1997ff stretches the limits of current astronomical mensuration and yet is still consistent with a growing field of data that imply that the universe is accelerating at distances beyond 5-7 billion light years.

This observation is in conflict with the traditional theory for The Big Bang.  The Big Bang Theory, in general, is the scientifically accepted theory and model for the origin and evolution of our known universe.  The basic tenets of Big Bang Theory are that the observable universe results from the isotropic and homogeneous expansion, from a single point, approximately 10 to 20 billion years ago.  Given these restrictions in the model, change can occur only in the time domain, which is modeled as an explosion that is spherically symmetrical.  This expansion/explosion started the evolution of all matter, energy, space and time, as we currently understand them.

There are numerous assumptions made in The Big Bang Model.  One of the basic assumptions is that the laws of thermodynamics and Newtonian physics apply.   The Big Bang Theory does not address dark matter, the internal dynamics of galactic formation, the distribution of mass and other stellar constructs.  On a related note, Einstein proposed The General Theory of Relativity that describes how the distribution of mass in the universe determines the geometry of the space.

The accuracy and verification of subsequent measurements and corroborating data will determine the impact that SN1997ff will have on cosmology theory.  There is a prevailing propensity to try and account for this "push" or force that causes this unexpected acceleration at the 5-7 billion light year horizon to be attributed to "dark matter ".  The concept of dark energy was first proposed, then discarded, by Albert Einstein.  The later sections give a discussion of the importance of "The Theory of General Relativity " and dark matter as they relate to cosmology theory.

The following sections will aid the reader in understanding how certain equations, that pertain to the Big Bang model, are generated from first principles and how these equations can reveal behaviors that transcend the original assumptions.

## 6. Constants of Proportionality :

Constants of proportionality are wonderful equalizers that are used by science to make the leap from observables to equations. These constants define our universe and the perspective, scale and interaction of the physics within the universe. It is the opinion of this author that a solid discussion on the method and philosophy of the constants of proportionality are sadly lacking at the undergraduate level.

From the fundamental viewpoint, as Newton and many other's have, we can observe that when we apply a force to a body it accelerates. The first order conclusion that can be made is that Force ($F$) is directly proportional to acceleration ($a$), where $F$ and $a$ are vectors. Using basic algebra we know that the equation for a line is of the form $y = sx + b$, where $F = y$ and $a = x$, $s$ is the slope of the line and $b$ is the y intercept when $x = 0$. In our case $b = 0$ because when $F = 0$, $a = 0$.

If the observer then applies various forces to the object and measures the resulting acceleration, an X-Y plot of the resulting data will show a line of the form $y = sx + b$ or in this case $F = sa$. Knowing that we have $F$ for each $a$, we can calculate $s$ by dividing $F$ by $a$; $s = F/a$. In this case the calculation of the slope of the line is relatively straightforward and yields the result $s = m$; where $m$ is equal to the mass of the object. The mass ($m$) is the constant of proportionality between any specific $F$ and $a$.

The resulting equation $F = ma$ is used to predict classical rectilinear non-relativistic motion. This is how Newton did it and this is how it is still done. This is one of the most fundamental procedures in utilizing the scientific method to predict behavior by generating the relevant equations.

This same method is applied to determine the equations that pertain to the weak, strong, nuclear and gravitational forces. These necessary 'constants of proportionality' are derived by observation in electromagnetics, thermodynamics, nuclear physics and astrodynamics. The resultant constants of proportionality have names such as mass, charge, voltage, frequency, Boltzman's constant, the gravitational constant, Plank's constant, Hubble's constant and many others.

Now let us see how this method is used to predict the large scale behavior of the universe and in particular how it leads us to conclude that SN1997 is a clue to a behavior in the universe that is in need of theory. To begin with we will need to concern ourselves with frequency and Hubble's constant.

405

## 7. In the Beginning There Was Hubble :

Assuming that all matter originated at a point in space and has been continuously expanding outwards from that point in space, the farther an object is from any given reference point, the older it is. This applies because of the assumption that the universe is an isotropic, homogeneous system and that change can occur only with respect to time. Like Newton, we observe also that if the velocity of matter in the universe is directed radially outward from the origin and proportional to the distance " **r** " from the origin, the following equation applies:

$$\mathbf{v} = H \, \mathbf{r} \quad \text{(equation 1)} .$$

" H " is referred to as Hubble's constant (after the astronomer). Hubble's constant " H ", is assumed to be independent of **r** and possibly a function of time. The next step involves a standard observation where we have a given **v**1 and **r**1 at an arbitrary point in space. We can perform a classic Galilean coordinate transformation on the system with respect to the origin **r** in the following manner:

$$\mathbf{R} = \mathbf{r} - \mathbf{r}1 \quad \text{(equation 2)} ,$$

where **R** & **V** are the relative vectors between an arbitrary point in space and the observational data. An example would be that **v**1 and **r**1 pertain to the Earth and **v** and **r** pertain to SN 1997ff. In this case then **R** would be the distance between Earth and SN 1997ff. Noting that $\mathbf{r} = \mathbf{v} / H$ from equation 1, we can substitute this value **r** into equation 2. This yields $(\mathbf{V} / H) = (\mathbf{v} / H) - (\mathbf{v}1 / H)$. Multiplying both sides by the constant H yields $\mathbf{V} = \mathbf{v} - \mathbf{v}1$ as expected. Noting that $\mathbf{v} / H = \mathbf{r}$ & $\mathbf{v}1 / H = \mathbf{r}1$ , and substituting these into equation 2 we get $\mathbf{R} = \mathbf{v} / H - \mathbf{v}1 / H$ . Multiplying both sides by H yields the following:

$$\mathbf{V} = H \, \mathbf{R} \quad \text{(equation 3)} ,$$

The point of the above manipulations is to show that under the given assumptions, any arbitrary point in space will appear to be moving away from any other given point. Therefore the Doppler shift of all objects that are observed from any given point will reflect a recession with respect to that point. This brings us to a basic discussion of Doppler.

Doppler shift is defined as the amount of change in the observed frequency of a wave due to the fact that the wave has a relative motion with respect to the observer. This phenomena translates well as a tool to measure velocities of stellar objects. When applied in such a manner it is commonly referred to as the cosmological redshift. This is the effect that happens when light appears to be shifted towards the red end of the spectrum because of the expansion of space-time itself.

Waves emitted by a moving object as received by an observer will be blue-shifted (compressed) if approaching and red-shifted (elongated) if receding. The Doppler effect is observed in both sound and light. How much the frequency changes depends on how fast the object is moving toward or away from the receiver.

Another method of calculating the distance of a given supernova from earth is by utilizing the measurement of luminosity. If the absolute maximum light is known, from a given type, one can derive the supernova distance and hence the distance to any associated supernova can also be derived. The light extinction of a supernova (or any other optical phenomenon) with distance is;

$$5 \, \text{Log}(D/10) = Mv(max) - mv(max) + Av ,$$

where D is the distance to the supernova in parsecs, Mv is the absolute magnitude, mv the apparent magnitude and Av the extinction. Therefore from observations of the mv for any given supernova and measurements of Av from a variety of sources the distance can be obtained, providing one knows the ' type ' of the supernova. The decrease in the luminosity for any supernova is known as it depends on the half-life of the various isotopes produced. Therefore the recorded rate of decay can be used to give the type, and hence the distance for a given supernova. These methods show some of the most basic steps in calculating distance to a supernova.

Hubble's constant was the first cosmological constant. The fact that the universe seems to be accelerating beyond 5-7 billion light years leads to an acknowledgement of Einstein's cosmological constant or a refinement of Hubble's. Some theories have refined the Hubble constant to 60-70 km per second per mparsec. In the past the universe was modeled as steadily decelerating with a zero or even negative cosmological constant. The fact that the universe is accelerating beyond the 5-7 billion light year threshold has led to a new estimate for the age of the universe. The latest estimate that accounts for this phenomena, predicts the age of the universe to be 12-13.5 billion years old.

Based on theoretical studies by Hawking-Penrose and empirical data by Freeman-Sandage it is being argued that the universe will never be able to contract or oscillate eternally. The idea that space-time is infinite is considered to be dead.

" The end is where we start from. "

- T. S. Eliot (1888-1965) -
" Four Quartets "

## 8. Some Fundamental Background on ' The Big Bang Theory ':

Some of the basic equations that have been used to describe the essence of the universe are based on the assumption that " **R** " can be defined as the scale factor for the universe, " **P** " is the total pressure arising from all sources, " p " is considered the density of matter and radiation, " k " is a constant describing the geometry of the universe, " c " is the speed of light and " G " is the gravitational constant.

Therefore d**R**/dt and d 2 **R**/dt 2 represent the velocity and acceleration of the universe.

$$2(d\,2\,\mathbf{R}/dt\,2\,)/\mathbf{R} + [(d\mathbf{R}/dt)/\mathbf{R}]\,2 + kc\,2\,/\mathbf{R}\,2 = -\,(8\,Pi\,G\,p)/c\,2 \quad \text{(equation 4)}$$

$$[(d\mathbf{R}/dt)/\mathbf{R}\,2]\,2 + kc\,2/\mathbf{R}\,2 = (8\,Pi\,G\,p/\,3) \quad \text{(equation 5)}$$

If we subtract equation 5 from equation 4 we get the following:

$$2(d\,2\,\mathbf{R}/dt\,2\,)/\mathbf{R} = -[8\,Pi\,G\,(p + 3P/c\,2\,)]/3 \quad \text{(equation 6)} \quad (17)$$

Equation 6 has been used to note that the acceleration of the universe based on the assumptions made is negative and therefore decelerating. Gravity slows the expansion of the universe. If the universe is dense enough, the expansion of the universe will eventually reverse and the universe will collapse. If the density is not high enough, then the expansion will continue forever.

The distribution of matter in the universe is nearly uniform. This assumption is confirmed by galaxy surveys and by the low level of fluctuations in the cosmic microwave background radiation.

The expansion of the universe also cools the microwave background radiation. Today the cosmic microwave background radiation has a temperature of 2.728 Kelvin. It is presumed that the background radiation was hotter in the past. Models that include this variation in background radiation temperature have been called ' Hot Big Bang Models."

The ' Hot Big Bang Theory ' is consistent with a number of important observations:

1. The observed expansion of the universe,
2. The observed abundance's of helium, deuterium and lithium, three elements
     thought to be synthesized primarily in the first three minutes of the universe,
3. The thermal spectrum of the cosmic microwave background radiation,
4. The cosmic microwave background radiation appears hotter in distant clouds
     of gas. Since light travels at a finite speed, we see these distant clouds at an
     early time in the history of the universe, when it was more dense and thus
     hotter.

In addition, another recent discovery has helped bolster Einstein's original assertions regarding 'dark matter' and his 'cosmological constant'. On 29 March 2001 it was announced that "the first complete "Einstein ring" ever spotted in the universe, had been discovered. Such cosmic mirages were predicted more than 60 years ago by the theory of general relativity."

## The First Image of a Complete Einstein Ring

The Hubble Institute's press release went on to say "The Einstein Ring" is an optical effect created when a

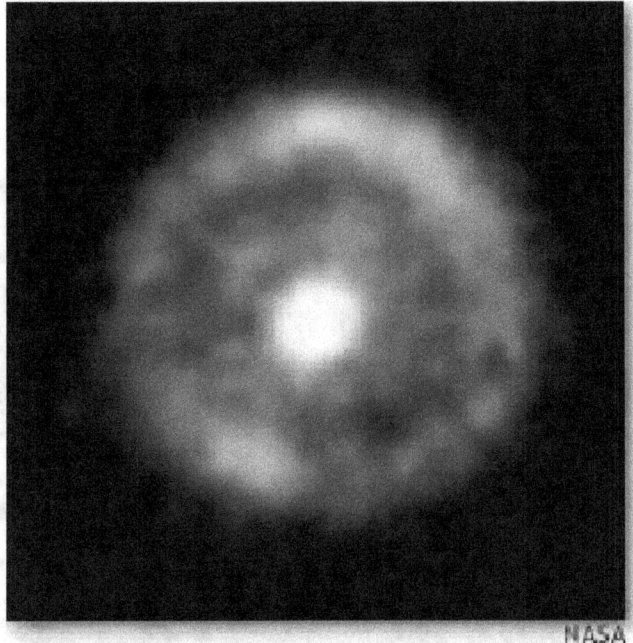

distant light source is directly behind a massive galaxy, as seen from Earth. Albert Einstein predicted that the galaxy in the middle would act as a "gravitational lens": The galaxy's gravitational pull would bend the light rays from the farther object so that a distant observer would see a halo or ring, with the galactic "lens" in the very center of the picture. (4)

# Einstein Ring Dynamics

About 25 gravitational lenses have been detected, but in virtually every case the distant light source, either a galaxy or a quasar, is not exactly lined up with the galactic lens. In such a case, the ring is incomplete, and there are multiple images of the distant source.

A British team of astronomers has been conducting a survey of gravitational lenses for the past several years.

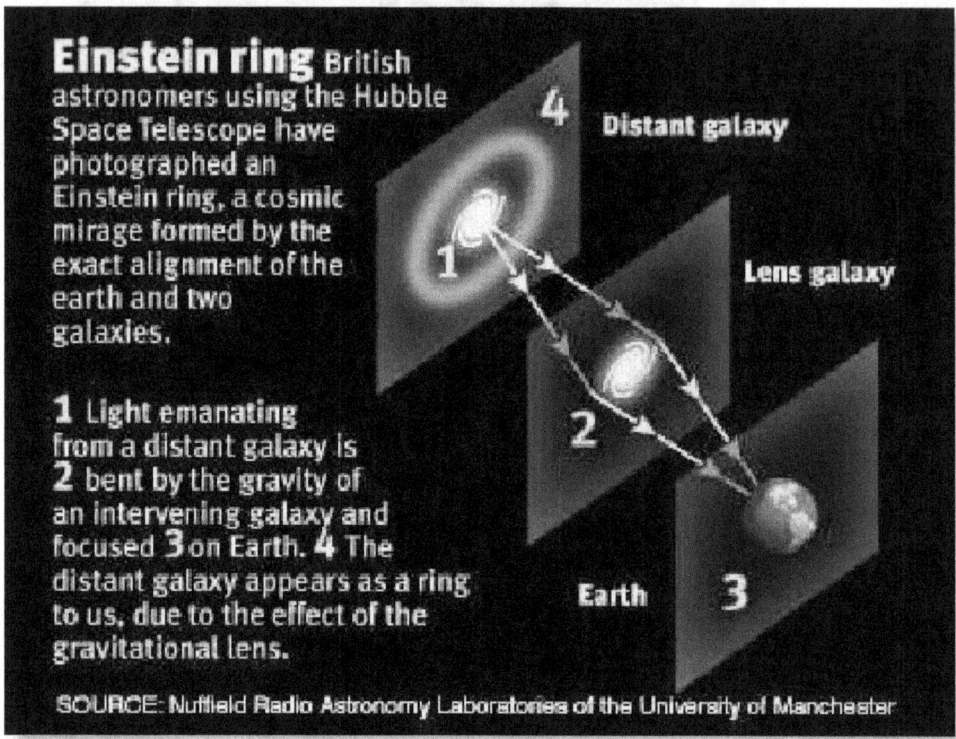

They selected an object known as B1938+666 in the constellation Draco for further study, based on radio observations from the Very Large Array of radio telescopes in New Mexico and the MERLIN radio array in Britain.

Astronomers examined B1938+666 using the Hubble Space Telescope's infrared camera and found the telltale circular signature with a galaxy in the middle. "At first sight it looks artificial, and we thought it was some sort of defect in the image," said Ian Browne of the University of Manchester, "but then we realized we were actually looking at a perfect Einstein ring."

In this case, the distant light source is another galaxy. Neal Jackson, another member of the research team from the University of Manchester, said astronomers don't know exactly how distant the galaxy is. But the object can't be seen with the naked eye — in fact, researchers said the observational feat was similar to examining a penny from a distance of 20 miles.

"MERLIN and the Hubble have scored a bull's eye," Bristol University astronomer Mark Birkinshaw said in a statement. The scientific results are being presented at this week's UK National Astronomy Meeting and appear in the April 1 issue of the Monthly Notices of the Royal Astronomical Society. Among the members of the research team are astronomers from the University of Oxford, the California Institute of Technology, the Netherlands Foundation for Radio Astronomy, the University of Groningen and the Paris Institute of Astrophysics.

Gravitational lenses could play an important role in solving some of the fundamental questions of the universe. For example, how much of the universe's mass is composed of "dark matter" that can't be detected directly by optical or radio telescopes? What is the total size of the universe? How old is it?

By observing a number of gravitational lenses that involve a variable light source, scientists can develop better estimates of the Hubble constant, which measures how fast the universe is expanding. The Hubble constant can be combined with other astronomical data to derive the age and total mass of the universe.

Jackson said B1938+666 probably won't help much with calculations of the Hubble constant, but he said that the continuing survey of gravitational lenses should go a long way toward settling the issue.
"We already have at least two lenses that are quite good candidates for the Hubble constant," Jackson said. "The idea in this game is to beat down the errors." The study of gravitational lenses also may help settle the controversy over the cosmological constant, a mathematical value that Einstein once included in his calculations but later labeled his "greatest blunder."

The cosmological constant describes a long-range "anti-gravitational" force. At first, Einstein thought the constant was required in order to make his theories match astronomical realities, but he was later convinced that it was unnecessary. In recent years, however, ever-more-detailed observations have led some physicists to suspect that the cosmological constant may be required after all.

Different values for the cosmological constant would yield different geometries of the universe, and physicists generally agree that counting the number of gravitational lenses is the best way to measure the universe's large-scale geometry. Astronomers should be able to come up with tighter limits on the value of the cosmological constant as a result of the survey of gravitational lenses, Jackson said. "That is something that will follow from the statistics once we've finished the survey," he said.

He said the survey would continue for another year or so, with researchers hoping to find about 10 more gravitational lenses." (from the Hubble Space Science Institute Press release on B1938+666 and the discovery of the first complete Einstein Ring.) (4)

" If Einstein didn't exist, we would<br/>
have to create him. "

## 9. Einstein's Theory of Gravity

(Portions reprinted from STScl- PR01-09: Blast from the Past: Farthest Supernova Ever Seen Sheds Light On Dark Universe)

Einstein's theory of gravity introduced some remarkable ideas such as black holes, curved space, dark matter and repulsive gravity. Even in Einstein's theory, matter always pulls. The repulsive aspect of gravity, a force that pushes matter apart, only arises in extraordinary circumstances.

A number of possibilities have been suggested for the source of the repulsive gravity - referred to as dark energy - that is driving the universe apart. They range from the energy of the quantum vacuum, to a mini-version of Guth's inflation, to the influence of hidden additional space dimensions. In some theories the dark energy eventually dissipates, with the slowing effect of ordinary matter taking over again. While there is no consensus as to what the dark energy is, astronomers and physicists agree that this mysterious force is extremely important. (Portions reprinted from STScl- PR01-09: Blast from the Past: Farthest Supernova Ever Seen Sheds Light On Dark Universe)

(The following is reprinted from the Center for Particle Astrophysics of the University of California at Berkeley.)

What is a white hole?

The equations of general relativity have an interesting mathematical property: they are symmetric in time. That means that you can take any solution to the equations and imagine that time flows backward rather than forward, and you'll get another valid solution to the equations. If you apply this rule to the solution that describes black holes, you get an object known as a white hole. Since a black hole is a region of space from which nothing can escape, the time-reversed version of a black hole is a region of space into which nothing can fall. In fact, just as a black hole can only suck things in, a white hole can only spit things out.

White holes are a perfectly valid mathematical solution to the equations of general relativity, but that doesn't mean that they actually exist in nature. In fact, they almost certainly do not exist, since there's no way to produce one. (Producing a white hole is just as impossible as destroying a black hole, since the two processes are time reversals of each other.)

What is a wormhole?

So far, we have only considered ordinary "vanilla" black holes. Specifically, we have been talking all along about black holes that are not rotating and have no electric charge. If we consider black holes that rotate and/or have charge, things get more complicated. In particular, it is possible to fall into such a black hole and not hit the singularity. In effect, the interior of a charged or rotating black hole can "join up" with a corresponding white hole in such a way that you can fall into the black hole and pop out of the white hole. This combination of black and white holes is called a wormhole.

The white hole may be somewhere very far away from the black hole; indeed, it may even be in a "different Universe" -- that is, a region of space-time that, aside from the wormhole itself, is completely disconnected from our own region. A conveniently located wormhole would therefore provide a convenient and rapid way to travel very large distances, or even to travel to another Universe. Maybe the exit to the wormhole would lie in the past, so that you could travel back in time by going through. All in all, they sound pretty cool.

But before you apply for that research grant to go search for them, there are a couple of things you should know. First of all, wormholes almost certainly do not exist. As we said above in the section on white holes, just because something is a valid mathematical solution to the equations doesn't mean that it actually exists in nature. In

412

particular, black holes that form from the collapse of ordinary matter (which includes all of the black holes that we think exist) do not form wormholes. If you fall into one of those, you're not going to pop out anywhere. You're going to hit a singularity, and that's all there is to it.

Furthermore, even if a wormhole were formed, it is thought that it would not be stable. Even the slightest perturbation (including the perturbation caused by your attempt to travel through it) would cause it to collapse. Finally, even if wormholes exist and are stable, they are quite unpleasant to travel through. Radiation that pours into the wormhole (from nearby stars, the cosmic microwave background, etc.) gets blue-shifted to very high frequencies.

One theory that I regard as particularly insightful and valuable, is one that takes into consideration the Schwarzschild geometry. The complete Schwarzschild geometry consists of a black hole, a white hole, and two universes connected at their horizons by a wormhole. The wormhole joining the two separate Universes is known as the Einstein-Rosen Bridge.

A white hole is then simply a black hole running backwards in time. Just as black holes swallow things irretrievably, so also do white holes spit them out. General relativity is time symmetric. It does not know about the second law of thermodynamics, and it does not know which way cause and effect go.

Do Schwarzschild wormholes really exist?

Schwarzschild wormholes certainly exist as exact solutions of Einstein's equations. However current models theorize that:

1. When a realistic star collapses to a black hole, it does not produce a wormhole;
2. The complete Schwarzschild geometry includes a white hole, which violates the second law of thermodynamics;
3. Even if a Schwarzschild wormhole were somehow formed, it would be unstable and fly apart.

(Portions of the above are reprinted from the Center for Particle Astrophysics of the University of California at Berkeley.)

In addition to the above models of the universe there are models based on the concept of string theory, twistor theory and an extension of string theory called M-theory. M-theory does not do away with the Big Bang. The evidence that everything emerged from a 'fireball' with a temperature of 10 billion degrees, expanding on a time scale of one second, is now very compelling and uncontroversial.

## 10.  Cosmological Observations :

As mentioned above, it is theorized, that a force must have resulted in the energy that moved SN1997ff from its predicted position to its actual position.  Scientists are asking themselves what force could this be?  The physical sciences recognize four forces; the strong (nuclear force), the weak (nuclear) force, the electromagnetic force and the gravitational force.

There has been a long standing effort to combine these forces into a single theory.  This theory has gone by various names such as "The Theory of Everything (TOE)' and "Grand Unification Theory (GUT)."  The idea of a fifth force has been theorized in the past.  This force, in essence is one that keeps pushing.  This proposed fifth force is considered to be a weak force that becomes noticeable only on large time scales.  This force would counteract the force of gravity, which has led many to theorize that the universe must eventually collapse under its own weight.

This predicament led Einstein to posit his version of ' the cosmological constant ' in the general theory of relativity.  Einstein's constant, which is represented by the upper case lambda, has led some to call it an ' anti-gravity ' term that would account for a stable universe.  Einstein's cosmological constant remained unchallenged until 1922 when the Russian mathematician Alexander Friedmann began producing cosmological models based on Einstein's equations but without the ' lambda '.  Einstein retracted his theoretical cosmological constant of ' Lambda ' in 1932.  This did not stop the development of theories that still postulate a similar ' Lambda cosmological constant.'

It is not often in science that such overwhelming predictions can be made before observations confirm the theory and/or equations.  Einstein's theory of general relativity predicted three fundamentally observable events.  These are as follows:

1. The path of light from stellar objects will be deflected by the sun.

2. Any body that orbits between the sun and the earth will experience an advance of it's perihelion. (The perihelion is the point in the orbit of a body where it is closest to the sun.)

3. A given spectral line will experience a red shift due to gravity. (17)

I have found it rare to find a reasonable assemblage of the theoretical and observed verifications of general relativity. As a result I have compiled the following;

1. In 1919 the British astronomer and mathematician Arthur Eddington led a team that verified that stellar light was bent by the sun's gravitational field by 1.8 +/- 0.2 arcseconds (one arcsecond is 1/3600th of a degree or about .06 % of the angular diameter of the moon.) Einstein had predicted 1.751 arcseconds.

2. General relativity predicted a 1.751 arc second deflection of starlight. 1.70 +/- .10 arcseconds has been observed.

3. General relativity predicted a 1.751 arc second deflection of starlight. 1.73 +/- .05 arcseconds has been observed.

4. General relativity predicted a 43.03 arcsecond centennial precession for the orbit of Mercury. A 43.11 +/- 0.45 arcseconds was observed.

5. General relativity predicted an 8.6 arcsecond centennial precession for the orbit of Venus. An 8.4 +/- 4.8 arcseconds was observed.

6. General relativity predicted a 3.8 arcsecond centennial precession for the orbit of Earth. A 5.0 +/- 1.2 arcseconds was observed.

7. General relativity predicted a 10.3 arcsecond centennial precession for the orbit of Icarus. A 9.8 +/- 0.8 arcseconds was observed.

8. General relativity predicted a 43.03 arcsecond perihelion advance for the orbit of Mercury. A 43.20 +/- 0.30 arcseconds was observed.

9. General relativity predicted a 4.2 +/- 0.3 degree/yr. advance for the periastron for the binary pulsar PSR 1913+16. A 4.225 +/- 0.002 degrees/yr. was observed.

10. The change in the orbital period of binary pulsar PSR 1913+16 due to gravitational radiation is 1.13 +/- 0.19 times the value predicted by general relativity.

11. General relativity predicted a beta parameter of 1.0 for the echo delay of laser signals from a moon-based corner cube reflector placed by an Apollo mission. A 1.003 +/- 0.005 beta was observed.

12. General relativity predicted a gamma parameter of 1.0 for the echo delay of laser signals from a moon-based corner cube reflector placed by an Apollo mission. A 1.008 +/- 0.008 beta was observed.

13. A gravitational red shift of spectral lines on the earth's surface, known as the Mossbauer effect, was 0.9970 +/- 0.0076 times the value predicted by general relativity.

14. General relativity predicted a gamma parameter of 1.0 for the gravitational retardation of radio signals. A 1.000 +/- 0.001 was observed.

15. A gravitational red shift of the neutral hydrogen spectral line was 1.000000 +/- 0.000070 times the value predicted by general relativity.

16. The gravitational lens effect that was predicted by general relativity was observed on quasar images. (17)

17. On 29 March 2001 analysis of Hubble Space Telescope photographs show the first complete "Einstein Ring" from the galaxy B1938+666. (4)

## 11. Conclusion :

Symmetry !

I happen to revel in this discovery because for me it could validate an intuition that others and I have had for decades; that a form of 'antigravitational' energy exists. Therefore, in my opinion, cosmological models that represent a one-sided coin, and no antonym for " entropy ", still require a "fudge factor" – negentropy is one such theory.

In my opinion there are sources in the universe that represent sources of replenishment. I use the term " extropy " to define the process. In addition we have seen how a new discovery has led to the revitalization of some old theories on cosmology and how Einstein's "dark matter " and "cosmological constant " maintain a place in current theory. The discovery of SN1997ff is a significant event, especially if it withstands further scrutiny.

It has been stated that white holes cannot exist because they violate the second law of thermodynamics. I will take this opportunity to posit some of my own theories. In my opinion, the laws of thermodynamics and most laws of our present day scientific body of knowledge, are universe specific, and therefore ultimately relative. They are defined and based on assumptions that are limited by our observational universe. It is my opinion when we think of the ultimate symmetry of the universe we must imagine how our universe interacts with higher dimensions and the possibility of other physically defined universes. Specifically models of The Schwarzschild geometry come to mind. When the higher order terms of these laws are considered and modeled based on interdimensional coupling, a new perspective is possible that accounts for white holes and the concept of
" extropy ".

## 12. Bibliography :

1.)    Hubble Space Science Institute Press Release : 01 - 58, dated 2 April 2001, Entitled: "Farthest Supernova Ever Seen Sheds Light On Dark Universe" ; Donald Savage,Headquarters, Washington, DC & Ray Villard, Hubble Space Telescope Science Institute, Baltimore, MD

2.)    Hubble Space Science Institute Press Release : STScI- PR01-09, dated 2 April 2001, Entitled: " Blast from the Past: Farthest Supernova Ever ...

3.)    Working notes on interview with Mr. Paul Preuss of Lawrence Berkeley National Laboratory on 24 April 2001; release 1-5 ref: SN1997ff.

4.)    Hubble Space Science Institute Press Release :   2 April 2001, Entitled: "First Complete Einstein Ring Discovered" ; Hubble Space Telescope Science Institute, Baltimore, MD

5.)    ``Wormholes in space-time and their use for interstellar travel: A tool for teaching general relativity", American Journal of Physics, 56, 395-412, M. S. Morris & K. S. Thorne (1988).

6.)    "The Case for the Relativistic Hot Big Bang Cosmology", Nature, 352, 769 -- 776,  Peebles, P.J.E., Schramm, D.N., Turner, E.L. \& R.G. Kron 1991.

7.)    "The Evolution of the Universe", Scientific American, 271, 29 -- 33, Peebles, P.J.E., Schramm, D.N., Turner, E.L. \& R.G. Kron 1994,

8.)    "Was Einstein Right?", Will, Clifford,

9.)    "Supernovae & Nucleosynthesis", David Arnett

10.)   "The Meaning of Quantum Theory", Baggott

11.)   "Theories of Everything", Barrow

12.)   "Fundamentals of Physics", David Halliday

13.)   "The Symbiotic Universe", Greenstein

14.)   "Black Holes & Baby Universes", Hawking

15.)   "The Nature of Space & Time", Hawking, Penrose

16.)   "Astronomy", Menzel

17.)   "The Fingerprint of God", Hugh Ross

www.ingramcontent.com/pod-product-compliance
Lightning Source LLC
Chambersburg PA
CBHW080757180526
45168CB00006B/2234